ADVANCES IN
MOLECULAR AND CELL BIOLOGY
VOLUME 38

Chloride Movements Across Cellular Membranes

ADVANCES IN
MOLECULAR AND CELL BIOLOGY
VOLUME 38

Chloride Movements Across Cellular Membranes

Series Editor

E. Edward Bittar
University of Wisconsin – Madison
Madison, Wisconsin

Volume Editor

Michael Pusch
Istituto di Biofisica, CNR
Genova, Italy

2007

ELSEVIER

AMSTERDAM • BOSTON • HEIDELBERG • LONDON • NEW YORK • OXFORD
PARIS • SAN DIEGO • SAN FRANCISCO • SINGAPORE • SYDNEY • TOKYO

Elsevier
525 B Street, Suite 1900, San Diego, California 92101-4495, USA
84 Theobald's Road, London WC1X 8RR, UK

This book is printed on acid-free paper. ∞

Copyright © 2007, Elsevier B. V. All Rights Reserved.

No part of this publication may be reproduced or transmitted in any form or by any means, electronic or mechanical, including photocopy, recording, or any information storage and retrieval system, without permission in writing from the Publisher.

The appearance of the code at the bottom of the first page of a chapter in this book indicates the Publisher's consent that copies of the chapter may be made for personal or internal use of specific clients. This consent is given on the condition, however, that the copier pay the stated per copy fee through the Copyright Clearance Center, Inc. (www.copyright.com), for copying beyond that permitted by Sections 107 or 108 of the U.S. Copyright Law. This consent does not extend to other kinds of copying, such as copying for general distribution, for advertising or promotional purposes, for creating new collective works, or for resale. Copy fees for pre-2007 chapters are as shown on the title pages. If no fee code appears on the title page, the copy fee is the same as for current chapters.
$35.00

Permissions may be sought directly from Elsevier's Science & Technology Rights Department in Oxford, UK: phone: (+44) 1865 843830, fax: (+44) 1865 853333, E-mail: permissions@elsevier.co.uk. You may also complete your request on-line via the Elsevier homepage (http://elsevier.com), by selecting "Support & Contact" then "Copyright and Permission" and then "Obtaining Permissions."

For all information on all Academic Press publications
visit our Web site at www.books.elsevier.com

ISBN-13: 978-0-444-52872-8
ISBN-10: 0-444-52872-5
ISSN: 1569-2558

PRINTED IN THE UNITED STATES OF AMERICA
07 08 09 9 8 7 6 5 4 3 2 1

Working together to grow libraries in developing countries

www.elsevier.com | www.bookaid.org | www.sabre.org

ELSEVIER   BOOK AID International   Sabre Foundation

Transferred to digital printing in 2008.

Contents

Contributors ix
Preface xi

1. Chloride-Transporting Proteins in Mammalian Organisms:
   An Overview

   Michael Pusch

       References 6

2. Physiological Functions of the CLC Chloride Transport Proteins

   Tanja Maritzen, Judith Blanz, and Thomas Jentsch

       I. Introduction 10
      II. Physiological Function of CLC-0 and Mammalian CLC Proteins 15
     III. CLC Proteins in Model Organisms 43
     IV. Summary and Outlook 46
        References 46

3. Structure and Function of CLC Chloride Channels and Transporters

   Alessio Accardi

        I. The Structure of CLC-ec1, a Bacterial CLC Transporter 60
       II. Structural Conservation Among the CLCs 62
      III. Ion Selectivity, Permeation and Binding 63
      IV. A Multi-Ion Pore 64
       V. Channels and Transporters: Where Is the Difference? 65
      VI. Function of the CLC Transporters 66
     VII. Speculative Modeling 68
    VIII. CLC Channel Gating 69
      IX. The Role of the C-Terminal Domain of CLC Channels:
          Recent Developments 77
       X. Conclusions and Outlook 78
         References 79

4. Pharmacology of CLC Chloride Channels and Transporters

   *Michael Pusch, Antonella Liantonio, Annamaria De Luca,
   and Diana Conte Camerino*

       I. Introduction 83
      II. The Pharmacology of the Macroscopic Skeletal Muscle
         Cl^- Conductance gCl 84

	III.	New Molecules Targeting ClC-1 Identified Using Heterologous Expression	88
	IV.	Mechanism of Block of Muscle Type CLC Channels by Clofibric Acid Derivatives	91
	V.	The Binding Site for 9-AC and CPA on ClC-1 and ClC-0 and Use of CPA as a Tool to Explore the Mechanisms of Gating of ClC-0	96
	VI.	Pharmacology of CLC-K Channels	98
	VII.	Other CLC Channels and Other Blockers	101
	VIII.	Potential of Having Blockers of ClC-2, ClC-3, ClC-5, ClC-7: Outlook	102
		References	103

5. The Physiology and Pharmacology of the CFTR Cl$^-$ Channel

    Zhiwei Cai, Jeng-Haur Chen, Lauren K. Hughes, Hongyu Li, and David N. Sheppard

	I.	Introduction	109
	II.	The Molecular Physiology of CFTR	110
	III.	The Cellular Physiology of CFTR	116
	IV.	The Role of CFTR in Epithelial Physiology	117
	V.	The Pathophysiology of CFTR	120
	VI.	The Pharmacology of CFTR	123
	VII.	Conclusions	137
		References	137

6. Gating of Cystic Fibrosis Transmembrane Conductance Regulator Chloride Channel

    Zhen Zhou and Tzyh-Chang Hwang

	I.	CFTR Overview	145
	II.	Regulation of CFTR by Phosphorylation/Dephosphorylation	148
	III.	CFTR Is an ATPase	152
	IV.	Structural Biology of ABC Transporters	153
	V.	Methods Used to Study CFTR Gating	157
	VI.	CFTR Gating by ATP	160
	VII.	Unsettled Issues and Future Directions	169
		References	172

7. Functional Properties of Ca^{2+}-Dependent Cl$^-$ Channels and Bestrophins: Do They Correlate?

    Jorge Arreola and Patricia Pérez-Cornejo

	I.	Summary	181
	II.	Introduction	182
	III.	Functional Properties of CaCCs	182
	IV.	Molecular Candidates of CaCCs	187

V.	Functional Properties of Bestrophins	188
VI.	Concluding Remarks	191
	References	193

8. Cell Volume Homeostasis: The Role of Volume-Sensitive Chloride Channels

Alessandro Sardini

I.	Introduction	199
II.	Cell Volume Homeostasis	200
III.	Volume-Sensitive Cl^- Channels	203
IV.	Concluding Remarks	210
	References	211

9. GABAergic Synaptic Transmission

Andreas Draguhn and Kristin Hartmann

I.	Postsynaptic GABA Responses	216
II.	Molecular Composition and Function of GABAergic/Glycinergic Synapses	221
III.	Modulation of Synaptic Functions Through Endogenous Substances and Drugs	228
IV.	Perspective: Specificity of GABAergic and Glycinergic Signaling in Defined Networks	232
	References	233

10. Physiology of Cation-Chloride Cotransporters

Christian A. Hübner and Marco B. Rust

I.	Introduction	241
II.	The Cation-Chloride Cotransporter Family	243
III.	Cation-Chloride Cotransporters in the Nervous System	250
IV.	Cation-Chloride Cotransporters in the Inner Ear	256
V.	Cation-Chloride Cotransporters in the Kidney	259
VI.	Cation-Chloride Cotransporters in RBCs	262
VII.	Cation-Chloride Cotransporter and Cancer	265
VIII.	Cation-Chloride Cotransport and Hypertension	266
IX.	Concluding Remarks	267
	References	267

11. Plasma Membrane Cl^-/HCO_3^- Exchange Proteins

Haley J. Shandro and Joseph R. Casey

I.	Introduction	280
II.	The Cl^-/HCO_3^- Exchangers	280
III.	Disputed Cl^-/HCO_3^- Exchangers	300

IV.	Structure	301
V.	Regulation and Inhibition of Cl^-/HCO_3^- Exchange	303
VI.	Pathophysiology	309
	References	312

12. Orchestration of Vectorial Chloride Transport by Epithelia

Peying Fong and Michael A. Gray

I.	Introduction	329
II.	Epithelial Cell Structure and Organization	330
III.	Junctional Proteins: An Overview	334
IV.	Chloride Transporting Proteins in Epithelia: Channels, Transporters, and All Things in Between	336
V.	PDZ Proteins: What Are They? Where Are They? What Do They Do?	347
VI.	Two Diseases Involving Epithelial Chloride Transport	351
VII.	Summary	356
	References	357

Index 369

Contributors

Alessio Accardi *Department of Biochemistry, HHMI/Brandeis University, Waltham, Massachusetts 02454*

Jorge Arreola *Institute of Physics, University of San Luis Potosí, San Luis Potosí, SLP 78000, México*

Judith Blanz *Zentrum für Molekulare Neurobiologie Hamburg (ZMNH), Hamburg University, Falkenried 94, D-20251 Hamburg, Germany*

Zhiwei Cai *Department of Physiology, School of Medical Sciences, University of Bristol, Bristol BS8 1TD, United Kingdom*

Joseph R. Casey *Membrane Protein Research Group, Department of Biochemistry, and Department of Physiology, University of Alberta, Edmonton, Alberta, Canada T6G 2H7*

Jeng-Haur Chen *Department of Physiology, School of Medical Sciences, University of Bristol, Bristol BS8 1TD, United Kingdom*

Diana Conte Camerino *Sezione di Farmacologia, Dipartimento Farmacobiologico, Facoltà di Farmacia, Università di Bari, Italy*

Annamaria De Luca *Sezione di Farmacologia, Dipartimento Farmacobiologico, Facoltà di Farmacia, Università di Bari, Italy*

Andreas Draguhn *Institut für Physiologie und Pathophysiologie, University of Heidelberg, Germany*

Peying Fong *The Department of Physiology, The Johns Hopkins University School of Medicine, Baltimore, Maryland 21205*

Michael A. Gray *Institute for Cell and Molecular Biosciences and School of Biomedical Sciences, University Medical School, Newcastle-Upon-Tyne, United Kingdom*

Kristin Hartmann *Institut für Physiologie und Pathophysiologie, University of Heidelberg, Germany*

Christian A. Hübner Institut für Humangenetik, Universitätskrankenhaus Hamburg-Eppendorf, Butenfeld 42, D-22529 Hamburg, Germany; Zentrum für Molekulare Neurobiologie, Universität Hamburg, Falkenried 94, 20246 Hamburg, Germany

Lauren K. Hughes Department of Physiology, School of Medical Sciences, University of Bristol, Bristol BS8 1TD, United Kingdom

Tzyh-Chang Hwang Department of Medical Pharmacology and Physiology, Dalton Cardiovascular Research Center, University of Missouri-Columbia, Columbia, Missouri 65211

Thomas Jentsch Zentrum für Molekulare Neurobiologie Hamburg (ZMNH), Hamburg University, Falkenried 94, D-20251 Hamburg, Germany

Hongyu Li Department of Physiology, School of Medical Sciences, University of Bristol, Bristol BS8 1TD, United Kingdom

Antonella Liantonio Sezione di Farmacologia, Dipartimento Farmacobiologico, Facoltà di Farmacia, Università di Bari, Italy

Tanja Maritzen Zentrum für Molekulare Neurobiologie Hamburg (ZMNH), Hamburg University, Falkenried 94, D-20251 Hamburg, Germany

Patricia Pérez-Cornejo School of Medicine, University of San Luis Potosí, San Luis Potosí, SLP 78000, México

Michael Pusch Istituto di Biofisica, CNR, Via De Marini, 6, I-16149 Genova, Italy

Marco B. Rust Zentrum für Molekulare Neurobiologie, Universität Hamburg, Falkenried 94, 20246 Hamburg, Germany

Alessandro Sardini MRC Clinical Sciences Centre, Faculty of Medicine, Imperial College, London W12 0NN, United Kingdom

Haley J. Shandro Membrane Protein Research Group, Department of Biochemistry, University of Alberta, Edmonton, Alberta, Canada T6G 2H7

David N. Sheppard Department of Physiology, School of Medical Sciences, University of Bristol, Bristol BS8 1TD, United Kingdom

Zhen Zhou Department of Medical Pharmacology and Physiology, Dalton Cardiovascular Research Center, University of Missouri-Columbia, Columbia, Missouri 65211

Preface

Chloride Movements Across Cellular Membranes presents the state of the art of a rapidly expanding and interest-gaining field of membrane transport. Traditionally, chloride-transporting proteins have somehow the reputation of being the boring parents of the truly exciting sodium, calcium, and potassium channels. But things have changed in the recent years and the physiological importance of chloride-transporting proteins is more evident than ever. Even the classical view of GABA and glycine receptors as pure "dampeners" of neural excitability cannot be sustained any more. Recent developments, including the molecular identification of numerous novel anion-transporting proteins, the generation of knockout animal models, the elucidation of three-dimensional structures of bacterial homologues of mammalian transporters, the detailed biophysical analysis of transport mechanisms, and the increasing availability of pharmacological tools, have clearly established the anion transport systems as physiologically highly relevant.

Completely unexpected and novel functions of chloride transport have been discovered, all of which are described in the present volume. In excitable cells like neurons and muscle fibers, the activation of chloride channels and the activity of chloride transporters directly and indirectly affect the membrane potential, the most important parameter in neural signaling. The large array of substances that target postsynaptic GABA receptor channels for treating epilepsy and other neurological disorders highlights the importance of chloride in the nervous system. In epithelial cells, chloride transport is one of the major, and long-studied tasks.

One of the major driving forces behind the recent developments has been the involvement of chloride channels and transporters in human genetic diseases. The most prominent example is cystic fibrosis, a frequent life-threatening disease, caused by mutations in cystic fibrosis transmembrane conductance regulator (CFTR), a kinase and ATP-regulated chloride channel. While cystic fibrosis is a relatively common disease, most pathologies associated with mutations in genes coding for transporters or channels are rare. However, as relatively simple model systems, their study may help also to increase our understanding of complex, multifactorial diseases.

The topic of this book, *Chloride Movements Across Cellular Membranes*, has arrived at a relatively mature state, making it interesting for a broad medically, physiologically, biologically, and biophysically oriented readership. From the current literature it is, however, relatively difficult to obtain a comprehensive overview of the field because the proteins involved are studied by a variety of different scientific communities with little overlap. I hope that the compilation of the 12 chapters of the book helps both newcomers as well as experts to obtain an up-to-date and comprehensive insight into the subject.

Michael Pusch

Chapter 1

Chloride-Transporting Proteins in Mammalian Organisms: An Overview

Michael Pusch

Istituto di Biofisica, CNR, Via De Marini, 6, I-16149 Genova, Italy

"*Chloride Movements Across Cellular Membranes*" presents the state of the art of a rapidly expanding and interest-gaining field of membrane transport. Traditionally, Cl^--transporting proteins have somehow the reputation of being the boring parents of the truly exciting Na, Ca, and K channels that are of prime importance for neural computing (Miller, 2006). Even in the nervous system, the activity of postsynaptic GABA and glycine receptor anion channels was associated mostly with a "dampening" of the actually important, excitatory mechanisms of neural computing.

But things have changed in the recent years and the physiological importance of Cl^--transporting proteins is more evident than ever. Even the classical view of GABA and glycine receptors as pure "dampeners" of neural excitability cannot be sustained any more [see Chapters 9 and 10 and Marty and Llano (2005)]. Recent developments, including the molecular identification of numerous novel anion-transporting proteins, the generation of knockout animal models, the elucidation of three-dimensional structures of bacterial homologues of mammalian transporters, the detailed biophysical analysis of transport mechanisms, and the increasing availability of pharmacological tools, have clearly established the anion transport systems as physiologically highly relevant.

The most significant progress has been made possible after the cloning of the first members of the various gene families that code for Cl^--transporting proteins (Table I). This, subsequently, allowed the application of the powerful molecular biology methods such as gene knockout in mice. With these techniques, completely unexpected and novel functions have been discovered, all of which are described in the present volume. But before going into detail, let us step back and get an overview of why and where Cl^- transport is important: Cl^- is the most abundant anion in the extracellular fluid, where its concentration is above 100 mM. Only in some specialized tissues Cl^- is replaced by other anions. In the pancreatic ducts, for example, Cl^- is almost completely replaced by HCO_3^- (Steward *et al.*, 2005). How the various Cl^-- and HCO_3^--transporting

Table I
Gene Families of Cl$^-$-Transporting Proteins[a]

Gene family	Architecture	Number of genes coding for subunits	Localization	Associated disease(s)	Structure of homologues	Chapters
Cation-chloride cotransporters	Monomer	9	PLM	Yes	No	10, 12
Cl$^-$/HCO$_3^-$ exchangers, SLC4	Monomer	4	PLM	Yes	No	11, 12
Cl$^-$/HCO$_3^-$ exchangers, SLC26	Monomer	5	PLM	Yes	No	11, 12
GABA/glycine receptors	Pseudosymmetric pentamers	>20	PLM	Yes	Torpedo ACh receptor	9, 12
CFTR	Monomer	1	PLM	Yes	Bacterial ABC transporters	5, 6, 12
CLC proteins	Homodimers (possibly heterodimers)	9	PLM/organelles	Yes	Bacterial CLC homologues	2–4, 12

[a]PLM, plasma membrane; CFTR, cystic fibrosis transmembrane conductance regulator.

proteins in the pancreatic epithelial cells achieve this remarkably high concentration is still a matter of debate (see Chapters 11 and 12). In fact, the relatively recent discovery of a novel family of Cl^-/HCO_3^- transporters, some of which may be electrogenic, has opened new perspectives regarding epithelial Cl^- and HCO_3^- transport, and many questions are still unresolved (see Chapters 11 and 12).

The intracellular Cl^- concentration can vary substantially between different cell types and this is an important parameter, subject to regulation. Clearly, in excitable cells like neurons and muscle fibers, the activation of Cl^- channels and the activity of Cl^- transporters directly and indirectly affect the membrane potential, the most important parameter in neural signaling (see Chapters 2, 4, 9, 10, and 12). The large array of substances that target postsynaptic GABA receptor channels for treating epilepsy and other neurological disorders highlight the importance of Cl^- in the nervous system (see Chapter 9). In epithelial cells, Cl^- transport is one of the major and long-studied tasks (see Chapters 2, 4–6, and 10–12). In fact, cystic fibrosis, a relatively common genetic disease with life-threatening symptoms, is caused by mutations in a gene coding for an epithelial Cl^- channel (see Chapters 5, 6, and 12). In contrast, the role of Cl^--transporting proteins in intracellular organelles is only beginning to emerge (see Chapters 2 and 12).

Single ion-transporting proteins, like all enzymes, do not work in isolation, but are functionally and sometimes also structurally, coupled to other proteins. The present collection of chapters provides a unique opportunity to understand the functional physiological relationships between the various transporter families. In particular, Chapters 9 and 12 highlight the functional interplay and the structural relationships between various transporters and channels in the nervous system and in epithelia, respectively.

Probably, one reason as why anion transporters have sometimes been considered as the poor cousins of cation transporters (Gadsby, 1996; Miller, 2006) is that they are often difficult to study for various reasons: many Cl^- channels have a small single channel conductance, like the muscle channel ClC-1 (see Chapter 3); some seem to be ubiquitous such that expression cloning is almost impossible (see Chapter 8); transporters that are not electrogenic, like many Cl^-/HCO_3^- exchangers, are more difficult to measure in functional detail than electrogenic transporters and channels (see Chapters 10–12); and few specific pharmacological tools are available.

In fact, the technical difficulties in studying Cl^- channels have led to numerous publications who claim that a certain protein represents a Cl^- channel, based mostly on heterologous expression (see Chapters 2, 7, 8, and references therein). However, for many of these, the identity is not widely accepted and thus, the gene families which are still not clearly established as coding for Cl^- transporters (e.g., pIcln, CLCA, CLIC, tweety, and others) are not covered in the present volume. Similarly, the Cl^--channel activity associated with various amino acid and phosphate transporters (Jentsch *et al.*, 2002; DeFelice, 2004) is not covered explicitly in this book, since the physiological role and relevance of this phenomenon is not yet established.

After the cloning of the gene families that code for GABA/glycine receptors, cation-chloride cotransporters, Cl^-/HCO_3^- exchangers, cystic fibrosis transmembrane conductance regulator (CFTR), and CLC proteins, we have seen a tremendous progress in the field. One of the major driving forces behind this development has been the involvement of these transporters in human genetic diseases. The most prominent example is cystic fibrosis, a frequent life-threatening disease, caused by mutations in CFTR, a kinase, and ATP-regulated Cl^- channel (see Chapters 5, 6, and 12). Generally, these transporter- or channel-associated pathologies are rare. However, as relatively simple model systems, their study may help also to increase our understanding of complex, multifactorial diseases.

The difficulties in studying Cl^- transport proteins of the plasma membrane are even greater for anion transport systems that are active in intracellular compartments, like endosomes, lysosomes, or Golgi. Several investigators study intracellular channels by reconstituting partially purified intracellular membranes into planar lipid bilayers. Currents with large unitary conductances have been described, but none of these is really selective for Cl^- and the molecular identity or physiological relevance of these currents are unknown. In fact, the full exploration of intracellular transport systems will undoubtedly be an active and interesting future field, but it will necessitate new technical advances. Several members of the CLC family are localized intracellularly and we know a great deal about the phenotypes of their functional knockout in mice (see Chapters 2 and 12). Among these, ClC-3, ClC-4, and ClC-5 can be expressed to reasonable degree in the plasma membrane in heterologous systems (Steinmeyer *et al.*, 1995; Friedrich *et al.*, 1999; Li *et al.*, 2000). The properties of these currents are such that it is almost *a priori* excluded that they could have been detected in bilayer reconstitution experiments. Thus, for these proteins, the genetic approach, using knockout, heterologous expression, and similar techniques, seems at the moment to be the only way to gain insight into their function.

Very little is known about the intracellularly located ClC-6, and there is practically no functional data available for ClC-7, mutations of which cause osteopetrosis, a bone disease (Cleiren *et al.*, 2001; Kornak *et al.*, 2001). Also, the known properties of ClC-3 to ClC-5 (see Chapters 2 and 3) cannot be easily reconciled with their proposed physiological role as shunting conductances in endosomes. In fact, it has been recently shown that these proteins mediate electrogenic chloride proton exchange, instead of being "simple" ion channels (Picollo and Pusch, 2005; Scheel *et al.*, 2005), similar to the bacterial ClC-ec1 (Accardi and Miller, 2004). It can also be speculated that several intracellularly active anion transporters await identification.

Among several others, two physiologically extremely important plasma membrane Cl^- channels still await their molecular identification: one is the ubiquitous swelling activated Cl^- channel that is important for regulatory volume decrease (Chapter 8), and the other is the Ca^{2+}-activated Cl^- channel (Chapter 7), that plays numerous physiological roles (Table II).

In order to gain detailed mechanistic insight into protein function, structural information is essential. Unfortunately, membrane proteins are difficult to analyze using X-ray crystallography, the technique that provides the highest, that is, atomic, resolution. In particular, eukaryotic membrane proteins are difficult to purify at high yields. Consequently, with a few exceptions (Toyoshima *et al.*, 2000; Long *et al.*, 2005;

Table II
Cl⁻ Channels of Unknown Molecular Identity

Current	Tissue	Viable candidate genes	Chapter/References
Volume-regulated anion channel (VRAC)	Ubiquitous	No	Chapter 8
$I_{Cl}(Ca^{2+})$	Smooth muscle, olfactory neurons, epithelia, other	Bestrophins	Chapter 7
Inwardly rectifying cyclic adenosine monophosphate (cAMP)-regulated Cl⁻ channel	Plexus choroidiae	No	Kibble et al. (1996)
Adenosine 5′-triphosphate (ATP)-activated Cl⁻ channel	Parotid acinar cells	No	Arreola and Melvin (2003)
H⁺-activated Cl⁻ channel	Several cells	No	Nobles et al. (2004)

Unwin, 2005), most structures of ion-transporting proteins have been obtained for prokaryotic homologues. In particular, for Cl⁻ transporters, direct structural data is available only for the CLC proteins with crystal structures from bacterial homologues (Dutzler et al., 2002, 2003). Even though for CFTR only the structure of quite distant homologues has been solved (Chang and Roth, 2001), the recent elucidation of the structure of the isolated nucleotide-binding domains has greatly advanced the understanding of the ATP-dependent gating of this channel (see Chapter 6). In the case of glycine and GABA$_A$ receptor channels, the structure of the Na⁺/K⁺-conducting *Torpedo* acetylcholine receptor at 4-Å resolution (Unwin, 2005) may serve as a guide to interpret functional measurements.

In these fast moving times with almost immediate information access via the World Wide Web, a classical book, like the present volume of the series *Advances in Molecular and Cellular Biology*, may seem obsolete. Reviews of the leaders in subfields are available for those who would like to gain an up-to-date overview of the state of the art in a specific matter. Why then compile a book with its intrinsically slower processing times and relatively high costs? The major reason is that such a book is much more comprehensive than a standard review in two aspects. First, the scope of the present book is much larger than any available single review. Thus, the book is useful for a much broader readership. For example, newcomers to any of the subfields addressed in this compilation, who would like to obtain an overview over anion transport in more general terms, can find a very useful broad overview about related, and thus immediately relevant, subjects. But also well-established scientists working on specific transport systems, but who are curious to recent developments in related subjects, will find an authorative, up-to-date overview. Finally, a comprehensive compilation of articles from leading scientists provides a valuable source of information for those who have a specific question regarding a specific problem in the area covered in the present book. In fact, even with the World Wide Web it is often difficult to find a reliable source of information that provides sufficient background. Most importantly, however, the difference between book publishing and journal publishing will undoubtedly blur in the near future. A first sign that the difference between books and journal articles will eventually disappear can be seen in the recent Google service called "Google Book Search" (http://books.google.com).

I believe that the topic of this book, *Chloride Movements Across Cellular Membranes*, has arrived at a relatively mature state, making it interesting for a broad medically, physiologically, biologically, and biophysically interested readership. Furthermore, the topic is generally relatively little represented in the biomedical literature and its importance is often underestimated.

Of course, the present book is not comprehensive but it covers the most exciting recent developments and presents the future challenges in the field.

REFERENCES

Accardi, A. and Miller, C. (2004). Secondary active transport mediated by a prokaryotic homologue of ClC Cl^- channels. *Nature* **427**, 803–807.

Arreola, J. and Melvin, J. E. (2003). A novel chloride conductance activated by extracellular ATP in mouse parotid acinar cells. *J. Physiol.* **547**, 197–208.

Chang, G. and Roth, C. B. (2001). Structure of MsbA from *E. coli*: A homolog of the multidrug resistance ATP binding cassette (ABC) transporters. *Science* **293**, 1793–1800.

Cleiren, E., Benichou, O., Van Hul, E., Gram, J., Bollerslev, J., Singer, F. R., Beaverson, K., Aledo, A., Whyte, M. P., Yoneyama, T., deVernejoul, M. C., and Van Hul, W. (2001). Albers-Schonberg disease (autosomal dominant osteopetrosis, type II) results from mutations in the ClCN7 chloride channel gene. *Hum. Mol. Genet.* **10**, 2861–2867.

DeFelice, L. J. (2004). Transporter structure and mechanism. *Trends Neurosci.* **27**, 352–359.

Dutzler, R., Campbell, E. B., Cadene, M., Chait, B. T., and MacKinnon, R. (2002). X-ray structure of a ClC chloride channel at 3.0 Å reveals the molecular basis of anion selectivity. *Nature* **415**, 287–294.

Dutzler, R., Campbell, E. B., and MacKinnon, R. (2003). Gating the selectivity filter in ClC chloride channels. *Science* **300**, 108–112.

Friedrich, T., Breiderhoff, T., and Jentsch, T. J. (1999). Mutational analysis demonstrates that ClC-4 and ClC-5 directly mediate plasma membrane currents. *J. Biol. Chem.* **274**, 896–902.

Gadsby, D. C. (1996). Two-bit anion channel really shapes up. *Nature* **383**, 295–296.

Jentsch, T. J., Stein, V., Weinreich, F., and Zdebik, A. A. (2002). Molecular structure and physiological function of chloride channels. *Physiol. Rev.* **82**, 503–568.

Kibble, J. D., Trezise, A. E., and Brown, P. D. (1996). Properties of the cAMP-activated Cl^- current in choroid plexus epithelial cells isolated from the rat. *J. Physiol.* **496**, 69–80.

Kornak, U., Kasper, D., Bösl, M. R., Kaiser, E., Schweizer, M., Schulz, A., Friedrich, W., Delling, G., and Jentsch, T. J. (2001). Loss of the ClC-7 chloride channel leads to osteopetrosis in mice and man. *Cell* **104**, 205–215.

Li, X., Shimada, K., Showalter, L. A., and Weinman, S. A. (2000). Biophysical properties of ClC-3 differentiate it from swelling-activated chloride channels in Chinese hamster ovary-K1 cells. *J. Biol. Chem.* **275**, 35994–35998.

Long, S. B., Campbell, E. B., and Mackinnon, R. (2005). Crystal structure of a mammalian voltage-dependent Shaker family K^+ channel. *Science* **309**, 897–903.

Marty, A. and Llano, I. (2005). Excitatory effects of GABA in established brain networks. *Trends Neurosci.* **28**, 284–289.

Miller, C. (2006). ClC chloride channels viewed through a transporter lens. *Nature* **440,** 484–489.
Nobles, M., Higgins, C. F., and Sardini, A. (2004). Extracellular acidification elicits a chloride current that shares characteristics with ICl(swell). *Am. J. Physiol. Cell Physiol.* **287,** C1426–C1435.
Picollo, A. and Pusch, M. (2005). Chloride/proton antiporter activity of mammalian CLC proteins ClC-4 and ClC-5. *Nature* **436,** 420–423.
Scheel, O., Zdebik, A. A., Lourdel, S., and Jentsch, T. J. (2005). Voltage-dependent electrogenic chloride/proton exchange by endosomal CLC proteins. *Nature* **436,** 424–427.
Steinmeyer, K., Schwappach, B., Bens, M., Vandewalle, A., and Jentsch, T. J. (1995). Cloning and functional expression of rat CLC-5, a chloride channel related to kidney disease. *J. Biol. Chem.* **270,** 31172–31177.
Steward, M. C., Ishiguro, H., and Case, R. M. (2005). Mechanisms of bicarbonate secretion in the pancreatic duct. *Annu. Rev. Physiol.* **67,** 377–409.
Toyoshima, C., Nakasako, M., Nomura, H., and Ogawa, H. (2000). Crystal structure of the calcium pump of sarcoplasmic reticulum at 2.6 Å resolution. *Nature* **405,** 647–655.
Unwin, N. (2005). Refined structure of the nicotinic acetylcholine receptor at 4 Å resolution. *J. Mol. Biol.* **346,** 967–989.

Chapter 2

Physiological Functions of the CLC Chloride Transport Proteins

Tanja Maritzen,[1] Judith Blanz,[2] and Thomas Jentsch[3]

Zentrum für Molekulare Neurobiologie Hamburg (ZMNH), Hamburg University, Falkenried 94, D-20251 Hamburg, Germany

I. Introduction
   A. CLC Subfamilies
   B. CLC Structure and Impact of Mutations
   C. CLC Function
II. Physiological Function of CLC-0 and Mammalian CLC Proteins
   A. ClC-0: The *Torpedo* Cl$^-$ Channel
   B. ClC-1 and Myotonia
   C. ClC-2: A Ubiquitously Expressed CLC Channel with Many Proposed Functions
   D. ClC-Ka/Kb: Cl$^-$ Channels Expressed in Kidney and Inner Ear
   E. ClC-3: An Intracellular CLC Protein Located on Endosomes and Synaptic Vesicles
   F. ClC-4: A Vesicular CLC Protein of Unclear Function
   G. ClC-5 and Dent's Disease
   H. ClC-6: A Late Endosomal Protein Important for Lysosomes
   I. ClC-7 and Osteopetrosis
III. CLC Proteins in Model Organisms
   A. Yeast
   B. *Caenorhabditis elegans*
   C. Plants
IV. Summary and Outlook
   References

---

[1]Present address: Institut für Chemie und Biochemie, Freie Universität Berlin, Takustrasse 6, D-14195 Berlin, Germany.
[2]Present address: Institut für Biochemie, Universität Kiel, Otto Hahn Platz 9, D-24118 Kiel, Germany.
[3]Present address: FMP (Leibniz-Institut für Molekulare Pharmakologie) and MDC (Max-Delbrück-Centrum für Molekulare Medizin), Robert Rössle Strasse 10, D-13125 Berlin, Germany.

I. INTRODUCTION

The CLC proteins constitute a highly conserved family of Cl^- channels and transporters. They can be found in all phylae ranging from bacteria to mammals. However, although CLC proteins serve important physiological functions, they are not per se essential for life. In fact, some organisms, like *Helicobacter pylori*, can do very well without any CLC protein. Nevertheless, the genome of most species encodes at least one CLC and often more than that. In mammals nine distinct members of the CLC family are found. Their great physiological importance is underlined by the fact that mutations in their genes are associated with a variety of human hereditary diseases.

A. CLC Subfamilies

According to their degree of sequence identity, the mammalian CLCs can be grouped into three subfamilies as illustrated by the phylogenetic tree in Fig. 1. The first subfamily consists of the CLC channels ClC-1, ClC-2, ClC-Ka, and ClC-Kb. These four proteins reside mainly at the plasma membrane. This localization has greatly facilitated their electrophysiological characterization. All of them have been shown to function as Cl^- channels that either mediate transepithelial Cl^- transport or contribute to the stabilization of the plasma membrane resting potential. The second subfamily comprises ClC-3, ClC-4, and ClC-5 which in contrast to the members of the first subfamily reside mainly on intracellular vesicles and are implicated in vesicular acidification. Contrary to previous belief, ClC-4 and ClC-5 were recently shown to be electrogenic Cl^-/H^+ antiporters instead of Cl^- channels (Picollo and Pusch, 2005; Scheel *et al.*, 2005). The high degree of homology between ClC-3, ClC-4, and ClC-5 renders it likely that ClC-3 is an antiporter as well. However, due to technical problems, its mode of operation has not yet been established experimentally. The members of the third subfamily, ClC-6 and ClC-7, are also localized on intracellular vesicles. In contrast to the members of the second subfamily, ClC-6 and ClC-7 do not traffic to the plasma membrane even on overexpression. This has prevented their electrophysiological characterization so far. Thus it is currently not known whether these two CLC proteins are channels or transporters.

B. CLC Structure and Impact of Mutations

CLC proteins possess a very peculiar structure that sets them apart from other channel families. They are dimers in which each of the two subunits has its own pore (Weinreich and Jentsch, 2001; Dutzler *et al.*, 2002), an arrangement that has been named "double-barrel." It is believed that CLC proteins function mostly as homodimers, although there is some evidence from *in vitro* and coimmunoprecipitation studies that at least the members of the first (Lorenz *et al.*, 1996) and second (Mohammad-Panah *et al.*, 2003; Suzuki *et al.*, 2006) subfamily can heterodimerize within the respective subfamily. The physiological significance of these results, however,

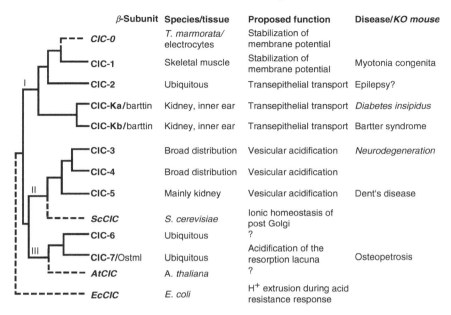

Figure 1. The CLC family of Cl⁻ channels and transporters in mammals and selected model organisms. Based on sequence identity, the nine mammalian CLC proteins can be grouped into three branches, as shown by the dendrogram. Sequences from *Torpedo marmorata*, *Saccharomyces cerevisiae*, and *Arabidopsis thaliana* also fit into this phylogenetic tree, while bacterial CLC proteins like the one from *Escherichia coli* comprise another more ancient evolutionary branch. Proteins of the first subfamily reside predominantly in the plasma membrane, while proteins from the two other branches are mainly located on vesicular membranes. The first column lists the β-subunits that are known to interact with CLC proteins. The second column indicates the most important features of the tissue distribution of the mammalian CLC proteins and states the names of the selected model organisms (italicized). The following columns list the presumed functions of the different CLC proteins and name the associated human disease or the phenotype of the corresponding knockout mouse model (italicized).

is unclear. Within the dimers, each pore can be opened by an individual gate. In addition, there is a common gating mechanism that closes both pores in parallel. The double-barrel structure of the CLC proteins combined with their gating mechanism has consequences for the impact of mutations on CLC function. Mutations that reduce single channel conductance or the gating of individual pores will generally not have dominant negative effects on coexpressed wild-type (WT) subunits, as the second pore will remain functional. Only mutations affecting the common gate or causing a missorting of the assembled dimer can exert dominant effects, and even the strongest dominant negative mutations are unlikely to decrease the current to less than 25% in the heterozygous state. This is in stark contrast to voltage-dependent K^+ channels in which the pore is assembled from four independent subunits, so that numerous mutations can reduce channel function down to 6% if coexpressed 1:1 with WT subunits.

Due to the comparably moderate dominant negative effects of CLC mutations, dominant forms of human CLC diseases like osteopetrosis (due to mutations in ClC-7) (Cleiren et al., 2001; Kornak et al., 2001) and myotonia congenita (due to mutations in ClC-1) (Steinmeyer et al., 1994; Meyer-Kleine et al., 1995; Pusch et al., 1995b) are clinically less severe than recessive variants in which the presence of two mutated alleles often leads to a complete loss of function.

The crystallization of bacterial CLC homologues by McKinnon and colleagues (Dutzler et al., 2002) revealed the three-dimensional structure of CLC proteins. Each CLC monomer consists of 18 α-helices, 17 of which are embedded in the plasma membrane. The N- as well as C-termini are located in the cytosol. In contrast to their prokaryotic homologues, mammalian CLC proteins have a long C-terminus which contains a pair of cystathionine-β-synthetase (CBS) domains. These domains which have recently been crystallized as part of the cytoplasmic domain of ClC-0 (Meyer and Dutzler, 2006) are implicated in the trafficking of CLC proteins (Schwappach et al., 1998; Carr et al., 2003) as well as in the binding of adenosine nucleotides (Scott et al., 2004).

C. CLC Function

The CLC-mediated anion transport over the plasma membrane serves either to modulate the electrical excitability of muscle and nerve cells by stabilizing their resting membrane potential (Fig. 2), to counteract cell swelling, or to allow salt and water movement across epithelia (Fig. 3). In the case of intracellularly located CLC proteins, the mediated Cl^- flux is supposed to enable the efficient acidification of vesicles such as endosomes (Figs. 4 and 5) and synaptic vesicles (Fig. 4). The luminal acidification of these vesicles is mainly achieved by V-type ATPases which transport H^+ into the vesicle lumen using the energy liberated by ATP hydrolysis. This import of H^+ is electrogenic, thus requiring a neutralizing current in order to prevent the generation of a large lumen-positive membrane potential that would limit the import of further H^+. This necessary charge neutralization is supposed to rely on the import of Cl^- (Sabolic and Burckhardt, 1986). CLC proteins are thought to provide the required charge neutralization thereby allowing the acidification to proceed. For a long time, the vesicular CLC proteins have been assumed to be Cl^- channels like the members of the first subfamily, in particular because ClC-4 and ClC-5 mediate Cl^- currents on heterologous expression (Steinmeyer et al., 1995; Friedrich et al., 1999). This view was challenged by the discovery that the prokaryotic CLC homologue ClC-ec1 is not an ion channel but rather a Cl^-/H^+ exchanger (Accardi and Miller, 2004). Subsequently, ClC-4 and ClC-5 were shown to function as Cl^-/H^+ exchangers as well (Picollo and Pusch, 2005; Scheel et al., 2005). Their highly electrogenic Cl^-/H^+ antiport remains compatible with the vesicular acidification concept. But the coupling of Cl^- transport to an H^+ counterflux means that more metabolic energy is needed for acidification. Unlike Cl^- channels, Cl^-/H^+ exchangers will directly couple Cl^- gradients to vesicular H^+ gradients. This suggests a physiological role for ClC-4 and

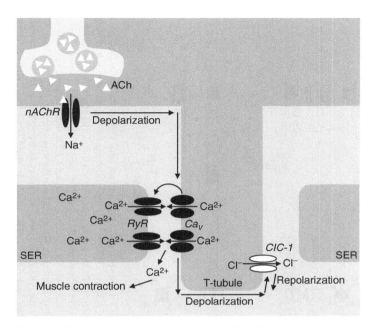

Figure 2. Stabilization of the muscle membrane potential by ClC-1. Skeletal muscle cells are innervated by motorneurons which release acetylcholine. Binding of the transmitter to nicotinic acetylcholine receptors on the muscle cell membrane causes a depolarizing Na^+ influx into the cells. Due to the t-tubule system, the depolarization spreads far down into the cells where it effects the opening of voltage-dependent Ca^{2+} channels (Ca_v) which in turn causes the opening of ryanodine-sensitive Ca^{2+} channels (RyR) in the membrane of the sarcoplasmatic reticulum (SER). The Ca^{2+} influx into the cytosol via Ca_v and RyR channels triggers muscle contraction. ClC-1 is located in the skeletal muscle membrane and stabilizes its voltage by allowing the influx of Cl^- ions already under resting conditions. It is further activated on depolarization and helps to repolarize the cell membrane after an action potential. Loss of ClC-1 results in myotonia, a disease associated with hyperexcitability of the muscle cells and repetitive action potential firing.

ClC-5 not only in facilitating endosomal acidification but also in regulating the endosomal Cl^- concentration which might influence enzymatic activities (Davis-Kaplan et al., 1998) or the osmotic regulation of vesicular volume. Additionally, ClC-4 and ClC-5 might directly acidify endosomes shortly after they pinch off from the plasma membrane by exchanging cytosolic H^+ for luminal Cl^- which initially is present at the high extracellular concentration. It was also speculated that the vesicular CLC proteins might be important for vesicle fusion (Picollo and Pusch, 2005) which is known to depend on Ca^{2+}. The local Ca^{2+} release from endosomes (Holroyd et al., 1999; Bayer et al., 2003) would create an inside negative potential which then could activate CLC proteins thereby facilitating endosomal acidification. The diverse physiological functions of CLC proteins will be discussed further in the following sections dealing with the different mammalian CLCs.

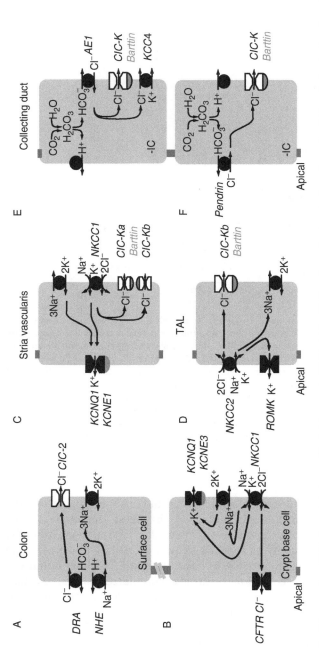

Figure 3. Diverse roles of Cl⁻ channels in transepithelial transport. (A) Model for NaCl reabsorption in colonic surface cells. Cells at the luminal surface of colonic epithelia possess an apical Cl⁻/HCO₃⁻ exchanger (Dra, Slc26a3) and an apical Na⁺/H⁺ (NHE3) exchanger, allowing for the reabsorption of NaCl. While Na⁺ leaves the cell via the basolateral Na⁺/K⁺-ATPase, Cl⁻ crosses the basolateral membrane probably in part via ClC-2. (B) Model for Cl⁻ secretion in colonic crypt base cells. Cells at the crypt base take up Cl⁻ across their basolateral membrane together with Na⁺ and K⁺ via NKCC1 using the Na⁺ gradient established by the basolateral Na⁺/K⁺-ATPase as driving force. Cl⁻ leaves the cell via the apical Cl⁻ channel CFTR thus being secreted into the colon. K⁺ is recycled at the basolateral membrane via the K⁺ channel KCNQ1/KCNE3. (C) Model for K⁺ secretion in the stria vascularis of the cochlea. The driving force for the secretion of K⁺ into the endolymph of the cochlea is provided by the NKCC1-mediated secondary active uptake of K⁺ together with Na⁺ and Cl⁻, which is powered by the Na⁺ gradient established by the Na⁺/K⁺-ATPase, and also by the K⁺ uptake by the ATPase itself. While K⁺ leaves the cell via the apical K⁺ channel KCNQ1/KCNE1 and is thus secreted, Cl⁻ ions are recycled via the basolateral Cl⁻ channels ClC-Ka/barttin and ClC-Kb/barttin. (D) Model for NaCl reabsorption in the TAL of Henle's loop. The steep Na⁺ gradient established by the basolateral Na⁺/K⁺-ATPase powers the apical NKCC2-mediated Na⁺-dependent uptake of Cl⁻ together with K⁺. While K⁺ is recycled via the apical K⁺ channel ROMK (Kir1.1), Cl⁻ and Na⁺ leave the cell at the basolateral site via the Cl⁻ channel ClC-Kb/barttin and the Na⁺/K⁺-ATPase, respectively, thus being reabsorbed from the primary urine. (E) Model for Cl⁻ recycling in α-intercalated cells of the collecting duct. α-intercalated cells acidify the primary urine by secreting protons via an H⁺-ATPase.

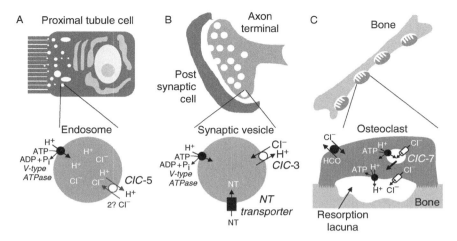

Figure 4. Role of CLC channels and transporters in acidification. The illustration depicts three cell types in which CLC proteins are important for acidification. (A) By shunting the current of the electrogenic H^+-ATPase, ClC-5 is crucial for the efficient acidification of endosomes of renal proximal tubular cells. While it has been established that ClC-5 is a Cl^-/H^+ antiporter rather than a channel, the stoichiometry of its electrogenic antiport is still elusive. (B) ClC-3 contributes to the acidification of synaptic vesicles. The electrochemical H^+ gradient is used to load the vesicles with neurotransmitter (NT). In addition, ClC-3 is also involved in endosomal acidification. ClC-3 is quite likely a transporter considering its high homology to ClC-5; however experimental evidence is still lacking. (C) ClC-7 provides the Cl^- conductance required for efficient proton pumping by the H^+-ATPase of the osteoclast-ruffled border, thus facilitating the acidification of the resorption lacuna which is essential for the dissolution of the mineral phase as well as the enzymatic degradation of the organic bone matrix. It is not known whether ClC-7 operates as channel or transporter.

II. PHYSIOLOGICAL FUNCTION OF CLC-0 AND MAMMALIAN CLC PROTEINS

A. ClC-0: The *Torpedo* Cl^- Channel

For a long time, the identification and purification of Cl^- channels from the CLC family had been hampered by the absence of high-affinity ligands, which were for example instrumental in the isolation of a different class of Cl^- channels, the postsynaptic glycine and GABA receptors (Pfeiffer *et al.*, 1982; Schofield *et al.*, 1987). The first member of the CLC family, ClC-0, was finally identified in 1990 by expression cloning

---

The remaining HCO_3^- ions are removed via the basolateral transporter AE1 which takes up Cl^- in exchange. Basolateral KCC4 transporters as well as the ClC-K/barttin channels are proposed to recycle the Cl^- taken up by AE1. (F) Model for Cl^- transport in β-intercalated cells of the collecting duct. β-intercalated cells serve to make the urine less acidic by secreting HCO_3^- via the transporter pendrin (SLC26a4) which takes up Cl^- in exchange. ClC-K/barttin channels have been suggested to allow Cl^- efflux at the basolateral site thus mediating Cl^- reabsorption.

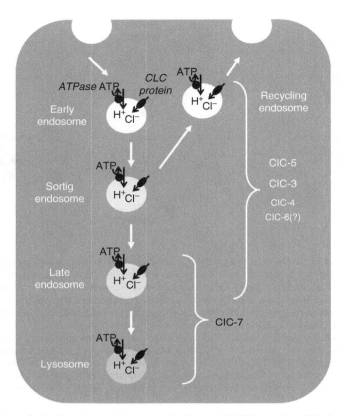

Figure 5. CLC proteins in the endosomal–lysosomal pathway. All CLC proteins of the intracellular branch of the CLC family have been shown to reside on vesicles of the endosomal–lysosomal pathway. Early endosomes on their progression to lysosomes become more and more acidic due to the activity of the vesicular H^+-ATPase. The voltage over the vesicular membrane that would be generated by this process is neutralized by Cl^- entering the vesicles via CLC proteins. While ClC-5 resides predominantly on early endosomes, ClC-7 is located on late endosomes and lysosomes. The precise localization of the remaining intracellular CLC proteins in regard to the endosomal–lysosomal pathway is less clear. The mechanisms determining the different vesicular localizations of the CLC proteins are not identified yet.

in *Xenopus* oocytes (Jentsch *et al.*, 1990). As the electric organ of the ray *Torpedo marmorata* was known to be an abundant source of voltage-gated Cl^- channels (White and Miller, 1979; Miller and White, 1984), Jentsch *et al.* used this organ as starting material to isolate mRNA for the expression cloning. In a depletion approach, single-stranded DNAs derived from a corresponding cDNA library were utilized to selectively deplete corresponding mRNAs. The total mRNA depleted of certain mRNAs was expressed in *Xenopus* oocytes which were analyzed by voltage clamping. In this way, the specific cDNA that caused the disappearance of the voltage-gated Cl^- current was finally identified, leading to the isolation of ClC-0 (Jentsch *et al.*, 1990).

1. Physiological Function of ClC-0

What is the physiological function of ClC-0? Marine rays like *Torpedo* stun their prey with short electric pulses which are generated in stacks of polarized cells, the electrocytes of the electric organ, a structure evolutionarily derived from skeletal muscle. In contrast to skeletal muscle cells, electrocytes from *Torpedo* lack voltage-gated Na^+ channels. The large depolarizing current which flows through the innervated membrane after activation of the electric organ is due to a Na^+ influx through acetylcholine receptors which are expressed at high levels in the innervated membrane. ClC-0 is found in the opposite, non-innervated membrane. Its Cl^- conductance stabilizes the membrane potential even in the face of the large transcellular currents that flow after acetylcholine receptor activation. ClC-0 thus allows the generation of a large transcellular potential, which amounts to about 90 mV across a single electrocyte. As electrocytes are organized in stacks equivalent to batteries arranged in series, the voltages across single electrocytes add up to more than 100 V during an electric shock.

ClC-0 has a single channel conductance of about 10 pS, which is much larger than that of ClC-1 (1.5 pS) for instance. Its two gates, the so-called fast and the slow gate, can be easily separated experimentally. ClC-0 thus provided a unique possibility to study the gating and structure–function relationship of CLC channels which are clearly voltage dependent, although they lack a classical voltage sensor like the one found in cation channels. The gating of CLC channels depends on the permeating Cl^- itself, which acts as the gating charge (Pusch *et al.*, 1995a). In ClC-0, fast gating is strongly influenced by extracellular Cl^- with a shift of the open probability curve to more positve voltages by about 50 mV per tenfold reduction in extracellular Cl^- concentration (Pusch *et al.*, 1995a). A detailed description of the biophysical properties of CLC channels is given in Chapter 3.

B. ClC-1 and Myotonia

Skeletal muscle differs from other mammalian tissues in that 70–80% of its resting membrane conductance is carried by Cl^- instead of K^+ (Rüdel and Lehmann-Horn, 1985). This difference is probably a consequence of the structure of the transverse t-tubule system which is essential for the coupling of excitation and contraction within the muscle fiber. In skeletal muscle, t-tubules invaginating from the plasma membrane propagate the electrical excitation deep into the muscle fiber, where the voltage-dependent activation of L-type Ca^{2+} channels eventually leads to the release of Ca^{2+} from intracellular stores and to muscle contraction (Fig. 2). If the repolarization of t-tubular membranes depended primarily on the efflux of K^+ via K^+ channels like in other tissues, the normally low extracellular K^+ concentration would increase significantly in the small space inside these tubules during prolonged muscle activity, thereby leading to a long-lasting moderate depolarization. To avoid this depolarization, evolution has chosen Cl^- channels to electrically stabilize and repolarize skeletal muscle membranes (Fig. 2). As the extracellular Cl^- concentration is 20-fold higher than that of K^+, the relative changes in t-tubular Cl^- concentration are much smaller than those in K^+ concentration for the same amount of repolarizing current.

1. Myotonia

Consistent with the importance of a Cl^- conductance for the proper functioning of skeletal muscle cells, the loss or reduction of this conductance causes myotonia, a disease affecting muscle relaxation. Indeed, muscle biopsies from myotonic goats and human patients revealed a reduced Cl^- conductance (Lipicky and Bryant, 1966; Lipicky et al., 1971), and myotonia could also be artificially induced in amphibians (Bretag et al., 1980) and goats (Franke et al., 1991; Iaizzo et al., 1991) for instance by applying the Cl^- channel blocker 9-anthracene-carboxylic acid (9-AC). Myotonia is a symptom found in several mouse mutants (Steinmeyer et al., 1991a; Gronemeier et al., 1994) and other animals (Rüdel and Lehmann-Horn, 1985; Bryant and Conte-Camerino, 1991; Beck et al., 1996). Patients with myotonia experience muscle stiffness after voluntary contraction which gradually improves with exercise ("warm-up" phenomenon). The disease is due to a hyperexcitability of the muscle cell membrane leading to series of repetitive action potentials, so-called myotonic runs, that can be detected in electromyograms. Which Cl^- channel mediates the high Cl^- conductance in the muscle membrane? Since the electric organ of rays is developmentally derived from skeletal muscle, it seemed likely that the Cl^- conductance of mammalian muscle might be mediated by a homologue of the *Torpedo* channel ClC-0. This was confirmed in 1991, when a Cl^- channel that is nearly exclusively expressed in skeletal muscle was cloned by its homology to ClC-0 (Steinmeyer et al., 1991b). This channel, which was named ClC-1, is 54% identical to the *Torpedo* channel ClC-0 and represents its closest mammalian orthologue. Soon after ClC-1 was cloned, it was shown that the open reading frame of ClC-1 was destroyed by a transposon insertion in the myotonic mouse strain *adr* (Steinmeyer et al., 1991a). This clearly demonstrated that ClC-1 is the major Cl^- channel in skeletal muscle and that it is essential for maintaining normal muscle excitability. Soon afterward, ClC-1 was identified as disease gene for the human disease myotonia congenita (Koch et al., 1992), which can be inherited in an autosomal recessive (Becker type) or an autosomal dominant form (Thomsen type). The Thomsen type was first described by Thomsen (1876) who himself suffered from this disease. The Thomsen mutation (P480L) exerts a strong dominant-negative effect on WT channels when coexpressed in *Xenopus* oocytes (Steinmeyer et al., 1994). By now, more than 80 different mutations in *CLCN1* (the gene encoding ClC-1), have been identified in patients with dominant or recessive myotonia (Pusch, 2002). These alterations are scattered throughout the *CLCN1* sequence and include nonsense, splice-site, and frame-shift mutations that truncate the channel protein. As explained in the introduction, recessive mutations are much more common in the case of CLC channels than dominant-negative mutations. Accordingly, only a few myotonia-causing mutations found in *CLCN1* are dominant negative (Pusch, 2002). Nearly all mutations identified in dominant myotonia cluster at the dimer interface (Pusch, 2002) and shift the voltage dependence of ClC-1 to more positive voltages (Pusch *et al.*, 1995b; Kubisch *et al.*, 1998). This means that the mutated Cl^- channels activate only at very positive voltages so that at physiological membrane potentials their contribution to the resting conductance is severly reduced. In light of the double-barrelled structure of CLC proteins, it is likely that these dominant mutations affect mainly the slow gate that acts on both pores simultaneously (Pusch, 2002).

2. ClC-1 Modulation by ATP

Recently, high concentrations of ATP were shown to shift the voltage dependence of ClC-1 to more positive voltages leading to channel closure (Bennetts *et al.*, 2005). The modulation of ClC-1 through ATP was speculated to be important during metabolic stress. Under such conditions of depleted ATP stores, a reduced excitability of muscle cells due to reduced ClC-1 activity might protect the cells from metabolic exhaustion which is thought to be the major factor in fatigue. The differential regulation of ClC-1 during metabolic stress via ATP might be a new mechanism linking muscle excitability to the metabolic state of the cells.

C. ClC-2: A Ubiquitously Expressed CLC Channel with Many Proposed Functions

Following the identification of ClC-0 and ClC-1, ClC-2, the third member of the first CLC subfamily, was cloned in 1992 by homology to ClC-1 (Thiemann *et al.*, 1992). Aside from ClC-1, ClC-2 is the only mammalian CLC protein for which single channel recordings could be obtained. ClC-2 has a relatively small single channel conductance of about 2–3 pS (Weinreich and Jentsch, 2001) and can be activated by various stimuli such as hyperpolarization (Thiemann *et al.*, 1992), cell swelling (Gründer *et al.*, 1992), and extracellular acidification (Jordt and Jentsch, 1997). Depending on the cell type, significant activation of ClC-2 starts at membrane potentials ranging from −30 to −80 mV (Thiemann *et al.*, 1992; Furukawa *et al.*, 1998; Park *et al.*, 1998; Schwiebert *et al.*, 1998; Xiong *et al.*, 1999; Cid *et al.*, 2000). Like in other CLC channels, the gating of ClC-2 is influenced by anions. But in contrast to ClC-0 and ClC-1, the gating of ClC-2 mainly depends on the intracellular Cl^- concentration rather than on extracellular Cl^-. Increases in intracellular Cl^- lead to channel activation (Pusch *et al.*, 1999; Niemeyer *et al.*, 2003; Catalán *et al.*, 2004). The inactivation of ClC-2 is probably mediated by an N-terminal domain (amino acids 16–61), the deletion of which leads to a constitutively open channel and thus to linear currents in two electrode voltage clamp recordings of *Xenopus* oocytes (Jordt and Jentsch, 1997), but neither in whole cell patch-clamp recordings of mammalian cells (Varela *et al.*, 2002) nor in excised patches (Pusch et al., 1999). These conflicting results were explained by a cell type-dependent diffusible factor which locks the mutated but not the WT channel in an open state. As mutations in the intracellular loop between transmembrane domains J and K also cause linear currents in Xenopus oocytes (Jordt and Jentsch, 1997) that are unaffected by cell swelling and pH, this region was suggested as a putative binding site for the N-terminal domain of the channel (Gründer et al., 1992; Jordt and Jentsch, 1997).

In contrast to its close homologues ClC-0 and ClC-1, which display a restricted tissue expression pattern, ClC-2 is broadly expressed. Its mRNA was found in all investigated tissues and cell lines (Thiemann et al., 1992). The most abundant expression of ClC-2 was found in brain and epithelia (Thiemann et al., 1992; Cid et al., 1995; Furukawa et al., 1995). Based on the expression of ClC-2 in epithelia of the fetal lung (Murray et al., 1995), intestine, kidney (Murray et al., 1996; Obermüller et al., 1998), and stomach (Malinowska et al., 1995) many functions for this channel like for example, a role in lung and kidney development (Murray et al., 1995; Blaisdell et al., 2000) and in gastric acid secretion (Malinowska et al., 1995) were proposed. The various stimuli modulating its activity invited speculations that ClC-2 might be involved in regulatory volume decrease (RVD), pH homeostasis, and regulation of the intracellular Cl^- concentration ($[Cl^-]_i$) of neurons (Jentsch et al., 2002). Before discussing possible roles of ClC-2, it should be mentioned that its reported subcellular localization was often based on immunohistochemistry that used antibodies which were not specific for ClC-2, as revealed later by control experiments on ClC-2 KO tissues. Thus ClC-2 immunohistochemistry and in particular hypothetical physiological roles of ClC-2 that have been based on such a localization should be interpreted with caution.

1. The ClC-2 KO Mouse

To elucidate the physiological function of ClC-2, a mouse model was generated in which the gene encoding ClC-2 was disrupted (Bösl et al., 2001). Rather unexpectedly, the analysis of ClC-2 KO mice revealed a degeneration of the testicular germ cells and of the retina causing male infertility and blindness. This testicular and retinal degeneration was also observed in a second ClC-2 KO mouse model (Nehrke et al., 2002). In testis and eye, ClC-2 is expressed in the plasma membrane of Sertoli cells and in the retinal pigment epithelium (RPE). Both cell types represent supportive tissue, in the case of the Sertoli cells for germ cells and in the case of the RPE for photoreceptor cells. Hence, loss of ClC-2 causes the degeneration of two cell types which depend on supporting cells that form the blood–testis and blood–retina barriers, respectively. In both instances, the supporting cells transport lactate and fluid. Therefore it is tempting to speculate that ClC-2 might be involved in the ionic homeostasis within the narrow subretinal space between RPE and photoreceptors as well as within the narrow cleft between Sertoli and germ cells (Bösl et al., 2001). As the transport of lactate results in an acidification of the narrow extracellular clefts between them, a tight pH regulation is needed to protect the vulnerable germ cells and photoreceptors from damage. This is mainly achieved by the activity of Cl^-/HCO_3^- exchangers which neutralize the acidic extracellular pH by secreting HCO_3^-. As ClC-2 is activated by low extracellular pH, it was hypothesized to support the pH neutralization in these cells by recycling the Cl^- necessary for the continued activity of the Cl^-/HCO_3^- exchangers. Thus the absence of ClC-2 might cause prolonged extracellular acidification finally resulting in the degeneration of the vulnerable germ cells and photoreceptors (Bösl et al., 2001).

2. ClC-2 in Transepithelial Cl^- Transport

The abundant expression of ClC-2 in epithelia and the degeneration of ClC-2-deficient tissues which depend on Cl^- transport suggest an important role for this channel in transepithelial Cl^- fluxes. To test whether the loss of ClC-2 affects currents across epithelia, Ussing chamber experiments were performed in which epithelial tissue is mounted as a flat sheet between two half-chambers, establishing a luminal and a serosal side. This technique allows the measurement of ion transport across an epithelium. In such experiments, the transepithelial voltage, resistance, and current across the RPE of ClC-2 KO mice was reduced. This underlines the severely compromised function of the RPE in ClC-2 KO mice (Bösl et al., 2001). The direction of transepithelial transport always depends on the localization of the transporting protein. At a voltage of −50 mV, which is typical for epithelial cells, the $[Cl^-]_i$ concentration (30–40 mM) is above its electrochemical equilibrium. Thus the opening of ClC-2 would lead to an efflux of Cl^-. Whether this Cl^- efflux results in Cl^- secretion or reabsorption depends on the localization of ClC-2, as apical Cl^- channels mediate Cl^- secretion while basolateral channels allow salt reabsorption. However, in many epithelia the precise localization of ClC-2 is uncertain: in intestinal epithelia and fetal lung tissue for instance the channel was reported to be expressed in the apical (Murray et al., 1995; Gyömörey et al., 2000; Mohammad-Panah et al., 2001), as well as in the basolateral (Lipecka et al., 2002; Catalán et al., 2004) membrane. However, recent experiments strengthen the case for a basolateral localization of ClC-2 in colon (Pena-Munzenmayer et al., 2005; Zdebik and Jentsch, unpublished data).

Due to the high expression of ClC-2 in fetal lung tissue, the channel was speculated to be involved in lung development (Murray et al., 1995; Blaisdell et al., 2000). However, ClC-2 KO mice did not exhibit any lung phenotype. The originally proposed involvement of ClC-2 in kidney development and gastric acid secretion was not confirmed by the analysis of ClC-2 KO mice either. Thus the role of ClC-2 in these epithelia is still unclear.

3. ClC-2 and CFTR

The expression of ClC-2 in epithelial tissues partially overlaps with that of another Cl^- channel, the cystic fibrosis transmembrane conductance regulator (CFTR) (Riordan et al., 1989) which is described in detail in Chapters 5 and 6. CFTR serves as a cAMP-activated Cl^- channel that mediates Cl^- secretion in many epithelia.

Mutations in CFTR are associated with cystic fibrosis (CF), the most common monogenic disease in Caucasians. Reduced CFTR activity affects mainly organs that depend on Cl^- and the associated salt and fluid transport for their function like lung, intestine, and pancreas. Because of its overlapping expression, ClC-2 has been suggested to function in parallel to CFTR, thus providing a second pathway for Cl^- secretion in these tissues. However, in the lung ClC-2 is expressed early on and declines after birth, whereas CFTR is only found postnatally. Therefore, in this organ a redundant function of the two channel proteins is unlikely. In the intestine, colonic cells residing at the crypt base take up Cl^- via the basolateral transporter NKCC1 and secrete this Cl^- through apical CFTR channels (Fig. 3B) (Welsh et al., 1982; Strong et al., 1994). In CFTR-deficient mice, the impaired intestinal fluid secretion results in a thickening of the intestinal content and intestinal obstruction. The fact that ClC-2 is expressed in the same epithelia as CFTR suggested that it might modulate the CF phenotype which was shown to depend on genetic background both in humans (Bronsveld et al., 2001; Salvatore et al., 2002) and in mice (Rozmahel et al., 1996; Gyömörey et al., 2000). It was hypothesized that an activation of ClC-2 should ameliorate the CF pathology, while the loss of ClC-2 in addition to CFTR was expected to yield a more severe CF phenotype. However, the phenotype of $Clcn2^{-/-}/CFTR^{\Delta F508}$ mice (Zdebik et al., 2004) was mainly a superimposition of the intestinal phenotype of the $CFTR^{\Delta F508}$ mice, which have a severely reduced membrane localization of CFTR (Clarke et al., 1992), and the degenerative phenotype of $Clcn2^{-/-}$ mice (Bösl et al., 2001). Quite contrary to expectations, the double transgenic mice survived even slightly better than the $CFTR^{\Delta F508}$ mice (Zdebik et al., 2004). From these data it was concluded that ClC-2 does not provide a parallel Cl^- secretion pathway to CFTR and thus cannot compensate for a loss of CFTR. The improved survival of $Clcn2^{-/-}/CFTR^{\Delta F508}$ double transgenic mice might rather be explained by an increased Cl^- secretion or decreased Cl^- absorption compatible with a basolateral instead of an apical localization of ClC-2 in the intestine. Indeed, recently it was shown that in murine colon, ClC-2 exclusively localizes to the basolateral membranes of surface epithelia (Pena-Munzenmayer et al., 2005) suggesting its involvement in Cl^- absorption (Fig. 3A). This basolateral ClC-2 staining of intestinal surface cells was also observed in our lab. The staining was completely abolished on ClC-2 KO sections, proving the specificity of the labeling (Zdebik and Jentsch, unpublished data). These findings and in particular the data of the $Clcn2^{-/-}/CFTR^{\Delta F508}$ double transgenic mice should end the speculations that ClC-2 activation might improve the CF phenotype.

4. ClC-2, a Candidate for $I_{Cl,swell}$?

In many cell types, hypotonic cell swelling activates a typical Cl^- current ($I_{Cl,swell}$) which participates in RVD. This current is described in detail in Chapter 8. As ClC-2 is stimulated by cell swelling (Gründer et al., 1992) and was shown to accelerate the RVD of insect cells (Xiong et al., 1999) on heterologous overexpression, it was suggested to play a role in RVD. However, ClC-2 cannot be the molecular correlate of $I_{Cl,swell}$ as there are great differences in its biophysical properties in respect to rectification and ion selectivity (Gründer et al., 1992; Jordt and Jentsch, 1997). ClC-2-deficient parotid acinar cells did not exhibit changes in their RVD (Nehrke et al., 2002) and ClC-2 KO mice did not show any morphological abnormalities in organs exposed to large osmotic changes like the kidney (Bösl et al., 2001; Nehrke et al., 2002). Hence, the physiological importance of the activation of ClC-2 by cell swelling is still not resolved.

5. ClC-2 in Brain Tissue

Aside from epithelia, ClC-2 was also found to be highly expressed in brain. ClC-2-like currents were obtained in neurons as well as glial cells (Staley, 1994; Smith et al., 1995; Ferroni et al., 1997; Nobile et al., 2000) such as cortical and hippocampal astrocytes (Makara et al., 2001, 2003). In the hippocampus, these currents correlated with the expression of ClC-2 transcripts and protein (Smith et al., 1995; Sik et al., 2000; Gulacsi et al., 2003). Immunoelectron microscopy analysis localized ClC-2 to pyramidal neurons of the CA1 and CA3 regions of the hippocampus as well as to astrocytes. In pyramidal neurons, ClC-2-specific antibodies stained the soma as well as dendrites which were located close to presumed GABAergic inhibitory synapses. ClC-2 was found in astrocytic endfeet that contact blood vessels, as well as in neurons close to inhibitory synapses. What might be the role of ClC-2 in these cell types? In the case of glial cells, ClC-2 could play a role in maintaining the extracellular ion composition and thus might be important for pH and volume regulation (Walz, 2002). The expression of ClC-2 in neurons which are close to inhibitory synapses suggested a role of ClC-2 in maintaining the low $[Cl^-]_i$ of adult neurons which is necessary for the inhibitory response of postsynaptic GABA and glycine receptors. These receptors are ligand-gated Cl^- channels that may yield hyperpolarizing or depolarizing currents on activation, depending on whether the $[Cl^-]_i$ is below or above its electrochemical equilibrium (Miles, 1999). Whereas hyperpolarizing currents result in the typical inhibitory response, depolarizing currents may be excitatory. Indeed, depolarizing excitatory responses due to low $[Cl^-]_i$ occur early in development, as well as in certain adult neurons like for instance in the spinal root ganglia. The excitation early in development gives later way to inhibition, as the neuronal $[Cl^-]_i$ decreases due to the expression of the cation-chloride cotransporter KCC2. This cotransporter is most likely the main player in lowering the $[Cl^-]_i$ (Rivera et al., 1999; Hübner et al., 2001) (see also Chapter 10), but other transport proteins like KCC3 (Böttger et al., 2003) and ClC-2 may also play a role. Indeed, the adenoviral transfection of ClC-2 into cultured spinal root ganglia changed their excitatory GABA response to an inhibitory response by decreasing $[Cl^-]_i$ (Staley et al., 1996). As ClC-2 is activated by high $[Cl^-]_i$, it was suggested to prevent the neuronal accumulation of Cl^- above its electrochemical equilibrium. Being a channel, ClC-2 can mediate Cl^- efflux only when Cl^- has accumulated above its equilibrium concentration. This might occur during high-frequency stimulation when extracellular K^+ increases and potentially causes a reversal of KCl cotransport (Kaila et al., 1997; DeFazio et al., 2000).

According to the proposed role of ClC-2 in lowering the $[Cl^-]_i$ of neurons, mutations in ClC-2 were suggested to give rise to epilepsy in humans (Haug et al., 2003). This idea was supported by the fact that a locus for multigenic idiopathic epilepsy (IGE) was mapped to 3q26, a chromosomal region harboring the human *CLCN2* gene (3q27) (Sander et al., 2000). Indeed in 2003, Haug et al. (2003) identified three *CLCN2* mutations by screening 46 unrelated families with IGE which had been previously mapped to 3q26. These mutations cosegregated in three pedigrees with epileptic syndromes in an autosomal-dominant fashion. One missense mutation (G715E) in the C-terminus of ClC-2 and two mutations which truncate the protein at helix F (Gins597) or lead to an increase in the skipping of exon 3 (IVS2–14del11) were described. The latter mutations were reported to be nonfunctional and to exert a dominant-negative effect on ClC-2 WT channels. However, in ClC-1, the closest homologue of ClC-2, the respective truncating mutations are associated with recessive myotonia (Pusch, 2002). The missense mutation G715E, which lies between CBS1 and CBS2, was reported to change the voltage dependence of ClC-2 to more positive voltages in a Cl^--dependent manner (Haug et al., 2003). This represents a gain of function contrasting with the loss of function caused by the truncating mutations. As the effects published by Haug et al. could neither be reproduced in our lab (Blanz and Jentsch, unpublished data) nor by Sepúlveda and colleagues (Niemeyer et al., 2004), it is puzzling how these mutations in *CLCN2* might contribute to the epilepsy phenotype observed in the patients. In the likely absence of a dominant-negative effect, the truncation mutations might act via haploinsufficiency with the remaining functional copy of the gene not sufficing to produce enough protein. This, however, contrasts with the lack of epilepsy in ClC-2 KO mice which totally lack ClC-2 protein (Bösl et al., 2001). In two additional screens for ClC-2 mutations in epileptic patients, only two polymorphisms were identified (D'Agostino et al., 2004; Marini et al., 2004). These do not alter the biophysical properties of ClC-2 in *Xenopus* oocytes (Blanz and Jentsch, unpublished data). Thus, at this point it is not clear yet whether ClC-2 mutations underlie monogenic epilepsy. Recently it was shown that CBS domains function as binding sites for nucleotides (Scott et al., 2004). The G715E mutation, even though it lies in between the CBS domains, was reported to impair the binding of AMP which changed the gating kinetics of WT but not of mutated ClC-2 proteins (Niemeyer et al., 2004). The physiological importance of ClC-2 modulation by AMP is presently unclear. The recently crystallized CBS domains of ClC-0, however, were reported not to bind ATP (Meyer and Dutzler, 2006).

D. ClC-Ka/Kb: Cl^- Channels Expressed in Kidney and Inner Ear

ClC-Ka and ClC-Kb (ClC-K1 and ClC-K2 in rodents) (Uchida et al., 1993; Adachi et al., 1994; Kieferle et al., 1994) are highly homologous Cl^- channels that are predominantly expressed in the kidney—hence their name ClC-K—and in the inner ear. Their high homology is probably the consequence of a recent gene duplication, since ClC-Ka and ClC-Kb are located very close to each other on chromosome

1p36 (Brandt and Jentsch, 1995). As the sequence of the ClC-K channel isoforms is nearly 90% identical within a single species, species orthologues could not be identified by sequence comparison. However, morphological and physiological data suggest that ClC-K1 corresponds to ClC-Ka and ClC-K2 to ClC-Kb.

1. ClC-K Expression

In the inner ear, both ClC-Ka and ClC-Kb are expressed in the marginal cells of the stria vascularis and in the dark cells of the vestibular organ (Ando and Takeuchi, 2000; Estévez et al., 2001). Both cell types are involved in K^+ secretion into the endolymph. In the kidney, ClC-Ka is found in the thin ascending limb of the loop of Henle, a highly Cl^- permeable nephron segment (Uchida et al., 1995). The subcellular localization of ClC-Ka is not yet clear, since it was reported to be expressed apically and basolaterally (Uchida et al., 1995) as well as exclusively basolaterally (Vandewalle et al., 1997). In contrast, ClC-Kb is clearly expressed in basolateral membranes of the distal convoluted tubule, of intercalated cells in the collecting duct, and in the thick ascending limb (TAL) of Henle's loop (Estévez et al., 2001).

2. ClC-K and Barttin

As the immunohistochemistry in the kidney showed a clear plasma membrane staining for ClC-K channels and because physiological data suggested a role in transepithelial transport, it was surprising that only the rodent ClC-K1 channel yielded Cl^- currents when expressed in Xenopus oocytes (Uchida et al., 1995; Waldegger and Jentsch, 2000). No currents were observed with either human ClC-Ka or ClC-Kb (Kieferle et al., 1994; Waldegger and Jentsch, 2000). These findings suggested that ClC-K channels may need a β-subunit for their functional expression. This, indeed, turned out to be true. In 2001, Hildebrandt and colleagues identified the small transmembrane protein barttin by positional cloning of the BSND gene that underlies Bartter syndrome type IV (Birkenhäger et al., 2001). Barttin is a small protein with a predicted molecular weight of around 50 kDa. It does not belong to a larger gene family. Barttin has two predicted transmembrane domains and intracellular N- and C-termini (Estévez et al., 2001). Its expression overlaps with that of ClC-K channels, suggesting that it might interact with them at the plasma membrane. When barttin was coexpressed with ClC-Ka und ClC-Kb in Xenopus oocytes, large Cl^- currents were observed, a consequence of the enhanced surface expression of ClC-K/barttin channels (Estévez et al., 2001). Accordingly, barttin seems to be necessary for the transport of ClC-K to the plasma membrane. So far, ClC-K channels are the only proteins that were reported to interact with barttin. The physiological function of Cl^- currents mediated by ClC-Ka/barttin and ClC-Kb/barttin in epithelia of the kidney and inner ear was gleaned from a KO mouse model for ClC-K1 (Matsumura et al., 1999) and by the hereditary disease Bartter syndrome which is caused by mutations in ClC-Kb and barttin (Simon et al., 1997; Birkenhäger et al., 2001).

3. Renal Functions of ClC-K/Barttin

In the nephron, many components of the primary filtrate like salts, organic solutes, and proteins are reabsorbed as the fluid passes the tubular epithelium. The salt reabsorption results from a complex interplay between ion channels, pumps, and transporters. Dysfunction of these proteins can lead to renal disorders like Bartter syndrome, a group of closely related hereditary tubulopathies characterized by renal salt wasting, hypokalemic metabolic acidosis, and hyperreninemic hyperaldosteronism with normal blood pressure. There are several different variants of Bartter syndrome, denoted Bartter syndrome type I–IV, which are caused by mutations in different proteins involved in the reabsorption process. All subtypes of Bartter syndrome show an autosomal recessive inheritance. The classical Bartter syndrome occurs during infancy or early childhood, but there are also two antenatal forms, one of which (type IV) is associated with deafness. The renal salt loss in Bartter syndrome is the consequence of a defective salt reabsorption in the TAL of Henle's loop. Four main proteins involved in this process have been identified (Fig. 3D). The uptake of NaCl is mediated by the Na^+-dependent K^+/Cl^- cotransporter NKCC2 (*SLC12A1*) which uses the Na^+ gradient established by the basolateral Na^+/K^+-ATPase to transport K^+ and Cl^- into the cell. Mutations in NKCC2 give rise to antenatal forms of Bartter syndrome type I (Simon et al., 1996a). To allow continued NKCC2 activity, the K^+ that has been taken up has to be recycled, as the luminal K^+ concentration is so low that K^+ would become limiting. This K^+ recycling occurs via the apical K^+ channel ROMK1/Kir1.1. Mutations in its gene (*KCNJ1*) underlie Bartter syndrome type II (Simon et al., 1996b). Whereas Na^+ leaves the cell via the basolateral Na^+/K^+-ATPase, the transported Cl^- crosses the basolateral plasma membrane through ClC-Kb channels which need to be associated with the β-subunit barttin to be functional. Thus mutations in either protein can cause Bartter syndrome. Mutations in the gene *CLCNKB* encoding ClC-Kb result in Bartter syndrome type III (Simon et al., 1997), whereas mutations in the barttin gene *BSDN* are associated with Bartter syndrome type IV (Birkenhäger et al., 2001), an antenatal severe form of Bartter syndrome accompanied by deafness.

The function of ClC-Kb/barttin in acid-secreting α-intercalated cells and base-secreting β-intercalated cells of the collecting duct is less well understood. In α-cells, ClC-Kb/barttin might serve to recycle Cl^- for the basolateral Cl^-/HCO_3^- exchanger AE1 (Fig. 3E). This role may also be performed by the K^+/Cl^--cotransporter KCC4, the disruption of which led to renal tubular acidosis in mice (Böttger et al., 2002). In the presence of the severe disturbance of renal salt handling in Bartter patients, an additional distal acidification defect may not be easily detectable. In β-cells, ClC-K/barttin channels have been suggested to allow the efflux of Cl^- taken up by the HCO_3^-/Cl^- exchanger pendrin thus mediating Cl^- reabsorption (Fig. 3F).

The physiological function of ClC-Ka was clarified by generating and analyzing a corresponding mouse model (Matsumura et al., 1999). ClC-K1 is the principal Cl^-

channel in the thin ascending limb of Henle's loop (Uchida et al., 1995; Matsumura et al., 1999). The high Cl$^-$ permeability of this nephron segment is essential for establishing the high osmolarity of the renal medulla which is used to drive water reabsorption in later nephron segments. In the absence of ClC-Ka, the solute accumulation in the renal medulla and consequently also the water reabsorption was impaired, resulting in renal water loss. The large increase in urinary volume in ClC-Ka-deficient mice (Matsumura et al., 1999) resembles nephrogenic diabetes insipidus. However, so far no mutations in ClC-Ka were identified in human patients with diabetes insipidus.

4. ClC-K Function in the Inner Ear

What is the role of ClC-K channels in the inner ear? Mutations in barttin, but not in ClC-Kb or ClC-Ka alone, are associated with deafness. This deafness might be the consequence of an altered composition of the cochlear endolymph of the scala media. This endolymph is normally very rich in K$^+$ (150 mM). This high K$^+$ concentration and the positive endocochlear potential are essential for the hearing process. The K$^+$ is secreted by the marginal cells of the stria vascularis by an interplay of different transporters and channels (Fig. 3C). Defects in any of the necessary components of the K$^+$ secretion system are associated with deafness. The basolateral cotransporter NKCC1 takes up K$^+$ and Cl$^-$ in a Na$^+$-dependent manner. K$^+$ leaves the cells apically through the K$^+$ channel KCNQ1 which is associated with the β-subunit KCNE1. Na$^+$ recycles basolaterally via the Na$^+$/K$^+$-ATPase and Cl$^-$ via ClC-Ka/barttin and ClC-Kb/barttin channels. In the inner ear, both ClC-K channels are thought to be coexpressed in the same cells. Consequently, mutations in ClC-Kb or ClC-Ka alone are not associated with deafness as the residual Cl$^-$ channel activity of the other isoform is sufficient to support K$^+$ secretion. However, when barttin is mutated, both ClC-K channels are affected and thus cannot compensate for each other. This model is further supported by the identification of a patient suffering from Bartter syndrome type IV as a consequence of mutations in ClC-Ka and ClC-Kb (Schlingmann et al., 2004).

E. ClC-3: An Intracellular CLC Protein Located on Endosomes and Synaptic Vesicles

ClC-3 which was first cloned by Kawasaki et al.(1994) and Borsani et al. (1995) is a rather broadly expressed member of the CLC family. It was detected in brain, retina, kidney, liver, epididymis, adrenal gland, skeletal muscle, heart, and pancreas (Stobrawa et al., 2001; Isnard-Bagnis et al., 2003). Within the brain, ClC-3 is abundantly expressed in the hippocampus, cerebellum, and bulbus olfactorius (Stobrawa et al., 2001). In the kidney, ClC-3 was reported to be highly expressed in acid-reabsorbing β-intercalated cells (Obermüller et al., 1998).

1. Subcellular Localization of ClC-3

Using subcellular fractionation experiments and immunofluorescence studies of cells transfected with epitope-tagged ClC-3, the protein was localized to endosomes, to glutamatergic and GABAergic synaptic vesicles, and to the related small synaptic-like microvesicles of PC12 cells (Stobrawa et al., 2001; Salazar et al., 2004; Hara-Chikuma et al., 2005b). Its localization on endosomes and synaptic vesicles is mutually consistent, as synaptic vesicles derive from endosomes and often also recycle through an endosomal compartment at the synaptic terminal (Südhof, 2000). The sorting of ClC-3 to endosomes and synaptic vesicles is thought to be mediated by the adaptor complex AP3 (Salazar et al., 2004), which exists in a neuronal and a ubiquitous isoform. The localization of ClC-3 on endosomes and/or synaptic vesicles seems to depend on recognition by both AP3 isoforms (Seong et al., 2005). The ubiquitous isoform may sort it to endosomes and the neuronal isoform to synaptic vesicles. However, there is no biochemical evidence so far for an interaction between ClC-3 and AP3. Although ClC-3 is mainly located on intracellular vesicles, the protein can reach the plasma membrane on overexpression (Weylandt et al., 2001), which has allowed its electrophysiological characterization. The currents published by Weinman and colleagues show the Cl^- over I^- conductance and strong outward rectification (Li et al., 2000) previously reported for ClC-4 and ClC-5 (Steinmeyer et al., 1995; Friedrich et al., 1999). The different currents which were attributed to ClC-3 earlier on (Kawasaki et al., 1994, 1995; Duan et al., 1997) were probably endogenous to the expression system. In some cell types, overexpression of ClC-3 resulted in artificial, large intracellular vesicles that were acidic and stained for ClC-3 and lysosomal markers (Li et al., 2002).

2. ClC-3 KO Mouse Models

The importance of ClC-3 became apparent from the analysis of ClC-3 KO mouse models. Three ClC-3 KO lines have been generated (Stobrawa et al., 2001; Dickerson et al., 2002; Yoshikawa et al., 2002). Previously, ClC-3 had been suggested to mediate the swelling-activated Cl^- current $I_{Cl,swell}$ that is present in most tissues and has an I^- over Cl^- conductance (Duan et al., 1997). This, however, contrasted with the Cl^- over I^- conductance reported by Weinman (Li et al., 2000) which agreed well with the ion conductance of the close homologues ClC-4 and ClC-5. In fact, the currents measured by Weinman and colleagues later proved to be indeed mediated by ClC-3, as the mutation of its "gating glutamate" had the same effect on the properties of these currents (Li et al., 2002) as the corresponding mutation in ClC-4 and ClC-5 on their currents (Friedrich et al., 1999).

The analysis of ClC-3-deficient mice clearly established that ClC-3 is not the molecular counterpart of $I_{Cl,swell}$ (see also Chapter 8). This outwardly rectifying current that inactivates at positive voltages and displays an I^- over Cl^- conductivity was still present in ClC-3-deficient hepatocytes, pancreatic acinar cells (Stobrawa et al., 2001), salivary gland cells (Arreola et al., 2002) as well as cardiac myocytes

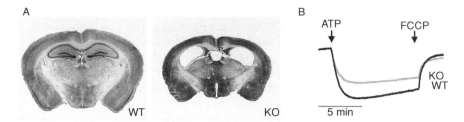

Figure 6. Neuronal phenotype of ClC-3-deficient mice. (A) Nissl-stained sections of adult WT and ClC-3 KO brains illustrate the complete degeneration of the hippocampus in ClC-3 KO mice. (B) *In vitro* acidification assays using WT and ClC-3 KO synaptic vesicle preparations indicate a 50% reduction in the acidification of ClC-3-deficient vesicles. Acidification was induced by addition of ATP to the vesicle preparations in the presence of 60-mM KCl and quantified by the pH-dependent trapping and quenching of the fluorescent dye acridine orange in the vesicular lumen. At the end of the experiment, the pH gradient was collapsed by addition of the protonophore FCCP. Modified from Stobrawa *et al.* (2001).

(Yamamoto-Mizuma *et al.*, 2004). Thus the previously reported currents that resembled $I_{Cl,swell}$ were most likely endogenous to the expression system.

The disruption of ClC-3 caused a severe neurodegeneration that resulted in a complete loss of the hippocampus (Fig. 6A) and of photoreceptors (Stobrawa *et al.*, 2001). The hippocampal degeneration became apparent at 2 weeks of age and affected initially only pyramidal cells of the CA1 region. Afterward it spread to the other regions of the hippocampus. At 3 months of age, the hippocampus was replaced by large cavities contiguous with the ventricular system. The neurodegeneration was accompanied by an activation of microglia and astrogliosis (Stobrawa *et al.*, 2001; Dickerson *et al.*, 2002). In spite of the severe neurodegeneration, ClC-3 KO mice were viable and able to survive for more than 1 year. However, there was an overall increase in lethality and a reduction in weight. Mice lacking the hippocampus showed abnormal folding of the hind limbs when held at the tail (Dickerson *et al.*, 2002), and exhibited increased motor activity and performance deficits on hanging wire tests and in rotarod assays (Stobrawa *et al.*, 2001; Yoshikawa *et al.*, 2002). They were still able to acquire new motor skills as judged by rotarod assay (Stobrawa *et al.*, 2001). In general, the hippocampus is assumed to be important for learning. An in depth analysis of the spatial learning capabilities of ClC-3 KO mice which is known to depend on vision, however, was precluded by their blindness. Besides, the broad distribution of the neurodegeneration should render the analysis of a hippocampus-specific learning deficit in the constitutive ClC-3 KO mice difficult.

3. Involvement of ClC-3 in Vesicular Acidification

Vesicular acidification depends on a counterion conductance to achieve the necessary charge neutralization. As ClC-3 is present on endosomes and synaptic vesicles and mediates Cl^- fluxes, it was a likely candidate for the shunt pathway of these vesicles.

Indeed, *in vitro* experiments revealed a slower acidification rate of ClC-3-deficient synaptic vesicles (Stobrawa *et al.*, 2001) (Fig. 6B), suggesting a contribution of ClC-3 to the charge neutralization required for the efficient operation of the electrogenic H^+-ATPase that acidifies the vesicles (Fig. 4B). The ATPase-mediated H^+ import into the vesicles generates a pH gradient as well as a membrane potential. Both gradients contribute to the uptake of neurotransmitters into synaptic vesicles. The import of monoamines and acetylcholine depends predominantly on the pH gradient (Johnson, 1987; Parsons *et al.*, 1993), whereas the potential is the decisive factor in the uptake of glutamate (Maycox *et al.*, 1988; Bellocchio *et al.*, 2000). An impaired charge neutralization due to the loss of ClC-3 should cause an increased membrane potential and a reduced pH gradient. Accordingly, glutamatergic vesicles should contain more neurotransmitter. Instead, ClC-3 KO synaptic vesicle preparations displayed reduced glutamate uptake (Stobrawa *et al.*, 2001). However, these synaptic vesicle preparations stemmed from animals already affected by neurodegeneration which seems to have caused the preferential loss of glutamatergic vesicles, as suggested by the detection of reduced amounts of the glutamate transporters in the preparation (Stobrawa *et al.*, 2001). Electrophysiological analysis of hippocampal slices from animals not yet suffering from overt neurodegeneration did not reveal prominent changes in synaptic transmission (Stobrawa *et al.*, 2001) except for a slight increase in the amplitudes of miniature excitatory postsynaptic currents which would agree with an enhanced glutamate uptake into ClC-3-deficient glutamatergic vesicles.

The mechanism of the neurodegeneration is still unclear. If glutamatergic vesicles indeed contain a greater amount of glutamate, the observed neurodegeneration in the brain might partially be due to glutamate toxicity. However, this explanation probably does not hold true for photoreceptor cells, as these cells are quite insensitive to activators of glutamate receptors (Facci *et al.*, 1990). On the other hand, cellular trafficking defects originating from disturbed endosomal function might be involved, as ClC-3 not only contributes to the efficient acidification of synaptic vesicles, but also to that of endosomes (Fig. 5). The initial observation of an elevated pH in a ClC-3-deficient hepatic vesicle fraction mainly representing endosomes (Yoshikawa *et al.*, 2002) has since been confirmed by *in vivo* measurements of endosomal pH in ClC-3 KO hepatocytes as well as ClC-3-transfected CHO-K1 cells (Hara-Chikuma *et al.*, 2005b). While in the absence of ClC-3 the pH of early endosomes was more alkaline by ~0.4 units, it was ~0.4 units more acidic in cells which overexpressed ClC-3 suggesting the ion conductance (rather than the ATPase) to be rate limiting in endosomal acidification. In the case of an endosomal acidification defect in ClC-3-deficient neurons, there might be a mislocalization of crucial membrane proteins or changes in luminal receptor ligand interactions or enzymatic activities which could contribute to the observed neuronal cell death.

ClC-3 has also been implicated in insulin secretion lately. It was reported to reside on the large dense core vesicles of β-cells which contain insulin and to participate in their acidification (Barg *et al.*, 2001). A sorting of ClC-3 to this compartment would be unrelated to its presence on endosomes and SVs, as the large dense core vesicles are not related to the endosomal compartment, but rather derive from the trans golgi network

(TGN). However, this result hinges on the quality of the antibody. A detailed evaluation of the potential involvement of ClC-3 in insulin secretion using the available ClC-3 KO mouse models has not yet been published. A first analysis of glucose homeostasis in ClC-3-deficient mice revealed that they were not hyperglycemic (Dickerson *et al.*, 2002). However, insulin levels following a glucose load were not measured.

4. ClC-3 Splice Variants

Apart from the originally cloned ClC-3, now sometimes called ClC-3A, a splice variant denoted ClC-3B has been described (Ogura *et al.*, 2002). In this splice variant, a stretch of 76bp is inserted into the C-terminus. As a result, the 29 C-terminal amino acids of ClC-3A are replaced with 77 different amino acids in ClC-3B. This new extended C-terminus contains a PDZ-binding motif. ClC-3B is predominantly expressed in epithelial cells, where it localizes to the Golgi apparatus (Gentzsch *et al.*, 2003). It was shown to interact with EBP50, PDZK1, and GOPC via its PDZ-binding motif (Gentzsch *et al.*, 2003). The interaction between ClC-3B and PDZK1 was confirmed in our lab (Maritzen and Jentsch, unpublished data). PDZK1 was found to promote interactions between ClC-3B and CFTR, as the two proteins bind to distinct PDZ domains of PDZK1 (Gentzsch *et al.*, 2003). However, the physiological importance of this finding is obscure, as the bulk of ClC-3B is located intracellularly, while CFTR resides mainly at the plasma membrane.

F. **ClC-4: A Vesicular CLC Protein of Unclear Function**

ClC-4 was identified 10 years ago during analysis of the chromosomal region Xp22.3 (van Slegtenhorst *et al.*, 1994) and by homology cloning (Jentsch *et al.*, 1995). However, the physiological role of ClC-4 is still not resolved. Like ClC-3, ClC-4 is quite broadly expressed. In mice, it was detected in skeletal muscle, heart, liver, and kidney. It is also abundantly expressed in brain, especially in the hippocampus (Adler *et al.*, 1997; Stobrawa *et al.*, 2001). There may be differences in the tissue distribution of ClC-4 between humans (van Slegtenhorst *et al.*, 1994) and rats (Jentsch *et al.*, 1995).

1. Involvement of ClC-4 in Endosomal Acidification?

Like its close relatives ClC-3 and ClC-5, ClC-4 resides primarily on intracellular membranes. In HEK293 cells, the localization of overexpressed epitope-tagged ClC-4 closely resembled that of ClC-3 and ClC-5, which are all present in endosomal compartments (Suzuki *et al.*, 2006) (Fig. 5). Recently, ClC-4 and ClC-5 were reported to be colocalized in endosomes of the renal proximal tubules (Mohammad-Panah *et al.*, 2003). In this study, the disruption of ClC-4 by antisense cDNA impaired endosomal acidification and transferrin receptor recycling to the same degree as the disruption of its homologue ClC-5, which is known to be crucial for renal proximal tubular endosomal acidification (Günther et al., 2003). Defects in endocytosis due to mutations in ClC-5 cause the X-chromosomal Dent's disease which is characterized by proteinuria (Section II.G). Although ClC-4 has been implicated in endosomal acidification as well, ClC-4 and ClC-5 are obviously not redundant in the kidney, as ClC-4 cannot compensate for the loss of ClC-5. ClC-5 KO mice and Dent's disease patients display proteinuria although expressing ClC-4. Due to the high degree of sequence identity between ClC-4 and ClC-5, their similar electrophysiological properties (Friedrich et al., 1999), their ability to form heterodimers *in vitro* (Mohammad-Panah et al., 2003; Suzuki et al., 2006), and the X-chromosomal location of the human *CLCN4* gene, ClC-4 has been suggested as an additional candidate gene for Dent's disease. Indeed, there are patients for this disease that do not harbor any mutations in ClC-5 (Ludwig et al., 2003). However, a study screening such patients for mutations in ClC-4 did not confirm this hypothesis (Ludwig and Utsch, 2004). In another study, overexpression of ClC-4 in cell lines increased copper incorporation into ceruloplasmin (Wang and Weinman, 2004), resembling the function of the yeast homologue Gef1p in iron metabolism (Greene et al., 1993; Davis-Kaplan et al., 1998; Schwappach et al., 1998) (Section III.A). Although no disruption of ClC-4 by homologous recombination has been reported to date, ClC-4-deficient mice are available. The surprising finding that the mouse *Clcn4* gene is located on the X chromosome in the mouse strain *Mus spretus*, while it resides on chromosome 7 in the laboratory mouse strain C57BL/6J enabled Rugarli et al. (1995) to generate ClC-4-deficient mice by crossing these two different strains. No obvious phenotype was found except for infertility, which, however, is expected when crossing two distantly related strains without the need to invoke a role of ClC-4. As the involvement of ClC-4 in endosomal acidification and copper incorporation into ceruloplasmin has not yet been verified by analysis of the ClC-4-deficient mice, the physiological importance of these findings remains unclear.

G. ClC-5 and Dent's Disease

Although identified later than ClC-3 and ClC-4 (Fisher et al., 1994, 1995; Steinmeyer et al., 1995), ClC-5 is the most thoroughly studied CLC protein within the second branch of the CLC family. ClC-5 was first identified by positional cloning as a candidate gene for the X-chromosomal Dent's disease by Thakker and colleagues (Fisher et al., 1994). This renal disorder was first described in 1964 by Dent and Friedman (1964). Since the cloning of ClC-5, many patients suffering from Dent's disease were shown to harbor mutations in the gene encoding ClC-5 (Lloyd et al., 1996). Historically, several other names (X-linked recessive nephrolithiasis, X-linked recessive hypophosphatemic rickets) were given to this disorder (Lloyd et al., 1996). As these diseases are now known to share the same genetic mechanism, the term Dent's disesase is used for all of these.

1. ClC-5 Expression and Subcellular Localization

ClC-5 displays a more restricted expression pattern than ClC-3 and ClC-4. While moderate levels of ClC-5 have been detected in intestine, brain, liver, and testis (Fisher et al., 1995; Steinmeyer et al., 1995; Vandewalle et al., 2001), ClC-5 is most abundantly expressed in the kidney (Günther et al., 1998; Devuyst et al., 1999; Sakamoto et al., 1999). In this organ, ClC-5 expression is restricted to the epithelial cells of all three segments of the proximal tubule (Günther et al., 1998), the acid-secreting α-intercalated cells (Günther et al., 1998; Obermüller et al., 1998; Devuyst et al., 1999; Sakamoto et al., 1999), and the base-secreting β-intercalated cells (Günther et al., 1998; Sakamoto et al., 1999) of the collecting duct (Günther et al., 1998). A weaker expression is also found in the TAL of Henle's loop (Devuyst et al., 1999).

Like ClC-3 and ClC-4, ClC-5 resides predominantly on endosomes (Fig. 5). This has been demonstrated in a variety of cell types. In transfected fibroblasts, only a minority of ClC-5 was detected at the plasma membrane even on overexpression. The bulk of the protein resided in small vesicles where it colocalized with endocytosed protein (Günther et al., 1998). On cotransfection with a constitutively active rab5 mutant, ClC-5 was targeted to the arising enlarged early endosomes (Günther et al., 1998). In the rat small intestine and colon ClC-5 has been detected on apical vesicles, where it partially colocalizes with transcytosed polymeric immunoglobulin receptor (Vandewalle et al., 2001). On fractionation of intestinal homogenates, ClC-5 copurified with rab4, rab5a, and the H^+-ATPase. Thus it is assumed that ClC-5 is present in the endocytotic and transcytotic pathways of intestinal epithelial cells. In α-intercalated cells of the kidney, ClC-5 colocalizes also with the H^+-ATPase that is present in apical vesicles and can be inserted into the plasma membrane by regulated exocytosis. However, it was still in apical vesicles in β-intercalated cells, where the H^+-ATPase is present in basolateral vesicles. It is still unclear whether ClC-5 plays an important role in these cells. This is also true for the TAL of Henle's loop, where some intracellular staining was observed by sensitive immunohistochemical methods (Devuyst et al., 1999). Immunofluorescence analysis of ClC-5 in proximal tubules revealed high expression right underneath the brush border of the proximal tubular cells. ClC-5 colocalized with the V-type ATPase (Günther et al., 1998; Sakamoto et al., 1999). When the uptake of labeled protein was simultaneously analyzed, ClC-5 was found to colocalize with the endocytosed material only at early time points (2 min) arguing for its presence on the apical early endosomes of proximal tubular cells (Günther et al., 1998) (Fig. 4A). This subcellular localization of ClC-5 suggested an important role for ClC-5 in proximal tubular endocytosis, most probably by providing the neutralizing current for the H^+-ATPase-mediated endosomal acidification—analogous to the role of ClC-3 in hepatic endosomes that was discovered later.

About 99% of all the low-molecular-weight proteins that pass the glomerular sieve are resorbed in the proximal tubule. In the presence of proximal tubular dysfunction, commonly denoted as Fanconi syndrome, filtered proteins are not efficiently resorbed any more and thus accumulate in the urine, a state called proteinuria. Low-molecular-weight (LMW) proteinuria is in fact the most common symptom of Dent's disease, strengthening the case for a potential role of ClC-5 in proximal tubular endocytosis.

2. The Mechanism of Dent's Disease

To understand the physiological function of ClC-5 and the pathophysiological mechanism behind Dent's disease, the groups of Jentsch and Guggino independently generated ClC-5 KO mice by homologous recombination (Piwon et al., 2000; Wang et al., 2000). Proximal tubular endocytosis was studied by following the fate of fluorescently labeled proteins that were injected into the bloodstream under anesthesia by confocal microscopy (Piwon et al., 2000). As the gene encoding ClC-5 is located on the X chromosome, female mice that are heterozygous for the disrupted allele display mosaic expression, that is, some cells express ClC-5, whereas other cells of the same tubule lack it. This mosaic pattern can clarify which effects of ClC-5 are cell-autonomous, rather than being due to secondary systemic changes (e.g., in hormones). The analysis of heterozygous females revealed that the uptake of labeled protein into the proximal tubular cells was reduced by $\sim$70% in a cell-autonomous manner (Piwon et al., 2000) (Fig. 7A). Likewise, the uptake of fluorescently labeled dextrans was impaired, indicating that receptor-mediated as well as fluid phase endocytosis is affected by the loss of ClC-5 (Piwon et al., 2000). In vitro experiments using preparations of cortical endosomes demonstrated a slower acidification rate and lower steady state levels of acidification of vesicles derived from ClC-5 KO mice (Piwon et al., 2000; Günther et al., 2003). These results have been confirmed more recently by in vivo pH measurements in cultured proximal tubular cells. In these experiments, the initial pH of transferrin-labeled endosomes was $\sim$7.2. In WT cells, the pH had decreased to $\sim$6.0 after 15 min, whereas it reached only $\sim$6.5 in ClC-5-deficient cells (Hara-Chikuma et al., 2005a). The increase in vesicular Cl^- concentration that accompanies acidification was reduced in the absence of ClC-5 as well. Late endosomes on the other hand were not affected. Thus ClC-5 is indeed involved in the acidification of early endosomes in the proximal tubule by a Cl^- shunt mechanism. However, the inhibition of endosomal acidification and Cl^- accumulation in ClC-5 KO proximal tubule cells by the Cl^- channel blocker NPPB suggests the presence of additional endosomal Cl^- channels that are not yet identified (Hara-Chikuma et al., 2005a). Although the contribution of ClC-5 to the endosomal acidification is clearly established, it is still difficult to reconcile its function in endosomes with its electrophysiological properties. Analyses of ClC-5 expressed heterologously in Xenopus oocytes showed strongly outwardly rectifying currents that were only measurable at voltages $>+20$ mV (Steinmeyer et al., 1995; Lloyd et al., 1996; Friedrich et al., 1999), and decreased on acidification of the extracellular medium (Friedrich et al., 1999). The physiological significance of the extremely strong outward rectification is unclear, as voltages more positive than $+20$ mV are unlikely to be reached in the plasma membrane of nonexcitable cells or intracellular vesicles under physiological conditions. Accordingly ClC-5 should mediate hardly any Cl^- flux in endosomes because of their acidic pH and their lumenal potential that is likely to be positive. To reconcile electrophysiological data and physiological function one has to assume that vesicularly located ClC-5 behaves differently from ClC-5 measured at the plasma membrane of oocytes. This might be due to a modulation of its properties via the interaction with so

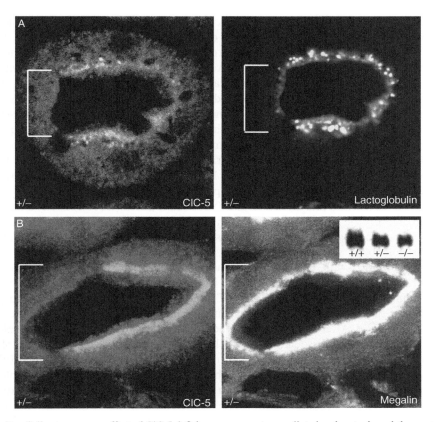

Figure 7. Cell-autonomous effect of ClC-5 deficiency on receptor-mediated endocytosis and the expression of the endocytic receptor megalin. Female mice that are heterozygous for the disrupted ClC-5 allele express ClC-5 in a mosaic pattern due to random inactivation of one of their X chromosomes. Cells lacking ClC-5 are indicated by the brackets. (A) Receptor-mediated endocytosis was investigated by injecting fluorescently labeled lactoglobulin in heterozygous female mice. Kidneys were fixed by perfusion a few minutes after lactoglobulin injection and analyzed by confocal microscopy. The uptake of lactoglobulin is specifically reduced in the ClC-5-deficient cells indicating a cell-autonomous endocytosis defect. (B) The expression of the endocytic receptor megalin was analyzed by confocal immunocytochemistry of kidney slices from heterozygous females. The cells lacking ClC-5 express less megalin. This result was confirmed by Western blot analysis of megalin in WT, heterozygous, and KO kidneys (inset). Modified from Piwon *et al.* (2000).

far unknown cellular factors. However, this is not the only question remaining unanswered. While the impairment of vesicular acidification as well as the endocytosis defect have been unambiguously shown, the link between these two defects is only incompletely understood. The first step in receptor-mediated endocytosis, the generation of clathrin-coated vesicles from the plasma membrane, is independent of pH. Drugs abrogating vacuolar acidification do not affect the rate of endocytic uptake (Cupers *et al.*, 1997; Tyteca *et al.*, 2002). However, they inhibit recycling (Basu *et al.*, 1981) or arrest transfer to lysosomes (Aniento *et al.*, 1996; Cupers *et al.*, 1997). Thus the progression along the endocytic apparatus depends on endosomal acidification. This is also evident from the pH-dependent formation of transport vesicles destined for late endosomes (Aniento *et al.*, 1996). According to a hypothesis put forward by Marshansky *et al.* (2002), the intravesicular acidification might not only provide the optimal conditions for receptor–ligand dissociation, but also present a signaling mechanism for coat recruitment. The progressively lower endosomal pH is assumed to cause a conformational change in an as yet unknown vesicular receptor protein enabling the receptor to recruit signaling molecules from the cytosol. This hypothesis is strengthened by the finding that several regulatory proteins involved in vesicle transport, such as β-COPs (Aniento *et al.*, 1996) and the small Arf GTPases which are involved in the recruitment of coat complexes associate with endosomes in a pH-dependent manner (Gu and Gruenberg, 2000; Maranda *et al.*, 2001). However, it is also possible that it is the rise in Cl^- rather than the acidification which affects the vesicular trafficking, as there are numerous proteins possessing binding sites for Cl^- (Faundez and Hartzell, 2004).

In accordance with the pH-dependence of endosomal trafficking, not only the endosomal uptake of proteins was found to be impaired in ClC-5-deficient mice, but also the recycling of membrane proteins. In WT animals, the multiligand receptors megalin, a member of the LDL family of proteins, and cubilin are predominantly localized at the plasma membrane. After ligand binding, the receptors are endocytosed into apical early endosomes. Following dissociation of the ligands, the receptors recycle to the plasma membrane. There is a reduced amount of megalin in ClC-5-deficient proximal tubular cells, as Jentsch and colleagues first showed by immunofluorescence and Western blotting (Piwon *et al.*, 2000) (Fig. 7B). Christensen *et al.* (2003) confirmed this result and extended it to cubilin. Using subcellular fractionation studies and EM, they also showed that the remaining receptors are redistributed to endosomes (Christensen *et al.*, 2003). Only very little protein is left at the plasma membrane suggesting that ClC-5 might play a role in recycling megalin back to the apical brush border membrane. These results might also explain the reduced amount of megalin in the urine of patients with Dent's disease. The cell-autonomous decrease in megalin suggests that receptor-mediated endocytosis is more severely impaired by a lack of ClC-5 than fluid phase endocytosis. As megalin and cubilin play an important role in the uptake of numerous ligands, their impaired recycling contributes to the proteinuria observed in the ClC-5 KO mice. The importance of megalin for proximal tubular endocytosis can also be gleaned from the fact that megalin KO mice show severe LMW proteinuria (Leheste *et al.*, 1999).

3. Hypercalciuria and Hyperphosphaturia: A Consequence of Impaired Endocytosis

While proteinuria is the symptom which is most often encountered in Dent's disease, it is clinically not the most important one. Patients often present with hyperphosphaturia and hypercalciuria which can result in rickets, nephrocalcinosis, nephrolithiasis (kidney stones), and finally progressive kidney failure (Wrong et al., 1994; Scheinman, 1998). Most male patients exhibit hypercalciuria, two-thirds present with nephrocalcinosis and about half of them develop kidney stones (Cebotaru et al., 2005), the most relevant clinical symptom of the disease. The analysis of the ClC-5 KO mouse models revealed that these additional symptoms are probably entirely secondary to the primary defect in endocytosis (Piwon et al., 2000). The endocytosis defect entails that filtered proteins pass later nephron segments in an elevated concentration. In the case of signaling proteins that bind to apical receptors of tubular cells, this can have pathological consequences.

A well-investigated example of such a signaling molecule is the parathyroid hormone (PTH) (Fig. 8). Being only 9.4 kD in size, PTH is able to pass the glomerular filter to a large extent. In ClC-5 KO animals (Piwon et al., 2000), as well as in Dent's disease patients (Norden et al., 2001), the urinary concentration of PTH was increased. One function of PTH is the regulation of renal phosphate (P_i) uptake. By binding to receptors located at the apical tubule membrane, PTH stimulates the internalization of

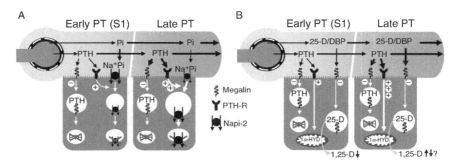

Figure 8. Mechanism underlying hyperphosphaturia and hypercalciuria in Dent's disease. (A) Model for hyperphosphaturia. The apical Na^+/phosphate cotransporter NaPi-2 which mediates the bulk of phosphate resorption in the proximal tubule is regulated by parathyroid hormone (PTH). This hormone stimulates the endocytosis and subsequent lysosomal degradation of NaPi-2. PTH is readily filtered into the nephron lumen from where it is normally removed already at the beginning of the proximal tubule by megalin-dependent endocytosis and subsequent lysosomal degradation. The endocytosis defect resulting from mutations in ClC-5 causes an accumulation of PTH in adjacent nephron segments and hence a greater stimulation of apical PTH receptors in these segments, which in turn triggers the enhanced endocytosis and degradation of NaPi-2. Due to decreased amounts of NaPi-2 at the plasma membrane, less phosphate is resorbed resulting in elevated phosphate levels in the urine. (B) Model for hypercalciuria. The augmented stimulation of apical PTH receptors does not only affect NaPi-2, but causes also a transcriptional up-regulation of the mitochondrial enzyme 1α-hydroxylase that catalyzes the conversion of the vitamin D precursor 25(OH)-vitamin D (25-D) into the active hormone 1,25(OH)$_2$-vitamin D (1,25-D). This increased enzymatic activity should in principle result in higher levels of the active hormone which would cause hypercalciuria as a consequence of elevated intestinal Ca^{2+} uptake. However, the vitamin D precursor which is filtered into the primary urine complexed to its binding protein DBP needs to be taken up into the proximal tubular cells by megalin-mediated endocytosis. Thus, the availability of the precursor is severely reduced in the ClC-5-deficient cells, counteracting the elevated enzyme transcription. The outcome of these two opposing influences is hard to predict and seems in fact to be variable, as there is a ClC-5-deficient mouse model with decreased vitamin D levels (Piwon et al., 2000) as well as one with increased vitamin D concentrations and thus hypercalciuria (Wang et al., 2000) and also great clinical variability between Dent's disease patients (Wrong et al., 1994).

the Na/phosphate cotransporter NaPi-2 (Bacic et al., 2003), thus decreasing the P_i uptake. In ClC-5-deficient animals, this signaling pathway is hyperactivated due to the elevated luminal PTH concentration and results in an increased internalization of NaPi-2 (Piwon et al., 2000). Immunocytochemistry revealed that the majority of NaPi-2 had shifted to intracellular vesicles in KO mice, whereas it resided mostly in the brush border of WT proximal tubules (PTs). The study of heterozygous females revealed that this effect was not cell-autonomous, strengthening the hypothesis that the increased internalization is due to alterations in hormonal regulation. Further support came from the fact that NaPi-2 was still predominantly apical in early segments of KO PTs where PTH is not yet concentrated. Due to the lack of sufficient amounts of the transporter at the plasma membrane of subsequent segments, less P_i is resorbed resulting in hyperphosphaturia, one of the additional symptoms of Dent's disease (Fig. 8A).

At the same time, the elevated PTH concentration results in an increased transcription of the 25(OH)-vitamin D 1α-hydroxylase (Günther et al., 2003). This enzyme converts the vitamin D precursor 25(OH)-vitamin D into the active metabolite 1,25(OH)$_2$-vitamin D. In a recent gene expression profiling of ClC-5 KO kidneys, the up-regulation of the vitamin D-activating enzyme was found to be accompanied by a marked down-regulation of the inactivating enzyme vitamin D 24-hydroxylase (Maritzen et al., 2006), most probably also due to the elevated PTH concentration, as PTH is known to decrease the stability of 24-hydroxylase mRNA. The increased capacity to generate active vitamin D is, however, partially counteracted by the scarcity of the precursor in ClC-5 KO animals. Following its filtration into the nephron, 25(OH)-vitamin D is endocytosed in complex with the vitamin D-binding protein (DBP) by proximal tubular cells in a megalin-dependent manner. In ClC-5 KO mice, less 25(OH)-vitamin D is taken up due to their impaired endocytosis. Which of the opposing aspects (increased enzyme amount, scarcity of the precursor, and loss of active metabolite into the urine) exerts the strongest influence on the final serum 1,25(OH)$_2$-vitamin D concentration seems to be variable and probably depends on nutritional factors like the amount of vitamin D in the diet as well as genetic factors (Fig. 8B). In Dent's disease patients, serum 1,25-(OH)$_2$-vitamin D$_3$ levels were mostly increased (Wrong et al., 1994; Scheinman, 1998), as they were in the ClC-5 KO mice from the Guggino lab (Wang et al., 2000). In contrast, serum levels of the precursor and of the active metabolite were decreased in the Jentsch mouse model (Piwon et al., 2000; Günther et al., 2003). In spite of the reduced serum vitamin D levels of the Jentsch mouse model, a number of vitamin D-dependent genes, such as Txnip and Calbindin D9k, were found to be up-regulated in the recent gene expression profiling of ClC-5 KO kidneys (Maritzen et al., 2006). This up-regulation is most likely caused by the luminal accumulation of vitamin D in distal nephron segments and its action on distally located target genes from within the nephron lumen. Thus, the findings from the gene expression profiling support the notion that the luminal accumulation of vitamins and hormones resulting from the impaired endocytosis influences signal cascades within the kidney and thus has important physiological consequences.

1,25-$(OH)_2$-vitamin D plays an important role in the regulation of Ca^{2+} homeostasis by stimulating renal and intestinal Ca^{2+} reabsorption and influencing bone mineralization (Brown et al., 1999). The elevated 1,25-$(OH)_2$-vitamin D of Dent's disease patients may be the cause for their hypercalciuria by increasing intestinal Ca^{2+} uptake which is then balanced by a greater renal Ca^{2+} excretion. However, it could be likewise caused by alterations in the Ca^{2+} homeostasis of kidney or bone, the other two important vitamin D target organs. An analysis of the overall Ca^{2+} handling in the hypercalciuric ClC-5 KO mice from the Guggino lab suggests that their hypercalciuria is rather of bone and renal origin instead of being caused by increased intestinal Ca^{2+} absorption (Silva et al., 2003). Especially the finding that the KO mice have elevated bone turnover markers points to a use of their bone Ca^{2+} stores.

The two ClC-5 KO mouse models recapitulate the symptomatic variability of Dent's disease, as ClC-5 KO mice from the Guggino lab display elevated 1,25-$(OH)_2$-vitamin D and hypercalciuria (Wang et al., 2000; Silva et al., 2003), while the KO mice generated in the Jentsch lab have reduced 1,25-$(OH)_2$-vitamin D levels and accordingly no hypercalciuria (Piwon et al., 2000). In those patients affected by hypercalciuria, the urine can become saturated with Ca^{2+} salts which will precipitate in the renal papillae or within the collecting ducts forming kidney stones. Such kidney stones have not been observed in the ClC-5 KO mouse models. However, old mice of the Guggino ClC-5 KO strain were reported to have some Ca^{2+} deposits in their kidneys and to develop renal insufficiency (Cebotaru et al., 2005). In these mice, a high-citrate diet was able to slow the progression of renal insufficiency, as the citrate decreases the free Ca^{2+} in the urine by complexing it (Cebotaru et al., 2005).

4. Additional Symptoms

In addition to the discussed Dent's disease symptoms (LMW proteinuria, hyperphosphaturia, hypercalciuria, nephrocalcinosis, kidney stones, and progressive renal failure), there are even more variable symptoms like aminoaciduria, glucosuria, kaliuresis, uricosuria, and impairment of urinary acidification which are not yet understood. However, the defect in endocytosis and vesicle trafficking may also impinge on other transporters and receptors in addition to NaPi-2 and megalin. It is tempting to speculate that the aminoaciduria and glucosuria observed in patients with Dent's disease might be due in part to altered trafficking of the respective transport proteins.

5. ClC-5 Mutations

More than 30 different human ClC-5 mutations are known comprising splice site, missense, and nonsense mutations. For reasons that are still unclear, the missense mutations cluster at the interface between the two subunits (Wu et al., 2003). Most of the missense mutations that have been studied in oocytes reduce or abolish ClC-5-mediated currents (Lloyd et al., 1996, 1997; Igarashi et al., 1998; Morimoto et al., 1998) without changing the biophysical properties of the current. Since the link between ClC-5 mutations and Dent's disease had been established, the disease had been consistently associated with ClC-5. Only recently the first evidence for genetic heterogeneity in Dent's disease was published. Two groups (Ludwig et al., 2003; Hoopes et al., 2004) reported patients with the typical symptoms of LMW proteinuria and hypercalciuria in which no mutations in ClC-5 could be found. In some of these patients mutations in OCRL1 were subsequently identified (Hoopes et al., 2005), a gene encoding a phosphatidylinositol 4,5-bisphosphate (PIP_2) 5-phosphatase which had been known to be mutated in the multisystem disease Lowe syndrome. Thus, there is at least one additional gene the mutation of which can cause a Dent's disease-like phenotype.

6. ClC-5 Sorting and Regulation

One level of regulation of ClC-5 involves ubiquitination. ClC-5 is the only CLC protein that carries a PY motif between its C-terminal CBS domains. This motif proved important for the internalization of ClC-5 from the plasma membrane. Destroying the PY motif by mutagenesis led to increased surface expression of the protein and thus to about twofold increased currents (Schwake et al., 2001). PY motifs are known to interact with WW domains of ubiquitin ligases. A peptide corresponding to the ClC-5 PY motif was shown to bind to the WW domain of the HECT ubiquitin ligase WWPII (Pirozzi et al., 1997) which is present in the kidney. Destruction of the PY motif and coexpression of a mutant of WWPII containing an inactivating mutation in its HECT ubiquitin ligase domain increased the surface expression as well in a PY motif-dependent manner (Schwake et al., 2001). Hence, the PY motif results in an increased rate of internalization of ClC-5 from the plasma membrane, an effect that is most likely triggered by ubiquitination. This resembles the model proposed for the regulation of the epithelial Na^+ channel (ENaC), the internalization and subsequent degradation of which is also triggered by a PY motif-dependent ubiquitination (Rotin et al., 2000). The involvement of ubiquitin ligases in ClC-5 regulation was recently confirmed by the finding that ClC-5 and the ubiquitin ligase Nedd4–2 interact as well, and that ClC-5 can indeed be ubiquitinated (Hryciw et al., 2004). The coexpression of Nedd4–2 and ClC-5 in oocytes decreased currents and cell surface expression of ClC-5 (Hryciw et al., 2004). It is not yet clear whether ubiquitination regulates ClC-5 function in vivo and, if so, whether it only influences the endocytosis from the plasma membrane or whether it also affects other intracellular transport steps.

H. ClC-6: A Late Endosomal Protein Important for Lysosomes

So far, very little is known about the physiological function of ClC-6, although it was cloned about 10 years ago in parallel to ClC-7 (Brandt and Jentsch, 1995). ClC-6 shares about 45% sequence identity with ClC-7, with which it forms a distinct branch of the CLC family. It is broadly expressed (Brandt and Jentsch, 1995; Kida et al., 2001) and is transcribed already early in mouse development (Brandt and Jentsch, 1995). Like ClC-7, ClC-6 is located on an intracellular vesicular compartment. On overexpression in COS or Chinese Hamster Ovary (CHO) cells, epitope-tagged ClC-6 was targeted to structures which were identified as endoplasmatic reticulum (Buyse et al., 1998). However, newer reports of overexpressed ClC-6 rather colocalize it with EEA1 and lamp-1, suggesting its presence on early and late endosomes (Suzuki et al., 2006) (Fig. 5). The predominantly intracellular localization of ClC-6 has precluded its electrophysiological characterization. As ClC-6 was able to complement the phenotype arising from the disruption of the single yeast CLC gene *Gef1*, it is assumed to be able to mediate Cl^- fluxes (Kida et al., 2001). Several splice variants of ClC-6 were identified by RT-PCR (Eggermont et al., 1997). As they severely truncate the protein, their physiological importance is obscure. In contrast to ClC-3 to ClC-5, ClC-6 could not be coimmunoprecipitated with any other intracellular CLC protein (Suzuki et al., 2006). Further clues as to the function of ClC-6 are expected from the analysis of ClC-6 KO mice.

Indeed, it was reported very recently that the disruption of ClC-6 in mice leads to a lysosomal storage disease that was exclusively observed in neurons (Poët et al., 2006). The ClC-6 protein—in contrast to its mRNA—was nearly exclusively found in neurons where it localized to late endosomes (Poët et al., 2006).

I. ClC-7 and Osteopetrosis

ClC-7 was cloned in 1995 and was found to have a broad tissue distribution including brain, testis, muscle, and kidney (Brandt and Jentsch, 1995). It has not yet been possible to measure ClC-7-mediated currents on its heterologous expression in oocytes or transfected cells, suggesting that ClC-7 is exclusively located intracellularly. Indeed, the protein resides predominantly on late endosomal/lysosomal vesicles where it colocalizes with lamp-1 (Kornak et al., 2001; Suzuki et al., 2006) (Fig. 5).

1. Phenotype of ClC-7 KO Mice

The physiological importance of ClC-7 became apparent after its disruption in mice (Kornak et al., 2001). ClC-7-deficient mice are severely sick animals that die within 6 weeks after birth and exhibit prominent skeletal abnormalities. Their teeth do not erupt and their bones being too calcified lack bone marrow cavities. Hence erythropoiesis has to take place extramedullarly (Kornak et al., 2001). Such dense, fragile bones devoid of bone marrow are the hallmark of osteopetrosis. ClC-7 mutations were also found in human osteopetrosis (Kornak et al., 2001). More than 30 ClC-7 mutations underlying osteopetrosis are known to date. Dominant-negative mutations in ClC-7 cause the less severe autosomal dominant osteopetrosis of the Albers Schönberg type, while recessive mutations are associated with the malignant infantile human osteopetrosis.

2. Mechanism of Osteopetrosis

How does the loss of ClC-7 cause osteopetrosis? The formation and maintenance of bone depends on the interaction between the bone-depositing osteoblasts and the bone-resorbing osteoclasts. Bone is not a static material, but rather exists in a dynamic equilibrium, being continually resorbed and rebuilt again. ClC-7 is prominently expressed in osteoclasts (Kornak et al., 2001). These specialists of bone degradation closely attach to the bone surface via integrin–vitronectin interactions and differentiate their bone facing plasma membrane into the so-called ruffled border by inserting H^+-ATPase containing vesicles of lysosomal origin (Boyle et al., 2003). The acidic space enclosed between ruffled border and bone surface is called resorption lacuna and contains lysosomal enzymes. It serves to enzymatically degrade the organic bone matrix and chemically dissolve the inorganic bone material, thus allowing the subsequent bone remodeling by osteoblasts. ClC-7 is coinserted into the ruffled border together with the H^+-ATPase and is assumed to serve as shunt for the acidification of the resorption lacuna just like ClC-3 and ClC-5 serve as shunt for endosomal acidification (Fig. 4C). Indeed, ClC-7-deficient osteoclasts were present in normal numbers and also attached to bone, but failed to acidify their resorption lacuna and thus were unable to degrade the bone material. The notion of an acidification defect which lies at the heart of the ClC-7-associated osteopetrosis is supported by the observation that mutations in the a3 subunit of the osteoclast H^+-ATPase cause osteopetrosis as well (Li et al., 1999; Frattini et al., 2000; Kornak et al., 2000; Scimeca et al., 2000).

3. ClC-7 and Neurodegeneration

However, the phenotype of ClC-7-deficient mice comprises also features not present in the a3 KO mice. In addition to their skeletal abnormalities, ClC-7 KO mice exhibit a rapidly progressing retinal degeneration that leads to blindness at about 4 weeks of age. There are only few photoreceptors left at that point, while the ganglion cells are still intact which argues against the retinal degeneration being a secondary effect of a narrowing of the optic canal due to osteopetrosis. Indeed, ClC-7-deficient mice in which the expression of ClC-7 was rescued in bone, did not exhibit osteopetrosis while still displaying retinal degeneration (Kasper et al., 2005). There is also a neurodegeneration of cortical and hippocampal neurons in the CNS of ClC-7 KO mice which is accompanied by microglial activation and astrogliosis (Kornak et al., 2000).

4. Lysosomal Storage

ClC-7-deficient neurons and renal proximal tubular cells contained electron dense storage material reminiscent of neuronal ceroid lipofuscinosis, a lysosomal storage disease (Kasper et al., 2005). These diseases typically result from defects in key lysosomal proteins causing an impaired lysosomal degradation which leads to an accumulation of undegraded material within the lysosome and finally to cell death. When this storage phenotype became apparent, the loss of ClC-7 was postulated to impair lysosomal acidification. As an acidic pH is necessary for the activity of many lysosomal hydrolases, this would reduce the degradative capacity of the lysosomes. However, no difference in the pH of lysosomes from ClC-7-deficient cultured neurons and fibroblasts could be detected, suggesting that it might not be impaired acidification at all that causes the phenotype. ClC-7 might for instance also play a role in the trafficking of late endosomal and lysosomal vesicles. This would fit to the observed impaired development of the ruffled border in ClC-7-deficient osteoclasts (Kornak et al., 2001).

Recently, ClC-7 was shown to need a β-subunit called Ostm1 to support bone resorption and lysosomal function (Lange et al., 2006). Like mutations in *CLCN7* also mutations in *OSTM1* cause severe osteopetrosis (Chalhoub et al., 2003). This overlapping phenotype suggested that Ostm1 and ClC-7 might be functionally linked in osteoclasts (Lange et al., 2006). Indeed, Ostm1 was found to colocalize with ClC-7 in late endosomes and lysosomes. Ostm1 needs ClC-7 to reach lysosomes, and coimmunoprecipitations revealed that ClC-7 and Ostm1 form a complex. As protein, but not RNA, levels of ClC-7 were greatly reduced in Ostm1 KO mice and *vice versa,* it was suggested that the interaction of both proteins is important for their stability and function. Mutations in Ostm1 likely cause osteopetrosis by destabilizing ClC-7 and thereby impairing the ClC-7-dependent acidification of the osteoclast resorption lacuna (Lange et al., 2006). The lysosomal storage found in Ostm1 KO brain and kidney (Lange et al., 2006) resembles that of ClC-7 KO mice (Kasper et al., 2005) and underlines the importance of Ostm1 for ClC-7 stability and function also in other organs.

III. CLC PROTEINS IN MODEL ORGANISMS

CLC proteins can be found in all phylae from bacteria to humans. As mentioned in the introduction, the mammalian CLC proteins can be grouped into three evolutionary categories. The identified CLC sequences from *Saccharomyces* (yeast), *Arabidopsis* (plant), *Drosophila* (fly), and *Caenorhabditis* (nematode worm) all fit into either of these three branches, while bacterial CLC proteins belong to another more ancient evolutionary branch (George et al., 2001) (Fig. 1). The existence of CLC proteins in species that are separated by billions of years of evolution underlines their importance for cellular function. In the following paragraphs the functions of CLC proteins in some common model organisms shall be discussed.

A. Yeast

In the yeast *Saccharomyces cerevisiae*, a single CLC gene denoted ScClC or Gef1 has been identified. It was isolated in a genetic screen for an iron-dependent defect in respiration (Greene et al., 1993). In this screen, yeast mutants were grown on a rich medium containing the nonfermentable carbon sources glycerol and ethanol and were checked for growth defects that could be suppressed by adding high concentrations of iron to the medium (hence the name *Gef* for *g*lycerol/*e*thanol, *F*e-requiring). As growth on a medium without fermentable carbon sources is dependent on an intact respiratory chain and as a number of enzymes of the respiratory chain need iron for proper functioning, such a screen can identify genes involved in iron transport and metabolism.

A key component of the high-affinity iron uptake system of yeast is the Fet3p oxidase which requires copper for its activity. The loading of copper onto Fet3p takes place in late Golgi vesicles and depends on the copper-transporting ATPase Ccc2p and an acidic lumen supplied by an H^+-ATPase. Gef1p colocalizes with Ccc2p in late Golgi (Gaxiola et al., 1998; Schwappach et al., 1998) and likely supports the electrogenic activity of Ccc2p as well as the H^+-ATPase by providing the necessary electric shunt allowing the efficient copper loading of Fet3p and thus iron uptake. This fits well with the fact that mutations in an H^+-ATPase subunit result in the same iron-dependent growth defect as Gef1p mutations.

The potential involvement of Gef1p in acidification was further supported by the impaired ability of the mutant strain to grow at neutral or alkaline pH (Gaxiola et al., 1998; Schwappach et al., 1998) or on media containing salts like $MnCl_2$ (Gaxiola et al., 1998), because salt tolerance is achieved in part by the intracellular Na^+/H^+ exchanger Nhx1p which likely relies on a pH gradient to sequester Na^+ (Gaxiola et al., 1999). However, there are also indications for functions of Gef1p which are independent of its potential ability to neutralize positive charges. For example, the import of Cl^- seems to be important in itself, that is, independent of its charge-neutralizing capacity, as Cl^- ions function as allosteric effectors of copper assembly for Fet3p (Davis-Kaplan et al., 1998). Additionally Gef1p was recently shown to interact with the ATPase Arr4p in a copper-dependent manner (Metz et al., 2006). Disruption of Gef1 also inhibited the elimination of misfolded receptors (Li et al., 1999) suggesting a function in quality control. Recently, it was reported that Gef1p is proteolytically processed within the first extracellular loop by the late Golgi Kex2p protease (Wächter and Schwappach, 2005). However, as a noncleavable mutant as well as the split protein are active, the functional significance of this processing is not clear.

B. *Caenorhabditis elegans*

The genome of the nematode worm *Caenorhabditis elegans* encodes six different CLC genes denoted CeClC-1 to CeClC-6 (Schriever et al., 1999) comprising all three branches of "mammalian-type" CLC proteins. While CeClC-1 to CeClC-4 represent the first branch of CLC proteins, CeClC-5 is most similar to ClC-3/ClC-4/ClC-5 and CeClC-6 finally belongs to the third branch sharing the highest amino acid identity with ClC-6/ClC-7. CeClC-1 to CeClC-4 seem to be expressed only in restricted subsets of cells. CeClC-4 for instance was found only in the H-shaped excretory cell (Schriever et al., 1999; Nehrke et al., 2000) where it was coexpressed with CeClC-3 which is also present in neurons, muscles, and epithelial cells. CeClC-5 and CeClC-6 showed broader expression patterns (Schriever et al., 1999). When CeClC-1 to CeClC-5 were tested in oocyte measurements only CeClC-1 to CeClC-3 gave currents (Schriever et al., 1999; Nehrke et al., 2000), arguing for a predominantly intracellular localization of the other CeClC proteins. The gene encoding CeClC-1, which is expressed in hypodermal cells like the seam cells that synthesize collagen for the cuticle, has been disrupted by transposon insertion. This causes a widening of the worm body. The wider body could be shrunk again by exposure to hyperosmolar solution, suggesting a role for CeClC-1 in whole animal osmoregulation (Petalcorin *et al.*, 1999). RNAi experiments reducing the expression of CeClC-3 revealed a slight difference in the timing of the contraction of gonadal sheath cells surrounding oocytes. In addition, a swelling-activated Cl^- current was abolished in CeClC-3 knockdown animals. Although the gating properties of CeClC-3 (Schriever *et al.*, 1999) differ from those of mammalian ClC-2 (Gründer *et al.*, 1992; Jordt and Jentsch, 1997; Pusch *et al.*, 1999), it was proposed to be the species orthologue of ClC-2 due to its activation on swelling and hyperpolarization (Rutledge *et al.*, 2001). Besides, there are two splice variants which might explain the differences observed in electrophysiological measurements.

C. Plants

Being the most abundant anion in higher plants, Cl^- plays an important role in plant nutrition and osmoregulation. It is especially plentiful in the plant vacuole which is the major storage site for ions. As the vacuole is crucial for turgor regulation, tonoplast anion channels are important for plant turgor as well as cell growth and development. In addition, anion channels are involved in various other plant-specific processes such as stomatal movement, nutrient transport and metal tolerance. Even the generation of action potentials depends on Cl^- channels, unlike in animals (Barbier-Brygoo *et al.*, 2000). While many plant Cl^- channels are characterized biophysically, only a subset is known at the molecular level. CLC proteins were the first, and still are the only, putative Cl^- channels cloned in plants (Hechenberger *et al.*, 1996; Lurin *et al.*, 1996). CLC homologues have been identified in *Nicotiana tabacum* (Lurin *et al.*, 1996) as well as *Arabidopsis thaliana* (Hechenberger *et al.*, 1996). All CLC proteins identified in plants belong to the third branch of the CLC family. They display about 35% sequence identity with human ClC-6 and ClC-7 and thus might be intracellularly located. This might explain the inability to obtain currents on heterologous expression of AtClC-a to AtClC-d in oocytes (Hechenberger *et al.*, 1996). In rescue experiments, in which AtClC-a to AtClC-d were tested, only AtClC-d rescued the iron-sensitive phenotype of the yeast Gef1 mutant (Section III.A). This is compatible with a function of AtClC-d as a Cl^- channel or transporter (Hechenberger *et al.*, 1996). The analysis of plants containing a mutation within the AtCLC-a gene revealed an unexpected role of AtClC-a in the regulation of the cellular nitrate status (Geelen *et al.*, 2000) as the mutants had a reduced capacity to accumulate nitrate under conditions of nitrate excess. The mechanism underlying this defect, however, is not yet clear. Analysis of the localization of the tobacco CLC-Nt1 protein by subcellular fractionation studies revealed its colocalization with markers of the mitochondrial inner membrane (Lurin *et al.*, 2000). Recently, a homologue of AtClC-f was reported to be located in the outer envelope of spinach chloroplasts, suggesting a role in photosynthesis (Teardo *et al.*, 2005). This result, however, hinges on the quality of the antibody. Very recent and elegant work showed that AtClC-a functions as a

$2NO_3^-/H^+$-exchanger in plant vacuoles (De Angeli et al., 2006), where it has a function in nitrate accumulation.

IV. SUMMARY AND OUTLOOK

The past 10 years have been very exciting in the still young CLC field. The knowledge about the physiology of CLC proteins exploded during this decade. Already shortly after the cloning of the first mammalian CLC protein ClC-1 (Steinmeyer et al., 1991b), its involvement in the human disease myotonia (Steinmeyer et al., 1991a) became apparent. Since then, the discovery of many more channelopathies associated with defects in CLC proteins provided amazing insights into the cellular functions of this protein family. The different pathologies (muscle stiffness, renal salt loss, deafness, blindness, neurodegeneration, male infertility, osteopetrosis, proteinuria, and kidney stones) revealed an unsuspected range of physiological roles of CLC proteins and demonstrated that these proteins are not only important for the Cl^- transport across the cellular plasma membrane, but also for the proper functioning of intracellular vesicles of the endocytic/lysosomal pathway.

The great progress made in the elucidation of CLC functions came primarily from two approaches: human genetics and mouse models. Human genetics was crucial for the determination of the roles of ClC-Kb in Bartter syndrome (renal salt loss) (Simon et al., 1997) and also in the case of ClC-5 and Dent's disease (Lloyd et al., 1996). Mouse models subsequently helped to solve the mechanism linking defects in ClC-5 and Dent's disease (Piwon et al., 2000; Wang et al., 2000). Besides, they were instrumental in identifying ClC-7 as disease gene for osteopetrosis (Kornak et al., 2001) and revealed the involvement of ClC-3 in neurodegeneration (Stobrawa et al., 2001). During the past 5 years it also became evident that some CLC proteins do not operate on their own, but rely on β-subunits for proper functioning. Thus the ClC-K channels were shown to need barttin for efficient cell surface localization (Birkenhäger et al., 2001), while ClC-7 protein stability depends crucially on Ostm1 expression (Lange et al., 2006). As many CLC proteins have overlapping expression patterns—in particular the vesicular CLCs that are located within the endosomal pathway (Fig. 5), and hence might have redundant functions in certain organs—future progress is expected from the generation of double knockout mouse models in conjunction with a broad spectrum of physiological, biophysical, and cell biological techniques.

REFERENCES

Accardi, A. and Miller, C. (2004). Secondary active transport mediated by a prokaryotic homologue of ClC Cl^- channels. *Nature* **427**, 803–807.

Adachi, S., Uchida, S., Ito, H., Hata, M., Hiroe, M., Marumo, F., and Sasaki, S. (1994). Two isoforms of a chloride channel predominantly expressed in thick ascending limb of Henle's loop and collecting ducts of rat kidney. *J. Biol. Chem.* **269**, 17677–17683.

Adler, D. A., Rugarli, E. I., Lingenfelter, P. A., Tsuchiya, K., Poslinski, D., Liggitt, H. D., Chapman, V. M., Elliott, R. W., Ballabio, A., and Disteche, C. M. (1997). Evidence of evolutionary up-regulation of the single active X chromosome in mammals based on *Clc4* expression levels in Mus spretus and Mus musculus. *Proc. Natl. Acad. Sci. USA* **94,** 9244–9248.

Ando, M. and Takeuchi, S. (2000). mRNA encoding 'ClC-K1, a kidney Cl^- channel' is expressed in marginal cells of the stria vascularis of rat cochlea: Its possible contribution to Cl^- currents. *Neurosci. Lett.* **284,** 171–174.

Aniento, F., Gu, F., Parton, R. G., and Gruenberg, J. (1996). An endosomal beta COP is involved in the pH-dependent formation of transport vesicles destined for late endosomes. *J. Cell Biol.* **133,** 29–41.

Arreola, J., Begenisch, T., Nehrke, K., Nguyen, H. V., Park, K., Richardson, L., Yang, B., Schutte, B. C., Lamb, F. S., and Melvin, J. E. (2002). Secretion and cell volume regulation by salivary acinar cells from mice lacking expression of the *Clcn3* Cl^- channel gene. *J. Physiol.* **545**(Pt. 1), 207–216.

Bacic, D., Schulz, N., Biber, J., Kaissling, B., Murer, H., and Wagner, C. A. (2003). Involvement of the MAPK-kinase pathway in the PTH-mediated regulation of the proximal tubule type IIa Na^+/Pi cotransporter in mouse kidney. *Pflügers Arch.* **446,** 52–60.

Barbier-Brygoo, H., Vinauger, M., Colcombet, J., Ephritikhine, G., Frachisse, J., and Maurel, C. (2000). Anion channels in higher plants: Functional characterization, molecular structure and physiological role. *Biochim. Biophys. Acta* **1465,** 199–218.

Barg, S., Huang, P., Eliasson, L., Nelson, D. J., Obermuller, S., Rorsman, P., Thevenod, F., and Renstrom, E. (2001). Priming of insulin granules for exocytosis by granular Cl^- uptake and acidification. *J. Cell Sci.* **114,** 2145–2154.

Basu, S. K., Goldstein, J. L., Anderson, R. G., and Brown, M. S. (1981). Monensin interrupts the recycling of low density lipoprotein receptors in human fibroblasts. *Cell* **24,** 493–502.

Bayer, M. J., Reese, C., Buhler, S., Peters, C., and Mayer, A. (2003). Vacuole membrane fusion: V0 functions after trans-SNARE pairing and is coupled to the Ca2+ -releasing channel. *J. Cell Biol.* **162,** 211–222.

Beck, C. L., Fahlke, C., and George, A. L., Jr. (1996). Molecular basis for decreased muscle chloride conductance in the myotonic goat. *Proc. Natl. Acad. Sci. USA* **93,** 11248–11252.

Bellochio, E. E., Reimer, R. J., Fremeau, R. T., Jr., and Edwards, R. H. (2000). Uptake of glutamate into synaptic vesicles by an inorganic phosphate transporter. *Science* **289,** 957–960.

Bennetts, B., Rychkov, G. Y., Ng, H. L., Morton, C. J., Stapleton, D., Parker, M. W., and Cromer, B. A. (2005). Cytoplasmic ATP-sensing domains regulate gating of skeletal muscle ClC-1 chloride channels. *J. Biol. Chem.* **280,** 32452–32458.

Birkenhäger, R., Otto, E., Schürmann, M. J., Vollmer, M., Ruf, E. M., Maier-Lutz, I., Beekmann, F., Fekete, A., Omran, H., Feldmann, D., Milford, D. V., Jeck, N., *et al.* (2001). Mutation of *BSND* causes Bartter syndrome with sensorineural deafness and kidney failure. *Nat. Genet.* **29,** 310–314.

Blaisdell, C. J., Edmonds, R. D., Wang, X. T., Guggino, S., and Zeitlin, P. L. (2000). pH-regulated chloride secretion in fetal lung epithelia. *Am. J. Physiol. Lung Cell Mol. Physiol.* **278,** L1248–L1255.

Borsani, G., Rugarli, E. I., Taglialatela, M., Wong, C., and Ballabio, A. (1995). Characterization of a human and murine gene (*CLCN3*) sharing similarities to voltage-gated chloride channels and to a yeast integral membrane protein. *Genomics* **27,** 131–141.

Bösl, M. R., Stein, V., Hübner, C., Zdebik, A. A., Jordt, S. E., Mukhophadhyay, A. K., Davidoff, M. S., Holstein, A. F., and Jentsch, T. J. (2001). Male germ cells and photoreceptors, both depending on close cell-cell interactions, degenerate upon ClC-2 Cl^--channel disruption. *EMBO J.* **20,** 1289–1299.

Böttger, T., Hübner, C., Maier, H., Rust, M. B., Beck, F. X., and Jentsch, T. J. (2002). Deafness and renal tubular acidosis in mice lacking the K-Cl cotransporter Kcc4. *Nature* **416,** 874–878.

Böttger, T., Rust, M. B., Maier, H., Seidenbecher, T., Schweizer, M., Keating, D. J., Faulhaber, J., Ehmke, H., Pfeffer, C., Scheel, O., Lemcke, B., Horst, J., *et al.* (2003). Loss of K-Cl co-transporter KCC3 causes deafness, neurodegeneration and reduced seizure threshold. *EMBO J.* **22,** 5422–5434.

Boyle, W. J., Simonet, W. S., and Lacey, D. L. (2003). Osteoclast differentiation and activation. *Nature* **423,** 337–342.

Brandt, S. and Jentsch, T. J. (1995). ClC-6 and ClC-7 are two novel broadly expressed members of the CLC chloride channel family. *FEBS Lett.* **377,** 15–20.

Bretag, A. H., Dawe, S. R., and Moskwa, A. G. (1980). Chemically induced myotonia in amphibia. *Nature* **286**, 625–626.
Bronsveld, I., Mekus, F., Bijman, J., Ballmann, M., de Jonge, H. R., Laabs, U., Halley, D. J., Ellemunter, H., Mastella, G., Thomas, S., Veeze, H. J., and Tummler, B. (2001). Chloride conductance and genetic background modulate the cystic fibrosis phenotype of Delta F508 homozygous twins and siblings. *J. Clin. Invest.* **108**, 1705–1715.
Brown, A. J., Dusso, A., and Slatopolsky, E. (1999). Vitamin D. *Am. J. Physiol.* **277**, F157–F175.
Bryant, S. H. and Conte-Camerino, D. (1991). Chloride channel regulation in the skeletal muscle of normal and myotonic goats. *Pflügers Arch.* **417**, 605–610.
Buyse, G., Trouet, D., Voets, T., Missiaen, L., Droogmans, G., Nilius, B., and Eggermont, J. (1998). Evidence for the intracellular location of chloride channel (ClC)-type proteins: Co-localization of ClC-6a and ClC-6c with the sarco/endoplasmic-reticulum Ca^{2+} pump SERCA2b. *Biochem. J.* **330**, 1015–1021.
Carr, G., Simmons, N., and Sayer, J. (2003). A role for CBS domain 2 in trafficking of chloride channel CLC-5. *Biochem. Biophys. Res. Commun.* **310**, 600–605.
Catalán, M., Niemeyer, M. I., Cid, L. P., and Sepúlveda, F. V. (2004). Basolateral ClC-2 chloride channels in surface colon epithelium: Regulation by a direct effect of intracellular chloride. *Gastroenterology* **126**, 1104–1114.
Cebotaru, V., Kaul, S., Devuyst, O., Cai, H., Racusen, L., Guggino, W. B., and Guggino, S. E. (2005). High citrate diet delays progression of renal insufficiency in the ClC-5 knockout mouse model of Dent's disease. *Kidney Int.* **68**, 642–652.
Chalhoub, N., Benachenhou, N., Rajapurohitam, V., Pata, M., Ferron, M., Frattini, A., Villa, A., and Vacher, J. (2003). Grey-lethal mutation induces severe malignant autosomal recessive osteopetrosis in mouse and human. *Nat. Med.* **9**, 399–406.
Christensen, E. I., Devuyst, O., Dom, G., Nielsen, R., Van der Smissen, P., Verroust, P., Leruth, M., Guggino, W. B., and Courtoy, P. J. (2003). Loss of chloride channel ClC-5 impairs endocytosis by defective trafficking of megalin and cubilin in kidney proximal tubules. *Proc. Natl. Acad. Sci. USA* **100**, 8472–8477.
Cid, L. P., Montrose-Rafizadeh, C., Smith, D. I., Guggino, W. B., and Cutting, G. R. (1995). Cloning of a putative human voltage-gated chloride channel (ClC-2) cDNA widely expressed in human tissues. *Hum. Mol. Genet.* **4**, 407–413.
Cid, L. P., Niemeyer, M. I., Ramírez, A., and Sepúlveda, F. V. (2000). Splice variants of a ClC-2 chloride channel with differing functional characteristics. *Am. J. Physiol. Cell Physiol.* **279**, C1198–C1210.
Clarke, L. L., Grubb, B. R., Gabriel, S. E., Smithies, O., Koller, B. H., and Boucher, R. C. (1992). Defective epithelial chloride transport in a gene-targeted mouse model of cystic fibrosis. *Science* **257**, 1125–1128.
Cleiren, E., Benichou, O., Van Hul, E., Gram, J., Bollerslev, J., Singer, F. R., Beaverson, K., Aledo, A., Whyte, M. P., Yoneyama, T., deVernejoul, M. C., and Van Hul, W. (2001). Albers-Schönberg disease (autosomal dominant osteopetrosis, type II) results from mutations in the *ClCN7* chloride channel gene. *Hum. Mol. Genet.* **10**, 2861–2867.
Cupers, P., Veithen, A., Hoekstra, D., Baudhuin, P., and Courtoy, P. J. (1997). Three unrelated perturbations similarly uncouple fluid, bulk-membrane, and receptor endosomal flow in rat fetal fibroblasts. *Biochem. Biophys. Res. Commun.* **236**, 661–664.
D'Agostino, D., Bertelli, M., Gallo, S., Cecchin, S., Albiero, E., Garofalo, P. G., Gambardella, A., St Hilaire, J. M., Kwiecinski, H., Andermann, E., and Pandolfo, M. (2004). Mutations and polymorphisms of the CLCN2 gene in idiopathic epilepsy. *Neurology* **63**, 1500–1502.
Davis-Kaplan, S. R., Askwith, C. C., Bengtzen, A. C., Radisky, D., and Kaplan, J. (1998). Chloride is an allosteric effector of copper assembly for the yeast multicopper oxidase Fet3p: An unexpected role for intracellular chloride channels. *Proc. Natl. Acad. Sci. USA* **95**, 13641–13645.
De Angeli, A., Monachello, D., Ephritikhine, G., Frachisse, J. M., Thomine, S., Gambale, F., and Barbier-Brygoo, H. (2006). The nitrate/proton antiporter AtCLCa mediates nitrate accumulation in plant vacuoles. *Nature* **442**, 939–942.
DeFazio, R. A., Keros, S., Quick, M. W., and Hablitz, J. J. (2000). Potassium-coupled chloride cotransport controls intracellular chloride in rat neocortical pyramidal neurons. *J. Neurosci.* **20**, 8069–8076.

Dent, C. E. and Friedman, M. (1964). Hypercalcuric rickets associated with renal tubular damage. *Arch. Dis. Child.* **39,** 240–249.
Devuyst, O., Christie, P. T., Courtoy, P. J., Beauwens, R., and Thakker, R. V. (1999). Intra-renal and subcellular distribution of the human chloride channel, CLC-5, reveals a pathophysiological basis for Dent's disease. *Hum. Mol. Genet.* **8,** 247–257.
Dickerson, L. W., Bonthius, D. J., Schutte, B. C., Yang, B., Barna, T. J., Bailey, M. C., Nehrke, K., Williamson, R. A., and Lamb, F. S. (2002). Altered GABAergic function accompanies hippocampal degeneration in mice lacking ClC-3 voltage-gated chloride channels. *Brain Res.* **958,** 227–250.
Duan, D., Winter, C., Cowley, S., Hume, J. R., and Horowitz, B. (1997). Molecular identification of a volume-regulated chloride channel. *Nature* **390,** 417–421.
Dutzler, R., Campbell, E. B., Cadene, M., Chait, B. T., and MacKinnon, R. (2002). X-ray structure of a ClC chloride channel at 3.0 Å reveals the molecular basis of anion selectivity. *Nature* **415,** 287–294.
Eggermont, J., Buyse, G., Voets, T., Tytgat, J., De Smedt, H., Droogmans, G., and Nilius, B. (1997). Alternative splicing of ClC-6 (a member of the ClC chloride-channel family) transcripts generates three truncated isoforms one of which, ClC-6c, is kidney-specific. *Biochem. J.* **325,** 269–276.
Estévez, R., Böttger, T., Stein, V., Birkenhäger, R., Otto, M., Hildebrandt, F., and Jentsch, T. J. (2001). Barttin is a Cl⁻-channel β-subunit crucial for renal Cl⁻-reabsorption and inner ear K⁺-secretion. *Nature* **414,** 558–561.
Facci, L., Leon, A., and Skaper, S. D. (1990). Excitatory amino acid neurotoxicity in cultured retinal neurons: Involvement of N-methyl-D-aspartate (NMDA) and non-NMDA receptors and effect of ganglioside GM1. *J. Neurosci. Res.* **27,** 202–210.
Faundez, V. and Hartzell, H. C. (2004). Intracellular chloride channels: Determinants of function in the endosomal pathway. *Sci. STKE* **2004**(233), re8.
Ferroni, S., Marchini, C., Nobile, M., and Rapisarda, C. (1997). Characterization of an inwardly rectifying chloride conductance expressed by cultured rat cortical astrocytes. *Glia* **21,** 217–227.
Fisher, S. E., Black, G. C., Lloyd, S. E., Hatchwell, E., Wrong, O., Thakker, R. V., and Craig, I. W. (1994). Isolation and partial characterization of a chloride channel gene which is expressed in kidney and is a candidate for Dent's disease (an X-linked hereditary nephrolithiasis). *Hum. Mol. Genet.* **3,** 2053–2059.
Fisher, S. E., van Bakel, I., Lloyd, S. E., Pearce, S. H., Thakker, R. V., and Craig, I. W. (1995). Cloning and characterization of CLCN5, the human kidney chloride channel gene implicated in Dent disease (an X-linked hereditary nephrolithiasis). *Genomics* **29,** 598–606.
Franke, C., Iaizzo, P. A., Hatt, H., Spittelmeister, W., Ricker, K., and Lehmann-Horn, F. (1991). Altered Na⁺ channel activity and reduced Cl⁻ conductance cause hyperexcitability in recessive generalized myotonia (Becker). *Muscle Nerve* **14,** 762–770.
Frattini, A., Orchard, P. J., Sobacchi, C., Giliani, S., Abinun, M., Mattsson, J. P., Keeling, D. J., Andersson, A. K., Wallbrandt, P., Zecca, L., Notarangelo, L. D., Vezzoni, P., et al. (2000). Defects in TCIRG1 subunit of the vacuolar proton pump are responsible for a subset of human autosomal recessive osteopetrosis. *Nat. Genet.* **25,** 343–346.
Friedrich, T., Breiderhoff, T., and Jentsch, T. J. (1999). Mutational analysis demonstrates that ClC-4 and ClC-5 directly mediate plasma membrane currents. *J. Biol. Chem.* **274,** 896–902.
Furukawa, T., Horikawa, S., Terai, T., Ogura, T., Katayama, Y., and Hiraoka, M. (1995). Molecular cloning and characterization of a novel truncated from (ClC-2 beta) of ClC-2 alpha (ClC-2G) in rabbit heart [published erratum appears in *FEBS Lett.* February 10, 1997; **403**(1), 111]. *FEBS Lett.* **375,** 56–62.
Furukawa, T., Ogura, T., Katayama, Y., and Hiraoka, M. (1998). Characteristics of rabbit ClC-2 current expressed in Xenopus oocytes and its contribution to volume regulation. *Am. J. Physiol.* **274,** C500–C512.
Gaxiola, R. A., Yuan, D. S., Klausner, R. D., and Fink, G. R. (1998). The yeast CLC chloride channel functions in cation homeostasis. *Proc. Natl. Acad. Sci. USA* **95,** 4046–4050.
Gaxiola, R. A., Rao, R., Sherman, A., Grisafi, P., Alper, S. L., and Fink, G. R. (1999). The *Arabidopsis thaliana* proton transporters, AtNhx1 and Avp1, can function in cation detoxification in yeast. *Proc. Natl. Acad. Sci. USA* **96,** 1480–1485.

Geelen, D., Lurin, C., Bouchez, D., Frachisse, J. M., Lelievre, F., Courtial, B., Barbier-Brygoo, H., and Maurel, C. (2000). Disruption of putative anion channel gene *AtCLC-a* in *Arabidopsis* suggests a role in the regulation of nitrate content. *Plant J.* **21**, 259–267.

Gentzsch, M., Cui, L., Mengos, A., Chang, X. B., Chen, J. H., and Riordan, J. R. (2003). The PDZ-binding chloride channel ClC-3B localizes to the Golgi and associates with CFTR-interacting PDZ proteins. *J. Biol. Chem.* **278**, 6440–6449.

George, A. L., Jr., Bianchi, L., Link, E. M., and Vanoye, C. G. (2001). From stones to bones: The biology of ClC chloride channels. *Curr. Biol.* **11**, R620–R628.

Greene, J. R., Brown, N. H., DiDomenico, B. J., Kaplan, J., and Eide, D. J. (1993). The *GEF1* gene of *Saccharomyces cerevisiae* encodes an integral membrane protein; mutations in which have effects on respiration and iron-limited growth. *Mol. Gen. Genet.* **241**, 542–553.

Gronemeier, M., Condie, A., Prosser, J., Steinmeyer, K., Jentsch, T. J., and Jockusch, H. (1994). Nonsense and missense mutations in the muscular chloride channel gene Clc-1 of myotonic mice. *J. Biol. Chem.* **269**, 5963–5967.

Gründer, S., Thiemann, A., Pusch, M., and Jentsch, T. J. (1992). Regions involved in the opening of ClC-2 chloride channel by voltage and cell volume. *Nature* **360**, 759–762.

Gu, F. and Gruenberg, J. (2000). ARF1 regulates pH-dependent COP functions in the early endocytic pathway. *J. Biol. Chem.* **275**, 8154–8160.

Gulacsi, A., Lee, C. R., Sik, A., Viitanen, T., Kaila, K., Tepper, J. M., and Freund, T. F. (2003). Cell type-specific differences in chloride-regulatory mechanisms and GABA(A) receptor-mediated inhibition in rat substantia nigra. *J. Neurosci.* **23**, 8237–8246.

Günther, W., Lüchow, A., Cluzeaud, F., Vandewalle, A., and Jentsch, T. J. (1998). ClC-5, the chloride channel mutated in Dent's disease, colocalizes with the proton pump in endocytotically active kidney cells. *Proc. Natl. Acad. Sci. USA* **95**, 8075–8080.

Günther, W., Piwon, N., and Jentsch, T. J. (2003). The ClC-5 chloride channel knock-out mouse—an animal model for Dent's disease. *Pflügers Arch.* **445**, 456–462.

Gyömörey, K., Yeger, H., Ackerley, C., Garami, E., and Bear, C. E. (2000). Expression of the chloride channel ClC-2 in the murine small intestine epithelium. *Am. J. Physiol. Cell Physiol.* **279**, C1787–C1794.

Hara-Chikuma, M., Wang, Y., Guggino, S. E., Guggino, W. B., and Verkman, A. S. (2005a). Impaired acidification in early endosomes of ClC-5 deficient proximal tubule. *Biochem. Biophys. Res. Commun.* **329**, 941–946.

Hara-Chikuma, M., Yang, B., Sonawane, N. D., Sasaki, S., Uchida, S., and Verkman, A. S. (2005b). ClC-3 chloride channels facilitate endosomal acidification and chloride accumulation. *J. Biol. Chem.* **280**, 1241–1247.

Haug, K., Warnstedt, M., Alekov, A. K., Sander, T., Ramirez, A., Poser, B., Maljevic, S., Hebeisen, S., Kubisch, C., Rebstock, J., Horvath, S., Hallmann, K., et al. (2003). Mutations in CLCN2 encoding a voltage-gated chloride channel are associated with idiopathic generalized epilepsies. *Nat. Genet.* **33**, 527–532.

Hechenberger, M., Schwappach, B., Fischer, W. N., Frommer, W. B., Jentsch, T. J., and Steinmeyer, K. (1996). A family of putative chloride channels from *Arabidopsis* and functional complementation of a yeast strain with a *CLC* gene disruption. *J. Biol. Chem.* **271**, 33632–33638.

Holroyd, C., Kistner, U., Annaert, W., and Jahn, R. (1999). Fusion of endosomes involved in synaptic vesicle recycling. *Mol. Biol. Cell* **10**, 3035–3044.

Hoopes, R. R., Jr., Raja, K. M., Koich, A., Hueber, P., Reid, R., Knohl, S. J., and Scheinman, S. J. (2004). Evidence for genetic heterogeneity in Dent's disease. *Kidney Int.* **65**, 1615–1620.

Hoopes, R. R., Jr., Shrimpton, A. E., Knohl, S. J., Hueber, P., Hoppe, B., Matyus, J., Simckes, A., Tasic, V., Toenshoff, B., Suchy, S. F., Nussbaum, R. L., and Scheinman, S. J. (2005). Dent disease with mutations in OCRL1. *Am. J. Hum. Genet.* **76**, 260–267.

Hryciw, D. H., Ekberg, J., Lee, A., Lensink, I. L., Kumar, S., Guggino, W. B., Cook, D. I., Pollock, C. A., and Poronnik, P. (2004). Nedd4-2 functionally interacts with ClC-5: Involvement in constitutive albumin endocytosis in proximal tubule cells. *J. Biol. Chem.* **279**, 54996–55007.

Hübner, C., Stein, V., Hermanns-Borgmeyer, I., Meyer, T., Ballanyi, K., and Jentsch, T. J. (2001). Disruption of KCC2 reveals an essential role of K-Cl-cotransport already in early synaptic inhibition. *Neuron* **30,** 515–524.
Iaizzo, P. A., Franke, C., Hatt, H., Spittelmeister, W., Ricker, K., Rudel, R., and Lehmann-Horn, F. (1991). Altered sodium channel behaviour causes myotonia in dominantly inherited myotonia congenita. *Neuromuscul. Disord.* **1,** 47–53.
Igarashi, T., Günther, W., Sekine, T., Inatomi, J., Shiraga, H., Takahashi, S., Suzuki, J., Tsuru, N., Yanagihara, T., Shimazu, M., Jentsch, T. J., and Thakker, R. V. (1998). Functional characterization of renal chloride channel, CLCN5, mutations associated with Dent's Japan disease. *Kidney Int.* **54,** 1850–1856.
Isnard-Bagnis, C., Da Silva, N., Beaulieu, V., Yu, A. S., Brown, D., and Breton, S. (2003). Detection of ClC-3 and ClC-5 in epididymal epithelium: Immunofluorescence and RT-PCR after LCM. *Am. J. Physiol. Cell Physiol.* **284,** C220–C232.
Jentsch, T. J., Steinmeyer, K., and Schwarz, G. (1990). Primary structure of *Torpedo marmorata* chloride channel isolated by expression cloning in *Xenopus* oocytes. *Nature* **348,** 510–514.
Jentsch, T. J., Günther, W., Pusch, M., and Schwappach, B. (1995). Properties of voltage-gated chloride channels of the ClC gene family. *J. Physiol. (Lond.)* **482,** 19S–25S.
Jentsch, T. J., Stein, V., Weinreich, F., and Zdebik, A. A. (2002). Molecular structure and physiological function of chloride channels. *Physiol. Rev.* **82,** 503–568.
Johnson, R. G., Jr. (1987). Proton pumps and chemiosmotic coupling as a generalized mechanism for neurotransmitter and hormone transport. *Ann. N. Y. Acad. Sci.* **493,** 162–177.
Jordt, S. E. and Jentsch, T. J. (1997). Molecular dissection of gating in the ClC-2 chloride channel. *EMBO J.* **16,** 1582–1592.
Kaila, K., Lamsa, K., Smirnov, S., Taira, T., and Voipio, J. (1997). Long-lasting GABA-mediated depolarization evoked by high-frequency stimulation in pyramidal neurons of rat hippocampal slice is attributable to a network-driven, bicarbonate-dependent K+ transient. *J. Neurosci.* **17,** 7662–7672.
Kasper, D., Planells-Cases, R., Fuhrmann, J. C., Scheel, O., Zeitz, O., Ruether, K., Schmitt, A., Poet, M., Steinfeld, R., Schweizer, M., Kornak, U., and Jentsch, T. J. (2005). Loss of the chloride channel ClC-7 leads to lysosomal storage disease and neurodegeneration. *EMBO J.* **24,** 1079–1091.
Kawasaki, M., Uchida, S., Monkawa, T., Miyawaki, A., Mikoshiba, K., Marumo, F., and Sasaki, S. (1994). Cloning and expression of a protein kinase C-regulated chloride channel abundantly expressed in rat brain neuronal cells. *Neuron* **12,** 597–604.
Kawasaki, M., Suzuki, M., Uchida, S., Sasaki, S., and Marumo, F. (1995). Stable and functional expression of the ClC-3 chloride channel in somatic cell lines. *Neuron* **14,** 1285–1291.
Kida, Y., Uchida, S., Miyazaki, H., Sasaki, S., and Marumo, F. (2001). Localization of mouse CLC-6 and CLC-7 mRNA and their functional complementation of yeast CLC gene mutant. *Histochem. Cell Biol.* **115,** 189–194.
Kieferle, S., Fong, P., Bens, M., Vandewalle, A., and Jentsch, T. J. (1994). Two highly homologous members of the ClC chloride channel family in both rat and human kidney. *Proc. Natl. Acad. Sci. USA* **91,** 6943–6947.
Koch, M. C., Steinmeyer, K., Lorenz, C., Ricker, K., Wolf, F., Otto, M., Zoll, B., Lehmann-Horn, F., Grzeschik, K. H., and Jentsch, T. J. (1992). The skeletal muscle chloride channel in dominant and recessive human myotonia. *Science* **257,** 797–800.
Kornak, U., Schulz, A., Friedrich, W., Uhlhaas, S., Kremens, B., Voit, T., Hasan, C., Bode, U., Jentsch, T. J., and Kubisch, C. (2000). Mutations in the a3 subunit of the vacuolar H^+-ATPase cause infantile malignant osteopetrosis. *Hum. Mol. Genet.* **9,** 2059–2063.
Kornak, U., Kasper, D., Bösl, M. R., Kaiser, E., Schweizer, M., Schulz, A., Friedrich, W., Delling, G., and Jentsch, T. J. (2001). Loss of the ClC-7 chloride channel leads to osteopetrosis in mice and man. *Cell* **104,** 205–215.
Kubisch, C., Schmidt-Rose, T., Fontaine, B., Bretag, A. H., and Jentsch, T. J. (1998). ClC-1 chloride channel mutations in myotonia congenita: Variable penetrance of mutations shifting the voltage dependence. *Hum. Mol. Genet.* **7,** 1753–1760.

Lange, P. F., Wartosch, L., Jentsch, T. J., and Fuhrmann, J. C. (2006). ClC-7 requires Ostm1 as a β-subunit to support bone resorption and lysosomal function. *Nature* **440**, 220–223.
Leheste, J. R., Rolinski, B., Vorum, H., Hilpert, J., Nykjaer, A., Jacobsen, C., Aucouturier, P., Moskaug, J. O., Otto, A., Christensen, E. I., and Willnow, T. E. (1999). Megalin knockout mice as an animal model of low molecular weight proteinuria. *Am. J. Pathol.* **155**, 1361–1370.
Li, X., Shimada, K., Showalter, L. A., and Weinman, S. A. (2000). Biophysical properties of ClC-3 differentiate it from swelling-activated chloride channels in Chinese hamster ovary-K1 cells. *J. Biol. Chem.* **275**, 35994–35998.
Li, X., Wang, T., Zhao, Z., and Weinman, S. A. (2002). The ClC-3 chloride channel promotes acidification of lysosomes in CHO-K1 and Huh-7 cells. *Am. J. Physiol. Cell Physiol.* **282**, C1483–C1491.
Li, Y., Kane, T., Tipper, C., Spatrick, P., and Jenness, D. D. (1999). Yeast mutants affecting possible quality control of plasma membrane proteins. *Mol. Cell. Biol.* **19**, 3588–3599.
Lipecka, J., Bali, M., Thomas, A., Fanen, P., Edelman, A., and Fritsch, J. (2002). Distribution of ClC-2 chloride channel in rat and human epithelial tissues. *Am. J. Physiol. Cell Physiol.* **282**, C805–C816.
Lipicky, R. J. and Bryant, S. H. (1966). Sodium, potassium, and chloride fluxes in intercostal muscle from normal goats and goats with hereditary myotonia. *J. Gen. Physiol.* **50**, 89–111.
Lipicky, R. J., Bryant, S. H., and Salmon, J. H. (1971). Cable parameters, sodium, potassium, chloride, and water content, and potassium efflux in isolated external intercostal muscle of normal volunteers and patients with myotonia congenita. *J. Clin. Invest.* **50**, 2091–2103.
Lloyd, S. E., Pearce, S. H., Fisher, S. E., Steinmeyer, K., Schwappach, B., Scheinman, S. J., Harding, B., Bolino, A., Devoto, M., Goodyer, P., Rigden, S. P., Wrong, O., *et al.* (1996). A common molecular basis for three inherited kidney stone diseases. *Nature* **379**, 445–449.
Lloyd, S. E., Günther, W., Pearce, S. H., Thomson, A., Bianchi, M. L., Bosio, M., Craig, I. W., Fisher, S. E., Scheinman, S. J., Wrong, O., Jentsch, T. J., and Thakker, R. V. (1997). Characterisation of renal chloride channel, CLCN5, mutations in hypercalciuric nephrolithiasis (kidney stones) disorders. *Hum. Mol. Genet.* **6**, 1233–1239.
Lorenz, C., Pusch, M., and Jentsch, T. J. (1996). Heteromultimeric CLC chloride channels with novel properties. *Proc. Natl. Acad. Sci. USA* **93**, 13362–13366.
Ludwig, M. and Utsch, B. (2004). Dent disease-like phenotype and the chloride channel ClC-4 (CLCN4) gene. *Am. J. Med. Genet. A* **128**, 434–435.
Ludwig, M., Waldegger, S., Nuutinen, M., Bokenkamp, A., Reissinger, A., Steckelbroeck, S., and Utsch, B. (2003). Four additional CLCN5 exons encode a widely expressed novel long CLC-5 isoform but fail to explain Dent's phenotype in patients without mutations in the short variant. *Kidney Blood Press. Res.* **26**, 176–184.
Lurin, C., Geelen, D., Barbier-Brygoo, H., Guern, J., and Maurel, C. (1996). Cloning and functional expression of a plant voltage-dependent chloride channel. *Plant Cell* **8**, 701–711.
Lurin, C., Güclü, J., Cheniclet, C., Carde, J. P., Barbier-Brygoo, H., and Maurel, C. (2000). CLC-Nt1, a putative chloride channel protein of tobacco, co-localizes with mitochondrial membrane markers. *Biochem. J.* **348**, 291–295.
Makara, J. K., Petheo, G. L., Toth, A., and Spat, A. (2001). pH-sensitive inwardly rectifying chloride current in cultured rat cortical astrocytes. *Glia* **34**, 52–58.
Makara, J. K., Rappert, A., Matthias, K., Steinhauser, C., Spat, A., and Kettenmann, H. (2003). Astrocytes from mouse brain slices express ClC-2-mediated Cl$^-$ currents regulated during development and after injury. *Mol. Cell. Neurosci.* **23**, 521–530.
Malinowska, D. H., Kupert, E. Y., Bahinski, A., Sherry, A. M., and Cuppoletti, J. (1995). Cloning, functional expression, and characterization of a PKA-activated gastric Cl$^-$ channel. *Am. J. Physiol.* **268**, C191–C200.
Maranda, B., Brown, D., Bourgoin, S., Casanova, J. E., Vinay, P., Ausiello, D. A., and Marshansky, V. (2001). Intra-endosomal pH-sensitive recruitment of the Arf-nucleotide exchange factor ARNO and Arf6 from cytoplasm to proximal tubule endosomes. *J. Biol. Chem.* **276**, 18540–18550.
Marini, C., Scheffer, I. E., Crossland, K. M., Grinton, B. E., Phillips, F. L., McMahon, J. M., Turner, S. J., Dean, J. T., Kivity, S., Mazarib, A., Neufeld, M. Y., Korczyn, A. D., *et al.* (2004). Genetic architecture of idiopathic generalized epilepsy: Clinical genetic analysis of 55 multiplex families. *Epilepsia* **45**, 467–478.

Maritzen, T., Rickheit, G., Schmitt, A., and Jentsch, T. J. (2006). Kidney-specific unregulation of vitamin D_3 target genes in ClC-5 KO mice. *Kidney Int.* **70**, 79–87.
Marshansky, V., Ausiello, D. A., and Brown, D. (2002). Physiological importance of endosomal acidification: Potential role in proximal tubulopathies. *Curr. Opin. Nephrol. Hypertens.* **11**, 527–537.
Matsumura, Y., Uchida, S., Kondo, Y., Miyazaki, H., Ko, S. B., Hayama, A., Morimoto, T., Liu, W., Arisawa, M., Sasaki, S., and Marumo, F. (1999). Overt nephrogenic diabetes insipidus in mice lacking the CLC-K1 chloride channel. *Nat. Genet.* **21**, 95–98.
Maycox, P. R., Deckwerth, T., Hell, J. W., and Jahn, R. (1988). Glutamate uptake by brain synaptic vesicles. Energy dependence of transport and functional reconstitution in proteoliposomes. *J. Biol. Chem.* **263**, 15423–15428.
Metz, J., Wachter, A., Schmidt, B., Bujnicki, J. M., and Schwappach, B. (2006). The yeast Arr4p ATPase binds the chloride transporter Gef1p when copper is available in the cytosol. *J. Biol. Chem.* **281**, 410–417.
Meyer, S. and Dutzler, R. (2006). Crystal Structure of the cytoplasmic domain of the chloride channel ClC-0. *Structure* **14**, 299–307.
Meyer-Kleine, C., Steinmeyer, K., Ricker, K., Jentsch, T. J., and Koch, M. C. (1995). Spectrum of mutations in the major human skeletal muscle chloride channel gene (*CLCN1*) leading to myotonia. *Am. J. Hum. Genet.* **57**, 1325–1334.
Miles, R. (1999). Neurobiology. A homeostatic switch. *Nature* **397**, 215–216.
Miller, C. and White, M. M. (1984). Dimeric structure of single chloride channels from Torpedo electroplax. *Proc. Natl. Acad. Sci. USA* **81**, 2772–2775.
Mohammad-Panah, R., Gyomorey, K., Rommens, J., Choudhury, M., Li, C., Wang, Y., and Bear, C. E. (2001). ClC-2 contributes to native chloride secretion by a human intestinal cell line, Caco-2. *J. Biol. Chem.* **276**, 8306–8313.
Mohammad-Panah, R., Harrison, R., Dhani, S., Ackerley, C., Huan, L. J., Wang, Y., and Bear, C. E. (2003). The chloride channel ClC-4 contributes to endosomal acidification and trafficking. *J. Biol. Chem.* **278**, 29267–29277.
Morimoto, T., Uchida, S., Sakamoto, H., Kondo, Y., Hanamizu, H., Fukui, M., Tomino, Y., Nagano, N., Sasaki, S., and Marumo, F. (1998). Mutations in *CLCN5* chloride channel in Japanese patients with low molecular weight proteinuria. *J. Am. Soc. Nephrol.* **9**, 811–818.
Murray, C. B., Morales, M. M., Flotte, T. R., McGrath-Morrow, S. A., Guggino, W. B., and Zeitlin, P. L. (1995). CLC-2: A developmentally dependent chloride channel expressed in the fetal lung and down-regulated after birth. *Am. J. Respir. Cell Mol. Biol.* **12**, 597–604.
Murray, C. B., Chu, S., and Zeitlin, P. L. (1996). Gestational and tissue-specific regulation of ClC-2 chloride channel expression. *Am. J. Physiol.* **271**, L829–L837.
Nehrke, K., Begenisich, T., Pilato, J., and Melvin, J. E. (2000). Into ion channel and transporter function. *Caenorhabditis elegans* ClC-type chloride channels: Novel variants and functional expression. *Am. J. Physiol. Cell Physiol.* **279**, C2052–C2066.
Nehrke, K., Arreola, J., Nguyen, H. V., Pilato, J., Richardson, L., Okunade, G., Baggs, R., Shull, G. E., and Melvin, J. E. (2002). Loss of hyperpolarization-activated Cl^- current in salivary acinar cells from *Clcn2* knockout mice. *J. Biol. Chem.* **26**, 23604–23611.
Niemeyer, M. I., Cid, L. P., Zuniga, L., Catalán, M., and Sepúlveda, F. V. (2003). A conserved pore-lining glutamate as a voltage- and chloride-dependent gate in the ClC-2 chloride channel. *J. Physiol.* **553**, 873–879.
Niemeyer, M. I., Yusef, Y. R., Cornejo, I., Flores, C. A., Sepúlveda, F. V., and Cid, L. P. (2004). Functional evaluation of human ClC-2 chloride channel mutations associated with idiopathic generalized epilepsies. *Physiol. Genomics* **19**, 74–83.
Nobile, M., Pusch, M., Rapisarda, C., and Ferroni, S. (2000). Single-channel analysis of a ClC-2-like chloride conductance in cultured rat cortical astrocytes. *FEBS Lett.* **479**, 10–14.
Norden, A. G., Lapsley, M., Lee, P. J., Pusey, C. D., Scheinman, S. J., Tam, F. W., Thakker, R. V., Unwin, R. J., and Wrong, O. (2001). Glomerular protein sieving and implications for renal failure in Fanconi syndrome. *Kidney Int.* **60**, 1885–1892.

Obermüller, N., Gretz, N., Kriz, W., Reilly, R. F., and Witzgall, R. (1998). The swelling-activated chloride channel ClC-2, the chloride channel ClC-3, and ClC-5, a chloride channel mutated in kidney stone disease, are expressed in distinct subpopulations of renal epithelial cells. *J. Clin. Invest.* **101,** 635–642.

Ogura, T., Furukawa, T., Toyozaki, T., Yamada, K., Zheng, Y. J., Katayama, Y., Nakaya, H., and Inagaki, N. (2002). ClC-3B, a novel ClC-3 splicing variant that interacts with EBP50 and facilitates expression of CFTR-regulated ORCC. *FASEB J.* **16,** S63–S65.

Park, K., Arreola, J., Begenisich, T., and Melvin, J. E. (1998). Comparison of voltage-activated Cl⁻ channels in rat parotid acinar cells with ClC-2 in a mammalian expression system. *J. Membr. Biol.* **163,** 87–95.

Parsons, S. M., Prior, C., and Marshall, I. G. (1993). Acetylcholine transport, storage, and release. *Int. Rev. Neurobiol.* **35,** 279–390.

Pena-Munzenmayer, G., Catalan, M., Cornejo, I., Figueroa, C. D., Melvin, J. E., Niemeyer, M. I., Cid, L. P., and Sepulveda, F. V. (2005). Basolateral localization of native ClC-2 chloride channels in absorptive intestinal epithelial cells and basolateral sorting encoded by a CBS-2 domain di-leucine motif. *J. Cell Sci.* **118,** 4243–4252.

Petalcorin, M. I., Oka, T., Koga, M., Ogura, K., Wada, Y., Ohshima, Y., and Futai, M. (1999). Disruption of *clh*-1, a chloride channel gene, results in a wider body of Caenorhabditis elegans. *J. Mol. Biol.* **294,** 347–355.

Pfeiffer, F., Graham, D., and Betz, H. (1982). Purification by affinity chromatography of the glycine receptor of rat spinal cord. *J. Biol. Chem.* **257,** 9389–9393.

Picollo, A. and Pusch, M. (2005). Chloride/proton antiporter activity of mammalian CLC proteins ClC-4 and ClC-5. *Nature* **436,** 420–423.

Pirozzi, G., McConnell, S. J., Uveges, A. J., Carter, J. M., Sparks, A. B., Kay, B. K., and Fowlkes, D. M. (1997). Identification of novel human WW domain-containing proteins by cloning of ligand targets. *J. Biol. Chem.* **272,** 14611–14616.

Piwon, N., Günther, W., Schwake, M., Bösl, M. R., and Jentsch, T. J. (2000). ClC-5 Cl⁻-channel disruption impairs endocytosis in a mouse model for Dent's disease. *Nature* **408,** 369–373.

Poët, M., Kornak, U., Schweizer, M., Zdebik, A. A., Scheel, O., Hoelter, S., Wurst, W., Schmitt, A., Fuhrmann, J. C., Planells-Cases, R., Mole, S. E., Hübner, C. A., and Jentsch, T. J. (2006). Lysosomal storage disease upon disruption of the neuronal chloride transport protein ClC-6. *Proc. Natl. Acad. Sci. USA* **103,** 13854–13859.

Pusch, M. (2002). Myotonia caused by mutations in the muscle chloride channel gene CLCN1. *Hum. Mutat.* **19,** 423–434.

Pusch, M., Ludewig, U., Rehfeldt, A., and Jentsch, T. J. (1995a). Gating of the voltage-dependent chloride channel ClC-0 by the permeant anion. *Nature* **373,** 527–531.

Pusch, M., Steinmeyer, K., Koch, M. C., and Jentsch, T. J. (1995b). Mutations in dominant human myotonia congenita drastically alter the voltage dependence of the ClC-1 chloride channel. *Neuron* **15,** 1455–1463.

Pusch, M., Jordt, S. E., Stein, V., and Jentsch, T. J. (1999). Chloride dependence of hyperpolarization-activated chloride channel gates. *J. Physiol. (Lond.)* **515,** 341–353.

Riordan, J. R., Rommens, J. M., Kerem, B., Alon, N., Rozmahel, R., Grzelczak, Z., Zielenski, J., Lok, S., Plavsic, N., Chou, J. L., Drumm, M. L., Iannuzzi, M. C., *et al.* (1989). Identification of the cystic fibrosis gene: Cloning and characterization of complementary DNA. *Science* **245,** 1066–1073.

Rivera, C., Voipio, J., Payne, J. A., Ruusuvuori, E., Lahtinen, H., Lamsa, K., Pirvola, U., Saarma, M., and Kaila, K. (1999). The K⁺/Cl⁻ co-transporter KCC2 renders GABA hyperpolarizing during neuronal maturation. *Nature* **397,** 251–255.

Rotin, D., Staub, O., and Haguenauer-Tsapis, R. (2000). Ubiquitination and endocytosis of plasma membrane proteins: Role of Nedd4/Rsp5p family of ubiquitin-protein ligases. *J. Membr. Biol.* **176,** 1–17.

Rozmahel, R., Wilschanski, M., Matin, A., Plyte, S., Oliver, M., Auerbach, W., Moore, A., Forstner, J., Durie, P., Nadeau, J., Bear, C., and Tsui, L. C. (1996). Modulation of disease severity in cystic fibrosis transmembrane conductance regulator deficient mice by a secondary genetic factor. *Nat. Genet.* **12,** 280–287.

Rüdel, R. and Lehmann-Horn, F. (1985). Membrane changes in cells from myotonia patients. *Physiol. Rev.* **65,** 310–356.

Rugarli, E. I., Adler, D. A., Borsani, G., Tsuchiya, K., Franco, B., Hauge, X., Disteche, C., Chapman, V., and Ballabio, A. (1995). Different chromosomal localization of the *Clcn4* gene in Mus spretus and C57BL/6J mice. *Nat. Genet.* **10**, 466–471.

Rutledge, E., Bianchi, L., Christensen, M., Boehmer, C., Morrison, R., Broslat, A., Beld, A. M., George, A. L., Greenstein, D., and Strange, K. (2001). CLH-3, a ClC-2 anion channel ortholog activated during meiotic maturation in *C. elegans* oocytes. *Curr. Biol.* **11**, 161–170.

Sabolic, I. and Burckhardt, G. (1986). Characteristics of the proton pump in rat renal cortical endocytotic vesicles. *Am. J. Physiol.* **250**, F817–F826.

Sakamoto, H., Sado, Y., Naito, I., Kwon, T. H., Inoue, S., Endo, K., Kawasaki, M., Uchida, S., Nielsen, S., Sasaki, S., and Marumo, F. (1999). Cellular and subcellular immunolocalization of ClC-5 channel in mouse kidney: Colocalization with H^+-ATPase. *Am. J. Physiol.* **277**, F957–F965.

Salazar, G., Love, R., Styers, M. L., Werner, E., Peden, A., Rodriguez, S., Gearing, M., Wainer, B. H., and Faundez, V. (2004). AP-3-dependent mechanisms control the targeting of a chloride channel (ClC-3) in neuronal and non-neuronal cells. *J. Biol. Chem.* **279**, 25430–25439.

Salvatore, F., Scudiero, O., and Castaldo, G. (2002). Genotype-phenotype correlation in cystic fibrosis: The role of modifier genes. *Am. J. Med. Genet.* **111**, 88–95.

Sander, T., Schulz, H., Saar, K., Gennaro, E., Riggio, M. C., Bianchi, A., Zara, F., Luna, D., Bulteau, C., Kaminska, A., Ville, D., Cieuta, C., *et al.* (2000). Genome search for susceptibility loci of common idiopathic generalised epilepsies. *Hum. Mol. Genet.* **9**, 1465–1472.

Scheel, O., Zdebik, A., Lourdel, S., and Jentsch, T. J. (2005). Voltage-dependent electrogenic chloride proton exchange by endosomal CLC proteins. *Nature* **436**, 424–427.

Scheinman, S. J. (1998). X-linked hypercalciuric nephrolithiasis: Clinical syndromes and chloride channel mutations. *Kidney Int.* **53**, 3–17.

Schlingmann, K. P., Konrad, M., Jeck, N., Waldegger, P., Reinalter, S. C., Holder, M., Seyberth, H. W., and Waldegger, S. (2004). Salt wasting and deafness resulting from mutations in two chloride channels. *N. Engl. J. Med.* **350**, 1314–1319.

Schofield, P. R., Darlison, M. G., Fujita, N., Burt, D. R., Stephenson, F. A., Rodriguez, H., Rhee, L. M., Ramachandran, J., Reale, V., Glencorse, T. A., Seeburg, P. H., and Barnard, E. A. (1987). Sequence and functional expression of the GABA A receptor shows a ligand-gated receptor super-family. *Nature* **328**, 221–227.

Schriever, A. M., Friedrich, T., Pusch, M., and Jentsch, T. J. (1999). CLC chloride channels in *Caenorhabditis elegans*. *J. Biol. Chem.* **274**, 34238–34244.

Schwake, M., Friedrich, T., and Jentsch, T. J. (2001). An internalization signal in ClC-5, an endosomal Cl^--channel mutated in Dent's disease. *J. Biol. Chem.* **276**, 12049–12054.

Schwappach, B., Stobrawa, S., Hechenberger, M., Steinmeyer, K., and Jentsch, T. J. (1998). Golgi localization and functionally important domains in the NH_2 and COOH terminus of the yeast CLC putative chloride channel Gef1p. *J. Biol. Chem.* **273**, 15110–15118.

Schwiebert, E. M., Cid-Soto, L. P., Stafford, D., Carter, M., Blaisdell, C. J., Zeitlin, P. L., Guggino, W. B., and Cutting, G. R. (1998). Analysis of ClC-2 channels as an alternative pathway for chloride conduction in cystic fibrosis airway cells. *Proc. Natl. Acad. Sci. USA* **95**, 3879–3884.

Scimeca, J. C., Franchi, A., Trojani, C., Parrinello, H., Grosgeorge, J., Robert, C., Jaillon, O., Poirier, C., Gaudray, P., and Carle, G. F. (2000). The gene encoding the mouse homologue of the human osteoclast-specific 116-kDa V-ATPase subunit bears a deletion in osteosclerotic (*oc/oc*) mutants. *Bone* **26**, 207–213.

Scott, J. W., Hawley, S. A., Green, K. A., Anis, M., Stewart, G., Scullion, G. A., Norman, D. G., and Hardie, D. G. (2004). CBS domains form energy-sensing modules whose binding of adenosine ligands is disrupted by disease mutations. *J. Clin. Invest.* **113**, 274–284.

Seong, E., Wainer, B. H., Hughes, E. D., Saunders, T. L., Burmeister, M., and Faundez, V. (2005). Genetic analysis of the neuronal and ubiquitous AP-3 adaptor complexes reveals divergent functions in brain. *Mol. Biol. Cell* **16**, 128–140.

Sik, A., Smith, R. L., and Freund, T. F. (2000). Distribution of chloride channel-2-immunoreactive neuronal and astrocytic processes in the hippocampus. *Neuroscience* **101**, 51–65.

Silva, I. V., Cebotaru, V., Wang, H., Wang, X. T., Wang, S. S., Guo, G., Devuyst, O., Thakker, R. V., Guggino, W. B., and Guggino, S. E. (2003). The ClC-5 knockout mouse model of Dent's disease has renal hypercalciuria and increased bone turnover. *J. Bone Miner. Res.* **18,** 615–623.
Simon, D. B., Karet, F. E., Hamdan, J. M., DiPietro, A., Sanjad, S. A., and Lifton, R. P. (1996a). Bartter's syndrome, hypokalaemic alkalosis with hypercalciuria, is caused by mutations in the Na-K-2Cl cotransporter NKCC2. *Nat. Genet.* **13,** 183–188.
Simon, D. B., Karet, F. E., Rodriguez-Soriano, J., Hamdan, J. H., DiPietro, A., Trachtman, H., Sanjad, S. A., and Lifton, R. P. (1996b). Genetic heterogeneity of Bartter's syndrome revealed by mutations in the K^+ channel, ROMK. *Nat. Genet.* **14,** 152–156.
Simon, D. B., Bindra, R. S., Mansfield, T. A., Nelson-Williams, C., Mendonca, E., Stone, R., Schurman, S., Nayir, A., Alpay, H., Bakkaloglu, A., Rodriguez-Soriano, J., Morales, J. M., *et al.* (1997). Mutations in the chloride channel gene, *CLCNKB*, cause Bartter's syndrome type III. *Nat. Genet.* **17,** 171–178.
Smith, R. L., Clayton, G. H., Wilcox, C. L., Escudero, K. W., and Staley, K. J. (1995). Differential expression of an inwardly rectifying chloride conductance in rat brain neurons: A potential mechanism for cell-specific modulation of postsynaptic inhibition. *J. Neurosci.* **15,** 4057–4067.
Staley, K. (1994). The role of an inwardly rectifying chloride conductance in postsynaptic inhibition. *J. Neurophysiol.* **72,** 273–284.
Staley, K., Smith, R., Schaack, J., Wilcox, C., and Jentsch, T. J. (1996). Alteration of $GABA_A$ receptor function following gene transfer of the CLC-2 chloride channel. *Neuron* **17,** 543–551.
Steinmeyer, K., Klocke, R., Ortland, C., Gronemeier, M., Jockusch, H., Gründer, S., and Jentsch, T. J. (1991a). Inactivation of muscle chloride channel by transposon insertion in myotonic mice. *Nature* **354,** 304–308.
Steinmeyer, K., Ortland, C., and Jentsch, T. J. (1991b). Primary structure and functional expression of a developmentally regulated skeletal muscle chloride channel. *Nature* **354,** 301–304.
Steinmeyer, K., Lorenz, C., Pusch, M., Koch, M. C., and Jentsch, T. J. (1994). Multimeric structure of ClC-1 chloride channel revealed by mutations in dominant myotonia congenita (Thomsen). *EMBO J.* **13,** 737–743.
Steinmeyer, K., Schwappach, B., Bens, M., Vandewalle, A., and Jentsch, T. J. (1995). Cloning and functional expression of rat CLC-5, a chloride channel related to kidney disease. *J. Biol. Chem.* **270,** 31172–31177.
Stobrawa, S. M., Breiderhoff, T., Takamori, S., Engel, D., Schweizer, M., Zdebik, A. A., Bösl, M. R., Ruether, K., Jahn, H., Draguhn, A., Jahn, R., and Jentsch, T. J. (2001). Disruption of ClC-3, a chloride channel expressed on synaptic vesicles, leads to a loss of the hippocampus. *Neuron* **29,** 185–196.
Strong, T. V., Boehm, K., and Collins, F. S. (1994). Localization of cystic fibrosis transmembrane conductance regulator mRNA in the human gastrointestinal tract by in situ hybridization. *J. Clin. Invest.* **93,** 347–354.
Südhof, T. C. (2000). The synaptic vesicle cycle revisited. *Neuron* **28,** 317–320.
Suzuki, T., Rai, T., Hayama, A., Sohara, E., Suda, S., Itoh, T., Sasaki, S., and Uchida, S. (2006). Intracellular localization of ClC chloride channels and their ability to form hetero-oligomers. *J. Cell. Physiol.* **206,** 792–798.
Teardo, E., Frare, E., Segalla, A., De Marco, V., Giacometti, G. M., and Szabo, I. (2005). Localization of a putative ClC chloride channel in spinach chloroplasts. *FEBS Lett.* **579,** 4991–4996.
Thiemann, A., Gründer, S., Pusch, M., and Jentsch, T. J. (1992). A chloride channel widely expressed in epithelial and non-epithelial cells. *Nature* **356,** 57–60.
Thomsen, J. (1876). Tonische Krämpfe in willkürlich beweglichen Muskeln in Folge von ererbter psychischer Disposition. *Arch. Psychiatr. Nervenkr.* **6,** 702–718.
Tyteca, D., Van Der Smissen, P., Mettlen, M., Van Bambeke, F., Tulkens, P. M., Mingeot-Leclercq, M. P., and Courtoy, P. J. (2002). Azithromycin, a lysosomotropic antibiotic, has distinct effects on fluid-phase and receptor-mediated endocytosis, but does not impair phagocytosis in J774 macrophages. *Exp. Cell Res.* **281,** 86–100.
Uchida, S., Sasaki, S., Furukawa, T., Hiraoka, M., Imai, T., Hirata, Y., and Marumo, F. (1993). Molecular cloning of a chloride channel that is regulated by dehydration and expressed predominantly in kidney medulla [published erratum appears in *J. Biol. Chem.* July 22, 1994, **269**(29), 19192]. *J. Biol. Chem.* **268,** 3821–3824.

Uchida, S., Sasaki, S., Nitta, K., Uchida, K., Horita, S., Nihei, H., and Marumo, F. (1995). Localization and functional characterization of rat kidney-specific chloride channel, ClC-K1. *J. Clin. Invest.* **95,** 104–113.
van Slegtenhorst, M. A., Bassi, M. T., Borsani, G., Wapenaar, M. C., Ferrero, G. B., de Conciliis, L., Rugarli, E. I., Grillo, A., Franco, B., Zoghbi, H. Y., and Ballabio, A. (1994). A gene from the Xp22.3 region shares homology with voltage-gated chloride channels. *Hum. Mol. Genet.* **3,** 547–552.
Vandewalle, A., Cluzeaud, F., Bens, M., Kieferle, S., Steinmeyer, K., and Jentsch, T. J. (1997). Localization and induction by dehydration of ClC-K chloride channels in the rat kidney. *Am. J. Physiol.* **272,** F678–F688.
Vandewalle, A., Cluzeaud, F., Peng, K. C., Bens, M., Lüchow, A., Günther, W., and Jentsch, T. J. (2001). Tissue distribution and subcellular localization of the ClC-5 chloride channel in rat intestinal cells. *Am. J. Physiol. Cell Physiol.* **280,** C373–C381.
Varela, D., Niemeyer, M. I., Cid, L. P., and Sepulveda, F. V. (2002). Effect of an N-terminus deletion on voltage-dependent gating of the ClC-2 chloride channel. *J. Physiol.* **544,** 363–372.
Wächter, A. and Schwappach, B. (2005). The yeast CLC chloride channel is proteolytically processed by the furin-like protease Kex2p in the first extracellular loop. *FEBS Lett.* **579,** 1149–1153.
Waldegger, S. and Jentsch, T. J. (2000). Functional and structural analysis of ClC-K chloride channels involved in renal disease. *J. Biol. Chem.* **275,** 24527–24533.
Walz, W. (2002). Chloride/anion channels in glial cell membranes. *Glia* **40,** 1–10.
Wang, S. S., Devuyst, O., Courtoy, P. J., Wang, X. T., Wang, H., Wang, Y., Thakker, R. V., Guggino, S., and Guggino, W. B. (2000). Mice lacking renal chloride channel, CLC-5, are a model for Dent's disease, a nephrolithiasis disorder associated with defective receptor-mediated endocytosis. *Hum. Mol. Genet.* **9,** 2937–2945.
Wang, T. and Weinman, S. A. (2004). Involvement of chloride channels in hepatic copper metabolism: ClC-4 promotes copper incorporation into ceruloplasmin. *Gastroenterology* **126,** 1157–1166.
Weinreich, F. and Jentsch, T. J. (2001). Pores formed by single subunits in mixed dimers of different CLC chloride channels. *J. Biol. Chem.* **276,** 2347–2353.
Welsh, M. J., Smith, P. L., Fromm, M., and Frizzell, R. A. (1982). Crypts are the site of intestinal fluid and electrolyte secretion. *Science* **218,** 1219–1221.
Weylandt, K. H., Valverde, M. A., Nobles, M., Raguz, S., Amey, J. S., Díaz, M., Nastrucci, C., Higgins, C. F., and Sardini, A. (2001). Human ClC-3 is not the swelling-activated chloride channel involved in cell volume regulation. *J. Biol. Chem.* **276,** 17461–17467.
White, M. M. and Miller, C. (1979). A voltage-gated anion channel from the electric organ of Torpedo californica. *J. Biol. Chem.* **254,** 10161–10166.
Wrong, O. M., Norden, A. G., and Feest, T. G. (1994). Dent's disease; a familial proximal renal tubular syndrome with low-molecular-weight proteinuria, hypercalciuria, nephrocalcinosis, metabolic bone disease, progressive renal failure and a marked male predominance. *QJM* **87,** 473–493.
Wu, F., Roche, P., Christie, P. T., Loh, N. Y., Reed, A. A., Esnouf, R. M., and Thakker, R. V. (2003). Modeling study of human renal chloride channel (hCLC-5) mutations suggests a structural-functional relationship. *Kidney Int.* **63,** 1426–1432.
Xiong, H., Li, C., Garami, E., Wang, Y., Ramjeesingh, M., Galley, K., and Bear, C. E. (1999). ClC-2 activation modulates regulatory volume decrease. *J. Membr. Biol.* **167,** 215–221.
Yamamoto-Mizuma, S., Wang, G. X., Liu, L. L., Schegg, K., Hatton, W. J., Duan, D., Horowitz, B., Lamb, F. S., and Hume, J. R. (2004). Altered properties of volume-sensitive osmolyte and anion channels (VSOACs) and membrane protein expression in cardiac and smooth muscle myocytes from $Clcn3^{-/-}$ mice. *J. Physiol.* **557,** 439–456.
Yoshikawa, M., Uchida, S., Ezaki, J., Rai, T., Hayama, A., Kobayashi, K., Kida, Y., Noda, M., Koike, M., Uchiyama, Y., Marumo, F., Kominami, E., et al. (2002). CLC-3 deficiency leads to phenotypes similar to human neuronal ceroid lipofuscinosis. *Genes Cells* **7,** 597–605.
Zdebik, A. A., Cuffe, J., Bertog, M., Korbmacher, C., and Jentsch, T. J. (2004). Additional disruption of the ClC-2 Cl$^-$ channel does not exacerbate the cystic fibrosis phenotype of CFTR mouse models. *J. Biol. Chem.* **279,** 22276–22283.

Chapter 3

Structure and Function of CLC Chloride Channels and Transporters

Alessio Accardi

Department of Biochemistry, HHMI/Brandeis University, Waltham, Massachusetts 02454

I. The Structure of CLC-ec1, a Bacterial CLC Transporter
II. Structural Conservation Among the CLCs
III. Ion Selectivity, Permeation and Binding
    A. The Channels
    B. The Transporters
IV. A Multi-Ion Pore
V. Channels and Transporters: Where Is the Difference?
VI. Function of the CLC Transporters
VII. Speculative Modeling
VIII. CLC Channel Gating
    A. Voltage, Cl^- and H^+ Gating of a CLC Channel
    B. Molecular Identity of the Fast Gate
    C. The Slow Gate
    D. Fast and Slow Gating in Other CLC Channels
IX. The Role of the C-Terminal Domain of CLC Channels: Recent Developments
X. Conclusions and Outlook
    References

The chloride channels (CLCs) form a large and widespread family of integral membrane proteins that allow Cl^- movement across biological membranes. Until a few years ago, all members of this family were believed to be Cl^--selective ion channels. However, recent developments have shown that two functionally distinct but structurally similar subclasses populate this protein family: ion channels and exchange-transporters (Accardi and Miller, 2004; Picollo and Pusch, 2005; Scheel *et al.*, 2005). Members of these two subgroups achieve the same basic purpose—moving Cl^- across a lipid bilayer—but in two completely different ways: the channels form an aqueous pore through which Cl^- ions freely diffuse down their electrochemical gradient, while the transporters stoichiometrically swap Cl^- from one side of the membrane for H^+ from the other, using the energy stored in the proton's electrochemical potential to drive uphill Cl^- movement (and vice versa). This distinction in function seems to correlate well with the cellular localization of the CLCs; the channels are found in the plasma membrane, while the transporters are generally segregated into intracellular compartments where they play a key role in acidification (Piwon et al., 2000; Kornak et al., 2001; Iyer et al., 2002).

In order to fulfill their physiological roles, channels must selectively allow ion movement and open and close (gate) their pores according to the cell's needs. An additional layer of complexity has to be considered for CLC proteins: ion permeation through the channels will obey the laws of free diffusion, while ion transport mediated by the Cl^-/H^+ exchangers will be tightly regulated by a fixed stoichiometry rule.

In this chapter, I will focus on how the structure of a bacterial CLC transporter, ClC-ec1 (Dutzler et al., 2002, 2003), has guided our understanding of these fundamental processes for both channels and transporters. The starting point will thus be the structure of ClC-ec1 and the few known things about its transport mechanism. I will then discuss ion permeation and CLC gating as seen from this structural viewpoint, trying to speculate on where the differences between CLC channels and transporters are hidden.

Despite their thermodynamically opposite mechanisms, both CLC subtypes share a common structural fold, quaternary organization, and ionic selectivity, making this family a case study in how evolution has fine-tuned one basic structure to achieve profoundly different functions.

I. THE STRUCTURE OF ClC-ec1, A BACTERIAL CLC TRANSPORTER

The identification of genes encoding members of the CLC family in bacteria led to the cloning, overexpression, and purification of ClC-ec1, one of the 3 CLC homologues from *Escherichia coli* (Maduke et al., 1999). This bacterial homologue is mostly membrane embedded and lacks the large C-terminal domain that is characteristically found in all eukaryotic CLCs, such that ClC-ec1 constitutes the minimal functional unit of a CLC protein. Immediately it became a prime target for structural investigations: its projection structure was first determined at 6.5-Å resolution through electron microscopy (Mindell et al., 2001) and eventually the MacKinnon lab successfully solved its three-dimensional structure to 2.5-Å resolution through X-ray crystallography (Dutzler et al., 2002, 2003). In agreement with previous biochemical and electrophysiological data (Middleton et al., 1994, 1996; Ludewig et al., 1996), the structure revealed that ClC-ec1 is a homodimer with twofold symmetry (Fig. 1A). Each monomer forms a well-defined Cl^- permeation pathway that is not straight and perpendicular to the plane of the membrane, as seen for K^+ channels, but rather is kinked in the middle such that Cl^- ions have an arched trajectory when crossing the membrane. The structure also revealed a weakly conserved internal repeat within ClC-ec1 that had not been previously identified based solely on the amino acid sequence (Fig. 1B). The N-terminal half (α-helices B–I, in black) runs antiparallel and is structurally related to the C-terminal half (α-helices J–Q, in white) forming a pseudo twofold symmetry axis

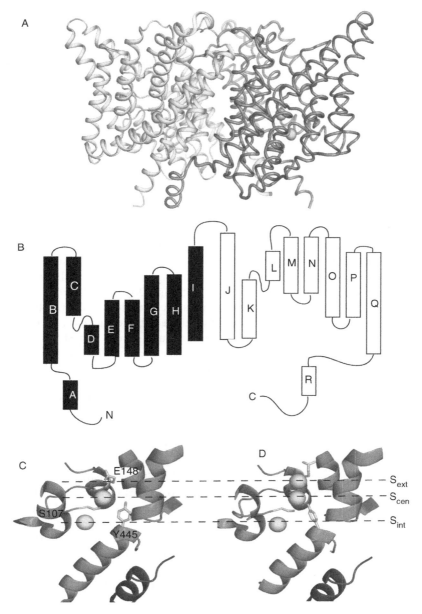

Figure 1. Structure of ClC-ec1. (A) View of the ClC-ec1 dimer from within the membrane with the extracellular side above. (B) Topological representation of a ClC-ec1 subunit where the α-helices are drawn as cylinders. The two internally repeated halves are colored in white and black. (C) The Cl^--binding region in WT ClC-ec1 and (D) in the E148Q mutant. This figure was prepared with Pymol.

within the membrane. A similar antiparallel architecture is also found in the individual subunits of aquaporin channels, even though their topology is completely unrelated to that of the ClCs. Each subunit in ClC-ec1 is formed by 18 α-helices that either cross the membrane with tilted angles or penetrate the membrane only partway and then loop back and emerge from the same side. This complex topology brings regions of the protein that are far apart in the sequence close together to form the Cl^--coordinating regions. In the wild-type (WT) structure two Cl^- ions are bound, the Cl^- in the inner site, S_{int}, is partly water exposed, while the ion in the central site, S_{cen}, is completely encased in the protein and likely to be dehydrated (Fig. 1C, see Section III.A for a detailed description of the binding sites).

II. STRUCTURAL CONSERVATION AMONG THE CLCs

Sequence conservation between members of the CLC family is low ($\sim$20%) and mostly localized to short stretches scattered throughout the protein that are involved in Cl^- binding (Dutzler *et al.*, 2002). Unlike K^+ channels, where the structure of KcsA was in agreement with and expanded the preexisting picture of K^+ channels based on over 40 years of electrophysiological investigations (Doyle *et al.*, 1998; Hille, 2001), our knowledge of CLC architecture in the prestructural days was primitive; even the number of pores in a CLC channel was hotly debated (Ludewig *et al.*, 1996; Middleton *et al.*, 1996; Fahlke *et al.*, 1998). Thus, several groups set out to test whether eukaryotic channels, such as ClC-0 and ClC-1, share the same complex topology as ClC-ec1. The overall agreement between the structure-based predictions and the experimental results clearly indicates that they do.

The first suggestion of a conserved structural architecture between channels and transporters came from the observation that many mutations in the eukaryotic ClC-1 channel that affect a process involving both subunits simultaneously (see Section VIII.C) are localized near the intersubunit interface of the prokaryotic ClC-ec1 (Estévez and Jentsch, 2002; Duffield *et al.*, 2003). So far, however, the main focus of these investigations has been the intracellular vestibule lining the Cl^- permeation pathway. Several laboratories used different yet complementary approaches to investigate the structural conservation. Estévez *et al.* (2003) identified a single residue crucial for the different binding affinities of 9-anthracene-carboxylic acid (9-AC) and chlorophenoxy-acetic acid (CPA), two CLC blockers, to ClC-0, ClC-1, and ClC-2. They then used the structure of ClC-ec1 to successfully predict several other residues that are far in the sequence but close in the three-dimensional structure and line the binding sites. Two other groups used cysteine-scanning mutagenesis to probe the conservation of the Cl^- pathway between ClC-0 and ClC-ec1 (Chen and Chen, 2003; Chen *et al.*, 2003; Lin and Chen, 2003; Engh and Maduke, 2005). Chen and collaborators showed that the accessibility pattern of the last 13 residues in the transmembrane region of ClC-0 is consistent with an α-helix that has the same orientation as helix R in ClC-ec1 (Lin and Chen, 2003). They also showed that altering the charges at residues predicted to line the Cl^- permeation pathway based on the ClC-ec1 structure alters the electrostatic properties of the ClC-0 pore

(Chen and Chen, 2003; Chen et al., 2003). Engh and Maduke (2005) performed an extensive cysteine scan of helices R and J, the C-D loop, and residues on helices C, D, E, F, and M and found that all of the MTSES-sensitive residues appear to be nearer to the bound Cl^- ions than the insensitive ones. They also showed that two residues that are in close proximity in ClC-ec1 are also close in ClC-0.

The striking conservation pattern observed at the intracellular end of the Cl^- pathway also extends to the extracellular side. Picollo et al. (2004) successfully used the crystal structure to identify two residues in the extracellular vestibule of CLC-K channels that line the 4,4′-diisothiocyanato-2,2′-stilbenedisulfonic acid (DIDS)-binding site.

This corpus of experimental results clearly indicates that the overall architecture of the CLCs is conserved and that, despite profound functional differences, ClC-ec1 is a surprisingly good structural template on which to model the eukaryotic CLC channels.

III. ION SELECTIVITY, PERMEATION AND BINDING

A. The Channels

CLC chloride channels are very selective for anions over cations, although their ability to discriminate among different ions with the same charge is weak compared to that seen in K^+ or Na^+ channels. This lack of interanionic selectivity probably reflects the fact that Cl^- is the only major anion present in physiological conditions, while the ability to selectively regulate passage of cations as similar as Na^+ and K^+ is vital for proper cellular function. Unlike K^+ channels, there is no unique selectivity sequence for all CLC channels, likely due to small differences in their permeation pathways. Investigations of CLC selectivity are further complicated by the effects that the permeant anions exert on the channel's open probability (see Section VIII.A), so that great care is needed to dissect effects on permeation from those on gating (Rychkov et al., 1998, 2001). Despite these differences, all CLCs are permeable to halides ($Cl^- > Br^- > I^- \gg F^-$) and to inorganic anions ($SCN^- \sim ClO_4^- > NO_3^- \sim ClO_3^-$). Generally the permeability decreases with the ion's size, with some exceptions; I^- and NO_3^- are less permeable than the larger ClO_4^- and SCN^- (Rychkov et al., 1998). Surprisingly, CLC channels not only let inorganic ions through, but are also permeable to large hydrophobic compounds such as benzoate or hexanoate (Rychkov et al., 1998). This suggests that the pore of the channels must be larger than 4.3 Å in diameter at its narrowest point to accommodate these ions.

B. The Transporters

The selectivity of the transporters has been characterized in far less detail than that of channels, but a similar pattern has emerged with $SCN^- > Cl^- > Br^- > NO_3^- > SO_4^{2-}$ permeating (Steinmeyer et al., 1995; Maduke et al., 1999; Accardi et al., 2004; Nguitragool and Miller, unpublished result). This suggests that the basic mechanism underlying anion selectivity in the ClCs is similar between channel and transporter subclasses. The crystal structure of the bacterial transporter, ClC-ec1, offers us a molecular view of Cl^- coordination in the selectivity filter (Dutzler et al., 2002, 2003). This region is highly conserved in ClCs belonging to both the channel and transporter subclass. In the WT structure, there are two Cl^--binding sites (Fig. 1C). The ion in the inner site, S_{int}, is partially exposed to water and is coordinated by the backbone amides of G106 and S107. The ion in S_{cen} is dehydrated and completely encased in the protein. The negatively charged side chain of E148 blocks access to and from the extracellular solution, while the side chains of S107 and Y445 form a constriction that sterically isolates the ion from the intracellular milieu. The energetic balance of a negatively charged ion in the middle of the lipid membrane is achieved through a complex network of interactions: the hydroxyl moieties from the side chains of S107 and Y445 directly coordinate the Cl^- ion, as do the backbone amides from I356 and F357. Finally, the positive dipole moments from the N-termini of helices D, F, and N have been postulated to help stabilize the Cl^- ion. However, recent theoretical calculations have proposed that the energetic contribution of the helical dipole moment to the central Cl^- ion is small (Faraldo-Gomez and Roux, 2004).

A third Cl^--binding site, S_{ext}, is revealed in ClC-ec1 on neutralization of the charge on E148, from now on referred to as Glu_{ext} (Dutzler et al., 2003) (Fig. 1D). The ion in S_{ext} is almost in Van der Waals contact with the one in S_{cen}, suggesting that the protein strongly favors the presence of a negative charge at that position and balances the electrostatic repulsion between the two ions. A recent measurement suggests that Cl^- affinity for these binding sites in ClC-ec1 is in the 5- to 50-mM range (Lobet and Dutzler, 2006), which is consistent with values deduced from conductance measurements for the ClC-0 and ClC-1 channels (Rychkov et al., 1998; Chen and Chen, 2003).

IV. A MULTI-ION PORE

Early experiments suggested that ClC-0 has a single-ion pore (White and Miller, 1981), based on the observation that the conductance followed an ideal Michaelis–Menten curve as a function of Cl^- concentration. However, more careful examination of the conduction properties of ClC-0 and ClC-1 led to the suggestion that these channels have a multi-ion pore (Pusch et al., 1995; Rychkov et al., 1998). The evidence supporting this idea is that when two anions with different permeabilities are mixed the conductance does not decrease linearly when the concentration of the less permeant one is increased, but rather has a biphasic behavior reaching a minimum and then increasing again. This behavior is generally referred to as an anomalous mole fraction effect and is a hallmark of a multi-ion pore containing two or more interacting ions (Hille, 2001).

This prediction was beautifully confirmed by the crystallographic analysis of WT and mutant structures of ClC-ec1 where three distinct binding sites define the Cl^- permeation pathway. Recently, a crystallographic study and theoretical calculations have independently suggested that these three sites can be populated simultaneously and with nearly full occupancy (Faraldo-Gomez and Roux, 2004; Lobet and Dutzler, 2006). The mutual repulsion experienced by the chlorides in the permeation pathway could foster rapid conduction in the ion channels (Dutzler et al., 2003; Lobet and Dutzler, 2006). However, some care must be exercised when inferring details of the permeation mechanism in the ion channels which is based on free diffusion, from results obtained on a transporter in a partially occluded state, which is built to prevent free diffusion.

V. CHANNELS AND TRANSPORTERS: WHERE IS THE DIFFERENCE?

While there is a lot of evidence supporting the idea that the overall structural architecture of the membrane portions of CLC channels and transporters are similar, their functions are profoundly different. The first striking discrepancy is the rate at which Cl^- is moved through the membrane: a Cl^- ion diffusing through a CLC channel crosses the membrane in 0.1–1 µs, while a CLC transporter will typically move the ion in about 0.1–10 ms, 1000–10,000 times slower. A second and related difference lies in the mechanism underlying Cl^- transport in the two systems. In a channel, opening and closing of a single gate allows or prevents Cl^- diffusion across the membrane. On the other hand, in a transporter two gates must be coordinated so that they are never simultaneously open in order to prevent free diffusion. Until the crystal structure of a CLC channel becomes available, we can only speculate how the same structural architecture accommodates the necessary changes.

In the crystal structure of ClC-ec1, S107 and Y445 form a constriction that almost completely isolates the ion in S_{cen} from the intracellular solution and is likely to form a barrier to free diffusion. Although channels and transporters have a high degree of homology in the Cl^--binding region, subtle structural changes could possibly widen this opening and facilitate free diffusion. This hypothesis is in agreement with the observation that mutating the conserved tyrosine (Y445 in ClC-ec1, Y512 in ClC-0) to alanine in ClC-ec1 increases the Cl^-/H^+ ratio per conformational cycle (Accardi et al., unpublished), while it has a small effect on the channel conductance (Accardi and Pusch, 2003; Chen and Chen, 2003; Estévez et al., 2003). This is particularly surprising in light of the fact that in the bacterial transporter this mutation severely destabilizes ion binding to the S_{cen} site (Accardi et al., unpublished). If the Cl^--binding profile in the channels is the same as in the transporters, elimination of one of the binding sites should produce a drastic effect on the conduction properties, as is observed in the KcsA K^+ channel (Zhou and MacKinnon, 2004). In the transporters, however, this mutation does not seem to greatly alter the overall turnover rate, but this might simply reflect that other steps in the conformational cycle are rate limiting. A widening of the permeation pathway in the channels would also account for the observed permeability of large hydrophobic anions through the CLC channels (Rychkov et al., 1997). No studies are currently available on the permeability of these compounds through the CLC transporters. A last suggestion that the Cl^- pathway might be wider for channels in comparison to ClC-ec1 comes from an accessibility study (Zhang and Chen, 2006) which indicates that E166 in ClC-0 (the glutamate corresponding to E148 in ClC-ec1) is accessible to modification from the intracellular side. In ClC-ec1 the Cl$^-$ pathway is too constricted to allow for passage of MTS reagents, suggesting that either in the E166C mutant the cysteine side chain can point inward, and is thus accessible to the MTS reagents, or that the pathway is larger and thus the reagents can reach it.

In the CLC channels it has been proposed that Cl$^-$ can bind to a site that is more extracellular than S_{ext} (Chen and Chen, 2001; Chen et al., 2003), and several computational studies have suggested various possibilities for its location (Bostick and Berkowitz, 2004; Faraldo-Gomez and Roux, 2004; Yin et al., 2004). However, crystallographic or other direct evidence for a fourth ion-binding site is still lacking.

VI. FUNCTION OF THE CLC TRANSPORTERS

Very little is known about the molecular details by which the CLC transporters catalyze the exchange of Cl$^-$ for H$^+$ across biological membranes, mostly because the discovery of the functional divide among the CLC proteins itself is extremely recent. Here I will briefly recapitulate the experimental evidence that led to the conclusion that some CLCs are transporters and not channels as had been generally assumed. Then I will discuss the recent structural and functional insights in our understanding of the transport cycle.

ClC-ec1 was the first member of the transporter subclass to be identified. When this protein is purified and reconstituted in synthetic lipid bilayers, it mediates anion-selective currents that display no visible voltage or time dependence (Accardi and Miller, 2004; Accardi et al., 2004). However, Cl$^-$ is not the sole charge carrier through ClC-ec1. In a Cl$^-$ gradient the reversal potential is ~2/3 of the expected Nernstian value and when applying a pH gradient, a shift in the reversal potential is observed, identifying protons as the other charge carriers. The behavior of the reversal potentials measured in the presence of a pH or a Cl$^-$ gradient is inconsistent with the predictions of a channel permeable to both ionic species, but is in quantitative agreement with the predictions of a transporter that catalyzes the stoichiometric exchange of two Cl$^-$ for one H$^+$. The proverbial "nail in the coffin" was the demonstration that ClC-ec1 can use the energy stored in a Cl$^-$ gradient to drive H$^+$ movement against its own electrochemical gradient (and vice versa). An ion channel is a passive device that cannot drive the electrochemically uphill movement of its substrates. A single point mutation, E148A, completely abolishes H$^+$/Cl$^-$ coupling by eliminating H$^+$ transport and the remaining currents become nearly ideally Cl$^-$ selective and lose their pH sensitivity. It is important to remark that the apparent turnover rate is very similar in the WT and E148A mutant, suggesting that the simple act of neutralizing the side chain of Glu$_{ext}$ does not turn ClC-ec1 into a Cl$^-$-selective ion channel and that Cl$^-$ transport is still rate limited by the conformational cycle of the transporter.

Soon after this initial discovery, two independent groups showed that transporters also populate the eukaryotic CLC branch: two human CLCs, ClC-4 and ClC-5, are not ion channels as had been generally believed but Cl^-/H^+ exchangers, just like their distant bacterial relative (Picollo and Pusch, 2005; Scheel et al., 2005). Using complementary approaches, they demonstrated that both proteins catalyze the exchange of Cl^- and H^+ with a similar stoichiometry to ClC-ec1; a Cl^- gradient can be used to actively drive H^+ uphill and neutralization of Glu_{ext} (E224 and E211, respectively, for ClC-4 and ClC-5) eliminates H^+ transport and yields pH-insensitive currents.

Although our picture of the molecular mechanism that allows the CLC transporters to exchange Cl^- for H^+ across biological membranes is still very cloudy, a few key players are starting to emerge. The crucial role played by Glu_{ext} in both Cl^- and H^+ transport strongly suggests that this residue lies on both pathways. More recently, a scan of the carboxyl-bearing side chains facing the intracellular side of ClC-ec1 has led to the identification of an inward-facing glutamate, E203 (thereby referred to as Glu_{int}), which is the intracellular H^+ acceptor (Accardi et al., 2005). Neutralization of this glutamate residue, E203Q, leads to a complete loss of H^+ transport, while the protein still retains the ability to facilitate transmembrane Cl^- movement in a pH-dependent manner. Computational studies had proposed that this residue could play a key role in tuning the electrostatic potential in the Cl^- pathway in ClC-ec1 (Faraldo-Gomez and Roux, 2004; Yin et al., 2004). While Glu_{ext} directly lines the Cl^- permeation pathway, Glu_{int} does not, instead it is found in a strongly conserved region commonly known as the "PIGG-pen" located at the intersubunit interface about 10 Å away from the Cl^- pathway. The distal location of Glu_{int} from the bound anions suggests that Cl^- and H^+ move through ClC-ec1 along two separate pathways that become congruent at the extracellular end and diverge at the intracellular side. Glu_{ext} lies at the confluence of the two trajectories and acts as a molecular switch regulating both Cl^- and H^+ movement when its side chain changes protonation state.

How a H^+ moves from Glu_{int} to Glu_{ext} remains a puzzle. The distances between these two residues in the available crystal structures of the unprotonated and protonated forms are too large for a H^+ to directly hop from one to the other (Dutzler et al., 2003; Accardi et al., 2005). Also, the interior of the protein between the two glutamates is very hydrophobic and there are no obvious transient H^+ acceptor candidates (Fig. 2). One possibility is that during the transport cycle a conformational change in the protein brings the two residues closer, allowing a H^+ to directly transfer from one to the other. This hypothesis encounters two difficulties: first, no conformational change was observed when Glu_{int} was neutralized and, second when the glutamate was replaced with an aspartate placing the carboxylates further away no disruption was seen in H^+ coupling (Accardi et al., 2005). Another possibility is that the H^+ and the Cl^- pathways converge at a more intracellular location than Glu_{ext}. In this scenario, a H^+ could directly protonate a residue lining the Cl^- pathway and then proceed on to one of the glutamates. However, the only directly protonatable residue between the inner and the central Cl^--binding sites is a tyrosine (Y445 in ClC-ec1) that does not get protonated/deprotonated during the transport cycle (Accardi et al., unpublished). Once again, we need more experiments and crystal structures to clarify how the protons bridge these last 10 Å while crossing the membrane.

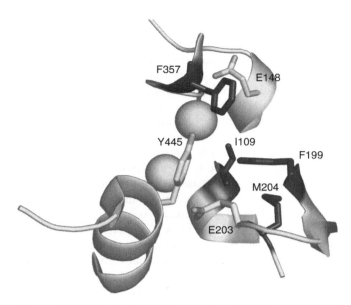

Figure 2. A hydrophobic patch separates E148 and E203. Side view of a single ClC-ec1 subunit. The hydrophobic side chains are in black and the hydrophilic ones in light gray. This figure was prepared with Pymol.

VII. SPECULATIVE MODELING

Although we have made some headway into improving our mechanistic understanding of a CLC transporter, our knowledge is still extremely limited. Any attempt at building a quantitative or even qualitative model for the transport cycle is thus doomed to fail. However, the available results put enough constraints on the system that we can at least sketch out a primitive outline for a model that captures the essential features of the system (at least those known so far). This attempt is based on many assumptions that I will try to explicitate, and its sole purpose is to help guide our thinking of future experiments aimed at understanding how these proteins achieve Cl^-/H^+ coupling. The experimental results that our model must account for are:

1. Cl^-/H^+ exchange with a 2:1 stoichiometry.
2. The crystal structures of WT and mutant ClC-ec1.
3. The involvement of E148 and E203 in H^+ transport.
4. E148 is the extracellular Cl^- gate.
5. The direct correlation between the occupancy of the central binding site and H^+ transport.
6. When E148 is protonated, three Cl^- ions are simultaneously bound to ClC-ec1.
7. H^+ and Cl^- ions can be simultaneously bound to ClC-ec1.

The assumptions that I am making in formulating the model are the following:

1. In order to prevent free diffusion, ClC-ec1 has two gates along the Cl^- pathway that are alternatively open and the intracellular one is located close to Y445.
2. Opening of the intracellular gate is associated with a conformational change in the protein. We have no idea of the magnitude of this change and it could well be limited to a side chain rearrangement (as in the case for the opening of the extracellular gate).
3. The WT structure represents the protein in a state with both gates closed.
4. There is an intermediate H^+ acceptor site whose protonation is favored by a Cl^- ion bound to S_{cen}.
5. At most, one transported H^+ can be bound to ClC-ec1 at any given moment.

When both gates in ClC-ec1 are in the closed state and there are two Cl^- ions bound, a H^+ binds to Glu_{int} from the intracellular side (Fig. 3A). A conformational change in the protein ensues, leading to an opening of the intracellular gate, H^+ movement from Glu_{int} to its intermediate site, and subsequently one of the two bound Cl^- ions moves into the internal solution (Fig. 3B). The H^+ bound to the intermediate site is destabilized and protonates E148 (Fig. 3C) and the intracellular gate closes. Protonation of E148 opens the extracellular gate and two Cl^- ions bind to ClC-ec1 from the extracellular side (Fig. 3D). After the H^+ is freed into the extracellular solution, the external gate closes and the internal one opens. Once a second Cl^- ion moves from the protein into the intracellular milieu (Fig. 3E), the intracellular gate closes and the transporter returns to its original state after having transported two Cl^- and one H^+ in opposite directions (Fig. 3F).

Clearly, this model is based on several assumptions and the order of events described could be changed (if they happen at all, that is). However, it could prove a useful tool in guiding future mechanistic investigations.

VIII. CLC CHANNEL GATING

Early functional analysis of single ClC-0 channels led to the proposal that two distinct gating mechanisms regulate channel opening: a slow gate which acts on both pores simultaneously and a fast gate which regulates opening and closing of each individual pore (Miller, 1982). At the single channel level, the distinction of these two processes is evident (Fig. 4): long periods of complete silence are interrupted by bursts of activity where the current is fluctuating between three different levels, reflecting the opening and closing of each individual pore. Within each burst, the probability distribution of the three conduction levels strictly follows a binomial distribution, indicating that once the slow gate is open the two pores gate independently.

Generally the CLCs have a low single channel conductance, starting at 1 pS (Saviane et al., 1999) for ClC-1 and topping out at 8 pS for ClC-0 (White and Miller, 1981). This has prevented an in-depth analysis at the single molecule level for all members aside from ClC-0. Thus, most of the properties that I will describe have been established for ClC-0 and then have been extended to other members of the CLC channel subfamily through analysis of macroscopic currents.

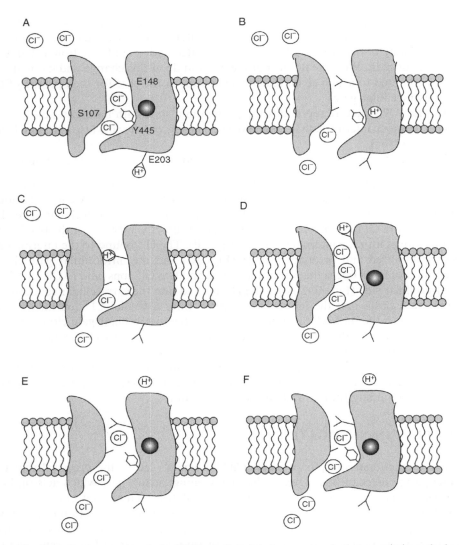

Figure 3. Tentative transport cycle for CLC-ec1. A 6-state transport cycle that recapitulates the known functional observations and structures is drawn for a single subunit of CLC-ec1. The assumptions and details of the model are described in the chapter.

A. Voltage, Cl^- and H^+ Gating of a CLC Channel

Generally the CLCs are referred to as voltage-gated ion channels; however, a more appropriate description would be to say that they are voltage-, H^+-, and Cl^--gated chloride channels. While pH and permeant ions modulate gating of most ion channels,

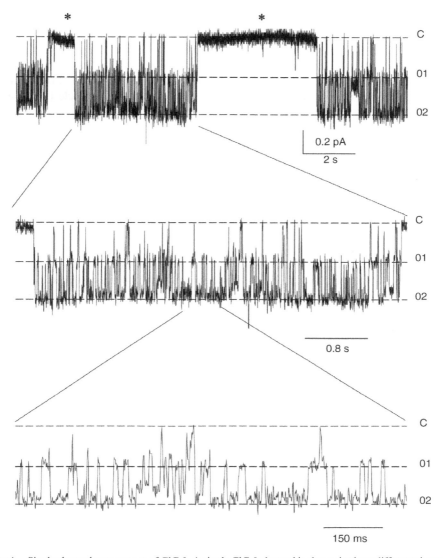

Figure 4. Single channel appearance of ClC-0. A single ClC-0 channel is shown in three different timescales. Dashed lines indicate the current levels associated to states with two open pores (O2), one open pore (O1), and both pores closed (C). The * indicate the closures of the slow gate. Trace recorded at −60 mV and in symmetrical 120 mM Cl⁻. This trace was a generous gift of Dr. T. Y. Chen.

the magnitude of the effects seen on the CLCs is unique: removing Cl^- or H^+ nearly abolishes channel activity. This functional oddity reflects a profound structural difference between conventional voltage-dependent cation channels and the CLCs: the former have a well-defined intrinsic voltage sensor (Yellen, 1998; Jiang et al., 2003a,b) while in the latter there is no identifiable domain that can play this role (Dutzler et al., 2002). So how does a CLC channel sense voltage? The consensus for the last 10 years or so has been that a permeating Cl^- ion is the gating charge carrier. This implies that channel gating is a nonequilibrium process: the ΔG associated with ion translocation through a CLC channel, an irreversible process, is an intrinsic part of the gating cycle (Richard and Miller, 1990; Pusch et al., 1995; Chen and Miller, 1996). Recently, the notion that the permeant ion is the charge carrier has been expanded to include not only Cl^- but also H^+ (Miller, 2006; Traverso et al., 2006).

Qualitatively, the voltage dependence of the fast gate of ClC-0 resembles that of many cation channels: depolarization enhances the open probability, the half activation potential is around -80 mV, and the maximal open probability is close to unity. However, these similarities are only superficial: the voltage dependence of activation is much weaker with a $z \sim 1$ (Chen and Miller, 1996) as opposed to the $z \sim 14$ observed for K^+ channels (Yellen, 1998) and there is no intrinsic voltage sensor. The first suggestion that Cl^- itself could be the voltage sensor came from the observation that when extracellular Cl^- is reduced, macroscopic activation of ClC-0 shifts towards more positive potentials (Pusch et al., 1995). However, in order to draw precise mechanistic correlations between channel gating and the conformational changes in the protein, a clear distinction between the opening and closing processes must be drawn. A careful and detailed analysis of single ClC-0 channels (Chen and Miller, 1996) revealed that while the closing rate decreases exponentially as the applied voltage increases, $\beta(V)=\beta_0\exp(z_{\beta 0}VF/RT)$, the opening rate has a biphasic voltage dependence with a minimum at negative voltages, $\alpha(V) = \alpha_0\exp(z_{\alpha 0}VF/RT) + \alpha_1\exp(z_{\alpha 1}VF/RT)$, where $z_{\alpha 0}$ and $z_{\alpha 1}$ have opposite signs. This was proposed to be due to the fact that channel opening can occur through two separate processes with opposite voltage dependences: a Cl^--dependent opening step that is strongly favored by depolarization, $\alpha_0(V)$, and a Cl^--independent pathway with a weaker voltage dependence that is enhanced at hyperpolarized voltages, $\alpha_1(V)$. Since the opening rate is nonvanishing at negative voltages, and increases with a slope similar to that of the closing rate, the open probability of the channel will remain finite at hyperpolarized voltages, in agreement with the experimental results. This model could also explain the strong modulatory effects that Cl^- and pH have on the fast gate of ClC-0. These four parameters affect the open probability in two qualitatively different ways: at low pH_{ext} and high $[Cl^-]_{int}$, the minimal open probability of the channel at negative voltages increases and the half activation potential is unchanged; conversely at low pH_{int} or high $[Cl^-]_{ext}$, channel opening is facilitated and the half activation potential shifts towards hyperpolarized voltages (Hanke and Miller, 1983; Pusch et al., 1995; Chen and Miller, 1996; Chen and Chen, 2001; Chen et al., 2003). This qualitative difference is mirrored at the microscopic level: pH_{ext} and $[Cl^-]_{int}$ affect mostly the opening rate while pH_{int} and $[Cl^-]_{ext}$ act on the closing rate. Interestingly, pH_{ext} and $[Cl^-]_{int}$ increase the opening rate of the fast gate of ClC-0 through two different microscopic mechanisms: pH_{ext} increases $\alpha_1(V)$, the Cl^--independent hyperpolarization-activated opening pathway, while increasing $[Cl^-]_{int}$ favors the depolarization-activated opening step, $\alpha_0(V)$.

Recently, two independent groups have proposed that the voltage dependence of CLC gating arises not from Cl^- movement but from a H^+ crossing the electric field (Miller, 2006; Traverso et al., 2006). In their models there is a single protonation site, the Glu_{ext} side chain, which in the CLC channels is accessible to H^+ from either side of the membrane (Miller, 2006; Traverso et al., 2006). A direct involvement of H^+ in voltage sensing had been initially rejected by Hanke and Miller (1983) who had proposed that the voltage dependence arose from a conformational change, not from the protonation step. Recent evidence requires a reavaluation of this initial assumption. Traverso et al. analyzed the ClC-0 mutant where the Glu_{ext} side chain is mutated to Asp (E166D). This conservative mutation has drastic effects on ClC-0 gating: channel opening is slowed down and the currents become outwardly rectifying. When pH_{int} is lowered, the activation profile of this mutant shifts leftward along the voltage axis and the half activation potential, $V_{1/2}$, varies linearly with pH_{int} in contrast with the demands of the Hanke and Miller model that instead predicts that the $V_{1/2}$ should saturate at acidic pH. Traverso et al. showed that if protonation of the Glu_{ext} side chain is the voltage-dependent step, then the linear dependence of $V_{1/2}$ on pH is accounted for. The second line of evidence supporting this proposal is a little more speculative. Starting from the recent finding that in ClC-ec1 there are two separate ion pathways for Cl^- and H^+, Miller (2006) proposed that a remnant of the H^+ pathway of the transporters is also found in the CLC channels and that H^+ transport along this pathway is voltage dependent. In his proposal he postulates the existence of a Cl^--binding site extracellular to Glu_{ext} that allosterically modulates the H^+ permeation pathway. Both of these models make one stunning prediction: all CLCs should transport protons. Fast gating in ClC-0 takes place on a timescale between 1–10 ms. Thus, if an H^+ is moved across the bilayer during each gating cycle, the turnover rate should be around 10^2-10^3 H^+ s^{-1}. This turnover value is entirely comparable to that observed in ClC-ec1 (Walden et al., unpublished) where H^+ movement can be readily measured (Accardi et al., 2004, 2005). However, so far no H^+ transport has been observed for WT or mutant ClC-0 or ClC-2 (Picollo and Pusch, 2005; Traverso et al., 2006). A small H^+ permeability has been reported for the skeletal muscle homologue, ClC-1, but its significance is still unclear (Picollo and Pusch, 2005). In the CLC transporters, the intracellular H^+ acceptor is thought to be a residue that is a strictly conserved glutamate, while at this position the channels have a strictly conserved valine (Accardi et al., 2005). Neutralization of this residue in ClC-ec1 completely abolishes all H^+ transport. This would then require that in the CLC channels either a new pathway for H^+ has been opened, or that the original one has been altered in the channels to bypass the missing Glu_{int}. Another possibility is that Glu_{ext} is directly accessible from the inside through a crevice that might or might not be different from the Cl^- conduction pathway. This latter hypothesis is supported by the observation that a cysteine at the Glu_{ext} position in ClC-0 is directly modifiable by MTS reagents when applied from the intracellular side (Zhang and Chen, 2006). While ascribing part or all of the voltage dependence of CLC fast gating to H^+ movement across the electric field is in agreement with the experimental data, no experimental result directly shows that Glu_{ext} needs to be protonated for the channel to open and consequently that an H^+ needs to completely cross the lipid bilayer.

Further experiments are clearly needed to test this novel and intriguing hypothesis and its structural implications.

B. Molecular Identity of the Fast Gate

The structure of ClC-ec1 has proven to be an invaluable tool in predicting several structural features of the CLC channels. Its most remarkable success to date has been the molecular identification of the fast gate of the CLC channels. The negatively charged side chain of a conserved glutamate (Glu_{ext}, E148 in ClC-ec1, E166 in ClC-0) isolates the Cl^- ion in S_{cen} from the extracellular solution. When this side chain is neutralized through mutagenesis to glutamine, no major backbone rearrangement takes place but the mutated side chain moves away from the pathway, thereby freeing it so that the central Cl^- ion has direct access to the external milieu. When the corresponding mutation is introduced in ClC-0, it results in constitutively open channels (Dutzler et al., 2003; Traverso et al., 2003). This result led to the proposal that fast gating of ClC-0 is limited to the movement of the side chain of E166 (Dutzler et al., 2003). In other words, the side chain of E166 acts as an intrinsic tethered blocker for ClC-0. This model accounts for the dependence on the extracellular pH of the fast gate: lowering pH_{ext} favors protonation of the E166 side chain which swings out unblocking the pathway and opening the channel. The dependence on $[Cl^-]_{int}$ can also be understood in the context of this model. The pore blocking side chain of E166 is displaced by the ions permeating through the pore in a classical "foot in the door" mechanism.

Is this the complete story? Probably not. Several lines of evidence suggest that fast gating in ClC-0 involves more than the simple conformational rearrangement associated with the movement of the side chain of E166. Side chain movements typically happen on a timescale much faster than the tens of milliseconds observed in fast gating, which would be more consistent with a larger allosteric transition. Second, extracellular and intracellular Cl^- act through two very different mechanisms: the former prevents channel closures while the latter favors voltage-dependent openings, suggesting that the two might act at different sites while the proposed model would require a common site of action. While the temperature dependence of the fast gate of ClC-0 (Pusch et al., 1997), with a Q_{10} of 2.2, is lower than that of transporter or channel proteins, where conformational changes larger than the simple bond rotation of a side chain take place, such as the K^+ channels, $Q_{10} > 4$ (Liu et al., 1997; Rodriguez et al., 1998; Yellen, 1998; Jiang et al., 2003a,b), or the Ca^{2+}-ATPase from the sarcoplasmatic reticulum, $Q_{10} = 3.1$ (Toyoshima and Nomura, 2002; Peinelt and Apell, 2004; Toyoshima and Mizutani, 2004), it is still much larger than that of free diffusion, $Q_{10} = 1.4$. This suggests that the conformational change associated to this process involves more than the few atoms of a single side chain but is not as extensive as the major rearrangements observed in these other transport proteins. Another inconsistency comes from the observation that the fast gates of ClC-0 or ClC-1 have opposite voltage dependence to that of ClC-2. How can the upward swinging motion of the Glu_{ext} side chain within a conserved pore in a rigid structure be alternatively favored by hyperpolarization or depolarization? Another puzzling result comes from the use of CPA, a small molecule pore blocker of ClC-0 that has a higher affinity for the closed pore than for the open. Several point mutations in ClC-0 selectively influence CPA binding to the open or to the closed state (Accardi and Pusch, 2003; Traverso et al., 2003). However, if channel gating were limited to the movement of the E166 side chain, then the affinity of the blocker for both states should be affected in a similar way. The most striking of these examples occurs with the mutation E166A, which results in constitutively open channels. If the open state of this mutant is identical to that of WT ClC-0, then the affinity of CPA for the mutant pore should be identical to that of the open channel. Instead, Traverso et al. saw an over 100-fold increase in affinity for the mutant pore. Should these results be taken to indicate that ClC-0 has another gate, maybe at the intracellular side? The answer is no. Lin and Chen (2003) demonstrated that the accessibility of the residues lining the intracellular vestibule (up to the Cl^--coordinating residue Y512) of ClC-0 does not show a pronounced state dependence. These residues have similar accessibility regardless of whether the channel is open or closed indicating that no tight constriction is present in ClC-0 at the intracellular side below Y512. However, ClC-0 channels never close 100% of the time (see above), but rather the open probability only changes approximately fivefold when the voltage is switched from -100 to $+100$ mV, from ~ 0.2 to ~ 1. This implies that the maximum expected difference in the reaction rate constant is fivefold. Thus, for ClC-0 this approach is much less sensitive than in the case of the Shaker channel, where the open probability changes over 1000-fold on a similar voltage change (Yellen, 1998) and the expected change in accessibility is comparably larger (Liu et al., 1997; del Camino et al., 2000). Lin and Chen (2003) observed up to a threefold change in the modification rates between open and closed ClC-0, indicating that while a "bundle-crossing" like gate is clearly not present in ClC-0, more subtle conformational changes might be taking place at the intracellular side.

In conclusion, I consider that fast gating of the CLC channels is likely to involve several residues, both at the intracellular and extracellular end of the pore, acting together as a team whose "M.V.P." is clearly the external glutamate.

C. The Slow Gate

In addition to the fast, single pore, gate discussed above, the CLC channels have another gating mechanism that acts on both pores simultaneously, referred to as the slow or common gate (Miller, 1982; Richard and Miller, 1990; Lin et al., 1999) (Fig. 4). While the fast gating process seems to be completely contained within the transmembrane portion of a single CLC subunit, slow gating appears to reflect a concerted structural change of both subunits. Not much is known about this process, mainly because its kinetics are so slow (ranging from tens of seconds to minutes) that extremely long recordings are required. In addition, the slow gate has an astonishingly high temperature dependence with a $Q_{10} \sim 40$ (Pusch et al., 1997), causing precise temperature control to be critical for quantitative measurements. This unusually strong temperature dependence probably indicates that this process involves a major conformational rearrangement of the protein.

In addition to temperature, slow gating is also modulated by voltage, pH, and Cl$^-$ concentration. The open probability of the slow gate increases at negative voltages and it has a half activation potential of approximately -80 mV. At positive voltages, the slow gate does not inactivate completely, leading to an offset of $p_{open}^{slow}(V)$ (Pusch et al., 1997; Lin et al., 1999). Lowering the extracellular pH favors slow gate opening (Miller and White, 1980; White and Miller, 1981), while lowering intracellular or extracellular Cl$^-$ promote its closing (Richard and Miller, 1990; Pusch et al., 1999). ClC-0 currents are also inhibited by extracellular Zn^{2+}, which acts by facilitating closure of the slow gate (Chen, 1998; Lin et al., 1999).

The conformational change associated with slow gating is likely to involve both subunits of the CLC channel and their transmembrane and cytosolic portions. This conclusion is supported by the drastic effects that several point mutations scattered throughout the protein have on slow gating. For example, the S132T mutant of ClC-0 alters the properties of the slow gate in homodimeric channels and also dominantly confers these alterations to heteromeric mutant-WT dimers (Ludewig et al., 1996). Moreover, mutations that cause dominant myotonia, and that are thought to affect the slow gate of ClC-1, are found at the interface between the two subunits (Estévez and Jentsch, 2002; Duffield et al., 2003).

Slow gating also involves both the transmembrane and cytosolic portions of a CLC channel, as highlighted by the number of mutations scattered throughout both regions that drastically alter its voltage dependence and kinetics. A most dramatic effect comes from a mutation of a conserved cysteine residue buried in the transmembrane region, C212 in ClC-0, which locks the slow gate in the open state (Lin et al., 1999), eliminates its temperature dependence and Zn^{2+} sensitivity.

D. Fast and Slow Gating in Other CLC Channels

From the limited single channel recordings available we can appreciate that ClC-1 and ClC-2 (Saviane et al., 1999; Weinreich and Jentsch, 2001) have the same double-barreled appearance as ClC-0: the channels fluctuate between one closed state and two, equally spaced, open states. However, the small conductance of these channels ($\sim$1.2 pS for ClC-1 and $\sim$2 pS for ClC-2) has prevented an in-depth characterization of their gating properties at the single molecule level. Thus, most of our understanding of their gating comes from macroscopic recordings and comparisons with the better characterized effects in ClC-0.

Channel gating in ClC-1 and ClC-2 is controlled by two separate mechanisms that closely resemble those found in ClC-0 (Rychkov et al., 1996; Accardi and Pusch, 2000; Accardi et al., 2001; Niemeyer et al., 2003; Zúñiga et al., 2004; de Santiago et al., 2005): each pore has its own intrinsic gate (fast gate) and there is a separate process that closes both pores simultaneously (slow gate).

In ClC-1, both the fast and the slow gates are activated by depolarization and are influenced by pH and Cl⁻ concentration in a manner similar to ClC-0 (Rychkov et al., 1996; Accardi and Pusch, 2000; Aromataris et al., 2001; Rychkov et al., 2001). At negative voltages these two processes have similar kinetics ($\tau \sim 10$ ms) so that their direct separation is harder than in ClC-0 (Rychkov et al., 1996; Saviane et al., 1999; Accardi and Pusch, 2000). At positive voltages, however, their activation kinetics differ by almost two orders of magnitude and are readily distinguishable (Accardi and Pusch, 2000).

Qualitatively, ClC-2 currents are very different from those of either ClC-0 or ClC-1. Hyperpolarization, not depolarization, opens this channel and the activation kinetics are extremely slow and are well described with the sum of two exponentials, with time constants $\tau_f \sim 30$ ms and $\tau_s \sim 200$ ms (Gründer et al., 1992; Thiemann et al., 1992; Pusch et al., 1999; Niemeyer et al., 2003; Zúñiga et al., 2004; de Santiago et al., 2005). Early studies showed that ClC-2 is also the only known member of the CLC family to be activated by hypotonicity (Gründer et al., 1992) and that its N-terminal region plays a key role in this process: deletion of residues 16–61 leads to channels that are insensitive to both voltage and osmotic stress.

The functional similarities between the gating processes found in the CLC channels are highlighted by the effects that two single point mutations have on the fast and slow gate, respectively. Neutralization of Glu_{ext} (E232 in ClC-1 and E217 in ClC-2) invariably yields constitutively open channels that have no time or voltage dependence (Dutzler et al., 2003; Estévez et al., 2003; Niemeyer et al., 2003; Traverso et al., 2003), and mutation of the cysteine corresponding to C212 in ClC-0 (C277 in ClC-1 and C256 in ClC-2) always eliminates the slow gating process and Zn^{2+} sensitivity (Lin et al., 1999; Accardi et al., 2001; Duffield et al., 2003; Zúñiga et al., 2004; Duffield et al., 2005).

Thus, despite the disparate kinetics, voltage and osmotic dependence, the molecular mechanisms underlying fast and slow gating are likely to be conserved in all CLC channels.

IX. THE ROLE OF THE C-TERMINAL DOMAIN OF CLC CHANNELS: RECENT DEVELOPMENTS

All eukaryotic CLCs have a large C-terminal domain that ranges from 155 to 398 amino acids and that contains two cystathione β-synthetase (CBS) domains (Estévez and Jentsch, 2002). Despite the clear involvement of these domains in CLC gating and regulation (Schmidt-Rose and Jentsch, 1997; Maduke et al., 1998), very little is actually known about their function in the context of the channel. Only recently it has been shown that ClC-1 is modulated by ATP through its binding to the CBS domains (Bennetts et al., 2005). It is not clear, however, if this is a general property of all the CLCs or if it is unique to ClC-1 (Meyer and Dutzler, 2006).

The molecular organization of the CBS domains within a CLC channel has been investigated by several groups, with contrasting results. The object of debate has been whether a CBS domain interacts only with its partner from the same monomer

(Hebeisen et al., 2004) or if there are also intersubunit interactions (Estévez et al., 2004). The recently published crystal structure of the C-terminal domain of ClC-0 (Meyer and Dutzler, 2006) has not solved this conundrum: in the crystal this domain is a monomer while in solution it is a dimer. Meyer and Dutzler argue that the monomeric state could be favored by the high salt concentrations needed for crystallization. These apparent contrasts could be reconciled by a recent study that suggests that the C-terminal domains are close to each other when the slow gate is in the open state and are far apart when the slow gate closes (Bykova et al., 2006). The relative mobility of the C-termini of the CLC channels is also in good agreement with the observation that several charged residues at the putative dimeric interface are substituted with hydrophobic residues in proteins where these domains stably form dimers (Meyer and Dutzler, 2006).

The crystal structure shows that the core of ClC-0's C-terminal domain forms a typical CBS fold: two $\beta 1-\alpha 1-\beta 2-\beta 3-\alpha 2$ repeats are related by a pseudo twofold axis of symmetry and the interdomain interactions are mediated by the β-sheets formed by $\beta 2-\beta 3$ (Meyer and Dutzler, 2006). This core domain is connected to the transmembrane portion of ClC-0 by a 22 amino acid linker and is followed by a 33 amino acid region (C-peptide), both these regions are disordered in the crystal structure. The two CBS domains are connected by a long linker region that is also mostly disordered. It is not clear at present if disorder is intrinsic to these domains or if they could acquire a more defined structure on dimerization. It is clear, however, that these regions play a key role in slow gating modulation (Hryciw et al., 1998; Hebeisen and Fahlke, 2005) and are likely to be involved in transducing the conformational changes taking place in the C-terminus to the transmembrane region. Until the crystal structure of a full-length CLC channel with a C-terminal domain becomes available, we can only speculate on the exact domain organization and how changes in one area can be transduced to the other.

X. CONCLUSIONS AND OUTLOOK

Since their initial discovery, almost 30 years ago, the CLCs have been a reliable source of unexpected functional behaviors and structural oddities, the latest of which has been the discovery that both ion channels and exchange transporters populate this family. How did nature adapt a single molecular architecture to achieve these two separate and opposite functions? The challenge that has to be met to answer this question is obvious: only the crystal structure of a bona fide CLC chloride channel will let us appreciate at molecular detail the differences underlying these thermodynamically opposite processes. A collateral question emerging from the same discovery is whether the channel or transporter functionality is determined by a limited number of residues or if it is more "diffuse"; in other words, is there a channel and transporter signature sequence? If so, can we turn one into the other by simply transplanting this sequence? However, in order to even think about achieving this transformation, we must first understand how CLC channels and transporters operate, and we do not.

Transporter function is still largely unexplored, only recently we have started to tap its outer shells. We are not yet able to satisfactorily address even the most basic questions such as where the ions are going through and how they are coupled. We still need to identify the different states visited by the protein during its gating cycle. Are there any major conformational changes involved or is coupled transport achieved through small movements of side chains?

We understand the channels a little more, or we think we do. There is a testable model for fast gating that accounts for most of the experimental results and glimpses of the molecular determinants of slow gating are starting to emerge. However, both pictures are still far from complete and we still ignore how these two processes are coupled.

Channels and transporters in the CLC family closely resemble each other, not just structurally but also functionally. The intimate coupling between channel opening and permeant ion(s) transport seen in the channels is strongly reminiscent of a transporter gating cycle, and a single point mutation can turn a tight exchanger into a leaky transporter, taking it a step closer to a channel. Thus, the divide between channels and transporters appears thin and easily breached. We can only wonder—and actively investigate—whether the separation between the two subgroups is sharp and discontinuous, or if there is a continuum of functions slowly morphing channels and transporters into each other.

ACKNOWLEDGMENTS

I thank Christopher Miller, Brittany Zadek, Hari Jayaram, Michael Walden, Carole Williams, and Carlo Bortolotti for helpful comments and discussions on the chapter. I wish to thank T. Y. Chen for the generous gift of the ClC-0 trace.

REFERENCES

Accardi, A. and Miller, C. (2004). Secondary active transport mediated by a prokaryotic homologue of ClC Cl^- channels. *Nature* **427**, 803–807.
Accardi, A. and Pusch, M. (2000). Fast and slow gating relaxations in the muscle chloride channel CLC-1. *J. Gen. Physiol.* **116**, 433–444.
Accardi, A. and Pusch, M. (2003). Conformational changes in the pore of CLC-0. *J. Gen. Physiol.* **122**, 277–293.
Accardi, A., Ferrera, L., and Pusch, M. (2001). Drastic reduction of the slow gate of human muscle chloride channel (ClC-1) by mutation C277S. *J. Physiol.* **534**, 745–752.
Accardi, A., Kolmakova-Partensky, L., Williams, C., and Miller, C. (2004). Ionic currents mediated by a prokaryotic homologue of CLC Cl^- channels. *J. Gen. Physiol.* **123**, 109–119.
Accardi, A., Walden, M., Nguitragool, W., Jayaram, H., Williams, C., and Miller, C. (2005). Separate ion pathways in a Cl^-/H^+ exchanger. *J. Gen. Physiol.* **126**, 563–570.
Aromataris, E. C., Rychkov, G. Y., Bennetts, B., Hughes, B. P., Bretag, A. H., and Roberts, M. L. (2001). Fast and slow gating of CLC-1: Differential effects of 2-(4-chlorophenoxy) propionic acid and dominant negative mutations. *Mol. Pharmacol.* **60**, 200–208.

Bennetts, B., Rychkov, G. Y., Ng, H. L., Morton, C. J., Stapleton, D., Parker, M. W., and Cromer, B. A. (2005). Cytoplasmic ATP-sensing domains regulate gating of skeletal muscle ClC-1 chloride channels. *J. Biol. Chem.* **280,** 32452–32458.
Bostick, D. L. and Berkowitz, M. L. (2004). Exterior site occupancy infers chloride-induced proton gating in a prokaryotic homolog of the ClC chloride channel. *Biophys. J.* **87,** 1686–1696.
Bykova, E., Zhang, X. D., Chen, T. Y., and Zheng, J. (2006). Fluorescence study of CLC-0 chloride channel structure and slow gating. *Biophys. J.* **90,** Pos-L78 (abstract).
Chen, M. F. and Chen, T. Y. (2001). Different fast-gate regulation by external Cl^- and H^+ of the muscle-type ClC chloride channels. *J. Gen. Physiol.* **118,** 23–32.
Chen, M. F. and Chen, T. Y. (2003). Side-chain charge effects and conductance determinants in the pore of ClC-0 chloride channels. *J. Gen. Physiol.* **122,** 133–145.
Chen, T. Y. (1998). Extracellular zinc ion inhibits ClC-0 chloride channels by facilitating slow gating. *J. Gen. Physiol.* **112,** 715–726.
Chen, T. Y. and Miller, C. (1996). Nonequilibrium gating and voltage dependence of the ClC-0 Cl^- channel. *J. Gen. Physiol.* **108,** 237–250.
Chen, T. Y., Chen, M. F., and Lin, C. W. (2003). Electrostatic control and chloride regulation of the fast gating of ClC-0 chloride channels. *J. Gen. Physiol.* **122,** 641–651.
del Camino, D., Holmgren, M., Liu, Y., and Yellen, G. (2000). Blocker protection in the pore of a voltage-gated K^+ channel and its structural implications. *Nature* **403,** 321–325.
de Santiago, J. A., Nehrke, K., and Arreola, J. (2005). Quantitative analysis of the voltage-dependent gating of mouse parotid ClC-2 chloride channel. *J. Gen. Physiol.* **126,** 591–603.
Doyle, D. A., Morais Cabral, J., Pfuetzner, R. A., Kuo, A., Gulbis, J. M., Cohen, S. L., Chait, B. T., and MacKinnon, R. (1998). The structure of the potassium channel: Molecular basis of K^+ conduction and selectivity. *Science* **280,** 69–77.
Duffield, M., Rychkov, G., Bretag, A., and Roberts, M. (2003). Involvement of helices at the dimer interface in ClC-1 common gating. *J. Gen. Physiol.* **121,** 149–161.
Duffield, M. D., Rychkov, G. Y., Bretag, A. H., and Roberts, M. L. (2005). Zinc inhibits human ClC-1 muscle chloride channel by interacting with its common gating mechanism. *J. Physiol.* **568,** 5–12.
Dutzler, R., Campbell, E. B., Cadene, M., Chait, B. T., and MacKinnon, R. (2002). X-ray structure of a ClC chloride channel at 3.0 Å reveals the molecular basis of anion selectivity. *Nature* **415,** 287–294.
Dutzler, R., Campbell, E. B., and MacKinnon, R. (2003). Gating the selectivity filter in ClC chloride channels. *Science* **300,** 108–112.
Engh, A. M. and Maduke, M. (2005). Cysteine accessibility in ClC-0 supports conservation of the ClC intracellular vestibule. *J. Gen. Physiol.* **125,** 601–617.
Estévez, R. and Jentsch, T. J. (2002). CLC chloride channels: Correlating structure with function. *Curr. Opin. Struct. Biol.* **12,** 531–539.
Estévez, R., Schroeder, B. C., Accardi, A., Jentsch, T. J., and Pusch, M. (2003). Conservation of chloride channel structure revealed by an inhibitor binding site in ClC-1. *Neuron* **38,** 47–59.
Estévez, R., Pusch, M., Ferrer-Costa, C., Orozco, M., and Jentsch, T. J. (2004). Functional and structural conservation of CBS domains from CLC chloride channels. *J. Physiol.* **557,** 363–378.
Fahlke, C., Rhodes, T. H., Desai, R. R., and George, A. L., Jr. (1998). Pore stoichiometry of a voltage-gated chloride channel. *Nature* **394,** 687–690.
Faraldo-Gomez, J. D. and Roux, B. (2004). Electrostatics of ion stabilization in a ClC chloride channel homologue from *Escherichia coli*. *J. Mol. Biol.* **339,** 981–1000.
Gründer, S., Thiemann, A., Pusch, M., and Jentsch, T. J. (1992). Regions involved in the opening of ClC-2 chloride channel by voltage and cell volume. *Nature* **360,** 759–762.
Hanke, W. and Miller, C. (1983). Single chloride channels from *Torpedo* electroplax. Activation by protons. *J. Gen. Physiol.* **82,** 25–45.
Hebeisen, S. and Fahlke, C. (2005). Carboxy-terminal truncations modify the outer pore vestibule of muscle chloride channels. *Biophys. J.* **89,** 1710–1720.
Hebeisen, S., Biela, A., Giese, B., Muller-Newen, G., Hidalgo, P., and Fahlke, C. (2004). The role of the carboxyl terminus in ClC chloride channel function. *J. Biol. Chem.* **279,** 13140–13147.

Hille, B. (2001). *Ion Channels of Excitable Membranes*, 3rd ed. Sinauer Associates, Sunderland, MA.
Hryciw, D. H., Rychkov, G. Y., Hughes, B. P., and Bretag, A. H. (1998). Relevance of the D13 region to the function of the skeletal muscle chloride channel, ClC-1. *J. Biol. Chem.* **273**, 4304–4307.
Iyer, R., Iverson, T. M., Accardi, A., and Miller, C. (2002). A biological role for prokaryotic ClC chloride channels. *Nature* **419**, 715–718.
Jiang, Y., Ruta, V., Chen, J., Lee, A., and Mackinnon, R. (2003a). The principle of gating charge movement in a voltage dependent K^+ channel. *Nature* **423**, 42–48.
Jiang, Y., Lee, A., Chen, J., Ruta, V., Cadene, M., Chait, B. T., and MacKinnon, R. (2003b). X-ray structure of a voltage-dependent K^+ channel. *Nature* **423**, 33–41.
Kornak, U., Kasper, D., Bösl, M. R., Kaiser, E., Schweizer, M., Schulz, A., Friedrich, W., Delling, G., and Jentsch, T. J. (2001). Loss of the ClC-7 chloride channel leads to osteopetrosis in mice and man. *Cell* **104**, 205–215.
Lin, C. W. and Chen, T. Y. (2003). Probing the pore of ClC-0 by substituted cysteine accessibility method using methane thiosulfonate reagents. *J. Gen. Physiol.* **122**, 147–159.
Lin, Y. W., Lin, C. W., and Chen, T. Y. (1999). Elimination of the slow gating of ClC-0 chloride channel by a point mutation. *J. Gen. Physiol.* **114**, 1–12.
Liu, Y., Holmgren, M., Jurman, M. E., and Yellen, G. (1997). Gated access to the pore of a voltage-dependent K^+ channel. *Neuron* **19**, 175–184.
Lobet, S. and Dutzler, R. (2006). Ion-binding properties of the ClC chloride selectivity filter. *EMBO J.* **25**, 24–33.
Ludewig, U., Pusch, M., and Jentsch, T. J. (1996). Two physically distinct pores in the dimeric ClC-0 chloride channel. *Nature* **383**, 340–343.
Maduke, M., Williams, C., and Miller, C. (1998). Formation of ClC-0 chloride channels from separated transmembrane and cytoplasmic domains. *Biochemistry* **37**, 1315–1321.
Maduke, M., Pheasant, D. J., and Miller, C. (1999). High-level expression, functional reconstitution, and quaternary structure of a prokaryotic ClC-type chloride channel. *J. Gen. Physiol.* **114**, 713–722.
Meyer, S. and Dutzler, R. (2006). Crystal structure of the cytoplasmic domain of the chloride channel ClC-0. *Structure* **14**, 299–307.
Middleton, R. E., Pheasant, D. J., and Miller, C. (1994). Purification, reconstitution, and subunit composition of a voltage-gated chloride channel from *Torpedo* electroplax. *Biochemistry* **33**, 13189–13198.
Middleton, R. E., Pheasant, D. J., and Miller, C. (1996). Homodimeric architecture of a ClC-type chloride ion channel. *Nature* **383**, 337–340.
Miller, C. (1982). Open-state substructure of single chloride channels from *Torpedo* electroplax. *Philos. Trans. R. Soc. Lond. B Biol. Sci.* **299**, 401–411.
Miller, C. (2006). CLC chloride channels viewed through a transporter lens. *Nature* **440**, 484–489.
Miller, C. and White, M. M. (1980). A voltage-dependent chloride conductance channel from *Torpedo* electroplax membrane. *Ann. NY Acad. Sci.* **341**, 534–551.
Mindell, J. A., Maduke, M., Miller, C., and Grigorieff, N. (2001). Projection structure of a ClC-type chloride channel at 6.5 Å resolution. *Nature* **409**, 219–223.
Niemeyer, M. I., Cid, L. P., Zúñiga, L., Catalán, M., and Sepúlveda, F. V. (2003). A conserved pore-lining glutamate as a voltage- and chloride-dependent gate in the ClC-2 chloride channel. *J. Physiol.* **553**, 873–879.
Peinelt, C. and Apell, H. J. (2004). Time-resolved charge movements in the sarcoplasmatic reticulum Ca-ATPase. *Biophys. J.* **86**, 815–824.
Picollo, A. and Pusch, M. (2005). Chloride/proton antiporter activity of mammalian CLC proteins ClC-4 and ClC-5. *Nature* **436**, 420–423.
Picollo, A., Liantonio, A., Didonna, M. P., Elia, L., Camerino, D. C., and Pusch, M. (2004). Molecular determinants of differential pore blocking of kidney CLC-K chloride channels. *EMBO Rep.* **5**, 584–589.
Piwon, N., Gunther, W., Schwake, M., Bösl, M. R., and Jentsch, T. J. (2000). ClC-5 Cl^--channel disruption impairs endocytosis in a mouse model for Dent's disease. *Nature* **408**, 369–373.
Pusch, M., Ludewig, U., Rehfeldt, A., and Jentsch, T. J. (1995). Gating of the voltage-dependent chloride channel ClC-0 by the permeant anion. *Nature* **373**, 527–531.

Pusch, M., Ludewig, U., and Jentsch, T. J. (1997). Temperature dependence of fast and slow gating relaxations of ClC-0 chloride channels. *J. Gen. Physiol.* **109**, 105–116.
Pusch, M., Jordt, S. E., Stein, V., and Jentsch, T. J. (1999). Chloride dependence of hyperpolarization-activated chloride channel gates. *J. Physiol.* **515**, 341–353.
Richard, E. A. and Miller, C. (1990). Steady-state coupling of ion-channel conformations to a transmembrane ion gradient. *Science* **247**, 1208–1210.
Rodriguez, B. M., Sigg, D., and Bezanilla, F. (1998). Voltage gating of Shaker K^+ channels. The effect of temperature on ionic and gating currents. *J. Gen. Physiol.* **112**, 223–242.
Rychkov, G. Y., Pusch, M., Astill, D. S., Roberts, M. L., Jentsch, T. J., and Bretag, A. H. (1996). Concentration and pH dependence of skeletal muscle chloride channel ClC-1. *J. Physiol.* **497**, 423–435.
Rychkov, G. Y., Pusch, M., Roberts, M. L., Jentsch, T. J., and Bretag, A. H. (1998). Permeation and block of the skeletal muscle chloride channel, ClC-1, by foreign anions. *J. Gen. Physiol.* **111**, 653–665.
Rychkov, G. Y., Pusch, M., Roberts, M. L., and Bretag, A. H. (2001). Interaction of hydrophobic anions with the rat skeletal muscle chloride channel ClC-1: Effects on permeation and gating. *J. Physiol.* **530**, 379–393.
Saviane, C., Conti, F., and Pusch, M. (1999). The muscle chloride channel ClC-1 has a double-barreled appearance that is differentially affected in dominant and recessive myotonia. *J. Gen. Physiol.* **113**, 457–468.
Scheel, O., Zdebik, A. A., Lourdel, S., and Jentsch, T. J. (2005). Voltage-dependent electrogenic chloride/proton exchange by endosomal CLC proteins. *Nature* **436**, 424–427.
Schmidt-Rose, T. and Jentsch, T. J. (1997). Reconstitution of functional voltage-gated chloride channels from complementary fragments of CLC-1. *J. Biol. Chem.* **272**, 20515–20521.
Steinmeyer, K., Schwappach, B., Bens, M., Vandewalle, A., and Jentsch, T. J. (1995). Cloning and functional expression of rat CLC-5, a chloride channel related to kidney disease. *J. Biol. Chem.* **270**, 31172–31177.
Thiemann, A., Gründer, S., Pusch, M., and Jentsch, T. J. (1992). A chloride channel widely expressed in epithelial and non-epithelial cells. *Nature* **356**, 57–60.
Traverso, S., Elia, L., and Pusch, M. (2003). Gating competence of constitutively open ClC-0 mutants revealed by the interaction with a small organic inhibitor. *J. Gen. Physiol.* **122**, 295–306.
Traverso, S., Zifarelli, G., Aiello, R., and Pusch, M. (2006). Proton sensing of ClC-0 mutant E166D. *J. Gen. Physiol.* **127**, 51–66.
Toyoshima, C. and Mizutani, T. (2004). Lumenal gating mechanism revealed in calcium pump crystal structures with phosphate analogues. *Nature* **432**, 361–368.
Toyoshima, C. and Nomura, H. (2002). Structural changes in the calcium pump accompanying the dissociation of calcium. *Nature* **418**, 605–611.
Weinreich, F. and Jentsch, T. J. (2001). Pores formed by single subunits in mixed dimers of different CLC chloride channels. *J. Biol. Chem.* **276**, 2347–2353.
White, M. M. and Miller, C. (1981). Probes of the conduction process of a voltage-gated Cl^- channel from Torpedo electroplax. *J. Gen. Physiol.* **78**, 1–18.
Yellen, G. (1998). The moving parts of voltage-gated ion channels. *Q. Rev. Biophys.* **31**, 239–295.
Yin, J., Kuang, Z., Mahankali, U., and Beck, T. L. (2004). Ion transit pathways and gating in ClC chloride channels. *Proteins* **57**, 414–421.
Zhang, X. D. and Chen, T. Y. (2006). Pore electrostatic potential in ClC-0 altered by binding of a charged inhibitor and by mutations of a charged residue. *Biophys. J.* **90**, 1502Plat (abstract).
Zhou, M. and MacKinnon, R. (2004). A mutant KcsA $K(^+)$ channel with altered conduction properties and selectivity filter ion distribution. *J. Mol. Biol.* **338**, 839–846.
Zúñiga, L., Niemeyer, M. I., Varela, D., Catalán, M., Cid, L. P., and Sepúlveda, F. V. (2004). The voltage-dependent ClC-2 chloride channel has a dual gating mechanism. *J. Physiol.* **555**, 671–682.

Chapter 4
Pharmacology of CLC Chloride Channels and Transporters

Michael Pusch,[1] Antonella Liantonio,[2] Annamaria De Luca,[2] and Diana Conte Camerino[2]

[1] *Istituto di Biofisica, CNR, Via De Marini, 6, I-16149 Genova, Italy*
[2] *Sezione di Farmacologia, Dipartimento Farmacobiologico, Facoltà di Farmacia, Università di Bari, Italy*

I. Introduction
II. The Pharmacology of the Macroscopic Skeletal Muscle Cl⁻ Conductance gCl
III. New Molecules Targeting ClC-1 Identified Using Heterologous Expression
IV. Mechanism of Block of Muscle Type CLC Channels by Clofibric Acid Derivatives
V. The Binding Site for 9-AC and CPA on ClC-1 and ClC-0 and Use of CPA as a Tool to Explore the Mechanisms of Gating of ClC-0
VI. Pharmacology of CLC-K Channels
VII. Other CLC Channels and Other Blockers
VIII. Potential of Having Blockers of ClC-2, ClC-3, ClC-5, ClC-7: Outlook
References

I. INTRODUCTION

The "pharmacology" of a protein comprises the description of its interactions with small organic or inorganic molecules that can bind to it and alter its function. Such pharmacological ligands constitute a tremendous resource, and the action of most medically useful drugs can be pinpointed to one or a few specific interactions with specific target proteins or other cellular macromolecules. But in addition to the obvious benefits as therapeutic agents, pharmacological ligands can be very useful tools for the exploration of the molecular mechanisms of function of enzymes. In fact, many inhibitors of enzymes are pseudosubstrates that bind to the active site and lock the enzyme in a certain state of the enzymatic cycle. Such ligands can be used in kinetic studies to explore the mechanism of function, or in structural studies to characterize an isolated kinetic state. For example, the X-ray structure of the sarcoplasmic Ca^{2+}-ATPase has been obtained in the presence (and absence) of various ligands, resulting in deep insights into the pumping mechanism (Obara et al., 2005). Another important application of pharmacological tools is to eliminate specifically the function of one or more components in a complex biological system. This is desirable if one particular protein shall be studied in isolation. For example, most neurons have Ca^{2+} currents that reflect the sum of the contributions of several distinct Ca^{2+} channel proteins. Specific inhibitors for practically each of the individual Ca^{2+} channels are available such that the contribution of each component can be estimated by applying appropriate combinations of inhibitors (Trimmer and Rhodes, 2004). Similarly, such specific inhibitors can be used to ask the specific question: is the activity of a certain protein essential for a certain physiological process? Often, a clear answer to such a question is provided by genetic knockout studies, mainly done in mice, in which the protein in question is genetically inactivated. However, compensation by up- and/or down-regulation of other genes may complicate the picture. Thus, the availability of specific and high-affinity inhibitors is potentially very useful to elucidate the physiological function of proteins.

Historically, for ion channels, high-affinity blockers have been essential for their biochemical, molecular identification because these proteins, as most membrane proteins, are of low abundance and difficult to purify from native tissues (Hille, 2001). Examples include the well-known tetrodotoxin (TTX), a specific blocker of voltage-gated Na channels (Hille, 2001), and α-bungarotoxin, an extremely potent blocker of nicotinic acetylcholine receptors (Katz and Miledi, 1973).

In contrast, relatively few high-affinity blockers generally exist for Cl^- channels. An exception to this are the postsynaptic gamma aminobutyric acid (GABA) and glycine receptors, for which some high-affinity blockers are historically well known, like strychnine that blocks glycine receptors (Young and Snyder, 1973). Recent progress has been made for the cystic fibrosis transmembrane conductance regulator (CFTR) Cl^- channel (see Chapters 5 and 6) for which relatively high-affinity inhibitors and activators have been developed (Galietta and Moran, 2004) (see Chapter 5). Also for some members of the CLC family of Cl^- channels and transporters, important insight into the mechanisms of action and binding sites of several drugs have been gained, and several new molecules have been developed. In the present chapter, we will provide a historical retrospective, describe this recent progress, and provide an overview about the potential benefits of small molecule ligands of CLC proteins in various physiological contexts.

II. THE PHARMACOLOGY OF THE MACROSCOPIC SKELETAL MUSCLE Cl^- CONDUCTANCE gCl

The sarcolemma is characterized by a larger resting permeability for Cl^- (gCl) than for K^+ (gK). The evidence that Cl^- permeation occurs through specific channels was soon derived from pharmacology, as it could be specifically blocked by inorganic (e.g., external Zn^{2+}) and organic molecules such as 9-anthracene-carboxylic acid (9-AC) (Fig. 1). The main physiological role for the large gCl is to maintain the electrical stability of the sarcolemma. In fact in pioneering studies, Bryant showed that the hyperexcitability recorded in the intercostal muscle of myotonic "fainting"

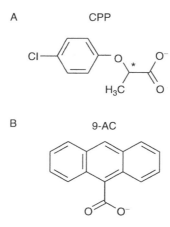

Figure 1. Chemical structure of classical inhibitors of the skeletal muscle Cl⁻ channel. In (A) is shown the structure of CPP, with the asterisk indicating the chiral carbon. In (B) is shown the structure of 9-AC.

goat was related to an abnormally low gCl, and could be reproduced by 9-AC, putting the basis for the discovery of a large series of genetic diseases due to mutations in membrane ion channels (Bryant and Morales-Aguilera, 1971).

Since then, the physiological and pharmacological properties of muscle gCl were actively studied by classical two microelectrode current-clamp recordings and were pivotal for the future studies on cloned channel proteins (see following paragraphs). For instance, the evidence that gCl increases age-dependently in rat EDL muscle during the first month of postnatal life, contributed, along with its sensitivity to 9-AC, to support that the ClC-1 protein was indeed the channel accounting for the macroscopic resting conductance (Conte Camerino *et al.*, 1989b; Steinmeyer *et al.*, 1991).

Other than 9-AC and the agents classically defined as "Cl⁻ channel blockers," as for example 4,4′-diisothiocyanatostilbene-2,2′-disulfonic acid (DIDS) and diphenylamine-2-carboxylate (DPC), other drugs can affect gCl (Camerino *et al.*, 1989) (Table I). The main finding for the identification of specific ClC-1 modulators was the observation that a hypolipidemic drug, clofibrate, was able to induce an "iatrogen" form of myotonia. Apart from an unspecific membrane effect, likely due to change in the lipid environment, it was rapidly demonstrated that clofibric acid, the active *in vivo* metabolite of clofibrate, could specifically block muscle gCl, in a concentration-dependent manner, when applied *in vitro* (Conte-Camerino *et al.*, 1984). This discovery opened the way toward an intense study of structure-activity relationship using a large number of clofibric acid derivatives, which turned out to be important pharmacological tools for studying various members of the CLC channel family (Pusch *et al.*, 2000) (see following paragraphs). The 2-*p*-chlorophenoxy propionic acid (CPP) (Fig. 1), a chiral molecule, allowed also to investigate the possible stereoselectivity of the drug-binding site on muscle Cl⁻ channels. In the native environment, the two enantiomers showed an opposite behavior. $S(-)$-CPP blocks gCl concentration-dependently and is one of the

Table I
Direct and Indirect Modulators of Muscle Cl⁻ Conductance

Modulators	Mechanism	Effect	Condition	Pathology
Clofibric acid derivatives	Direct high-affinity interaction	Block of gCl. The $R(+)$ isomer can increase gCl at low concentrations	The effects are always detectable with differences that are age related	Blockers can induce an iatrogen-myotonia
Taurine	Low-affinity interaction (exogenous)	Increase of gCl	The effects are more evident on fast than on slow muscle types and in condition of taurine depletion	Taurine supplementation can restore gCl in aged and dystrophic muscles
Phorbol esters	Activation of PKC	Block of gCl	The effects are always detectable, with differences that are age related or fiber-phenotype dependent	Overactivity can lead to a myotonic state
IGF-1	Activation of a phosphatase	Increase in gCl	The effects are more evident in condition of PKC overactivity or in slow-twitch fibers	IGF-1 has proved beneficial effect in aging and dystrophic conditions
GH	IGF-1-mediated activation of phosphatase?	Increase in gCl	The effects are more evident in aged subjects	GH has proved beneficial effects during aging
Ghrelin	Receptor-mediated activation of PKC	Reduction of gCl		
Statins	Direct or indirect (cholesterol pathways or PKC mediated)?	Reduction of gCl		The reduction of gCl may contribute to myopathies by statins
Niflumic acid	Both direct and PKC mediated	Reduction of gCl		The reduction of gCl may account for possible side effects by NSAID on muscle function on chronic use

most potent compounds with an IC_{50} of about 15 μM. $R(+)$-CPP is much less potent in blocking gCl, but shows at low concentrations (1–5 μM) the ability to increase gCl (Conte-Camerino et al., 1988). This behavior was well fitted with a model of two sites able to oppositely modulate gCl and on which the enantiomers can act with different affinity and intrinsic activity (De Luca et al., 1992). The "opener" activity of $R(+)$-CPP is not observed for ClC-1 expressed in heterologous systems, suggesting that for the native muscle Cl^- channel, some aspect of the native tissue plays an important role for modulating drug sensitivity (Aromataris et al., 1999; Pusch et al., 2000). However, it is worth noticing that a similar hypothesis, that is, the presence of both an "agonist" and an "antagonist" site, is now proposed for the renal CLC-K channels (Liantonio et al., 2006b).

The extensive structure–activity studies allowed to gain insight into the molecular requisites for modulating gCl, and, therefore, for drug–channel interactions. Structure modifications were conducted in all parts of the CPP molecule potentially involved in binding such as the chiral center, the aromatic moiety, the acid function, and the oxygen atom of the aryloxy group. It was demonstrated that CPP is the most active structure on muscle gCl and that—other than the chiral center—a pivotal role is played by the carboxylic function, ensuring a proper acidity, the halogens on the aromatic ring, ensuring the proper electronic clouds, and the oxygen nearby the aromatic ring (Liantonio et al., 2003). Based on experiments with cloned ClC-1, it could be shown that the binding site for CPP and derivatives is directly accessible only from the intracellular side (Pusch et al., 2000). Thus, assaying drug efficacy in intact skeletal muscle fibers bears the complication that the drug has to enter the cytoplasm (see below).

Muscle gCl is a highly sensitive index of muscle function, being generally one of the first parameters to be changed in many pathophysiological conditions, such as aging, denervation, and dystrophic degeneration, as possible consequence of changes in channel expression and/or function (Conte Camerino et al., 1989b; De Luca et al., 1990; Pierno et al., 1999; De Luca et al., 2003). Consequently, gCl can be directly or indirectly sensitive to the action of various pharmacologically modulated pathways.

For instance, muscle gCl is controlled by biochemical pathways involving a system of protein kinases and insulin-like growth factor-1 (IGF-1)-sensitive phosphatases. A phorbol ester-sensitive protein kinase C (PKC) can potently block gCl and the phosphorylation state may control the trafficking of ClC-1 to the sarcolemma, its expression in physiological conditions as well as its drug sensitivity (De Luca et al., 1994, 1998; Rosenbohm et al., 1999; Papponen et al., 2005). Nonetheless, such a mechanism can also play a role in the phenotypic-dependent difference in gCl between fast-twitch and slow-twitch muscles, as well as in its modulation in conditions as disuse and microgravity in which muscle plasticity is activated (Pierno et al., 2002; Desaphy et al., 2005). Interestingly, even growth hormone, likely through production of IGF-1, or ghrelin, through a direct modulation of a muscular receptor, can increase or decrease gCl, respectively, by acting through the biochemical modulatory pathways (De Luca et al., 1997; Pierno et al., 2003). As these latter require the native environment, their influence on the effect of direct channel modulators is not easy to study on heterologously expressed channels.

Another interesting modulator of ClC-1 is taurine, an osmolyte usually present in high concentrations in skeletal muscle. Pharmacological and structure–activity relationship studies support the ability of taurine to control gCl, acting on a low-affinity site (mM range) nearby the channel (Pierno et al., 1994). The main activity of taurine is to increase gCl. Preliminary two microelectrode voltage-clamp recordings showed that *in vitro* application of taurine modestly enhances the Cl$^-$ currents sustained by human ClC-1 heterologously expressed in *Xenopus* oocytes. In parallel, taurine slightly shifts the channel activation toward more negative potentials, an effect that possibly accounts for the increase in resting gCl observed in native fibers during current-clamp recordings (Conte Camerino et al., 2004). The low-affinity site may account for taurine effectiveness in some forms of myotonic states (Conte Camerino et al., 1989a).

Other than a pharmacological action, taurine can also exert a long-term physiological control on the function of muscle Cl$^-$ channels. In fact, a depletion of taurine content decreases gCl; this effect may be due to the ability of taurine to modulate the pathways (Ca homeostasis, kinase/phosphatase pathways) involved in the maintenance of ClC-1 in an active state (De Luca et al., 1996). Accordingly, the *in vivo* treatment with taurine, likely acting by restoring intracellular pools, may counteract the gCl impairment due to diseases, such as muscular dystrophy, or to physiological states as aging (Pierno et al., 1999; De Luca et al., 2003).

On the other hand, drugs with side effects on skeletal muscle can have gCl as a first target. For instance, statins with a lipophilic structure can reduce muscle gCl (Pierno et al., 1995), a cellular event that may account for some of the muscle effects described for this class of therapeutic compounds. Although the mechanism by which statins can act on Cl$^-$ channels is under investigation, possible hypotheses include the reduction of cholesterol synthesis and consequently the alteration of cholesterol-dependent pathways, as well as the drug activity on the biochemical events involved in ClC-1 modulation. Interestingly, even niflumic acid (NFA), a drug belonging to nonsteroidal anti-inflammatory drugs (NSAIDs), has been found to decrease muscle gCl both directly and through a PKC-mediated action due to the mobilization of intracellular Ca (Liantonio et al., 2006a). Also in this case, the mechanism can lead to unwanted muscular effects on chronic use of the drug. The indirect modulation of Cl$^-$ channels by drugs able to affect or rather improve skeletal muscle function may seem far from the direct action of specific tools, as CPP derivatives. Nonetheless, these lines of evidence suggest that ClC-1, and possibly other members of the CLC family, may undergo a strict control through not yet defined pathways, subunits, or enzymatic systems able to affect, in the native environment, the biophysical and pharmacological properties of the channel.

III. NEW MOLECULES TARGETING ClC-1 IDENTIFIED USING HETEROLOGOUS EXPRESSION

Although the evaluation of the effect of small organic molecules on native skeletal muscle Cl$^-$ conductance is of sure physiological relevance since in this system all biochemical constituents fundamental for channel activity are preserved (Conte-Camerino *et al.*, 1988; De Luca *et al.*, 1992, 1998), an invaluable contribution to the investigation of the pharmacological profile of the muscle Cl$^-$ channel derives from the use of heterologously expressed ClC-1 that opened the way for systematically studying established inhibitors but also previously untested or novel compounds. Indeed, taking into account the intracellular location of the CPP-binding site (Pusch *et al.*, 2000, 2001), the easy accessibility in the inside-out configuration of the patch-clamp technique (see below) allowed to study the interaction between the drugs and the amino acid residues involved in the binding site independently from the capability of the molecules to cross the plasma membrane (Liantonio *et al.*, 2003). Among the various ClC-1 inhibitors described in the literature (Jentsch *et al.*, 2002), such as 9-AC, DPC, NFA, and $S(-)$-CPP, 9-AC has the highest affinity (Estévez *et al.*, 2003). However, onset and wash of 9-AC block is very slow (timescale of minutes), rendering difficult a precise structure–activity study. $S(-)$-CPP exerts a specific and reasonably high-affinity action, blocking ClC-1 currents with a K_D of about 40 μM at -140 mV. Importantly, CPP block is quickly reversible and easily quantified using inside-out patch-clamp measurements and excised patch results can be well compared with gCl measurements. Starting from the CPP structure, the synthesis and the evaluation of the inhibitory effect on ClC-1 of a large array of derivatives, with modification at several strategic position of the molecule, allowed to perform a detailed structure-activity study as well as to develop potent and selective blockers. As summarized in Fig. 2, the modifications that have been accomplished were: (1) removal or substitution of the chlorine atom on the aromatic ring with other halogen atoms, or introduction of other substituents to evaluate the role of the electric cloud and of the steric hindrance of the ring; (2) isosteric substitution to evaluate the function of the oxygen atom of the phenoxy group; (3) introduction of a six- or five-membered ring to evaluate the effect of the increased molecular rigidity; (4) substitution of the methyl group of the chiral center with different alkyl or chlorophenoxy groups to evaluate the role of the asymmetric carbon atom as well as of the bulkiness in this part of the molecule; and (5) substitution of the carboxylic moiety with a bioisosteric phosphonate group to clarify the role of the acid function.

Several maneuvers, that is, modification of the substituent on the aromatic ring, isosteric substitution of the oxygen atom, elimination of the carboxylic group, or a change in molecular rigidity, strongly compromised drug-blocking activity. In contrast, the introduction of a second chlorophenoxy group on the chiral center of CPP significantly increases affinity toward the binding site. Particularly, these new CPP-like molecules, named bis-phenoxy derivatives, produced a block of heterologously expressed ClC-1 with a tenfold increased affinity with respect to $S(-)$-CPP showing a K_D value of about 4 μM at -140 mV. Thus, from a structural point of view, it was concluded that the presence of well-established chemical groups with an adequate spatial disposition are necessary to potently inhibit the muscle ClC-1 channel. First, as is the case for most of the standard Cl$^-$ channel inhibitors, a key role is played by the presence of a carboxylic group that confers to the molecule a negative charge allowing a competition between Cl$^-$ ions and drug and consequently the drug interference with channel permeation and/or gating. The acidic function should be carried by a chlorophenoxy group that represents the lead pharmacophore moiety which could

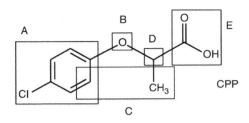

		Effect on muscle gCl IC$_{50}$ (μM)	Effect on ClC-1 expressed in oocytes K_D (−140 mV) (μM)
	S(−)-CPP	15	40
A	F-phenoxy-CH(CH$_3$)-COOH	≈	N.D.
	H$_3$C-phenoxy-CH(CH$_3$)-COOH	>	N.D.
	Cl-dimethylphenoxy-CH(CH$_3$)-COOH	>	N.D.
B	Cl-phenyl-NH-CH(CH$_3$)-COOH	≫	N.D.
	Cl-phenyl-S-CH(CH$_3$)-COOH	≈	N.D.
C	Cl-benzofuran-COOH	>	N.D.
	Cl-chroman-COOH	>	N.D.
D	Cl-phenoxy-CH(CH$_2$)$_3$-Cl-phenyl-COOH	>	<
	Cl-phenoxy-CH(CH$_2$)$_3$-OCH$_3$-phenyl-COOH	>	<
	Cl-phenoxy-CH(CH$_2$)$_3$-phenyl-COOH	>	<
E	Cl-phenoxy-CH$_2$-P(O)(OH)$_2$	>	>
	Cl-phenoxy-CH$_2$-P(O)(OCH$_3$)$_2$	−	−

Figure 2. Structure–activity study on ClC-1 using CPP derivatives (for details see chapter). The main structural modifications performed on the CPP structure were: (A) substitutions on the aromatic ring, (B) isosteric substitutions of the oxygen atom, (C) increase of molecular rigidity, (D) substitutions on the chiral center, (E) substitution of the carboxylic group. Symbols: ≈, IC$_{50}$ overlapping to that of CPP; >, 100 μM < IC$_{50}$ < 500 μM or K_D (−140 mV) > 50 μM; ≫, IC$_{50}$ < 1 mM; <, 5 μM < K_D (−140 mV) > 10 μM; −, completely ineffective; N.D., not determined.

interact with a hydrophobic pocket and at the same time realize a π-π interaction. Furthermore, the presence of an electron-attractive substituent in para position of the aromatic ring probably favors a dipole–dipole interaction with the binding site. The introduction of an additional phenoxy group on the chiral center stabilizes the interaction with the binding site, probably by an interaction with a second hydrophobic pocket. Particularly, the presence of a substituent in para position confers to this aromatic ring a bulkiness that could improve such an interaction.

Interestingly, bis-phenoxy derivatives of CPP turned out to inhibit also ClC-K1 and ClC-Ka channels when applied to the extracellular side (see below); thus, these compounds proved very useful tools to explore the pore structure also of these renal CLC members other than of the muscle ClC-1.

In a recent study, the effect of NFA, a molecule belonging to the class of fenamates normally used as nonsteroidal anti-inflammatory drugs, has been evaluated on ClC-1 (Liantonio et al., 2006a). Particularly, NFA inhibited native gCl with an IC_{50} of 42 µM and blocked ClC-1 through an interaction with an intracellular binding site. Although some common features shared by the two different class of inhibitors (CPP derivatives and NFA derivatives) either from a chemical view point (the presence of two aromatic rings and of a carboxylic function) or from a mechanistic view point (a voltage-dependent inhibition with an affinity in the micromolar range), the effect of NFA on the muscle ClC-1 current is somehow peculiar. In addition to a direct block of ClC-1, NFA was able to increase the basal intracellular Ca^{2+} concentration $[Ca^{2+}]_i$ in fura-2-loaded EDL muscle fibers by promoting a mitochondrial calcium efflux in an independent manner from cyclooxygenase and Cl^- channel inhibition. Considering that ClC-1 is down-regulated by the calcium-dependent PKC (De Luca et al., 1998; Rosenbohm et al., 1999) (see above), by using specific PKC inhibitors, the involvement of this kinase in NFA-mediated modulation of native gCl could be demonstrated. Thus, other than to produce a direct block of ClC-1 by an interaction with a blocking binding site located on the channel protein, NFA also indirectly modulates native gCl by increasing intracellular Ca levels which in turn produces a Ca-dependent PKC activation.

This class of inhibitors may be used to explore the physiological significance of the phosphorylation–dephosphorylation pathway in modulating ClC-1 activity in skeletal muscle fibers.

IV. MECHANISM OF BLOCK OF MUSCLE TYPE CLC CHANNELS BY CLOFIBRIC ACID DERIVATIVES

ClC-0, the curious double-barreled Cl^- channel (Miller and Richard, 1990), was cloned from the electric organ of *Torpedo marmorata* (Jentsch et al., 1990). This organ is related to skeletal muscle, and indeed, the first mammalian homologue of ClC-0 to be cloned was ClC-1, that is almost exclusively expressed in skeletal muscle and underlies its large gCl (Steinmeyer et al., 1991). Using heterologous expression in *Xenopus* oocytes and mammalian cells and application of voltage-clamp techniques it was thus possible to characterize the pharmacological properties of ClC-1 in great detail. The initial studies demonstrated that 9-AC potently blocks ClC-1 expressed in *Xenopus* oocytes (Steinmeyer *et al.*, 1991), confirming its identity as the channel that generates the muscle Cl$^-$ conductance. In these studies, block by 9-AC had a slow onset and was practically irreversible, suggesting that 9-AC has to enter the oocyte in order to block the channel from the intracellular side. Once inside the oocyte, 9-AC may be trapped, leading to an apparently irreversible inhibition. This hypothesis was tested directly more recently using inside-out and outside-out patch-clamp recordings (Estévez *et al.*, 2003). When applied to the extracellular side in outside-out patch-clamp recordings, 100 µM 9-AC had practically no effect, while the same concentration almost completely blocked ClC-1 when applied to the cytoplasmic side of inside-out patches (Estévez *et al.*, 2003). Thus, the binding site of 9-AC is directly accessible only from the intracellular side of the channel. However, 9-AC is a rather hydrophobic molecule (Fig. 1) and is thus able to diffuse across the lipid bilayer. In the very large *Xenopus* oocytes, this process takes a very long time and is practically irreversible because of the slow diffusion. In contrast, in small cells, like Sf9 cells or HEK 293 cells, the diffusion into and out of the cell is much faster. This may explain the fast onset and reversibility of block by extracellularly applied 9-AC in small, transfected cells (Astill *et al.*, 1996; Rychkov *et al.*, 1997).

CPP and derivatives represent the other class of compounds that were known to inhibit the skeletal muscle Cl$^-$ conductance, gCl (see above). CPP and derivatives (Fig. 1) were first tested on the cloned ClC-1 expressed in Sf9 cells (Aromataris *et al.*, 1999). Surprisingly, CPP blocked ClC-1 in a highly voltage-dependent manner: currents were strongly reduced at negative voltages, where the channels tend to close. In contrast, almost no block was seen at positive voltages. This effect was interpreted initially as a pure modulation of gating (Aromataris *et al.*, 1999): CPP renders opening of the channel more difficult, leading to apparent shifts of the voltage dependence of the open probability (Aromataris *et al.*, 1999). CPP is an optically active molecule with two enantiomers: $R(+)$-CPP and $S(-)$-CPP (Fig. 1). In agreement with earlier studies on the macroscopic skeletal muscle Cl$^-$ conductance, gCl (Conte-Camerino *et al.*, 1988), the $S(-)$ enantiomer was found to be much more effective as a blocker/gating modifier than the $R(+)$ enantiomer (Aromataris *et al.*, 1999). In these initial studies (Aromataris *et al.*, 1999), it could not be decided if CPP acts from the intracellular or from the extracellular side of the channel. Using excised patch-clamp recordings on ClC-1 expressed in *Xenopus* oocytes, Pusch *et al.* (2000) could show that CPP and derivatives act exclusively from the intracellular side of the channel. Additionally, the accurate inside-out patch-clamp measurements showed that $S(-)$-CPP does not simply cause a "shift" of the p_{open} (V) curve but that $S(-)$-CPP additionally decreases the minimal open probability at negative voltages (Pusch *et al.*, 2000) (Fig. 3). CPP or close analogues were tested also on other CLC channels and were found to be effective only on the plasma membrane channels ClC-1, ClC-0, ClC-2, and ClC-K1 (Pusch *et al.*, 2000; Estévez *et al.*, 2003; Pusch, unpublished result). Among these channels, ClC-1 shows the highest affinity, while ClC-2 has the lowest affinity. On all these channels, CPP and derivatives exclusively act from the intracellular side of the membrane, and the

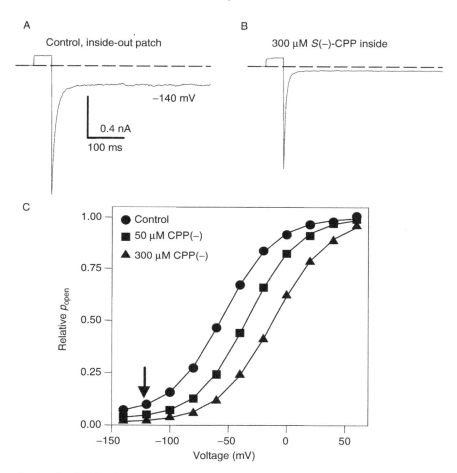

Figure 3. Block of ClC-1 by CPP from the inside. In (A) and (B) are shown inside-out patch-clamp recordings from an oocyte expressing human ClC-1 before (A) and after (B) the application of 300 μM S(−)-CPP to the intracellular side. In (C) is shown the apparent open probability in control and in the presence of various concentrations of S(−)-CPP. The arrow highlights the region of negative voltages at which a major CPP block occurs that cannot be described by a "shift" of the activation curve. Block at these voltages is physiologically most relevant.

effect is markedly voltage dependent with strong block at negative voltages and almost no block at positive voltages.

While the pharmacology of the skeletal muscle Cl⁻ channel is surely of direct physiological relevance, its complicated biophysical properties render the elucidation of the mechanism of action of CPP block quite difficult. For example, the single channel conductance of ClC-1 is very small (Pusch et al., 1994; Saviane et al., 1999) and its gating is complex (Accardi and Pusch, 2000). In this respect, the "model" CLC channel ClC-0 from the *Torpedo* electric organ has comparably more favorable properties (see Chapter 3). It can be studied at the single channel level and its gating can be quite easily separated in two kinetically vastly distinct components: a fast gate operates independently on each protopore of the dimeric double-barreled channel and a slow gate shuts both pores simultaneously (Pusch and Jentsch, 2005). Using ClC-0 as a model, the mechanism of clock by CPP derivatives has been elucidated in great detail (Pusch et al., 2001; Accardi and Pusch, 2003). First of all, it could be shown that CPP functionally acts on the individual protopores. This implies that the double-barreled channel has two binding sites, one in each pore. Next, it could be shown that intracellular Cl$^-$ ions compete with CPP binding, suggesting that CPP binds close to the ion-conducting pore (Pusch et al., 2001). The kinetics and steady state voltage and concentration dependence of CPP block could be quantitatively described by a four-state model in which channel opening/closing occurs with rate constants α and β, respectively, and CPP, at the concentration c, binds to the pore with a second order association constant k_{on} and a first order dissociation rate k_{off}.

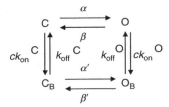

In the model C is the drug-free closed state, O the drug-free open state, C_B the drug-bound closed state, and O_B the drug-bound open state. The strong state dependence and the resulting voltage dependence of block (Fig. 3) is captured in Model 1 by the constraint that

$$K_D^C = \frac{k_{off}^C}{k_{on}^C} \ll K_D^O = \frac{k_{off}^O}{k_{on}^O}$$

that is, the fact that CPP binds with much higher affinity to closed channels than to open channels. Because a scheme like Model 1 must obey the principle of microscopic reversibility, the above relation implies that the "open probability" of drug-occupied channels must be much smaller than that of drug-free channels, that is,

$$\frac{\alpha'}{\beta'} \ll \frac{\alpha}{\beta}$$

A major difficulty in the analysis of CPP block is to find out if drug-bound channels that are in the "open" conformation, corresponding to state O_B in Model 1, are actually conducting or if they are blocked. The difficulty arises from the fact that, in either case, the state O_B has a low overall probability. This problem might seem to be esoteric at a first glance. However, from a mechanistic point of view, an answer to this question is important to distinguish between two principal modes of action of an ion channel inhibitor, as shown schematically in Fig. 4.

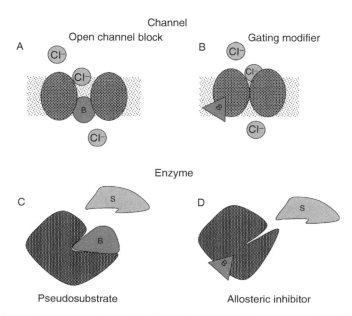

Figure 4. Difference between open channel blocker and gating modifier. An open channel blocker binds to the pore and, competing with permeant ions, blocks ion conduction (A). A gating modifier does not necessarily bind in the pore but reduces ion flow by promoting channel closure (B). In analogy, enzyme inhibitors can be pseudosubstrates (C) or allosteric inhibitors (D). "B" denotes blocker and "S" denotes substrate.

An open channel blocker binds to the pore and impedes ion flow by occluding the pore, while a pure gating modifier binds to a structure of the channel that promotes channel closure. A classical example of a gating modifier of voltage-dependent K^+ channels is hanatoxin that binds to the voltage sensor and thereby renders opening more difficult (Swartz and MacKinnon, 1997). Comparing channels to enzymes, a pore blocker is a pseudosubstrate (Fig. 4A and C) while a gating modifier is an allosteric modulator of enzyme function (Fig. 4B and D). Thus, if state O_B is conductive, CPP can be classified as a pure gating modifier that does not directly impede ion permeation when bound to the channel. In contrast, if state O_B is nonconductive, CPP is more likely to directly occlude the pore.

Direct information on this issue was obtained by Accardi and Pusch (2003) who used chlorophenoxy acetic acid (CPA), the simplest CPP derivative. In complete analogy to CPP, CPA also binds with higher affinity to closed channels. Importantly, however, it could be shown that CPA exerts a low affinity, flickery open channel block, with an affinity in the 10-mM range. The open channel block was increased at reduced extracellular Cl^- concentrations, suggesting that CPA binds to the pore of the channel. These results demonstrated that CPA is an open channel blocker that, in addition, has profound effects on channel gating. In the context of Model 1, Accardi and Pusch developed a robust assay to measure separately the affinity of CPA for closed and open ClC-0 channels (Accardi and Pusch, 2003), allowing to use CPA as a tool to explore the gating of ClC-0 (see below).

V. THE BINDING SITE FOR 9-AC AND CPA ON ClC-1 AND ClC-0 AND USE OF CPA AS A TOOL TO EXPLORE THE MECHANISMS OF GATING OF ClC-0

The determination of the crystal structure of bacterial CLC homologues (Dutzler *et al.*, 2002) opened a new era for the pharmacological investigation of CLC proteins. For the first time the blocking and modulating effect of small organic molecules could be mapped onto the structure (Estévez *et al.*, 2003; Picollo *et al.*, 2004) and computational methods could be employed to explore the interaction of small ligands with the protein (Moran *et al.*, 2003).

Even before the structure determination, Björn Schroeder, at that time a doctoral student in the laboratory of Thomas Jentsch in Hamburg, had identified amino acids that are critically involved in the inhibition of ClC-1 by 9-AC (Schroeder, 2000). The approach was based on the very different affinity of ClC-2 and ClC-1 for 9-AC: whereas the apparent K_D of ClC-1 for 9-AC is of the order of 10 μM, ClC-2 needs more than 500 μM 9-AC for half-maximal block (Estévez *et al.*, 2003). Using a chimeric strategy, Ser-537 in ClC-1, corresponding to Thr-518 in ClC-2, was identified as one of the most crucial differences between the two channel types. The S537T mutation in ClC-1 reduced the 9-AC affinity to the value measured for ClC-2, and the corresponding T518S mutation in ClC-2 significantly increased its affinity for 9-AC. When mapped onto the structure of the bacterial StClC protein, this crucial amino acid corresponds to Val-402 that is located in the very short loop between helices O and P. Using as a guide the structure of the bacterial *Salmonella* CLC homologue, Estévez *et al.* identified several amino acids in ClC-1 that, when mutated, drastically altered the apparent 9-AC or CPA affinity. When mapped onto the bacterial CLC protein, these "high-impact" residues clustered, without exception, in a region between the central Cl^--binding site, S_{cent}, and the critical Ser-537 (Estévez *et al.*, 2003). On one hand, this result demonstrated that the structure of StClC could be used as a valid guide for a structure-function analysis of mammalian proteins. On the other hand, the region found by Estévez *et al.* most likely defines the rough localization of the binding site for 9-AC and CPA on these channels. This putative binding site is completely buried within the protein and it is currently unclear how 9-AC and CPA may gain access to it (Maduke and Mindell, 2003). However, given the negative charge of the blockers and the competition of CPP block with intracellular Cl^- ions (Pusch *et al.*, 2001), the most likely access pathway is that used by the permeating Cl^- ions themselves. Building on these results, molecular modeling may in the future be useful to develop new drugs that are of therapeutic value. However, the structural conservation of the mammalian and the bacterial CLCs is not large enough, and homology models contain too much ambiguity, to allow currently a direct application of molecular modeling for rational drug design. The availability of the structure of a CLC channel (and not transporter) would certainly be a big advantage.

The inhibitors of ClC-0 or ClC-1, that is, CPP derivatives and 9-AC, are not really of high affinity with none of them blocking substantially at submicromolar concentrations. Nevertheless, these substances, and in particular CPP derivatives, have been used as molecular tools to explore the pore and the gating of the *Torpedo* ClC-0 channel. Starting from the fact that CPP derivatives show a strong state-dependent inhibition (see above) (Pusch *et al.*, 2001), Accardi and Pusch (2003) studied in a quantitative manner the effect of several point mutations of the ClC-0 channel on closed channel binding and open channel binding. The amino acids studied were S123, Y512, K519, and T481. The first two residues, S123 and Y512, turned out to be apparently crucially involved in the coordination of a Cl$^-$ ion in the crystal structure of StClC (Dutzler *et al.*, 2002), while K519 probably lines the intracellular pore entrance (Pusch *et al.*, 1995; Ludewig *et al.*, 1996; Middleton *et al.*, 1996; Ludewig *et al.*, 1997; Dutzler *et al.*, 2002) and T481 is critically involved in CPA and 9-AC binding, as described above. Interestingly, the mutation T481S exclusively altered binding of CPA to closed channels, while mutations S123T, Y512A, and K519Q mainly altered open channel block (Accardi and Pusch, 2003). These results suggested that the inner pore structure, where CPA binds, has a different conformation in the open and in the closed state. This conclusion was in disagreement with crystallographic results from Dutzler *et al.* (2003), who suggested that opening of the pore, that is, gating, only involves a small movement of the side chain of the conserved glutamate E166 (see Chapter 3), while the rest of the protein maintains a fixed structure. However, even assuming only a very small conformational change, it cannot be excluded that this alters drastically the pore occupancy by Cl$^-$ and the protonation state of titrable groups in the pore, leading indirectly to different open and closed channel affinity for CPA, and to the experimentally observed differential effects of mutations on open channel and closed channel block. Yet, further support for the hypothesis that fast gating of ClC-0 is associated with a significant conformational change was provided by additional experiments using CPA as a tool (Traverso *et al.*, 2003). Accardi and Pusch (2003) had found that the block of open ClC-0 pores by CPA is of very low affinity (in the 10-mM range) and is of a flickery, difficult to resolve nature. Thus, it was to be expected that a mutation that renders ClC-0 almost constitutively open by removing the charged side chain of E166 (mutant E166A) (Dutzler *et al.*, 2003; Traverso *et al.*, 2003) should show a low-affinity block by CPA. In complete contrast to this simple-minded expectation, E166A had a more than 200-fold higher affinity for CPA compared to the open channel affinity of wild-type (WT) ClC-0 (Traverso *et al.*, 2003). The affinity of E166A was even about 25-fold larger than the closed channel affinity of WT. Traverso *et al.* could explain these data by a simple three-state kinetic scheme in which one of the transitions had several biophysical properties in common with the regular opening step of WT ClC-0. Since the "gating" glutamate carboxylate side chain is absent in mutant E166A, these results thus suggested that, in addition to the swing-out of the glutamate side chain, the pore undergoes a conformational change that is associated with gating (Traverso *et al.*, 2003). This conclusion was in stark disagreement with the proposal of Dutzler *et al.* (2003) that opening of ClC-0 involves no structural rearrangement apart the "swing-out" of the side chain of Glu-166. It must be kept in mind, however, that these functional results are indirect, and that alternative hypotheses cannot be excluded. In particular, a more complex involvement of E166 in fast gating, related to the protonation from the intracellular side, must be considered (Traverso *et al.*, 2006). In fact, it may even be speculated (!) that the carboxylate group of a CPA molecule bound in the E166A pore substitutes somehow for the carboxylate group of E166, mimicking regular gating mediated via the voltage-dependent protonation from the intracellular side (Miller, 2006; Traverso *et al.*, 2006). There is definitely room for a further employment of CPA and other small molecule ligands as tools to explore the voltage-dependent gating of CLC channels.

VI. PHARMACOLOGY OF CLC-K CHANNELS

ClC-Ka and ClC-Kb, as their correspondent murine orthologues ClC-K1 and ClC-K2, are members of the CLC family that are selectively expressed in kidney and inner ear where they are essential for water and salt conservation and for the production of endolymph, respectively (Jentsch, 2005; Uchida and Sasaki, 2005). For a correct expression and function, these channels require the presence of the barttin β-subunit (Estévez *et al.*, 2001). Particularly, the ClC-Kb/ClC-K2 channel is expressed in the basolateral plasma membrane of the thick ascending limb of Henle's loop (TAL) where it plays a key role in Cl^-, Na, and Mg reabsorption. Mutations in the gene encoding ClC-Kb impair transepithelial NaCl transport in the TAL and in the distal convoluted tubule and are responsible for type III Bartter syndrome, a kidney disease characterized by severe salt wasting and hypokalemia (Jentsch, 2005) (see Chapter 2). The elucidation of the role of CLC-K channels in kidney salt reabsorption, obtained by using mouse models, human molecular genetics, and heterologous expression systems, has brought up a growing interest toward the identification of specific ligands that allow pharmacological interventions aimed to modulate CLC-K channel activity.

In contrast to the muscle ClC-1 channel, the pharmacological characterization of renal CLC-K channels in native systems is rather poor, mainly due to the technical difficulties in measuring the Cl^- conductance *in situ*. Intracellular application of DPC and DIDS in patch-clamp recordings on mouse microdissected distal-convoluted tubules produced a decrease of Cl^- currents. Based on the high CLC-K expression in this section of the nephron, a direct action of these two aspecific Cl^- channel blockers on ClC-K1 and ClC-K2 was hypothesized, even if the possibility of the presence of an unrelated, nonidentified channel could not be excluded (Nissant *et al.*, 2004; Teulon *et al.*, 2005). The limited information obtained from mouse native kidney was counterbalanced by a detailed pharmacological investigation performed on CLC-K channels expressed in heterologous systems. Indeed, after screening a variety of molecules belonging to different structural classes, it was demonstrated in the last years that CLC-K channels have two functionally different extracellular drug-binding sites: a blocking site and an activating site.

In a first attempt to find CLC-K inhibitors, a variety of derivatives of CPP, a specific ligand of ClC-1 (see above), were tested on chimeras of ClC-K1 and ClC-Kb (Liantonio *et al.*, 2002). This study pinpointed that bis-phenoxy derivatives of CPP were able to inhibit CLC-K chimeras from the extracellular side with an affinity in the 150-µM range. The discovery of barttin opened the way to intensive investigations of the pharmacological properties of each CLC-K isoform (Birkenhäger et al., 2001; Estévez et al., 2001). First, the inhibitory effect of the bis-phenoxy derivatives was confirmed for ClC-K1, which had a similar affinity as the previously studied chimeras (Liantonio et al., 2002, 2004). The mechanism of block of ClC-K1 by bis-phenoxy derivatives from the extracellular side was shown to be quite different from that of the block of ClC-1 that occurs from the intracellular side. For instance, the small voltage dependence of the block of ClC-K1, in contrast to the extremely voltage-dependent block of ClC-1, demonstrates that the block by extracellular bis-phenoxy derivatives is not state dependent. By performing an accurate structure-activity study it was concluded that a simpler structure than that of bis-chlorophenoxy derivatives is sufficient to bind and block ClC-K1 (Liantonio et al., 2004). Among several tested compounds, the CPP analogue carrying a benzyl group on the chiral center (3-phenyl-CPP) (Fig. 5) represented the minimal structure capable of stereoselectively inhibiting ClC-K1 currents with micromolar affinity (Liantonio et al., 2004). The rapid onset and reversibility of block together with the competition with extracellular Cl⁻ led to the hypothesis that the binding site is exposed to the extracellular side and is located close to the ion-conducting pore. This sidedness of the binding site was unequivocally demonstrated using excised patch-clamp measurements of ClC-K1 in which only extracellular drug application led to a current block (Liantonio et al., 2002). Among several "classical" Cl^- channel blockers, DIDS was capable of inhibiting ClC-K1 currents with a similar affinity as

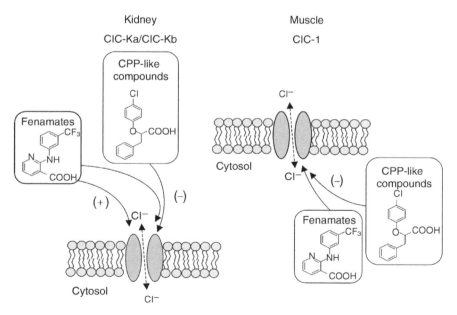

Figure 5. Summary of blocking and activating binding sites on renal CLC-K channels (left) and muscle ClC-1 (right).

3-phenyl-CPP, although in an apparently irreversible manner (Liantonio et al., 2004). At this regard, it is noteworthy that DIDS inhibits ClC-0 (Miller and White, 1984) and the bacterial transporter ClC-ec1 (Matulef and Maduke, 2005) in the micromolar range acting from the intracellular side. It seems to be a coincidence that the same substance inhibits channels of the same family from opposite sides of the membrane as in the case of bis-phenoxy derivatives of CPP (ClC-1 inside, ClC-K outside) and DIDS (ClC-0 inside, ClC-K outside). The extracellular modulation of CLC-K channels could be of high therapeutic interest, because the binding site is easily accessible, considering that the channels show specific basolateral, and possibly also apical, localization in renal epithelia.

Interestingly and unexpectedly, in view of the high homology between the two human isoforms, human ClC-Ka was found to be much more sensitive to the inhibitory activity of both 3-phenyl-CPP and DIDS than ClC-Kb (Picollo et al., 2004). At 60 mV, 3-phenyl-CPP is about fivefold more potent on ClC-Ka compared with ClC-Kb, with respective apparent K_D values of 80 µM (ClC-Ka) and 380 µM (ClC-Kb). The use of sequence comparison and of the crystal structure of the bacterial StClC (Dutzler et al., 2002) as a guide allowed to identify the structural basis responsible for this different pharmacological sensitivity (Picollo et al., 2004). Particularly, it was demonstrated that Asn-68 in ClC-Ka plays a pivotal role for the blocking activity of 3-phenyl-CPP because exchanging it with a negatively charged aspartate (N68D), as found in ClC-Kb, markedly reduced drug sensitivity. Furthermore, Asn-68, together with Gly-72, is also pivotal for the more pronounced activity of the unrelated stilbene blocker DIDS on ClC-Ka with respect to ClC-Kb. Both residues presumably expose their side chain to the extracellular pore mouth. Thus, it was speculated that the electrostatic interaction between these amino acids and the negatively charged group of the drug molecules might either permit, as in the case of ClC-Ka, or impede, as in the case of ClC-Kb, the interaction of the inhibitors. These results suggest that the binding site for 3-phenyl-CPP and DIDS is located close to the side chains of residues 68 and 72, that is, in the extracellular pore mouth of the channel, just above the side chain of residue 166, that is a glutamate in most CLC proteins with important roles for gating and proton transport (see Chapter 3).

More recently, the use of NFA derivatives gave the chance to confirm the presence of a blocking binding site on ClC-Ka and at the same time to unmask and to preliminarily characterize an activating binding site on both human CLC-K isoforms (Liantonio et al., 2006b), but not rat ClC-K1 (Liantonio et al., 2004). Depending on the chemical structure, fenamates are capable of blocking or opening ClC-Ka. Screening a series of NFA derivatives led to the definition of structural requisites to mediate the two opposite effects. NFA was the most potent opener, indicating that for an efficient activation of ClC-Ka, an acidic carboxylic group, two aromatic rings, one of which pyridinic, an anilinic moiety connecting the two rings, are required. Also the presence of an electronegative group, such as a CF_3, in meta position of the phenyl ring seemed to increase the drug affinity for the activating binding site. Nevertheless, some of these requisites, as the presence of the acidic function and of two aromatic rings, are also required for a blocking activity. In fact, all derivatives of flufenamic acid (FFA) are efficient blockers of ClC-Ka currents. Chemical modeling revealed that NFA shows a nearly planar conformation, whereas FFA derivatives and 3-phenyl-CPP are forced to assume a noncoplanar arrangement of the aromatic rings. Thus, the spatial geometry profile associated to each molecule seemed to be the main determinant of the final effect (activating or blocking).

Both NFA and FFA derivatives act as openers of ClC-Kb, further corroborating the previous finding that ClC-Kb is much less sensitive to blockers than ClC-Ka. In agreement with this, FFA derivatives had a reduced efficacy on the ClC-Ka mutant N68D strongly indicating that the mechanism of action of this class of inhibitors resembles that of 3-phenyl-CPP (Picollo *et al.*, 2004; Liantonio *et al.*, 2006b). On the other hand, it seems that the NFA-mediated current potentiation is caused by the interaction with a different binding site. Indeed, NFA induced a reproducible current increase on all mutants used for the identification of the blocking binding site (Picollo *et al.*, 2004; Liantonio *et al.*, 2006b). Nevertheless, considering that the activating receptor site is not yet identified and that it might be located within the pore, it can also be hypothesized that the two binding sites are partially overlapping.

It is well known that, secondary to the compromised ClC-Kb channel activity, type III Bartter syndrome patients showed a markedly elevated prostaglandin (PGE2) activity (Reinalter *et al.*, 2002). The direct action of NFA on CLC-K channels as an opener, together with its cyclooxygenase inhibition activity, makes NFA a lead starting point molecule on which to work for identifying drugs that might be therapeutically useful for this renal channelopathy. At the same time, considering the involvement of CLC-K channels in the mechanism of urine concentration, the reported inhibitors, such as 3-phenyl-CPP and FFA derivatives, could represent a new class of drugs with diuretic activity (Fong, 2004). At this regard, a selective action on the two CLC-K isoforms, that is to say a drug that specifically inhibits ClC-Ka but not ClC-Kb, could have significant advantages compared with a general CLC-K blocker.

VII. OTHER CLC CHANNELS AND OTHER BLOCKERS

All organic CLC blockers known so far are weak organic acids. Apart from 9-AC, CPP and derivatives (see above), a few other organic acids have been tried as blockers of CLC proteins. Among the CLC channels, the one that seems to be most sensitive to such organic inhibitors is the muscle channel, ClC-1. Apart from the block by CPP, 9-AC, NFA, and other molecules that bind from the inside (see above), some simple organic acids also affect ClC-1 currents from the outside. A detailed analysis of this phenomenon was performed by Rychkov *et al.* (2001) for the block of ClC-1 by extracellular hexanoate and similar compounds. These acids produced a paradoxical increase of block of instantaneous currents at negative voltages that could be quantitatively explained by an extracellular, superficial, binding site (Rychkov *et al.*, 2001). In retrospect, this site seems to be different from the crystallographically identified anion-binding sites (Dutzler *et al.*, 2003). No site-directed mutagenesis has been performed so far to identify this binding site.

A well-known organic acid acting as a "generic" Cl^- channel and Cl^- transporter blocker is DIDS (Miller and Richard, 1990). DIDS is quite reactive and can covalently bind to several groups, like for example to the side chain of lysines. In fact, the individual protopores of the *Torpedo* channel were irreversibly inhibited by intracellular DIDS (Miller and White, 1984). These elegant early experiments lent strong support to the hypothesis of a double-barreled channel with two physically distinct permeation pathways. These experiments with DIDS were never reproduced for the cloned ClC-0 channel and the "DIDS receptor" of ClC-0 is unknown. It is also unclear if the DIDS-binding site overlaps with the CPA/9-AC-binding site of ClC-0/ClC-1 (Estévez *et al.*, 2003). Interestingly, Matulef and Maduke (2005) showed recently that also the bacterial transporter ClC-ec1 is inhibited by intracellular DIDS in the micromolar range, however, in a reversible manner. For this transporter, the DIDS block (or other inhibitors) might be useful to functionally orient the protein in the lipid bilayer (Matulef and Maduke, 2005). However, the mechanism of action of DIDS on ClC-ec1 and the binding site are still little understood.

Divalent (or multivalent) heavy metal cations may interact at high affinity with proteins, often mediated by cysteine or histidine side chains. In fact, currents mediated by ClC-0, ClC-1, and ClC-2 are inhibited by various divalent cations (Rychkov *et al.*, 1997; Chen, 1998; Clark *et al.*, 1998; Kürz *et al.*, 1999). In all cases, block occurs from the extracellular side. The mechanism of inhibition has been thoroughly studied in ClC-0 by Chen (1998). They found that binding of Zn^{2+} ions favors the closure of the slow gate, that is, the gate that shuts off both protopores of the channel. Thus, Zn^{2+} can be regarded as a gating modifier of the common gate. Chen and colleagues went on to identify the possible target of Zn^{2+} on the channel by mutating several cysteine residues. They identified indeed a cysteine residue, C212, which greatly diminished Zn^{2+} effects when mutated to serine (C212S) (Lin *et al.*, 1999). This cysteine was therefore a good candidate for binding the divalent metal ions. However, besides reducing Zn^{2+} effects, the C212S mutation locked the slow, common, gate of the channel almost completely open, rendering it insensitive to voltage and temperature (Lin *et al.*, 1999). Thus, the lack of effect of Zn^{2+} on the mutant C212S does not demonstrate that C212 is part of the Zn^{2+}-binding site. Thus, while C212 is still a possible candidate for Zn^{2+} binding, the exact site remains to be identified. Recent studies have shown that the Zn^{2+} effects on the muscle channel ClC-1 are probably mediated by a very similar mechanism (Duffield *et al.*, 2005).

VIII. POTENTIAL OF HAVING BLOCKERS OF ClC-2, ClC-3, ClC-5, ClC-7: OUTLOOK

Our current understanding of the block of ClC-1 and ClC-Ks by several small organic molecules is summarized in Fig. 5. Unfortunately, practically no efficient organic inhibitor has been described for the other mammalian CLC proteins. Thus, no reasonably potent organic blocker of the ubiquitous plasma membrane channel ClC-2 is known, and the same holds true for the mostly intracellular proteins ClC-3—ClC-7. It is assumed that ClC-2 is involved in the control of ionic homeostasis in specialized cellular structures (Bösl et al., 2001), and specific inhibitors could be extremely useful to decipher its precise physiological role. Recently, McCarty and colleagues reported that crude scorpion venom inhibits ClC-2 (Thompson et al., 2005), but the ingredient responsible for the block remains to be identified.

Similarly, practically no small molecules are known to interfere with the mostly intracellular proteins ClC-3—ClC-7. The problem is especially hard for ClC-6 and ClC-7 because it has not yet been possible to study these two proteins in heterologous systems with electrophysiological techniques, impeding a direct assessment of the action of possible inhibitors. Drugs targeted at ClC-7 have been hypothesized to be of potential benefit to treat osteoporosis (Schaller et al., 2004). The argument is based on the fact that ClC-7 activity is needed for proper bone resorption (Kornak et al., 2001), and that osteoporosis is characterized by excessive bone resorption. Karsdal and colleagues have identified several compounds that seem to inhibit bone resorption (Henriksen et al., 2004; Schaller et al., 2004; Karsdal et al., 2005), and that are promising starting points to treat the disease. It remains, however, to be shown if these compounds act directly on ClC-7.

In summary, we can say that some progress has been made regarding the pharmacology of CLC channels and transporters, in particular regarding the identification of binding sites on ClC-1 and ClC-K channels, the CLC channels for which we have the most information. Nevertheless, we still lack really high-affinity (i.e., submicromolar) and specific inhibitors (or activators, in the case of ClC-K) for the plasma membrane channels. The situation is worse for the intracellular CLC proteins, ClC-3—ClC-7, for which practically no small molecule ligand has been identified so far. We believe that such ligands are of great potential utility, and that significant research activity should be devoted to the development of potent CLC inhibitors, and if possible, also activators.

ACKNOWLEDGMENTS

The financial support by Telethon Italy (grant GGP04018 to M.P. and grant GGP04140 to D.C.C.) and by MIUR COFIN (project 2005 to D.C.C.) is gratefully acknowledged. We thank Domenico Tricarico and Sabata Pierno for their contribution during the experimental work regarding the study on the native muscle Cl^- conductance and Alessandra Picollo, Elena Babini, Paola Didonna, Laura Elia, Sonia Traverso, and Alessio Accardi for their contribution to the elucidation of the pharmacology of expressed ClC-1 and CLC-K channels.

REFERENCES

Accardi, A. and Pusch, M. (2000). Fast and slow gating relaxations in the muscle chloride channel CLC-1. J. Gen. Physiol. **116**, 433–444.
Accardi, A. and Pusch, M. (2003). Conformational changes in the pore of CLC-0. J. Gen. Physiol. **122**, 277–293.

Aromataris, E. C., Astill, D. S., Rychkov, G. Y., Bryant, S. H., Bretag, A. H., and Roberts, M. L. (1999). Modulation of the gating of ClC-1 by S-(–) 2-(4-chlorophenoxy) propionic acid. *Br. J. Pharmacol.* **126**, 1375–1382.

Astill, D. S., Rychkov, G., Clarke, J. D., Hughes, B. P., Roberts, M. L., and Bretag, A. H. (1996). Characteristics of skeletal muscle chloride channel C1C-1 and point mutant R304E expressed in Sf-9 insect cells. *Biochim. Biophys. Acta* **1280**, 178–186.

Birkenhäger, R., Otto, E., Schurmann, M. J., Vollmer, M., Ruf, E. M., Maier-Lutz, I., Beekmann, F., Fekete, A., Omran, H., Feldmann, D., Milford, D. V., Jeck, N., et al. (2001). Mutation of BSND causes Bartter syndrome with sensorineural deafness and kidney failure. *Nat. Genet.* **29**, 310–314.

Bösl, M. R., Stein, V., Hübner, C., Zdebik, A. A., Jordt, S. E., Mukhopadhyay, A. K., Davidoff, M. S., Holstein, A. F., and Jentsch, T. J. (2001). Male germ cells and photoreceptors, both dependent on close cell-cell interactions, degenerate upon ClC-2 Cl(-) channel disruption. *EMBO J.* **20**, 1289–1299.

Bryant, S. H. and Morales-Aguilera, A. (1971). Chloride conductance in normal and myotonic muscle fibres and the action of monocarboxylic aromatic acids. *J. Physiol.* **219**, 367–383.

Camerino, D. C., De Luca, A., and Mambrini, M. (1989). The effect of diphenylamine-2-carboxylate on Cl^- channel conductance and on excitability characteristics of rat skeletal muscle. *J. Pharm. Pharmacol.* **41**, 42–45.

Chen, T. Y. (1998). Extracellular zinc ion inhibits ClC-0 chloride channels by facilitating slow gating. *J. Gen. Physiol.* **112**, 715–726.

Clark, S., Jordt, S. E., Jentsch, T. J., and Mathie, A. (1998). Characterization of the hyperpolarization-activated chloride current in dissociated rat sympathetic neurons. *J. Physiol.* **506**, 665–678.

Conte Camerino, D., De Luca, A., Mambrini, M., Ferrannini, E., Franconi, F., Giotti, A., and Bryant, S. H. (1989a). The effects of taurine on pharmacologically induced myotonia. *Muscle Nerve* **12**, 898–904.

Conte Camerino, D., De Luca, A., Mambrini, M., and Vrbova, G. (1989b). Membrane ionic conductances in normal and denervated skeletal muscle of the rat during development. *Pflügers Arch.* **413**, 568–570.

Conte Camerino, D., Tricarico, D., Pierno, S., Desaphy, J. F., Liantonio, A., Pusch, M., Burdi, R., Camerino, C., Fraysse, B., and De Luca, A. (2004). Taurine and skeletal muscle disorders. *Neurochem. Res.* **29**, 135–142.

Conte-Camerino, D., Tortorella, V., Ferranini, E., and Bryant, S. H. (1984). The toxic effects of clofibrate and its metabolite on mammalian skeletal muscle: An electrophysiological study. *Arch. Toxicol. Suppl.* **7**, 482–484.

Conte-Camerino, D., Mambrini, M., DeLuca, A., Tricarico, D., Bryant, S. H., Tortorella, V., and Bettoni, G. (1988). Enantiomers of clofibric acid analogs have opposite actions on rat skeletal muscle chloride channels. *Pflügers Arch.* **413**, 105–107.

De Luca, A., Mambrini, M., and Conte Camerino, D. (1990). Changes in membrane ionic conductances and excitability characteristics of rat skeletal muscle during aging. *Pflügers Arch.* **415**, 642–644.

De Luca, A., Tricarico, D., Wagner, R., Bryant, S. H., Tortorella, V., and Conte Camerino, D. (1992). Opposite effects of enantiomers of clofibric acid derivative on rat skeletal muscle chloride conductance: Antagonism studies and theoretical modeling of two different receptor site interactions. *J. Pharmacol. Exp. Ther.* **260**, 364–368.

De Luca, A., Tricarico, D., Pierno, S., and Conte Camerino, D. (1994). Aging and chloride channel regulation in rat fast-twitch muscle fibres. *Pflügers Arch.* **427**, 80–85.

De Luca, A., Pierno, S., and Camerino, D. C. (1996). Effect of taurine depletion on excitation-contraction coupling and Cl^- conductance of rat skeletal muscle. *Eur. J. Pharmacol.* **296**, 215–222.

De Luca, A., Pierno, S., Cocchi, D., and Conte Camerino, D. (1997). Effects of chronic growth hormone treatment in aged rats on the biophysical and pharmacological properties of skeletal muscle chloride channels. *Br. J. Pharmacol.* **121**, 369–374.

De Luca, A., Pierno, S., Liantonio, A., Camerino, C., and Conte Camerino, D. (1998). Phosphorylation and IGF-1-mediated dephosphorylation pathways control the activity and the pharmacological properties of skeletal muscle chloride channels. *Br. J. Pharmacol.* **125**, 477–482.

De Luca, A., Pierno, S., Liantonio, A., Cetrone, M., Camerino, C., Fraysse, B., Mirabella, M., Servidei, S., Ruegg, U. T., and Conte Camerino, D. (2003). Enhanced dystrophic progression in mdx mice by

exercise and beneficial effects of taurine and insulin-like growth factor-1. *J. Pharmacol. Exp. Ther.* **304,** 453–463.
Desaphy, J. F., Pierno, S., Liantonio, A., De Luca, A., Didonna, M. P., Frigeri, A., Nicchia, G. P., Svelto, M., Camerino, C., Zallone, A., and Camerino, D. C. (2005). Recovery of the soleus muscle after short- and long-term disuse induced by hindlimb unloading: Effects on the electrical properties and myosin heavy chain profile. *Neurobiol. Dis.* **18,** 356–365.
Duffield, M. D., Rychkov, G. Y., Bretag, A. H., and Roberts, M. L. (2005). Zinc inhibits human ClC-1 muscle chloride channel by interacting with its common gating mechanism. *J. Physiol.* **568,** 5–12.
Dutzler, R., Campbell, E. B., Cadene, M., Chait, B. T., and MacKinnon, R. (2002). X-ray structure of a ClC chloride channel at 3.0 Å reveals the molecular basis of anion selectivity. *Nature* **415,** 287–294.
Dutzler, R., Campbell, E. B., and MacKinnon, R. (2003). Gating the selectivity filter in ClC chloride channels. *Science* **300,** 108–112.
Estévez, R., Boettger, T., Stein, V., Birkenhäger, R., Otto, E., Hildebrandt, F., and Jentsch, T. J. (2001). Barttin is a Cl$^-$ channel beta-subunit crucial for renal Cl$^-$ reabsorption and inner ear K$^+$ secretion. *Nature* **414,** 558–561.
Estévez, R., Schroeder, B. C., Accardi, A., Jentsch, T. J., and Pusch, M. (2003). Conservation of chloride channel structure revealed by an inhibitor binding site in ClC-1. *Neuron* **38,** 47–59.
Fong, P. (2004). CLC-K channels: If the drug fits, use it. *EMBO Rep.* **5,** 565–566.
Galietta, L. J. and Moran, O. (2004). Identification of CFTR activators and inhibitors: Chance or design? *Curr. Opin. Pharmacol.* **4,** 497–503.
Henriksen, K., Gram, J., Schaller, S., Dahl, B. H., Dziegiel, M. H., Bollerslev, J., and Karsdal, M. A. (2004). Characterization of osteoclasts from patients harboring a G215R mutation in ClC-7 causing autosomal dominant osteopetrosis type II. *Am. J. Pathol.* **164,** 1537–1545.
Hille, B. (2001). *Ion Channels of Excitable Membranes.* Sinauer, Sunderland, MA.
Jentsch, T. J. (2005). Chloride transport in the kidney: Lessons from human disease and knockout mice. *J. Am. Soc. Nephrol.* **16,** 1549–1561.
Jentsch, T. J., Steinmeyer, K., and Schwarz, G. (1990). Primary structure of *Torpedo marmorata* chloride channel isolated by expression cloning in *Xenopus* oocytes. *Nature* **348,** 510–514.
Jentsch, T. J., Stein, V., Weinreich, F., and Zdebik, A. A. (2002). Molecular structure and physiological function of chloride channels. *Physiol. Rev.* **82,** 503–568.
Karsdal, M. A., Henriksen, K., Sørensen, M. G., Gram, J., Schaller, S., Dziegiel, M. H., Heegaard, A. M., Christophersen, P., Martin, T. J., Christiansen, C., and Bollerslev, J. (2005). Acidification of the osteoclastic resorption compartment provides insight into the coupling of bone formation to bone resorption. *Am. J. Pathol.* **166,** 467–476.
Katz, B. and Miledi, R. (1973). The effect of alpha-bungarotoxin on acetylcholine receptors. *Br. J. Pharmacol.* **49,** 138–139.
Kornak, U., Kasper, D., Bösl, M. R., Kaiser, E., Schweizer, M., Schulz, A., Friedrich, W., Delling, G., and Jentsch, T. J. (2001). Loss of the ClC-7 chloride channel leads to osteopetrosis in mice and man. *Cell* **104,** 205–215.
Kürz, L. L., Klink, H., Jakob, I., Kuchenbecker, M., Benz, S., Lehmann-Horn, F., and Rüdel, R. (1999). Identification of three cysteines as targets for the Zn^{2+} blockade of the human skeletal muscle chloride channel. *J. Biol. Chem.* **274,** 11687–11692.
Liantonio, A., Accardi, A., Carbonara, G., Fracchiolla, G., Loiodice, F., Tortorella, P., Traverso, S., Guida, P., Pierno, S., De Luca, A., Camerino, D. C., and Pusch, M. (2002). Molecular requisites for drug binding to muscle CLC-1 and renal CLC-K channel revealed by the use of phenoxy-alkyl derivatives of 2-(p-chlorophenoxy)propionic acid. *Mol. Pharmacol.* **62,** 265–271.
Liantonio, A., De Luca, A., Pierno, S., Didonna, M. P., Loiodice, F., Fracchiolla, G., Tortorella, P., Antonio, L., Bonerba, E., Traverso, S., Elia, L., Picollo, A., *et al.* (2003). Structural requisites of 2-(p-chlorophenoxy)propionic acid analogues for activity on native rat skeletal muscle chloride conductance and on heterologously expressed CLC-1. *Br. J. Pharmacol.* **139,** 1255–1264.
Liantonio, A., Pusch, M., Picollo, A., Guida, P., De Luca, A., Pierno, S., Fracchiolla, G., Loiodice, F., Tortorella, P., and Conte Camerino, D. (2004). Investigations of pharmacologic properties of the renal

CLC-K1 chloride channel co-expressed with barttin by the use of 2-(p-chlorophenoxy)propionic acid derivatives and other structurally unrelated chloride channels blockers. *J. Am. Soc. Nephrol.* **15**, 13–20.
Liantonio, A., Giannuzzi, V., Picollo, A., Babini, E., Pusch, M., and Conte Camerino, D. (2006a). Niflumic acid inhibits chloride conductance of rat skeletal muscle by directly inhibiting the CLC-1 channel and by increasing intracellular calcium. *Br. J. Pharmacol.* submitted for publication.
Liantonio, A., Picollo, A., Babini, E., Carbonara, G., Fracchiolla, G., Loiodice, F., Tortorella, V., Pusch, M., and Camerino, D. C. (2006b). Activation and inhibition of kidney CLC-K chloride channels by fenamates. *Mol. Pharmacol.* **69**, 165–173.
Lin, Y. W., Lin, C. W., and Chen, T. Y. (1999). Elimination of the slow gating of ClC-0 chloride channel by a point mutation. *J. Gen. Physiol.* **114**, 1–12.
Ludewig, U., Pusch, M., and Jentsch, T. J. (1996). Two physically distinct pores in the dimeric ClC-0 chloride channel. *Nature* **383**, 340–343.
Ludewig, U., Jentsch, T. J., and Pusch, M. (1997). Analysis of a protein region involved in permeation and gating of the voltage-gated *Torpedo* chloride channel ClC-0. *J. Physiol.* **498**, 691–702.
Maduke, M. and Mindell, J. A. (2003). The poststructural festivities begin. *Neuron* **38**, 1–3.
Matulef, K. and Maduke, M. (2005). Side-dependent inhibition of a prokaryotic ClC by DIDS. *Biophys. J.* **89**, 1721–1730.
Middleton, R. E., Pheasant, D. J., and Miller, C. (1996). Homodimeric architecture of a ClC-type chloride ion channel. *Nature* **383**, 337–340.
Miller, C. (2006). ClC chloride channels viewed through a transporter lens. *Nature* **440**, 484–489.
Miller, C. and Richard, E. A. (1990). The voltage-dependent chloride channel of *Torpedo* electroplax. Intimations of molecular structure from quirks of single-channel function. In: *Chloride Channels and Carriers in Nerve, Muscle and Glial Cells* (F. J. Alvarez-Leefmans and J. M. Russell, Eds.), pp. 383–405. Plenum, New York.
Miller, C. and White, M. M. (1984). Dimeric structure of single chloride channels from *Torpedo* electroplax. *Proc. Natl. Acad. Sci. USA* **81**, 2772–2775.
Moran, O., Traverso, S., Elia, L., and Pusch, M. (2003). Molecular modeling of p-chlorophenoxyacetic acid binding to the CLC-0 channel. *Biochemistry* **42**, 5176–5185.
Nissant, A., Lourdel, S., Baillet, S., Paulais, M., Marvao, P., Teulon, J., and Imbert-Teboul, M. (2004). Heterogeneous distribution of chloride channels along the distal convoluted tubule probed by single-cell RT-PCR and patch clamp. *Am. J. Physiol. Renal Physiol.* **287**, F1233–F1243.
Obara, K., Miyashita, N., Xu, C., Toyoshima, I., Sugita, Y., Inesi, G., and Toyoshima, C. (2005). Structural role of countertransport revealed in Ca^{2+} pump crystal structure in the absence of Ca^{2+}. *Proc. Natl. Acad. Sci. USA* **102**, 14489–14496.
Papponen, H., Kaisto, T., Myllyla, V. V., Myllyla, R., and Metsikko, K. (2005). Regulated sarcolemmal localization of the muscle-specific ClC-1 chloride channel. *Exp. Neurol.* **191**, 163–173.
Picollo, A., Liantonio, A., Didonna, M. P., Elia, L., Camerino, D. C., and Pusch, M. (2004). Molecular determinants of differential pore blocking of kidney CLC-K chloride channels. *EMBO Rep.* **5**, 584–589.
Pierno, S., De Luca, A., Huxtable, R. J., and Conte Camerino, D. (1994). Dual effects of taurine on membrane ionic conductances of rat skeletal muscle fibers. *Adv. Exp. Med. Biol.* **359**, 217–224.
Pierno, S., De Luca, A., Tricarico, D., Roselli, A., Natuzzi, F., Ferrannini, E., Laico, M., and Camerino, D. C. (1995). Potential risk of myopathy by HMG-CoA reductase inhibitors: A comparison of pravastatin and simvastatin effects on membrane electrical properties of rat skeletal muscle fibers. *J. Pharmacol. Exp. Ther.* **275**, 1490–1496.
Pierno, S., De Luca, A., Beck, C. L., George, A. L., Jr., and Conte Camerino, D. (1999). Aging-associated down-regulation of ClC-1 expression in skeletal muscle: Phenotypic-independent relation to the decrease of chloride conductance. *FEBS Lett.* **449**, 12–16.
Pierno, S., Desaphy, J. F., Liantonio, A., De Bellis, M., Bianco, G., De Luca, A., Frigeri, A., Nicchia, G. P., Svelto, M., Leoty, C., George, A. L., Jr., and Camerino, D. C. (2002). Change of chloride ion channel conductance is an early event of slow-to-fast fibre type transition during unloading-induced muscle disuse. *Brain* **125**, 1510–1521.
Pierno, S., De Luca, A., Desaphy, J. F., Fraysse, B., Liantonio, A., Didonna, M. P., Lograno, M., Cocchi, D., Smith, R. G., and Camerino, D. C. (2003). Growth hormone secretagogues modulate the electrical

and contractile properties of rat skeletal muscle through a ghrelin-specific receptor. *Br. J. Pharmacol.* **139,** 575–584.
Pusch, M. and Jentsch, T. J. (2005). Unique structure and function of chloride transporting CLC proteins. *IEEE Trans. Nanobioscience* **4,** 49–57.
Pusch, M., Steinmeyer, K., and Jentsch, T. J. (1994). Low single channel conductance of the major skeletal muscle chloride channel, ClC-1. *Biophys. J.* **66,** 149–152.
Pusch, M., Ludewig, U., Rehfeldt, A., and Jentsch, T. J. (1995). Gating of the voltage-dependent chloride channel CIC-0 by the permeant anion. *Nature* **373,** 527–531.
Pusch, M., Liantonio, A., Bertorello, L., Accardi, A., De Luca, A., Pierno, S., Tortorella, V., and Camerino, D. C. (2000). Pharmacological characterization of chloride channels belonging to the ClC family by the use of chiral clofibric acid derivatives. *Mol. Pharmacol.* **58,** 498–507.
Pusch, M., Accardi, A., Liantonio, A., Ferrera, L., De Luca, A., Camerino, D. C., and Conti, F. (2001). Mechanism of block of single protopores of the *Torpedo* chloride channel ClC-0 by 2-(p-chlorophenoxy) butyric acid (CPB). *J. Gen. Physiol.* **118,** 45–62.
Reinalter, S. C., Jeck, N., Brochhausen, C., Watzer, B., Nusing, R. M., Seyberth, H. W., and Komhoff, M. (2002). Role of cyclooxygenase-2 in hyperprostaglandin E syndrome/antenatal Bartter syndrome. *Kidney Int.* **62,** 253–260.
Rosenbohm, A., Rüdel, R., and Fahlke, C. (1999). Regulation of the human skeletal muscle chloride channel hClC-1 by protein kinase C. *J. Physiol.* **514,** 677–685.
Rychkov, G., Pusch, M., Roberts, M., and Bretag, A. (2001). Interaction of hydrophobic anions with the rat skeletal muscle chloride channel ClC-1: Effects on permeation and gating. *J. Physiol.* **530,** 379–393.
Rychkov, G. Y., Astill, D. S., Bennetts, B., Hughes, B. P., Bretag, A. H., and Roberts, M. L. (1997). pH-dependent interactions of Cd^{2+} and a carboxylate blocker with the rat ClC-1 chloride channel and its R304E mutant in the Sf-9 insect cell line. *J. Physiol.* **501,** 355–362.
Saviane, C., Conti, F., and Pusch, M. (1999). The muscle chloride channel ClC-1 has a double-barreled appearance that is differentially affected in dominant and recessive myotonia. *J. Gen. Physiol.* **113,** 457–468.
Schaller, S., Henriksen, K., Sveigaard, C., Heegaard, A. M., Helix, N., Stahlhut, M., Ovejero, M. C., Johansen, J. V., Solberg, H., Andersen, T. L., Hougaard, D., Berryman, M., *et al.* (2004). The chloride channel inhibitor NS3736 [corrected] prevents bone resorption in ovariectomized rats without changing bone formation. *J. Bone Miner. Res.* **19,** 1144–1153.
Schroeder, B. C. (2000). Klonierung und Charakterisierung von an Repolarisationsstörungen beteiligten Ionenkanälen. In: *Fachbereich Biochemie, Pharmazie und Lebensmittelchemie,* p. 138. Johann Wolfgang Goethe-Universität, Frankfurt, Germany.
Steinmeyer, K., Ortland, C., and Jentsch, T. J. (1991). Primary structure and functional expression of a developmentally regulated skeletal muscle chloride channel. *Nature* **354,** 301–304.
Swartz, K. J. and MacKinnon, R. (1997). Hanatoxin modifies the gating of a voltage-dependent K+ channel through multiple binding sites. *Neuron* **18,** 665–673.
Teulon, J., Lourdel, S., Nissant, A., Paulais, M., Guinamard, R., Marvao, P., and Imbert-Teboul, M. (2005). Exploration of the basolateral chloride channels in the renal tubule using. *Nephron. Physiol.* **99,** 64–68.
Thompson, C. H., Fields, D. M., Olivetti, P. R., Fuller, M. D., Zhang, Z. R., Kubanek, J., and McCarty, N. A. (2005). Inhibition of ClC-2 chloride channels by a peptide component or components of scorpion venom. *J. Membr. Biol.* **208,** 65–76.
Traverso, S., Elia, L., and Pusch, M. (2003). Gating competence of constitutively open CLC-0 mutants revealed by the interaction with a small organic inhibitor. *J. Gen. Physiol.* **122,** 295–306.
Traverso, S., Zifarelli, G., Aiello, R., and Pusch, M. (2006). Proton sensing of CLC-0 mutant E166D. *J. Gen. Physiol.* **127,** 51–66.
Trimmer, J. S. and Rhodes, K. J. (2004). Localization of voltage-gated ion channels in mammalian brain. *Annu. Rev. Physiol.* **66,** 477–519.
Uchida, S. and Sasaki, S. (2005). Function of chloride channels in the kidney. *Ann. Rev. Physiol.* **67,** 759–778.
Young, A. B. and Snyder, S. H. (1973). Strychnine binding associated with glycine receptors of the central nervous system. *Proc. Natl. Acad. Sci. USA* **70,** 2832–2836.

Chapter 5
The Physiology and Pharmacology of the CFTR Cl⁻ Channel

Zhiwei Cai, Jeng-Haur Chen, Lauren K. Hughes, Hongyu Li, and David N. Sheppard

Department of Physiology, School of Medical Sciences, University of Bristol, Bristol BS8 1TD, United Kingdom

- I. Introduction
- II. The Molecular Physiology of CFTR
  - A. Anion Flow Through the CFTR Pore: The MSDs
  - B. Regulation of CFTR Channel Gating by Intracellular ATP: The NBDs
  - C. Phosphorylation-Dependent Regulation of CFTR: The R Domain
- III. The Cellular Physiology of CFTR
- IV. The Role of CFTR in Epithelial Physiology
  - A. The Pancreas, Intestine, Hepatobiliary System, and Reproductive Tissues
  - B. The Sweat Gland
  - C. The Respiratory Airways
- V. The Pathophysiology of CFTR
  - A. Cystic Fibrosis
  - B. Autosomal Dominant Polycystic Kidney Disease
- VI. The Pharmacology of CFTR
  - A. Potentiators of the CFTR Cl⁻ Channel
  - B. Identification of CFTR Potentiators by HTS
  - C. Rescue of F508del-CFTR by Pharmacological Chaperones
  - D. Therapeutic Potential of CFTR Potentiators
  - E. Inhibitors of the CFTR Cl⁻ Channel
  - F. Inhibition of CFTR by a Peptide Toxin
  - G. Identification of CFTR Inhibitors by HTS
  - H. Therapeutic Potential of CFTR Inhibitors
- VII. Conclusions
  - References

I. INTRODUCTION

Cl⁻ channels regulated by $3',5'$-cyclic adenosine monophosphate (cAMP)-dependent phosphorylation are widely expressed in epithelial tissues (Welsh *et al.*, 2001). They are also found in some nonepithelial tissues, most notably cardiac myocytes

(Hume et al., 2000). In epithelia, these Cl⁻ channels play a fundamental role in fluid and electrolyte transport. They provide a pathway for Cl⁻ to exit from the cell and a key point at which to regulate transepithelial Cl⁻ secretion (Welsh et al., 2001). The discovery that cAMP fails to activate a Cl⁻ conductance in the apical membrane of cystic fibrosis (CF) epithelia led to the identification of the cystic fibrosis transmembrane conductance regulator (CFTR) and the demonstration that CFTR is the epithelial Cl⁻ channel regulated by cAMP-dependent phosphorylation (Quinton, 1983; Riordan et al., 1989; Anderson et al., 1991b).

CFTR has several distinguishing characteristics (for review see Sheppard and Welsh, 1999). First, CFTR has a small single-channel conductance (6–10 pS). Second, the current-voltage (I-V) relationship of CFTR is linear. Third, the anion permeability sequence of CFTR is $Br^- > Cl^- > I^- > F^-$. Fourth, CFTR shows time- and voltage-independent gating behavior. Fifth, the activity of CFTR is regulated by cAMP-dependent phosphorylation and intracellular nucleotides. These hallmarks are conferred on CFTR by the function of the different domains from which CFTR is assembled: the two membrane-spanning domains (MSDs) that are each composed of six transmembrane segments linked by intra- and extracellular loops, the two nucleotide-binding domains (NBDs) that each contain sequences, which interact with ATP (Walker A and B and LSGGQ motifs) and the regulatory (R) domain that contains multiple consensus phosphorylation sites and many charged amino acids. Here, we discuss how the molecular behavior of CFTR determines its function in cells and tissues, the consequences of CFTR malfunction for transepithelial ion transport and the implications of this knowledge for the development of rational new therapeutic strategies for CF and other diseases associated with the malfunction of CFTR.

II. THE MOLECULAR PHYSIOLOGY OF CFTR

Amino acid sequence analysis and comparison with other proteins placed CFTR in a large family of transport proteins called ATP-binding cassette (ABC) transporters (Holland et al., 2003). ABC transporters are composed of two motifs (see Fig. 1 for a model of CFTR). Each motif contains an MSD and an NBD. In CFTR, a unique domain called the R domain links the two MSD–NBD motifs. The vast majority of ABC transporters actively transport substrates across cell membranes. For example, P-glycoprotein (P-gp) confers cancer cells resistant to chemotherapeutic drugs; STE6 secretes the mating pheromone a-factor from yeast, and the histidine and maltose permeases transport nutrients into bacteria. Other ABC transporters, such as the sulphonylurea receptor that controls insulin secretion from pancreatic β-cells, regulate associated ion channels. However, only CFTR forms an ion channel.

In CFTR, the MSDs assemble to form an anion selective pore (Linsdell, 2006). Flow of anions through this pore is powered by cycles of ATP binding and hydrolysis at two ATP-binding sites located at the interface of an NBD1:NBD2 dimer (Gadsby et al., 2006). Phosphorylation of the R domain stimulates CFTR function by

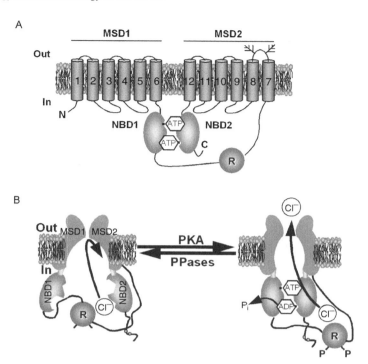

Figure 1. The CFTR Cl⁻ channel is regulated by phosphorylation and intracellular ATP. (A) Model showing the proposed domain structure of CFTR. Glycosylation sites located in the extracellular loop between transmembrane segment 7 (M7) and M8 are indicated by branched structures. (B) The simplified model shows a CFTR Cl⁻ channel under quiescent and activated conditions. MSD, membrane-spanning domain; NBD, nucleotide-binding domain; P, phosphorylation of the R domain; P_i, inorganic phosphate; PKA, cAMP-dependent protein kinase; PPase, protein phosphatase; R, regulatory domain. In and Out denote the intra- and extracellular sides of the membrane, respectively. See Section II for further information. Modified from Chen et al. (2006) with permission of S. Karger AG.

enhancing ATP-dependent channel gating at the NBDs (Ostedgaard et al., 2001). In this section, we consider the relationship between CFTR structure and function.

A. Anion Flow Through the CFTR Pore: The MSDs

In most ABC transporters, the MSDs assemble to form a pathway to shuttle substrates across the cell membrane. In contrast, the MSDs of CFTR form an anion-selective pore through which anions stream across the cell membrane driven by the transmembrane electrochemical gradient (Sheppard and Welsh, 1999; McCarty, 2000; Linsdell, 2006). Excitingly, Rosenberg et al. (2004) have determined the three-dimensional structure of CFTR by electron crystallography. Their data suggest several important conclusions: (1) the quaternary structure of CFTR is monomeric (for review see Riordan, 2005), (2) CFTR exists in different conformations with either single- or double-barreled central cavities (Rosenberg et al., 2004), and (3) the CFTR pore has a deep wide intracellular vestibule, but a shallow extracellular vestibule (Rosenberg et al., 2004). This image of the CFTR pore demonstrates notable similarity to that predicted by functional studies.

Anion permeation studies indicate that the narrowest part of the CFTR pore is ~ 0.53–0.60 nm in diameter (e.g., Cheung and Akabas, 1996), widening under certain circumstances to a diameter of ~ 1.3 nm (Linsdell and Hanrahan, 1998). On the intra- and extracellular sides of this constriction, the pore enlarges (Fig. 1B). The voltage-dependence of channel block by large organic anions suggests that the CFTR pore contains a wide intracellular vestibule that funnels blocking anions deep into the pore where they bind, occlude the pore, and block Cl^- permeation (for discussion see Section VI.E). Based on the inability of open-channel blockers to reach their binding sites when added to the extracellular solution and the short length of extracellular loops 3 and 6, McCarty (2000) proposed that the CFTR pore has a small extracellular vestibule. Consistent with this idea, CFTR exhibits an asymmetric permeability to large organic anions (Linsdell and Hanrahan, 1998) and current flow through the CFTR pore weakly inwardly rectifies (Cai et al., 2003b). Curiously, with some difference in drug-binding sites (Scott-Ward et al., 2004), the CFTR pore appears to be an inverted version of the Ca^{2+}-activated Cl^- channel (Qu and Hartzell, 2001).

The number, identity, and organization of the transmembrane segments that line the CFTR pore are unknown at the present time. Functional and structural studies of the ABC transporter P-gp suggest that the translocation pathway of P-gp is lined by transmembrane segment 6 (M6) and M12 (Loo et al., 2003; Rosenberg et al., 2005). Given the similarities in the crystal structures of CFTR and P-gp (Rosenberg et al., 2004), M6 and M12 might be predicted to be important determinants of the pore properties of the CFTR Cl^- channel. Numerous studies have demonstrated that amino acid residues in M6 line the CFTR pore (see following paragraph). However, the role of M12 and other transmembrane segments in MSD2 in determining CFTR's pore properties is much less certain (for discussion see Sheppard and Welsh, 1999).

M6 plays a crucial role in determining the pore properties of CFTR (Sheppard and Welsh, 1999; McCarty, 2000; Linsdell, 2006). Within M6, the residues arginine (R) 334, lysine (K) 335, phenylalanine (F) 337, threonine (T) 338, serine (S) 341, isoleucine (I) 344, and possibly R352 contribute to Cl^--binding sites. The narrowest part of the CFTR pore, the location of the selectivity filter, likely occurs in the region of F337 and T338 because mutation of these residues altered dramatically the anion permeability sequence of CFTR, whereas mutation of other residues in M6 had less marked effects on anion permeation (e.g., Gong et al., 2002). These data suggest that three Cl^--binding sites (S341, I344, and R352) might be located in a spacious intracellular vestibule, whereas two Cl^--binding sites (R334 and K335) might be located in a more confined extracellular vestibule. Two important caveats of this model are: first, present knowledge of the mechanism of anion permeation by the CFTR Cl^- channel is incomplete and it is quite likely that residues in other transmembrane segments contribute to Cl^--binding sites (Linsdell, 2005). Second, based on the weak anion selectivity of CFTR, Liu et al. (2003) speculate that the CFTR pore might lack a discrete selectivity filter.

CFTR is a multi-ion pore capable of holding multiple anions simultaneously (for review see Sheppard and Welsh, 1999; Linsdell, 2006). This hallmark of the CFTR pore is a key determinant of the channel's conduction properties because repulsive interactions between anions inside the pore ensure rapid anion permeation (Gong and Linsdell, 2004). As discussed by Liu *et al.* (2003), anion flow through the CFTR pore is determined by (1) anion permeability, the ease with which anions enter the CFTR pore and (2) anion binding, the tightness of the interaction between anions and the CFTR pore. Studies of polyatomic anions of known dimensions (e.g., Linsdell *et al.*, 1997b) demonstrate that the anion permeability sequence of CFTR follows a lyotropic sequence. This suggests that anion permeation is determined by the hydration energy of anions with large, weakly hydrated anions being most permeant (Linsdell *et al.*, 1997b). Similarly, anion binding exhibits a lyotropic sequence with large anions binding tighter to the CFTR pore. This tight binding of large anions within the CFTR pore (e.g., $Au(CN)_2^-$; Gong *et al.*, 2002) explains why these anions block Cl^- permeation avidly. Models with a series of binding sites (or wells) separated by energy barriers (or peaks) have been developed to describe the anion permeation properties of the CFTR Cl^- channel (Linsdell *et al.*, 1997a). While highly speculative, these models simulate accurately many of the characteristics of the CFTR pore.

B. Regulation of CFTR Channel Gating by Intracellular ATP: The NBDs

In ABC transporters, the NBDs are the site of ATP hydrolysis. In most ABC transporters, the energy released during the hydrolysis of ATP is used to actively transport substrates across the cell membrane (Holland *et al.*, 2003). However, for the reasons described in Section II.A, it was unclear why CFTR should have two domains that might hydrolyze ATP. Nevertheless, studies of channel regulation by ATP indicated that nonhydrolyzable ATP analogues and Mg^{2+}-free ATP were unable to support channel activity (for review see Gadsby and Nairn, 1999; Sheppard and Welsh, 1999; Gadsby *et al.*, 2006). The data argued that ATP hydrolysis is a prerequisite for CFTR function.

Crystal structures of several NBDs (e.g., NBD1 of CFTR; Lewis *et al.*, 2004, 2005) and a few complete ABC transporters (e.g., BtuCD, the vitamin B_{12} transporter of *Escherichia coli*; Locher *et al.*, 2002) have been elucidated. These data suggest strongly that the two NBDs of CFTR function as a head-to-tail dimer with the ATP-binding sites located at the interface of the two subunits (Fig. 1B). The data suggest that one ATP-binding site is formed by the Walker A and B motifs of NBD1 and the LSGGQ motif of NBD2 (termed site 1), while the other is formed by the Walker A and B motifs of NBD2 and the LSGGQ motif of NBD1 (termed site 2; Lewis *et al.*, 2004). Consistent with this model of the NBDs, functional studies demonstrated that the association of NBD1 and NBD2 is required for optimal ATPase activity and channel gating by CFTR (Kidd *et al.*, 2004; Vergani *et al.*, 2005).

Photolabeling studies argue that the two ATP-binding sites of CFTR are not equivalent in function: site 1 stably binds nucleotides, whereas site 2 rapidly hydrolyzes them (Aleksandrov *et al.*, 2002b). Because CFTR Cl^- channels transit between the

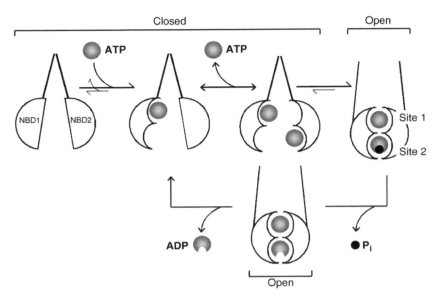

Figure 2. ATP-dependent NBD dimerization drives channel opening. Simplified scheme linking the interaction of ATP with ATP-binding sites located at the interface of the two NBDs with NBD dimerization and channel opening. See text for explanation. Modified from The Journal of General Physiology 2003, 120: 17–36. Copyright 2003 The Rockefeller University Press.

closed and open configurations in seconds, Vergani et al. (2003) interpreted the photolabeling data to suggest that CFTR channel gating is controlled by ATP binding and hydrolysis at site 2 driving cycles of NBD dimer assembly and disassembly. In the model developed by the authors (Fig. 2), MgATP binding to both ATP-binding sites drives conformational changes in the NBDs leading to the formation of an NBD dimer. During this period, the channel remains closed. Channel opening requires the formation of a prehydrolysis complex at site 2. Hydrolysis of this ATP molecule at site 2 drives a conformational change leading to closure of the channel and separation of the two NBDs. However, provided that ATP remains stably bound at site 1, cycles of NBD association and disassociation and hence, channel opening and closing are driven by ATP binding and hydrolysis at site 2 (Vergani et al., 2003; Fig. 2). For a detailed analysis of the ATP-driven NBD dimerization model of CFTR channel gating see Chapter 6.

C. Phosphorylation-Dependent Regulation of CFTR: The R Domain

The R domain contains multiple consensus phosphorylation sites for cAMP-dependent protein kinase (PKA) and protein kinase C (PKC) and many charged amino acids. Unlike the other domains from which CFTR is assembled, there is little sequence similarity across species with the exception of the phosphorylation sites and two regions of acidic amino acids (amino acids 725–733 and amino acids 817–838). Consistent with the lack of sequence conservation, Ostedgaard *et al.* (2000) demonstrated that a recombinant R domain (amino acids 708–831) was unstructured with the vast majority of the protein being random coil. Of note, phosphorylation with PKA failed to alter R domain structure (Ostedgaard *et al.*, 2000).

PKA is the most important kinase responsible for regulating CFTR. However, other kinases phosphorylate the R domain and stimulate the channel including PKC, the type II isotype of cyclic guanosine monophosphate (cGMP)-dependent protein kinase and tyrosine kinases (Gadsby and Nairn, 1999; Sheppard and Welsh, 1999). Dephosphorylation of the R domain is cell-type specific. In cardiac myocytes and sweat duct epithelia, protein phosphatase 2A (PP2A) plays an important role (Hwang *et al.*, 1993; Reddy and Quinton, 1996). However, in airway and intestinal epithelia, PP2C dephosphorylates CFTR and deactivates the channel (Travis *et al.*, 1997).

The most important phosphorylation sites for CFTR regulation by PKA are S660, S700, S737, S795, and S813. Cyclic AMP agonists phosphorylate these sites *in vivo* (e.g., Cheng *et al.*, 1991). Moreover, the largest decrement in CFTR activity occurred with the simultaneous mutation of S660, S737, S795, and S813 (e.g., Rich *et al.*, 1993). The data argue that PKA stimulation of CFTR is redundant with no individual phosphoserine residue being essential. They also suggest that there are functional interactions between different phosphoserines. However, it remains unclear how individual phosphoserines control channel activity. For example, while most phosphoserines are stimulatory, S737 and S768 (e.g., Wilkinson *et al.*, 1997) are inhibitory. Perhaps individual phosphoserines have differential sensitivity to protein kinases, phosphatases, and other regulatory molecules.

The R domain was originally proposed to regulate CFTR by keeping the channel closed at rest. In this model, the unphosphorylated R domain acts as an inhibitor maintaining the CFTR pore in a closed state; channel inhibition is relieved by either phosphorylation or deletion of the R domain (Sheppard and Welsh, 1999). However, the activity of phosphorylated CFTR greatly exceeded variants, which did not require phosphorylation to open in the presence of ATP (e.g., Rich *et al.*, 1993, but see Csanády *et al.*, 2000). These data suggest that the phosphorylated R domain might stimulate the activity of CFTR. Consistent with this idea, phosphorylation augments ATP binding and hydrolysis by the NBDs to accelerate the rate of channel opening (Li *et al.*, 1996; Winter and Welsh, 1997). Thus, the R domain does not function solely as an "on–off" switch, but rather as a tethered enzyme stimulating the interaction of ATP with the NBDs. Based on the model proposed by Ostedgaard *et al.* (2000), multiple phosphoserines situated on the outside of an unstructured R domain would interact with different sites on the NBDs and MSDs to stimulate channel function. In support of this model, using spectroscopic methods Grimard *et al.* (2004) observed gross changes in the structure of purified reconstituted CFTR protein following PKA-dependent phosphorylation.

The data reviewed in this section highlight the complexity of the specific domains from which CFTR is assembled. While current knowledge remains far from complete, our understanding is growing such that we can begin to understand how the function of individual domains determines the function of CFTR. In the following two sections, we consider the role of CFTR in cells and tissues. First, we consider how epithelial cells regulate CFTR function. Then, we review CFTR-driven fluid and electrolyte transport in different epithelial tissues.

III. THE CELLULAR PHYSIOLOGY OF CFTR

The transport of fluids and electrolytes across epithelial tissues involves the coordinated activation of ion channels and transporters located in the apical and basolateral membranes of individual epithelial cells. Tight spatial and temporal control of transepithelial ion transport is achieved by a network of intracellular signaling pathways that respond to a variety of extracellular cues. In part, the specificity and fidelity of intracellular signals is achieved by the association of transport proteins with protein kinases and phosphatases to form macromolecular signaling complexes.

The assembly of macromolecular signaling complexes is promoted by PDZ domain proteins that mediate protein–protein interactions (for review see Guggino and Stanton, 2006). At the C-terminus of CFTR, there is a PDZ-binding domain that mediates the interaction of CFTR with PDZ domain proteins such as the Na^+/H^+ exchanger regulatory factor isoform-1 [NHERF1, also termed ezrin-binding protein 50 kDa (EBP50); Guggino and Stanton, 2006]. Short *et al.* (1998) demonstrated that NHERF1 is localized to the apical membrane of human airway epithelia and interacts with CFTR in *in vitro* binding assays. Based on these and other data, Short *et al.* (1998) proposed a model of the CFTR macromolecular signaling complex. In this model (see Fig. 5 of Short *et al.*, 1998), NHERF1 binds the C-terminus of CFTR through a PDZ-binding domain at its N-terminus (termed PDZ1) and interacts with ezrin via an ezrin/radixin/moesin (ERM)-binding domain at its C-terminus. Because ezrin is an actin-binding protein, protein–protein interactions facilitated by NHERF1 tether CFTR to the actin cytoskeleton in the apical membrane of polarized epithelial cells (Short *et al.*, 1998). Of note, ezrin also acts as an anchoring protein for PKA, raising the possibility that the R domain of CFTR might be positioned in close proximity to PKA as a consequence of the interaction of CFTR with NHERF1 (Short *et al.*, 1998). Other studies have demonstrated that PP2A and PP2C that dephosphorylate the R domain and terminate channel activity interact directly with CFTR (Zhu *et al.*, 1999; Vastiau *et al.*, 2005). Taken together, these data suggest that the phosphorylation status of CFTR is tightly controlled by a macromolecular signaling complex that involves CFTR, scaffolding proteins, PKA, PP2C, and/or PP2A.

Physiologically, CFTR-driven transepithelial ion transport is regulated by multiple signaling pathways including β_2-adrenergic and A_{2B} adenosine receptors. Using biochemical and functional approaches, Naren *et al.* (2003) demonstrated that protein–protein interactions mediated by NHERF1 govern the regulation of CFTR by β_2-adrenergic receptors in airway epithelia. The data suggest that NHERF1 binds CFTR at PDZ1 and the β_2-adrenergic receptor at a second PDZ domain (termed PDZ2) (Naren et al., 2003). Thus, activation of the β_2-adrenergic receptor causes the G_s pathway to stimulate the adenylate cyclase leading to a local rise in the cAMP concentration and hence, the phosphorylation of the CFTR Cl$^-$ channel by PKA, positioned close to CFTR by ezrin (Naren et al., 2003). Huang et al. (2001) demonstrated that A_{2B} adenosine receptors are also coupled to CFTR through NHERF1-mediated protein–protein interactions. Of note, the data of Huang et al. (2001) suggest a compartmentalized autocrine-signaling mechanism linking physical stimuli (e.g., cilia beating) to CFTR-mediated Cl$^-$ secretion. In this signaling mechanism, physical stimuli cause the local release of ATP from airway epithelial cells which is metabolized to adenosine by ectonucleotidases and sensed by A_{2B} adenosine receptors (Huang et al., 2001).

Transepithelial ion transport imposes significant metabolic demands on epithelial cells. When Anderson et al. (1991a) demonstrated that nucleoside triphosphates regulate the CFTR Cl$^-$ channel, there was speculation that the ATP dependence of CFTR might provide a way of matching CFTR-mediated Cl$^-$ secretion to the availability of cellular ATP. However, subsequent studies suggest that the metabolic-sensing kinase AMP-activated protein kinase (AMPK) instead fulfills this important role. AMPK responds to changes in the [AMP]/[ATP] ratio during cell stress by inhibiting the activity of enzymes in crucial metabolic pathways. As a result, AMPK acts to conserve vital energy stores within cells. Using a yeast two-hybrid screen to identify proteins that interact with the C-terminus of CFTR, Hallows et al. (2000) identified the α_1 (catalytic) subunit of AMPK as a CFTR-interacting protein. Hallows et al. (2000) demonstrated that AMPK and CFTR colocalize at the apical membrane of epithelia. Moreover, they showed that phosphorylation of CFTR by AMPK inhibits CFTR Cl$^-$ currents by altering channel gating (Hallows et al., 2000, 2003). Thus, CFTR-interacting proteins play a crucial role in controlling the activity of the CFTR Cl$^-$ channel during transepithelial ion transport.

IV. THE ROLE OF CFTR IN EPITHELIAL PHYSIOLOGY

CFTR is principally expressed in epithelial tissues where it plays a crucial role in regulation of the quantity and composition of epithelial secretions. However, an examination of the physiology of the individual epithelia where CFTR is expressed reveals tissue-specific differences in CFTR function. A variety of factors contribute to this diversity including tissue architecture, interacting proteins, and the function of CFTR both as a Cl$^-$ channel and as a regulator of ion channels and transporters (Welsh et al., 2001; Guggino and Stanton, 2006). In the following paragraphs, we consider the role of CFTR in epithelial physiology. For further information see Welsh et al. (2001). Beyond the scope of this chapter is the function of CFTR in nonepithelial tissues such as cardiac myocytes (Hume et al., 2000).

A. The Pancreas, Intestine, Hepatobiliary System, and Reproductive Tissues

Ducts and tubes in these organs transport a variety of protein-rich cargoes. The efficient movement of these cargoes is facilitated by the lubrication of ducts and tubes by cAMP-stimulated fluid and electrolyte secretions from duct-lining epithelial cells. The pathophysiology of CF dramatically highlights the central role that the CFTR Cl^- channel plays in this secretion of fluids and electrolytes. The failure to lubricate ducts and tubes in CF causes the stasis of protein-rich cargoes leading to obstruction and ultimately the atrophy of ducts and tubes. Thus, the rate of movement of protein-rich cargoes along ducts and tubes is intimately related to tissue damage in CF (Forstner et al., 1987).

The cellular mechanism of cAMP-stimulated fluid and electrolyte secretion by epithelia involves the accumulation of Cl^- ions intracellularly at a concentration above the electrochemical equilibrium. This accumulation of Cl^- ions is achieved by a series of basolateral membrane transporters: the $Na^+/K^+/2Cl^-$-cotransporter; the Na^+/K^+-ATPase; and a K^+ channel. The extrusion of Na^+ ions from epithelial cells by the Na^+/K^+-ATPase provides the energy for transepithelial Cl^- secretion by establishing a steep electrochemical gradient for Na^+ entry into the cell. The $Na^+/K^+/2Cl^-$-cotransporter exploits this electrochemical gradient for Na^+ ions to power the entry of Cl^- ions. K^+ ions that enter the cell via the $Na^+/K^+/2Cl^-$-cotransporter and the Na^+/K^+-ATPase are recycled by a K^+ channel. K^+ exit across the basolateral membrane plays the additional important role of hyperpolarizing the basolateral membrane to maintain the driving force for Cl^- secretion. On activation by PKA-dependent phosphorylation, CFTR provides a pathway for Cl^- ions to exit passively from the cell across the apical membrane moving down a favorable electrochemical gradient. Apical Cl^- exit is followed by paracellular Na^+ flow and the secretion of water ensues.

Some epithelia, most notably the pancreas, secrete copious amounts of HCO_3^--rich fluids. This is achieved by the coordinated activity of CFTR Cl^- channels and Cl^-/HCO_3^- exchangers located in the apical membrane of duct-lining epithelial cells (Novak and Greger, 1988). Recent studies have demonstrated that members of the SLC26 family of anion transporters mediate Cl^-/HCO_3^- exchange in pancreatic duct epithelia (e.g., Ko et al., 2004). They also suggest that PDZ domain proteins mediate the close association of CFTR and SLC26 transporters at the apical membrane (Ko et al., 2004). Of special note, Ko et al. (2004) demonstrated that CFTR and SLC26 transporters coordinate their activities through the interaction of CFTR's phosphorylated R domain with the sulphate transporter and antisigma antagonist (STAS) domain of SLC26 transporters. These data provide a molecular mechanism for HCO_3^- secretion by epithelial tissues.

B. The Sweat Gland

To generate normal hypotonic sweat, the sweat gland contains two distinct regions: a secretory coil that generates an isotonic NaCl secretion and a water-impermeable duct that absorbs NaCl to leave a hypotonic fluid in the lumen of the duct which emerges onto the surface of the skin (see Fig. 201-21 of Welsh *et al.*, 2001). In the secretory coil, Ca^{2+}-activated Cl^- channels stimulated by cholinergic agonists predominantly mediate the secretion of Cl^- into the lumen of the coil with Na^+ ions and water following passively through the paracellular pathways between cells (Welsh *et al.*, 2001). In the sweat duct, the electrochemical gradient for Na^+ movements established by the Na^+/K^+-ATPase located in the basolateral membrane drives the influx of Na^+ ions across the apical membrane through the epithelial Na^+ channel (ENaC). This influx of Na^+ is removed from the cell across its basolateral membrane via the Na^+/K^+-ATPase with K^+ channels located in the basolateral membrane recycling K^+ ions. In contrast to other epithelial tissues, CFTR is located in both the apical and basolateral membranes of sweat duct epithelia where it provides pathways for the passive transport of Cl^- ions across the epithelium following the active transport of Na^+ ions (Reddy and Quinton, 1992). In CF, loss of CFTR function prevents NaCl reabsorption from the sweat duct leading to salty sweat, a hallmark used clinically to diagnose the disease (Welsh *et al.*, 2001).

C. The Respiratory Airways

The respiratory airways are lined by a pseudostratified columnar epithelium with a variety of distinct cell types including ciliated and goblet cells that play crucial roles in mucociliary clearance and host defense. Overlying the epithelium is a thin layer of airway surface liquid (ASL) composed of a watery sol that washes over the protruding cilia overlaid by a mucus gel that traps debris in the inhaled air. In normal mucociliary clearance, beating of the cilia in the watery sol propels mucus up the airways, removing debris from the lungs. Defective epithelial ion transport in CF prevents normal mucociliary clearance, leading to the accumulation of thick sticky mucus and airway obstruction. Distinct from other organs affected by CF, there is also a localized failure of the host defense system in the respiratory airways with the result that CF airway disease is characterized by persistent bacterial infections (Welsh *et al.*, 2001).

Like sweat duct epithelia, airway epithelia possess both CFTR and ENaC channels in their apical membrane. They are also endowed with a similar repertoire of basolateral membrane transporters to those found in secretory epithelia (e.g., the intestine). As a result, airway epithelia are capable of both absorbing or secreting fluids and electrolytes depending on the prevailing electrochemical gradients and neurohormonal signals (Welsh *et al.*, 2001). CF airway epithelia are characterized by two defects in transepithelial ion transport (Wine, 1999). First, the Cl^- permeability of the apical membrane is greatly reduced and unresponsive to cAMP agonists. Second, the Na^+ permeability of the apical membrane is enhanced dramatically. Two different hypotheses have been proposed to explain the pathogenesis of CF lung disease. The high salt hypothesis is based on the role of CFTR as a regulated Cl^- channel (Smith *et al.*, 1996), whereas the low-volume hypothesis emphasizes the function of CFTR as a regulator of ENaC (Matsui *et al.*, 1998; Guggino and Stanton, 2006). In the high salt hypothesis, the NaCl concentration of ASL remains elevated as a consequence of the failure of CFTR-driven salt and water reabsorption. The high salt content of ASL inactivates antimicrobial substances causing a localized failure of the host defense system in the respiratory airways (Smith *et al.*, 1996). By contrast, in the low-volume hypothesis, CFTR dysfunction disinhibits ENaC leading to the unrestrained absorption of fluids and electrolytes. This accelerated absorption depletes ASL volume and dehydrates mucus causing the failure of mucociliary clearance (Matsui *et al.*, 1998). Strikingly, the high salt and low-volume hypotheses suggest opposing treatment regimes for CF lung disease.

The proximal airways also contain submucosal glands with serous, mucous, and cuboidal epithelial cells. Engelhardt *et al.* (1992) demonstrated that serous cells of the submucosal glands are the predominant site of CFTR expression in the respiratory airways. CFTR-driven fluid and electrolyte secretion by these cells flushes mucins, glycoconjugates, and antimicrobial substances released from the mucus and serous cells out of the submucosal glands and into the lumen of the airway where they play important roles in mucociliary clearance and host defense (Welsh *et al.*, 2001). Malfunction of CFTR in CF causes the blockage of submucosal gland ducts, preventing the delivery of substances to the airway lumen. CFTR dysfunction might also reduce the volume and/or alter the composition of secreted fluid with the result that the viscoelastic properties of mucus and the composition of ASL are perturbed (Welsh *et al.*, 2001). To summarize, CFTR plays a number of crucial roles in the respiratory airways, its dysfunction causes the complex problem of CF lung disease.

V. THE PATHOPHYSIOLOGY OF CFTR

Malfunction of the CFTR Cl^- channel is associated with a wide spectrum of human disease. The genetic disease CF is caused by mutations that abolish the function of CFTR (Welsh *et al.*, 2001). CF affects approximately 1 in 2500 live births and is the most common fatal autosomal recessive disease to affect Caucasian populations (Welsh *et al.*, 2001). Other diseases such as autosomal dominant polycystic kidney disease (ADPKD) and secretory diarrhea involve unphysiologic activation of the CFTR Cl^- channel (Gabriel *et al.*, 1994; Sullivan *et al.*, 1998). ADPKD affects more than 1 in 1000 live births and is the most common single gene disorder associated with the loss of kidney function (Wilson, 2004). Secretory diarrhea annually kills millions of infants in Africa, Asia, and Latin America (Guerrant *et al.*, 1990). In the following paragraphs, we discuss briefly the molecular basis of CFTR dysfunction in CF and ADPKD; see Thiagarajah and Verkman (2003) for review of secretory diarrhea. For further discussion see Chapter 12.

A. Cystic Fibrosis

To date, over 1400 disease-causing mutations have been identified in the CFTR gene (see http://www.genet.sickkids.on.ca/cftr/). They are located throughout the entire coding sequence and include deletions, missense, and nonsense mutations. Most mutations are very rare. The exception is the deletion of a phenylalanine residue at position 508 of the CFTR sequence (termed F508del). This mutation, located in NBD1, accounts for about 70% of CF mutations worldwide and is associated with a severe disease phenotype (Welsh et al., 2001). In general, there are two mechanisms by which CF mutations cause a loss of CFTR function. First, the mutant protein might not be delivered to the apical membrane of epithelial cells. Second, the mutant protein might be present at the apical membrane, but its functional properties might be altered. The F508del mutation exemplifies both mechanisms.

The molecular basis for the F508del defect is retention of the mutant protein in the endoplasmic reticulum (ER) because the "quality control" machinery of the cell recognizes F508del as abnormal (for review see Hanrahan et al., 2003; Amaral, 2005). It is neither processed through the Golgi apparatus, where complex sugars are added, nor delivered to the plasma membrane. Crystal structures of NBD1 of human CFTR reveal that F508 is located on the surface of the domain in a position where it might interact with other parts of the protein or CFTR-interacting proteins (Lewis et al., 2005). They also demonstrate that the loss of F508 is without gross effect on the folding of NBD1, but perturbs the local surface topography of NBD1 in the region of F508 (Lewis et al., 2005). Based on these data, Lewis et al. (2005) speculated that the principal effect of F508del is to disrupt interdomain interactions within CFTR. Consistent with this idea, Du et al. (2006) demonstrated that F508del impairs interactions between NBD1 and NBD2 and arrests the posttranslational folding of CFTR. Misfolded F508del-CFTR is retained in the ER by chaperone proteins and degraded by the ubiquitin-proteasome system (Amaral, 2005).

Despite the defect in protein folding, some F508del-CFTR reaches the cell membrane where it forms Cl^- channels with many properties in common with those of wild-type CFTR (e.g., Dalemans et al., 1991; Denning et al., 1992). However, there is one notable exception, the pattern of channel gating. The gating behavior of wild-type CFTR is characterized by frequent bursts of channel activity that are interrupted by brief flickery closures and separated by longer closures between bursts (Fig. 3). In contrast, the gating behavior of F508del is characterized by infrequent bursts of channel openings separated by long closures of prolonged duration (Fig. 3). Surprisingly, the biophysical basis of this defect in channel gating is a reduced sensitivity of F508del-CFTR to PKA-dependent channel activation (Wang et al., 2000). Thus, the F508del mutation perturbs severely both the biosynthesis and function of CFTR leading to the loss of CFTR-mediated Cl^- transport in affected epithelial tissues and the blockage of ducts and tubes by thick sticky mucus (see Section IV).

B. Autosomal Dominant Polycystic Kidney Disease

ADPKD is caused by mutations in the polycystin proteins (polycystin-1 and polycystin-2) that lead to the formation of multiple fluid-filled epithelial cysts within the kidney, which grow insidiously destroying kidney function leading to renal failure

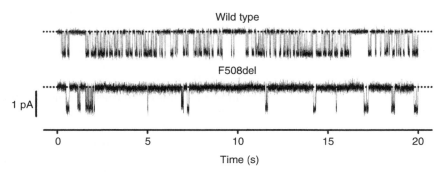

Figure 3. F508del disrupts CFTR channel gating. Representative recordings of wild-type and F508del-CFTR Cl⁻ channels in excised inside-out membrane patches from C127 cells expressing recombinant human CFTR. ATP (1 mM) and PKA (75 nM) were continuously present in the intracellular solution, voltage was −50 mV and there was a large Cl⁻ concentration gradient across the membrane patch (internal [Cl⁻] = 147 mM; external [Cl⁻] = 10 mM). Dotted lines indicate where channels are closed and downward deflections of the traces correspond to channel openings. Modified from Cai and Sheppard (2002) and Cai et al. (2006) with permission of The American Society for Biochemistry and Molecular Biology.

(Wilson, 2004). Polycystin-1 is a unique G-protein–coupled receptor that activates multiple intracellular signaling pathways (Delmas et al., 2002; Lakkis and Zhou, 2003), whereas polycystin-2 is a member of the transient receptor potential (TRP) family of Ca^{2+} channels that control the intracellular Ca^{2+} concentration (González-Perrett et al., 2001; Koulen et al., 2002). Polycystins are expressed in a variety of locations within renal epithelial cells including the primary cilium that protrudes into the nephron from the apical membrane and acts as a flow sensor (Nauli et al., 2003). The data suggest that polycystins might signal changes in luminal flow rates to control cell proliferation and differentiation in the developing kidney (Boletta et al., 2000; Nauli et al., 2003). This suggests that polycystin dysfunction in ADPKD leads to loss of important morphological cues with the result that tubules become cysts.

The location of polycystins on the primary cilium raises the possibility that polycystins might also regulate fluid movement across renal epithelia. Fluid movement across epithelia is driven by transepithelial ion flow with the apical membrane channels ENaC and CFTR controlling fluid absorption and secretion, respectively (Welsh et al., 2001). This suggests that fluid accumulation within ADPKD cysts might be achieved by (1) inhibition of ENaC-mediated Na^+ absorption, (2) stimulation of CFTR-driven Cl⁻ secretion, or (3) a combination of both mechanisms. The role of ENaC in ADPKD is uncertain. In contrast, immunocytochemical (Brill et al., 1996; Hanaoka et al., 1996), functional (Hanaoka et al., 1996), and pharmacological (Hanaoka and Guggino, 2000; Li et al., 2004) evidence argue that the CFTR Cl⁻ channel plays a key role in fluid accumulation within ADPKD cysts. Thus, the CFTR Cl⁻ channel contributes significantly to the pathogenesis of ADPKD.

VI. THE PHARMACOLOGY OF CFTR

CFTR Cl⁻ channels have a distinct pharmacological profile. CFTR Cl⁻ currents are enhanced by flavonoids (e.g., genistein; Wang *et al.*, 1998), substituted benzimidazolones (e.g., NS004; Gribkoff *et al.*, 1994), and xanthines (e.g., IBMX; Chappe *et al.*, 1998). By contrast, CFTR Cl⁻ currents are inhibited by sulphonylureas (e.g., glibenclamide; Sheppard and Welsh, 1992), arylaminobenzoates (e.g., NPPB; Zhang *et al.*, 2000), and the thiazolidinone $CFTR_{inh}$-172 (Ma *et al.*, 2002a). Importantly, disulfonic stilbenes (e.g., DIDS; Tabcharani *et al.*, 1990) and calixarenes (e.g., calix[4] arene; Singh *et al.*, 1996) are without effect when added to the extracellular side of the membrane. The failure of extracellular disulfonic stilbenes and calixarenes to inhibit the CFTR Cl⁻ channel distinguishes CFTR from other Cl⁻ channels in epithelia that are blocked by these agents (Singh *et al.*, 1995; Droogmans *et al.*, 1998; Schultz *et al.*, 1999).

The pharmacology of the CFTR Cl⁻ channel has attracted very significant interest in recent years. Much of this attention has been fueled by the search for rational new therapies for diseases caused by CFTR malfunction. Agents that enhance the activity of mutant Cl⁻ channels, present at the apical membrane of epithelia, might be of value in the treatment of CF and related diseases. By contrast, inhibitors of the CFTR Cl⁻ channel might be used in the treatment of secretory diarrhea and ADPKD and perhaps even for male contraception. They are also important in several other respects: (1) as probes to identify CFTR-dependent function and to investigate CFTR structure and function, (2) to study the pathogenesis of lung disease in CF, and (3) to develop animal models with which to evaluate new therapies for CF. In the search for modulators of CFTR, several pharmacological strategies to manipulate the activity of the CFTR Cl⁻ channel have been identified (Hwang and Sheppard, 1999; Schultz *et al.*, 1999). Some agents modulate the activity of the protein kinases and phosphatases that regulate CFTR, whereas others interact directly with CFTR to control its function.

In Section VI, we discuss the mechanism of action of agents that interact directly with CFTR to modulate channel activity and their therapeutic potential. Beyond the scope of this chapter is information about agents that (1) restore the cell surface expression of CF mutants (CFTR correctors), (2) modulate the activity of protein kinases and phosphatases that control CFTR activity, and (3) affect CFTR-mediated transport indirectly by regulating basolateral membrane channels and transporters. For further information about the pharmacology of the CFTR Cl⁻ channel see Schultz *et al.* (1999), Hwang and Sheppard (1999), and Cai *et al.* (2003a); for reviews of new therapies for CF see Rubenstein (2006) and Becq (2006).

A. Potentiators of the CFTR Cl⁻ Channel

CFTR potentiators are small molecules that interact directly with CFTR to augment channel activity, rescuing the defective gating of mutant Cl⁻ channels. Importantly, these agents do not open quiescent channels. Instead, they promote the gating and hence, increase the activity of CFTR Cl⁻ channels that have been phosphorylated by PKA. Some CFTR potentiators increase the frequency of channel opening (e.g., dichlorofluorescein; Cai and Sheppard, unpublished observation), other CFTR potentiators prolong the duration of channel openings (e.g., phloxine B; Cai and Sheppard, 2002), and yet others increase both the frequency and duration of channel openings (e.g., genistein; Wang et al., 1998). The effects of different CFTR potentiators on CFTR channel gating have been summarized in previous reviews (Hwang and Sheppard, 1999). In the following paragraphs, we use the ATP-driven NBD dimerization model of CFTR channel gating (Vergani et al., 2003) to discuss how different CFTR potentiators enhance the activity of wild-type and mutant CFTR Cl⁻ channels.

1. Genistein

Ai et al. (2004) first proposed that genistein and other CFTR potentiators might enhance CFTR channel gating by affecting NBD dimerization. The authors speculated first that the binding of genistein at the interface of the NBD dimer might lower the free energy of the transition state and hence, accelerate channel opening (Ai et al., 2004). Second, that genistein might slow the rate of channel closure by stabilizing the NBD dimer conformation (Ai et al., 2004). Third, that the binding site for genistein might be located at the dimer interface (Ai et al., 2004). Consistent with this latter idea, Moran et al. (2005) used a molecular model of the NBD dimer to localize at the dimer interface the binding site of genistein, apigenin, and a series of novel CFTR potentiators identified by high-throughput screening (HTS). As predicted by functional data (Wang et al., 1998), this drug-binding site is distinct from the two ATP-binding sites (sites 1 and 2). Moreover, sequences from both NBD1 (Walker A and B and LSGGQ) and NBD2 (LSGGQ) contribute to the drug-binding site, with those from NBD1 forming a cavity with which CFTR potentiators dock (Moran et al., 2005).

2. Phloxine B

Cai and Sheppard (2002) demonstrated that saturating (micromolar) concentrations of phloxine B potentiate wild-type CFTR by slowing the rate of channel closure without altering the opening rate. Using the model of Vergani et al. (2003), Cai et al. (2006) suggested that the interaction of phloxine B with the NBDs might strengthen the binding energy for stable dimer formation and hence, prolong the interaction of ATP with site 2. Because phloxine B did not potentiate the CFTR Cl⁻ channel in the absence of ATP (Cai and Sheppard, 2002), Cai et al. (2006) argued that phloxine B is unlikely to interact with site 1. Instead, Cai et al. (2006) proposed that the phloxine B-binding site might be located at the NBD dimer interface based on the data of Moran et al. (2005). The interaction of phloxine B with this site might, via a steric effect on the conformation of the NBDs, enhance the affinity of ATP binding and hence, the stability of the NBD dimer. This mechanism of action of phloxine B is similar to that proposed by Ai *et al.* (2004) to explain how genistein potentiates the CFTR Cl$^-$ channel (see Section VI.A.1).

3. Pyrophosphate (PP$_i$)

Gunderson and Kopito (1994) and Carson *et al.* (1995) demonstrated that the inorganic phosphate analogue PP$_i$ potentiates robustly wild-type CFTR Cl$^-$ channels by accelerating the rate of channel opening and slowing dramatically the rate of channel closure. Using the model of Vergani *et al.* (2003), Cai *et al.* (2006) suggested that PP$_i$ interacts with the NBD dimer at site 2 to disrupt the hydrolysis of ATP that determines the duration of channel openings. Consistent with this idea, G551D-CFTR markedly attenuated PP$_i$ potentiation of CFTR (Cai *et al.*, 2006). Moreover, G1349D-CFTR abolished the potentiation of CFTR Cl$^-$ currents by PP$_i$ (Cai *et al.*, 2006), suggesting that PP$_i$ might also bind to site 1 and accelerate channel opening by providing binding energy to drive NBD dimerization. However, because PP$_i$ cannot substitute for ATP in supporting channel gating (Carson *et al.*, 1995) and G1349D-CFTR has global effects on NBD dimer function (Moran *et al.*, 2005), Cai *et al.* (2006) suggested that the interaction of PP$_i$ with site 2 might energize NBD dimerization.

4. 2'-Deoxy-ATP

Like PP$_i$ (Carson *et al.*, 1995), 2'-deoxy-ATP (2'-dATP) augments wild-type CFTR channel gating by accelerating the rate of channel opening and slowing the closing rate (Aleksandrov *et al.*, 2002a; Cai *et al.*, 2006). Because 2'-dATP alone can open the CFTR Cl$^-$ channel and because it can substitute for ATP in supporting channel gating (Aleksandrov *et al.*, 2002a; Cai *et al.*, 2006), Cai *et al.* (2006) proposed that 2'-dATP can interact with both sites 1 and 2. Moreover, to explain the faster rate of channel opening in the presence of 2'-dATP (Aleksandrov *et al.*, 2002a; Cai *et al.*, 2006), Cai *et al.* (2006) suggested that tight binding of 2'-dATP to sites 1 and 2 drives NBD dimerization. Similarly, a slower rate of 2'-dATP hydrolysis compared to that of ATP might account for the prolonged channel openings observed in the presence of 2'-dATP (Aleksandrov *et al.*, 2002a; Cai *et al.*, 2006). Thus, different CFTR potentiators enhance CFTR channel gating by distinct mechanisms.

B. Identification of CFTR Potentiators by HTS

In recent years, the discovery of lead compounds for therapy development has been revolutionized by HTS. HTS exploits a reliable, sensitive, cost-effective assay to screen rapidly libraries of compounds for novel active small molecules. In the search for drugs to treat CF, several groups have developed HTS assays to identify novel

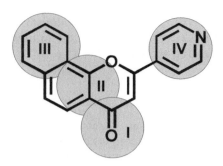

Figure 4. A four-point pharmacophore model of CFTR potentiators. The pharmacophore model developed by Springsteel et al. (2003) is denoted by labeled gray spheres and is overlaid onto the chemical structure of UC$_{CF}$-029. I, hydrogen bond acceptor; II, aromatic region; III, hydrophobic region 1; IV, hydrophobic region 2; UC$_{CF}$-029, 2-(4-pyridinium)benzo[*h*]4H-chromen-4-one bisulfate. Modified from Springsteel et al. (2003) with permission of Elsevier.

CFTR potentiators. For example, Alan Verkman (UCSF, San Francisco) used Fischer rat thyroid (FRT) cells coexpressing recombinant human CFTR and a green fluorescent protein (GFP) with ultra high halide sensitivity (Ma et al., 2002b), whereas Vertex Pharmaceuticals (San Diego) employed NIH-3T3 cells expressing recombinant CFTR and a fluorescence resonance energy transfer (FRET)-based voltage-sensitive assay (Van Goor et al., 2006). Both these automated HTS assays monitor the change in CFTR-mediated anion flux elicited by CFTR modulators in real time.

Verkman first applied his HTS assay to a small combinatorial library of compounds based on flavones and benzo[c]quinoliziniums because agents from these classes of compounds are known CFTR potentiators [e.g., genistein (Wang et al., 1998) and MPB-07 (Dormer et al., 2001)]. This screen revealed that 7,8-benzoflavones (e.g., UC$_{CF}$-029) are effective potentiators of wild-type CFTR (Galietta et al., 2001). Using UC$_{CF}$-029 as a scaffold for structure-activity relationships (SAR), Springsteel et al. (2003) identified UC$_{CF}$-339 (K_d 1.7 μM) as the most potent CFTR potentiator among the 7,8-benzoflavone analogues. By computational analysis of SAR data, Springsteel et al. (2003) developed a pharmacophore model to represent the interaction of CFTR potentiators with the NBD dimer (Fig. 4). Interestingly, this model, which consists of a hydrogen bond acceptor, an aromatic ring, and two hydrophobic regions, is applicable to other CFTR potentiators (Ma et al., 2002b).

Concurrently, Verkman applied his HTS assay to compound collections selected for chemical diversity and drug-like properties. Using FRT cells expressing wild-type CFTR, Ma et al. (2002b) identified 14 strong CFTR potentiators among a library of 60,000 compounds. The most efficacious agents (K_d down to 200 nM) had tetrahydrocarbazole (e.g., CFTR$_{act}$-09; Fig. 5), hydroxycoumarin, and thiazolidine structures. To identify CFTR potentiators that rescue the gating defect of F508del-CFTR, Yang et al. (2003) studied FRT cells expressing temperature-corrected F508del-CFTR. A screen of 100,000 compounds identified six novel classes of high-affinity F508del-CFTR potentiators. However, by screening additional structural analogues, Yang et al. (2003)

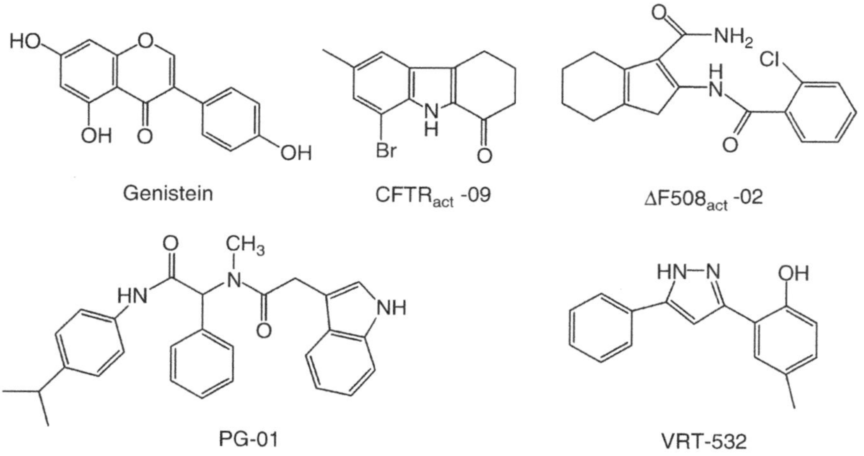

Figure 5. Chemical structures of some CFTR potentiators identified by HTS. CFTR$_{act}$-09, 8-bromo-6-methyl-3,4-dihydro-2H-carbazol-1(9H)-one; ΔF508$_{act}$-02, 2-(2-chlorobenzamido)-4,5,6,7-tetrahydro-3H-indene-1-carboxamide; PG-01, 2-[(2–1H-indol-3-yl-acetyl)methylamino]-N-(4-isopropylphenyl)-2-phenylacetamide; VRT-532, 4-methyl-2-(5-phenyl-1H-pyrazol-3-yl)phenol. For comparison, the chemical structure of genistein is shown.

discovered tetrahydrobenzothiophenes (e.g., ΔF508$_{act}$-02; Fig. 5), which potentiate F508del-CFTR with $K_d < 100$ nM. Most recently, Pedemonte et al. (2005) screened 50,000 compounds to search for further compounds that rescue the gating defect of F508del-CFTR. After secondary analyses, Pedemonte et al. (2005) identified phenylglycines (e.g., PG-01; Fig. 5) and sulphonamides that potentiate both F508del- and G551D-CFTR with nanomolar potencies. Thus, Verkman's HTS strategy has proven astonishingly successful at identifying novel CFTR potentiators. The challenge now for Verkman and his colleagues is to decide carefully which CFTR potentiators to pursue further in their quest to develop new therapies for CF.

To identify potentiators of F508del-CFTR, Vertex Pharmaceuticals screened 122,000 synthetic compounds from their compound collection using NIH-3T3 cells expressing temperature-corrected F508del-CFTR (Van Goor et al., 2006). After careful scrutiny, Van Goor et al. (2006) selected for further study 53 compounds consisting of 10 distinct chemical structures. One compound, the pyrazole VRT-532 (Fig. 5) rescued the gating defect of F508del-CFTR Cl$^-$ channels by enhancing the rate of channel opening and slowing the rate of channel closure (Van Goor et al., 2006). Critically, VRT-532 augmented robustly CFTR-mediated transepithelial Cl$^-$ current in F508del-CFTR human bronchial epithelia (HBE) (EC$_{50}$ 2.7 ± 0.2 μM; Van Goor et al., 2006). Moreover, the effects of VRT-532 on F508del-HBE were synergistic with a CFTR corrector (VRT-325) discovered by Vertex Pharmaceuticals (Van Goor et al., 2006). Together VRT-325 and VRT-532 generated levels of CFTR-mediated transepithelial Cl$^-$ current in F508del-HBE $> 20\%$ of those observed in HBE expressing wild-type CFTR. It is clear that HTS has revolutionized the pharmacology of the CFTR Cl⁻ channel. But, just as HTS has provided a wealth of tools to probe the CFTR Cl⁻ channel, the data have raised many new questions. Answers to these questions will provide important new understanding of the molecular pharmacology of CFTR.

C. Rescue of F508del-CFTR by Pharmacological Chaperones

The most common CF mutation F508del disrupts both the biosynthesis and function of the CFTR Cl⁻ channel (Cheng *et al.*, 1990; Dalemans *et al.*, 1991). Thus, to restore Cl⁻ channel function to F508del-CFTR requires the use of both CFTR correctors to deliver the mutant protein to its correct cellular location and CFTR potentiators to enhance channel gating. Clearly, a single agent with the properties of both types of drugs would be preferable. Excitingly, such drugs, termed pharmacological chaperones, have been identified for a number of misprocessed proteins including CFTR. Dormer *et al.* (2001) demonstrated that the benzo(c)quinolizinium MPB-07, an agent that augments the activity of F508del-CFTR at the cell surface, rescues CFTR biosynthesis in nasal epithelial cells from CF patients homozygous for the F508del mutation. Of note, Dormer *et al.* (2005) demonstrated that the phosphodiesterase type 5 (PDE5) inhibitor sildenafil (Viagra) also acts as a pharmacological chaperone. Because sildenafil is approved for clinical use, Dormer *et al.* (2005) speculate that their data might accelerate the development of new therapies for CF.

D. Therapeutic Potential of CFTR Potentiators

A plethora of agents that rescue the function of CF mutants in recombinant cells have been identified in recent years, most notably using HTS assays. The challenge now is to carefully decide which of the candidate drugs to pursue further in the quest to develop rational new therapies for CF. Two fundamental questions for the development of drug therapies for CF are first: how much CFTR function is required to rescue CF mutants? and second, is drug therapy for CF mutation specific? In the following paragraphs, we consider each question in turn.

1. How Much CFTR Function Is Required to Rescue CF Mutants?

The amount of function that must be restored to individual CF mutants to rescue the defect in transepithelial ion transport in CF epithelia is currently unknown. However, by analyzing published data on the relationship between genotype, phenotype, and CFTR Cl⁻ channel function, Van Goor *et al.* (2006) speculated that restoration of 5–30% of wild-type function to CF patients bearing the F508del mutation would be of therapeutic benefit.

Quantitative analyses of the effects of CFTR potentiators on CF mutants can be employed to evaluate how much function different therapeutic strategies restore to

Table I
Rescue of the Function of CF Mutants by CFTR Potentiators[a]

CFTR construct	CFTR potentiator	N (%)	i (%)	P_o (%)	$N \times i \times P_o$ (%)
Wild type	Control	100	100	100	100
F508del	Control	4	100	30	1.2
	Phloxine B (1 μM)	4	90	60	22
G551D	Control	100	100	2	2
	Phloxine B (5 μM)	100	88	19	17
	PP$_i$ (5 mM)	100	91	6	5
	2′-dATP (1 mM)	100	100	20	20
G1349D	Control	100	100	6	6
	Phloxine B (5 μM)	100	90	10	9
	PP$_i$ (5 mM)	100	95	10	9
	2′-dATP (1 mM)	100	100	23	23

[a] N, the number of CFTR Cl$^-$ channels in the cell membrane; i, single-channel current amplitude; P_o, open probability; $N \times i \times P_o$, predicted CFTR Cl$^-$ current. For all values, wild-type CFTR was assigned a value of 100%. Data for F508del-CFTR are from Denning et al. (1992) and those for G551D- and G1349D-CFTR are from Cai et al. (2006). For further information see Section VI.D.

CF mutants. Apical CFTR Cl$^-$ current [I^{CFTR}(apical)] is determined by the product of the number of CFTR Cl$^-$ channels in the apical membrane (N), the current amplitude (i) of an individual CFTR Cl$^-$ channel, and the probability (P_o) that a single CFTR Cl$^-$ channel is open: I^{CFTR}(apical) $= N \times i \times P_o$. Using biochemical (N) and functional (i and P_o) data, the apical CFTR Cl$^-$ current generated by CF mutants can be predicted. For this purpose, N, i, and P_o are set to 100% for wild-type CFTR. We previously successfully used this approach to provide a molecular explanation for the quantitative reduction in I^{CFTR}(apical) caused by CF mutants (Sheppard et al., 1995). In Table I, we provide some examples of the effects of different CFTR potentiators on the predicted values of $N \times i \times P_o$ for F508del-, G551D-, and G1349D-CFTR Cl$^-$ currents. Using the CFTR potentiator phloxine B (1 μM), Cai and Sheppard (2002) increased the P_o of F508del-CFTR to 60% that of wild-type CFTR. As a result, the predicted current for F508del-CFTR becomes 22% that of wild-type CFTR (Table I). Encouragingly, 2′-dATP (1 mM) increased the predicted current of both G551D- and G1349D-CFTR to approximately 20% that of wild-type CFTR (Table I). Importantly, it is currently unknown what level of CFTR function is required to promote ion transport in the respiratory airways of CF patients *in vivo*. Clearly, answers to the question "how much CFTR function is required to rescue CF mutants" are of fundamental importance for therapy development.

2. Is Drug Therapy for CF Mutation Specific?

With the establishment of a classification system for CF mutants (Welsh and Smith, 1993), it is tempting to speculate that CF mutants, which disrupt CFTR function by similar mechanisms might be treated with the same drug. However, studies

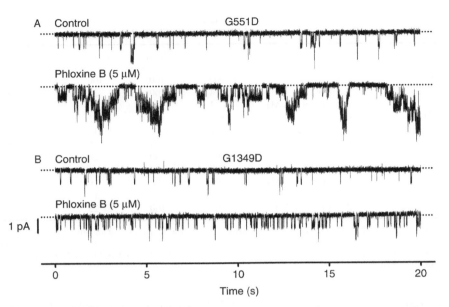

Figure 6. Phloxine B potentiates strongly the single-channel activity of G551D-CFTR. (A, B) Representative recordings show the effects of phloxine B (5 μM) on the activity of G551D- and G1349D-CFTR Cl⁻ channels, respectively. Other details as in Fig. 3.

of the CF mutants G551D and G1349D argue that this might not be the case. These CF mutants profoundly disrupt CFTR channel gating by affecting equivalent residues in the LSGGQ motifs of CFTR's two ATP-binding sites (G551D, site 2 and G1349D, site 1; Lewis et al., 2004; Cai et al., 2006; Fig. 6). Cai et al. (2006) demonstrated that phloxine B enhanced markedly G551D-CFTR channel gating, but was without effect on the G1349D-CFTR Cl⁻ channel (Fig. 6). Using genistein, Becq and his colleagues (Melin et al., 2004) reported similar results. Cai et al. (2006) also found that PP_i was without effect on both G551D- and G1349D-CFTR. Only 2′-dATP and the phenylglycine PG-01 rescued the function of G1349D-CFTR (Pedemonte et al., 2005; Cai et al., 2006). These data demonstrate that G551D- and G1349D-CFTR have distinct pharmacological profiles. They also raise the possibility that a single drug might not rescue the function of all CF mutants. Instead, therapeutic strategies might need to be tailored to specific CF mutants.

E. Inhibitors of the CFTR Cl⁻ Channel

A wide variety of diverse small molecules have been identified, which inhibit the CFTR Cl⁻ channel (Schultz et al., 1999; Cai et al., 2003a). Analysis of these molecules reveals several common themes: they are anions, most are lipophilic, and many are large in size. Two general mechanisms by which pharmacological agents inhibit the

CFTR Cl⁻ channel have been identified: open-channel and allosteric block. In the following paragraphs, we consider each mechanism in turn.

1. Open-Channel Blockers

These agents prevent Cl⁻ flow through the channel by occluding the CFTR pore (Hwang and Sheppard, 1999). The vast majority of open-channel blockers of CFTR identified to date bind within the deep wide vestibule at the intracellular end of the CFTR pore. However, at least one agent, the glycine hydrazide GlyH-101, has been identified, which occludes the extracellular end of the CFTR pore (see Section VI.G; Muanprasat et al., 2004).

Many of the agents that occlude the intracellular end of the CFTR pore are too large to permeate through the CFTR Cl⁻ channel. Thus, when added to the extracellular side of the membrane, these agents are prevented from passing through CFTR's selectivity filter because of their bulky size. As a result, open-channel blockers either fail to inhibit CFTR from the extracellular end of the pore (e.g., DIDS) or they reach their binding site by permeating through the lipid phase of the cell membrane by nonionic diffusion (e.g., glibenclamide). Once open-channel blockers reach the cytoplasm, they are driven into the deep wide intracellular vestibule of the CFTR pore by the electrochemical gradient where they bind, occlude the pore, and prevent Cl⁻ permeation (Fig. 7B).

Inhibition of the CFTR Cl⁻ channel by open-channel blockers is both voltage and Cl⁻ concentration dependent. Cl⁻ flow from the intra- to the extracellular side of the membrane is more strongly attenuated than that in the opposite direction by open-channel blockers. Channel block is relieved both by positive voltages and increasing the external Cl⁻ concentration, two maneuvers that flush open-channel blockers from the intracellular end of the CFTR pore (for discussion see Cai et al., 2004). These effects of voltage and external Cl⁻ concentration necessitate that experimental conditions are carefully controlled to maximize CFTR inhibition by open-channel blockers. In contrast to their effects on allosteric blockers (see Section VI.E.2), elevated concentrations of ATP and PP_i are without effect on open-channel blockers of CFTR.

Information about the location of drug-binding sites within the transmembrane electric field has emerged from analysis of the voltage-dependence of block using the Woodhull relationship (Woodhull, 1973). Many open-channel blockers cause a voltage-dependent block of the CFTR Cl⁻ channel and their binding sites are located approximately 20–60% of the way through the transmembrane electric field from the inside (Hwang and Sheppard, 1999). Although electrical distance does not necessarily indicate physical distance, surprisingly good correlation has been found between electrical distance and the physical distance along the length of M6 in two studies (Tabcharani et al., 1993; McDonough et al., 1994). For example, McDonough et al. (1994) found that diphenylamine-2-carboxylic acid (DPC) binds at a site approximately 40% of the electrical distance through the membrane from the intracellular side; this value is in excellent agreement with the predicted location of S341 to which DPC binds.

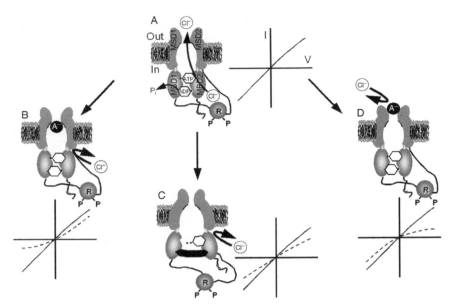

Figure 7. Mechanism of action of CFTR inhibitors. (A–D) Simplified models of the CFTR Cl⁻ channel and the effects of open-channel and allosteric blockers. Schematic representations of the I–V relationships of CFTR Cl⁻ currents in the absence (continuous line) and presence (dotted line) of blockers under conditions of symmetrical Cl⁻-rich solutions are shown beneath each model. (A) Topology of the CFTR pore. (B) Large anions (A⁻; e.g., glibenclamide) inhibit CFTR by occluding the intracellular vestibule. (C) Allosteric blockers (e.g., genistein) inhibit CFTR by interfering with channel gating. Because the NBDs function as a head-to-tail dimer, allosteric blockers might inhibit channel gating by preventing dimer formation. (D) The glycine hydrazide GlyH-101 (A⁻) inhibits CFTR by occluding the extracellular vestibule. See text for further information. Modified from The Journal of General Physiology 2004, 124: 109–113. Copyright 2004 The Rockefeller University Press.

In the case of glibenclamide, the best-studied CFTR inhibitor, channel block is weakened by mutation of two residues in M6 (F337 and T338) that play a key role in determining the anion selectivity of CFTR (Gupta and Linsdell, 2002). However, mutation of these residues did not abrogate the voltage-dependence of CFTR inhibition, suggesting that other residues likely contribute to the glibenclamide-binding site. Consistent with this idea, Linsdell (2005) demonstrated that K95 located toward the extracellular end of M1 acts as a crucial determinant of the glibenclamide-binding site. Site-directed mutagenesis of K95 revealed that positive charge at K95 is essential for channel block by glibenclamide and four other open-channel blockers of CFTR with unrelated chemical structures (Linsdell, 2005). Linsdell (2005) interpreted these data to propose a common mechanism of action of open-channel blockers: interaction with the positively charged side-chain of K95 located deep within the intracellular vestibule of the CFTR pore.

2. Allosteric Blockers

A number of agents that enhance CFTR Cl^- current at nanomolar and low micromolar concentrations inhibit the CFTR Cl^- channel at elevated concentrations. These agents block the CFTR Cl^- channel by interacting with the NBDs that control channel gating and slowing dramatically the rate of channel opening. As a result, in the presence of these agents the duration of long closures separating channel openings is increased greatly. Inhibition of CFTR by allosteric blockers is relieved by elevated concentrations of ATP and PP_i. However, inhibition of CFTR by allosteric blockers is voltage-independent and unaffected by altering the external Cl^- concentration. Thus, open-channel and allosteric blockers of CFTR can be distinguished by the effects of (1) voltage, (2) external Cl^- concentration, (3) ATP, and (4) PP_i (Cai *et al.*, 2004).

The observation that CFTR inhibition by allosteric blockers is relieved by elevated concentrations of ATP and PP_i has important implications for the mechanism of action of these agents. It suggests that allosteric blockers might either compete with ATP for a common binding site on the NBDs or interact with a closely related site. In the ATP-driven NBD dimerization model of CFTR channel gating (Vergani *et al.*, 2003), the interaction of allosteric blockers with the NBDs might impede strongly dimer formation (Fig. 7C). Because ATP binding is required at sites 1 and 2 for dimer formation and because ATP and PP_i relieve inhibition of CFTR by allosteric blockers, we speculate that allosteric blockers might inhibit CFTR by binding at site 2 and impeding NBD dimerization. Consistent with this idea, Dérand *et al.* (2002) demonstrated that the CF mutant G551D, which affects a key residue in site 2, abolishes the allosteric block of CFTR by genistein. We recommend that future studies explore further the role of site 2 in allosteric block of CFTR.

Despite the identification of many CFTR inhibitors using conventional assays of CFTR function, the potency and specificity of these agents has proved unsatisfactory (Schultz *et al.*, 1999; Cai *et al.*, 2003a). Few agents have been identified, which inhibit CFTR with nanomolar affinity. Worse, no specific blockers of the CFTR Cl^- channel have been identified. Open-channel blockers of CFTR invariably block other types of Cl^- channels, whereas allosteric blockers interact with other targets within cells at concentrations similar to those that inhibit CFTR. The lack of such blockers is a major obstacle for CFTR research. However, recent work raises the possibility that new agents to selectively abrogate CFTR activity will soon be available, which are a very significant improvement on those currently available.

F. Inhibition of CFTR by a Peptide Toxin

Nature has blessed investigators of cation channels with an armamentarium of peptide toxins to selectively eliminate different types of cation currents. Disappointingly, similar tools have, for a long time, been unavailable to investigators of anion channels. DeBin *et al.* (1993) purified a 4.1 kDa basic peptide from scorpion venom with sequence similarity to small insectotoxins. Because this peptide inhibited outwardly rectifying Cl⁻ channels reconstituted into planar lipid bilayers, DeBin et al. (1993) named this toxin chlorotoxin. Subsequently, Maertens et al. (2000) demonstrated that submicromolar concentrations of chlorotoxin were without effect on volume-regulated, Ca^{2+}-activated, and CFTR Cl⁻ channels. Based on their data, Maertens et al. (2000) concluded that chlorotoxin is not a general Cl⁻ channel inhibitor.

Undeterred by previous work, McCarty and his colleagues searched for peptide toxins that inhibit the CFTR Cl⁻ channel with the aim of identifying new tools to probe CFTR structure and function. In their initial study, Fuller et al. (2004) demonstrated that scorpion venom contains a low-molecular-weight peptide toxin that reversibly inhibits recombinant human CFTR expressed in *Xenopus* oocytes only when applied to the intracellular side of the membrane. Subsequently, Fuller et al. (2005) demonstrated that the peptide toxin preferentially inhibits CFTR when the channel is closed, suggesting that the toxin is a state-dependent blocker of the CFTR Cl⁻ channel. Of note, the off-rate for the interaction of the toxin with CFTR is one hundred times slower than that of glibenclamide (Fuller et al., 2005). This augurs well for the purification of a high-affinity peptide blocker of the CFTR Cl⁻ channel.

G. Identification of CFTR Inhibitors by HTS

To identify potent, specific blockers of the CFTR Cl⁻ channel, Verkman and his colleagues employed HTS of compound libraries. They modified their HTS assay for CFTR potentiators to identify agents that interact directly with CFTR to inhibit CFTR-mediated I⁻ influx into FRT cells coexpressing wild-type human CFTR and a GFP with ultrahigh halide sensitivity. Their initial screen of 50,000 compounds identified the thiazolidinone $CFTR_{inh}$-172 (Fig. 8), a CFTR inhibitor with several highly desirable properties (Ma et al., 2002a). First, potency: $CFTR_{inh}$-172 reversibly inhibited CFTR-mediated Cl⁻ currents with a half-maximal inhibitory concentration (K_i) of ~300 nM, an increase in potency of almost 500-fold compared with the widely used CFTR blocker glibenclamide (Ma et al., 2002a). Second, specificity: at relevant concentrations $CFTR_{inh}$-172 was without effect on several ion channels and transporters found in epithelial tissues including Ca^{2+}-activated and volume-regulated Cl⁻ channels and the ABC transporter P-gp (Ma et al., 2002a). Third, efficacy: $CFTR_{inh}$-172 inhibited cholera toxin-induced fluid secretion in the small intestine of mice, a striking demonstration of the therapeutic potential of CFTR blockers (Ma et al., 2002a). $CFTR_{inh}$-172, however, suffers from three drawbacks that constrain the drug's usefulness: (1) limited water solubility (~20 μM; Muanprasat et al., 2004), (2) reduced potency in intact cells and tissues (K_i~5 μM; Muanprasat et al., 2004), and (3) slow onset of action (Li et al., 2004, but see Pedemonte et al., 2005).

In a subsequent screen of 100,000 small molecules selected for chemical diversity and drug-like properties, Verkman and his colleagues (Muanprasat et al., 2004) identified a novel open-channel blocker of CFTR, the glycine hydrazide GlyH-101 (Fig. 8). The characteristic of channel block that sets GlyH-101 apart from other

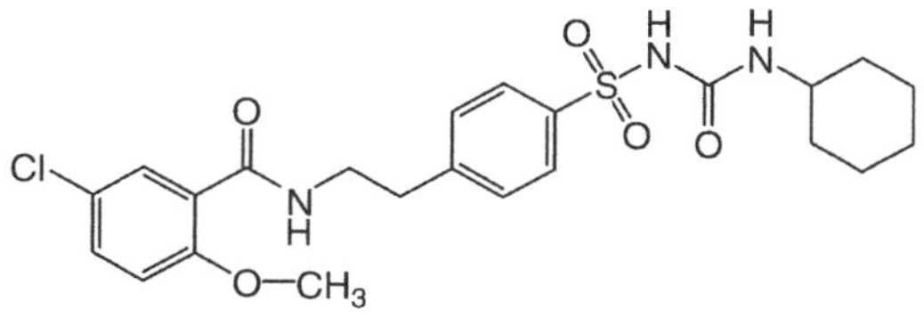

Figure 8. Chemical structures of CFTR inhibitors identified by HTS. CFTR$_{inh}$-172, 3-[(3-trifluoromethyl)phenyl]-5-[(4-carboxyphenyl)methylene]-2-thioxo-4-thiazolidinone; GlyH-101, N-(2-naphthalenyl)-[(3,5-dibromo-2,4-dihydroxyphenyl)methylene]glycine hydrazide. For comparison, the chemical structure of glibenclamide is shown.

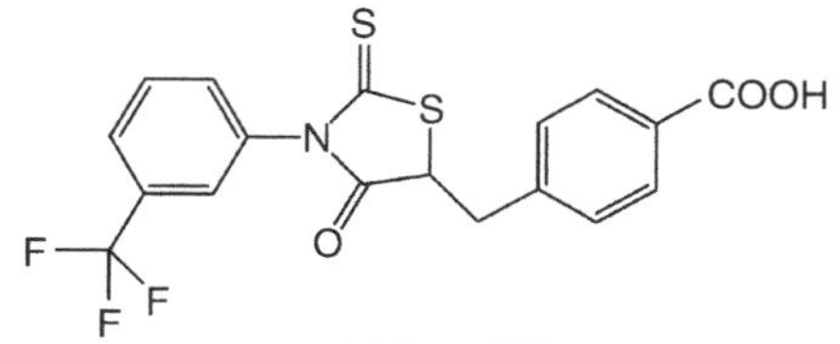

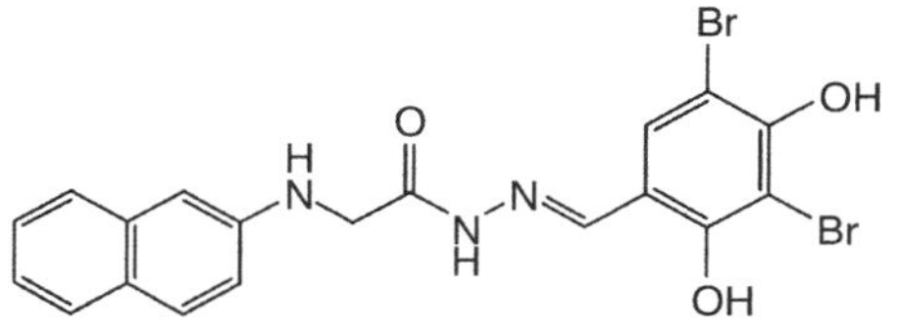

open-channel blockers of CFTR is the shape of the I–V relationship. As discussed in Section VI.E.1, typically, open-channel blockers of CFTR cause outward rectification of CFTR Cl⁻ currents. In contrast, submaximal concentrations of GlyH-101 (<30 µM) caused inward rectification of CFTR Cl⁻ currents, indicating that Cl⁻ flow from the extra- to the intracellular side of the membrane is more strongly attenuated than that in the opposite direction (Muanprasat et al., 2004). Taken together, the simplest interpretation of the data is that positive voltages drive the negatively charged GlyH-101 into the extracellular vestibule of the CFTR pore where it binds and occludes Cl⁻ permeation (Fig. 7D). Channel block by GlyH-101 is relieved both by negative voltages and reducing the extracellular Cl⁻ concentration; these maneuvers enable Cl⁻ flow from the intracellular to the extracellular side of the membrane to flush GlyH-101 from its binding site within the extracellular vestibule. Like CFTR$_{inh}$-172 (Ma et al., 2002a), GlyH-101 dramatically inhibited cholera toxin-induced fluid secretion (Muanprasat et al., 2004). However, in contrast to CFTR$_{inh}$-172, which must be administered intraperitoneally, GlyH-101 is active when added directly into the lumen of the small intestine (Muanprasat et al., 2004).

H. Therapeutic Potential of CFTR Inhibitors

Secretory diarrhea is the leading cause of death in young children worldwide (Guerrant et al., 1990). The disease results from the irreversible activation of fluid and electrolyte secretion in the intestine (Thiagarajah and Verkman, 2003). Ever since CFTR was identified as the apical membrane Cl^- channel responsible for cAMP-stimulated Cl^- secretion by intestinal epithelia, there has been speculation that CFTR blockers might be of value in the treatment of secretory diarrhea. However, it was not until the discovery of $CFTR_{inh}$-172 that proof of principle for this idea was obtained (Ma et al., 2002a). $CFTR_{inh}$-172 has several drawbacks that preclude its use as an antidiarrheal agent (see Section VI.G). In contrast, the demonstration that GlyH-101 inhibited the CFTR Cl^- channel by occluding the external mouth of the CFTR pore (Muanprasat et al., 2004) suggests that a nonabsorbable drug therapy for secretory diarrhea might be developed from the glycine hydrazide GlyH-101. Toward this goal, Sonawane et al. (2006) synthesized a series of malonic acid dihydrazides (MalH) linked to polar moieties and polyethylene glycol (PEG)-coupled butyric acid hydrazides [GlyH-(PEG)$_n$]. Importantly, the compounds developed by Sonawane et al. (2006) have a number of highly desirable properties including (1) high water solubility, (2) low toxicity, (3) potent inhibition of CFTR and cholera toxin-induced fluid secretion, (4) effective from the intestinal lumen, and (5) minimal absorption by the intestinal epithelium. Sonawane et al. (2006) speculate that nonabsorbable malonic acid dihydrazides and glycine hydrazides might be developed into an effective therapy for intestinal fluid loss in cholera.

Inappropriate CFTR-driven fluid accumulation contributes to the pathogenesis of ADPKD, the most common single gene disorder to affect the kidney (Wilson, 2004). Because CFTR is located on the lumen-facing membrane of epithelial cysts, there has been speculation that it is not a suitable target for ADPKD therapy. However, several lines of evidence raise the possibility that therapeutically active CFTR blockers might be of value in the treatment of ADPKD. First, Hanaoka and Guggino (2000) showed that DPC and glibenclamide, but not DIDS, retard the enlargement of ADPKD cysts. Second, Li et al. (2004) demonstrated that all drugs that inhibit CFTR retard cyst growth. These drugs include agents that directly inhibit the CFTR Cl^- channel by either open-channel (e.g., glibenclamide) or allosteric (e.g., genistein) mechanisms. They also include agents that act indirectly either by inhibiting cAMP-dependent phosphorylation of CFTR (e.g., H-89) or by interfering with basolateral membrane ion channels and transporters that accumulate Cl^- within epithelial cells (e.g., bumetanide). Importantly, $CFTR_{inh}$-172 retarded both cyst formation and diminished cyst growth (Li et al., 2004). Third, Torres et al. (2004) demonstrated that treatment of a mouse model of ADPKD with OPC31260 (a vasopressin V2 receptor antagonist) inhibited cyst formation and enlargement by blocking cAMP-stimulated cell proliferation and Cl^- secretion. Because related drugs are approved for clinical use, the data argue that V2 receptor antagonists might prove an effective therapy for ADPKD. Future studies should test this possibility.

VII. CONCLUSIONS

CFTR is a unique ABC transporter that plays a critical role in fluid and electrolyte transport across epithelial tissues. Its physiological role is determined by the individual domains from which it is assembled, but also by its interactions with a diverse repertoire of proteins. Malfunction of CFTR in human disease has profound consequences for transepithelial ion transport. In the search for new treatments for disease, pharmacological agents that interact directly with the CFTR Cl^- channel have been identified. Some agents potentiate CFTR by interacting with the NBDs that control channel gating, whereas others inhibit CFTR either by binding within the channel pore and preventing Cl^- permeation or by interfering with the control of channel gating by the NBDs. Knowledge of the pharmacology of CFTR-mediated transepithelial ion transport is leading to rational new therapies for diseases caused by the dysfunction of CFTR.

ACKNOWLEDGMENTS

We thank Drs. S. M. Husbands and O. Moran and our departmental colleagues for valuable discussions. J.-H.C. was supported by a scholarship from the University of Bristol and an ORS award from Universities UK. Work in the authors' laboratory is supported by the Biotechnology and Biological Sciences Research Council and the Cystic Fibrosis Trust.

REFERENCES

Ai, T., Bompadre, S. G., Wang, X., Hu, S., Li, M., and Hwang, T.-C. (2004). Capsaicin potentiates wild-type and mutant cystic fibrosis transmembrane conductance regulator chloride-channel currents. *Mol. Pharmacol.* **65,** 1415–1426.

Aleksandrov, A. A., Aleksandrov, L., and Riordan, J. R. (2002a). Nucleoside triphosphate pentose ring impact on CFTR gating and hydrolysis. *FEBS Lett.* **518,** 183–188.

Aleksandrov, L., Aleksandrov, A. A., Chang, X.-B., and Riordan, J. R. (2002b). The first nucleotide binding domain of cystic fibrosis transmembrane conductance regulator is a site of stable nucleotide interaction, whereas the second is a site of rapid turnover. *J. Biol. Chem.* **277,** 15419–15425.

Amaral, M. D. (2005). Processing of CFTR: Traversing the cellular maze-how much CFTR needs to go through to avoid cystic fibrosis? *Pediatr. Pulmonol.* **39,** 479–491.

Anderson, M. P., Berger, H. A., Rich, D. P., Gregory, R. J., Smith, A. E., and Welsh, M. J. (1991a). Nucleoside triphosphates are required to open the CFTR chloride channel. *Cell* **67,** 775–784.

Anderson, M. P., Gregory, R. J., Thompson, S., Souza, D. W., Paul, S., Mulligan, R. C., Smith, A. E., and Welsh, M. J. (1991b). Demonstration that CFTR is a chloride channel by alteration of its anion selectivity. *Science* **253,** 202–205.

Becq, F. (2006). On the discovery and development of CFTR chloride channel activators. *Curr. Pharm. Des.* **12,** 471–484.

Boletta, A., Qian, F., Onuchic, L. F., Bhunia, A. K., Phakdeekitcharoen, B., Hanaoka, K., Guggino, W., Monaco, L., and Germino, G. G. (2000). Polycystin-1, the gene product of PKD1, induces resistance to apoptosis and spontaneous tubulogenesis in MDCK cells. *Mol. Cell* **6,** 1267–1273.

Brill, S. R., Ross, K. E., Davidow, C. J., Ye, M., Grantham, J. J., and Caplan, M. J. (1996). Immunolocalization of ion transport proteins in human autosomal dominant polycystic kidney epithelial cells. *Proc. Natl. Acad. Sci. USA* **93**, 10206–10211.
Cai, Z. and Sheppard, D. N. (2002). Phloxine B interacts with the cystic fibrosis transmembrane conductance regulator at multiple sites to modulate channel activity. *J. Biol. Chem.* **277**, 19546–19553.
Cai, Z., Li, H., Scott-Ward, T. S., and Sheppard, D. N. (2003a). Pharmacology of the CFTR Cl$^-$ channel. Virtual repository of the cystic fibrosis European network. http://central.igc.gulbenkian.pt/cftr/vr/physiology.html
Cai, Z., Scott-Ward, T. S., and Sheppard, D. N. (2003b). Voltage-dependent gating of the cystic fibrosis transmembrane conductance regulator Cl$^-$ channel. *J. Gen. Physiol.* **122**, 605–620.
Cai, Z., Scott-Ward, T. S., Li, H., Schmidt, A., and Sheppard, D. N. (2004). Strategies to investigate the mechanism of action of CFTR modulators. *J. Cyst. Fibros.* 3(Suppl. 2), 141–147.
Cai, Z., Taddei, A., and Sheppard, D. N. (2006). Differential sensitivity of the cystic fibrosis (CF)-associated mutants G551D and G1349D to potentiators of the cystic fibrosis transmembrane conductance regulator (CFTR) Cl$^-$ channel. *J. Biol. Chem.* **281**, 1970–1977.
Carson, M. R., Winter, M. C., Travis, S. M., and Welsh, M. J. (1995). Pyrophosphate stimulates wild-type and mutant cystic fibrosis transmembrane conductance regulator Cl$^-$ channels. *J. Biol. Chem.* **270**, 20466–20472.
Chappe, V., Mettey, Y., Vierfond, J. M., Hanrahan, J. W., Gola, M., Verrier, B., and Becq, F. (1998). Structural basis for specificity and potency of xanthine derivatives as activators of the CFTR chloride channel. *Br. J. Pharmacol.* **123**, 683–693.
Chen, J.-H., Cai, Z., Li, H., and Sheppard, D. N. (2006). Function of CFTR protein: Ion transport. In: *Cystic Fibrosis in the 21st Century* (A. Bush, E. W. F. W. Alton, J. C. Davies, U. Griesenbach, and A. Jaffe, Eds.), pp. 38–44. Karger, Basel, Switzerland.
Cheng, S. H., Gregory, R. J., Marshall, J., Paul, S., Souza, D. W., White, G. A., O'Riordan, C. R., and Smith, A. E. (1990). Defective intracellular transport and processing of CFTR is the molecular basis of most cystic fibrosis. *Cell* **63**, 827–834.
Cheng, S. H., Rich, D. P., Marshall, J., Gregory, R. J., Welsh, M. J., and Smith, A. E. (1991). Phosphorylation of the R domain by cAMP-dependent protein kinase regulates the CFTR chloride channel. *Cell* **66**, 1027–1036.
Cheung, M. and Akabas, M. H. (1996). Identification of cystic fibrosis transmembrane conductance regulator channel-lining residues in and flanking the M6 membrane-spanning segment. *Biophys. J.* **70**, 2688–2695.
Csanády, L., Chan, K. W., Seto-Young, D., Kopsco, D. C., Nairn, A. C., and Gadsby, D. C. (2000). Severed channels probe regulation of gating of cystic fibrosis transmembrane conductance regulator by its cytoplasmic domains. *J. Gen. Physiol.* **116**, 477–500.
Dalemans, W., Barbry, P., Champigny, G., Jallat, S., Dott, K., Dreyer, D., Crystal, R. G., Pavirani, A., Lecocq, J.-P., and Lazdunski, M. (1991). Altered chloride ion channel kinetics associated with the ΔF508 cystic fibrosis mutation. *Nature* **354**, 526–528.
DeBin, J. A., Maggio, J. E., and Strichartz, G. R. (1993). Purification and characterization of chlorotoxin, a chloride channel ligand from the venom of the scorpion. *Am. J. Physiol.* **264**, C361–C369.
Delmas, P., Nomura, H., Li, X., Lakkis, M., Luo, Y., Segal, Y., Fernández-Fernández, J. M., Harris, P., Frischauf, A.-M., Brown, D. A., and Zhou, J. (2002). Constitutive activation of G-proteins by polycystin-1 is antagonized by polycystin-2. *J. Biol. Chem.* **277**, 11276–11283.
Denning, G. M., Anderson, M. P., Amara, J. F., Marshall, J., Smith, A. E., and Welsh, M. J. (1992). Processing of mutant cystic fibrosis transmembrane conductance regulator is temperature-sensitive. *Nature* **358**, 761–764.
Dérand, R., Bulteau-Pignoux, L., and Becq, F. (2002). The cystic fibrosis mutation G551D alters the non-Michaelis-Menten behaviour of the cystic fibrosis transmembrane conductance regulator (CFTR) channel and abolishes the inhibitory genistein binding site. *J. Biol. Chem.* **277**, 35999–36004.
Dormer, R. L., Dérand, R., McNeilly, C. M., Mettey, Y., Bulteau-Pignoux, L., Métayé, T., Vierfond, J.-M., Gray, M. A., Galietta, L. J. V., Morris, M. R., Pereira, M. M. C., Doull, I. J. M., *et al.* (2001).

Correction of delF508-CFTR activity with benzo(c)quinolizinium compounds through facilitation of its processing in cystic fibrosis airway cells. *J. Cell Sci.* **114**, 4073–4081.
Dormer, R. L., Harris, C. M., Clark, Z., Pereira, M. M. C., Doull, I. J. M., Norez, C., Becq, F., and McPherson, M. A. (2005). Sildenafil (Viagra) corrects ΔF508-CFTR location in nasal epithelial cells from patients with cystic fibrosis. *Thorax* **60**, 55–59.
Droogmans, G., Prenen, J., Eggermont, J., Voets, T., and Nilius, B. (1998). Voltage-dependent block of endothelial volume-regulated anion channels by calix[4]arenes. *Am. J. Physiol.* **275**, C646–C652.
Du, K., Sharma, M., and Lukacs, G. L. (2006). The ΔF508 cystic fibrosis mutation impairs domain-domain interactions and arrests post-translational folding of CFTR. *Nat. Struct. Mol. Biol.* **12**, 17–25.
Engelhardt, J. F., Yankaskas, J. R., Ernst, S. A., Yang, Y., Marino, C. R., Boucher, R. C., Cohn, J. A., and Wilson, J. M. (1992). Submucosal glands are the predominant site of CFTR expression in the human bronchus. *Nat. Genet.* **2**, 240–248.
Forstner, G., Kopelman, H., Durie, P., and Corey, M. (1987). Pancreatic and intestinal dysfunction in cystic fibrosis. In: *Genetics and Epithelial Cell Dysfunction in Cystic Fibrosis* (J. R. Riordan and M. Buchwald, Eds.), pp. 7–17. Alan R. Liss, New York.
Fuller, M. D., Zhang, Z.-R., Cui, G., Kubanek, J., and McCarty, N. A. (2004). Inhibition of CFTR channels by a peptide toxin of scorpion venom. *Am. J. Physiol.* **287**, C1328–C1341.
Fuller, M. D., Zhang, Z.-R., Cui, G., and McCarty, N. A. (2005). The block of CFTR by scorpion venom is state-dependent. *Biophys. J.* **89**, 3960–3975.
Gabriel, S. E., Brigman, K. N., Koller, B. H., Boucher, R. C., and Stutts, M. J. (1994). Cystic fibrosis heterozygote resistance to cholera toxin in the cystic fibrosis mouse model. *Science* **266**, 107–109.
Gadsby, D. C. and Nairn, A. C. (1999). Control of cystic fibrosis transmembrane conductance regulator channel gating by phosphorylation and nucleotide hydrolysis. *Physiol. Rev.* **79**, S77–S107.
Gadsby, D. C., Vergani, P., and Csanády, L. (2006). The ABC protein turned chloride channel whose failure causes cystic fibrosis. *Nature* **440**, 477–483.
Galietta, L. J. V., Springsteel, M. F., Eda, M., Niedzinski, E. J., By, K., Haddadin, M. J., Kurth, M. J., Nantz, M. H., and Verkman, A. S. (2001). Novel CFTR chloride channel activators identified by screening of combinatorial libraries based on flavone and benzoquinolizinium lead compounds. *J. Biol. Chem.* **276**, 19723–19728.
Gong, X. and Linsdell, P. (2004). Maximization of the rate of chloride conduction in the CFTR channel pore by ion-ion interactions. *Arch. Biochem. Biophys.* **426**, 78–82.
Gong, X., Burbridge, S. M., Cowley, E. A., and Linsdell, P. (2002). Molecular determinants of $Au(CN)_2^-$ binding and permeability within the cystic fibrosis transmembrane conductance regulator Cl^- channel pore. *J. Physiol.* **540**, 39–47.
González-Perrett, S., Kim, K., Ibarra, C., Damiano, A. E., Zotta, E., Batelli, M., Harris, P. C., Reisin, I. L., Arnaout, M. A., and Cantiello, H. F. (2001). Polycystin-2, the protein mutated in autosomal dominant polycystic kidney disease (ADPKD), is a Ca^{2+}-permeable nonselective cation channel. *Proc. Natl. Acad. Sci. USA* **98**, 1182–1187.
Gribkoff, V. K., Champigny, G., Barbry, P., Dworetzky, S. I., Meanwell, N. A., and Lazdunski, M. (1994). The substituted benzimidazolone NS004 is an opener of the cystic fibrosis chloride channel. *J. Biol. Chem.* **269**, 10983–10986.
Grimard, V., Li, C., Ramjeesingh, M., Bear, C. E., Goormaghtigh, E., and Ruysschaert, J.-M. (2004). Phosphorylation-induced conformational changes of cystic fibrosis transmembrane conductance regulator monitored by attenuated total reflection-fourier transform IR spectroscopy and fluorescence spectroscopy. *J. Biol. Chem.* **279**, 5528–5536.
Guerrant, R. L., Hughes, J. M., Lima, N. L., and Crane, J. (1990). Diarrhea in developed and developing countries: Magnitude, special settings, and etiologies. *Rev. Infect. Dis.* **12**, S41–S50.
Guggino, W. B. and Stanton, B. A. (2006). New insights into cystic fibrosis: Molecular switches that regulate CFTR. *Nat. Rev. Mol. Cell Biol.* **7**, 426–436.
Gunderson, K. L. and Kopito, R. R. (1994). Effects of pyrophosphate and nucleotide analogs suggest a role for ATP hydrolysis in cystic fibrosis transmembrane regulator channel gating. *J. Biol. Chem.* **269**, 19349–19353.

Gupta, J. and Linsdell, P. (2002). Point mutations in the pore region directly or indirectly affect glibenclamide block of the CFTR chloride channel. *Pflügers Arch.* **443,** 739–747.
Hallows, K. R., Raghuram, V., Kemp, B. E., Witters, L. A., and Foskett, J. K. (2000). Inhibition of cystic fibrosis transmembrane conductance regulator by novel interaction with the metabolic sensor AMP-activated protein kinase. *J. Clin. Invest.* **105,** 1711–1721.
Hallows, K. R., McCane, J. E., Kemp, B. E., Witters, L. A., and Foskett, J. K. (2003). Regulation of channel gating by AMP-activated protein kinase modulates cystic fibrosis transmembrane conductance regulator activity in lung submucosal cells. *J. Biol. Chem.* **278,** 998–1004.
Hanaoka, K. and Guggino, W. B. (2000). cAMP regulates cell proliferation and cyst formation in autosomal polycystic kidney disease cells. *J. Am. Soc. Nephrol.* **11,** 1179–1187.
Hanaoka, K., Devuyst, O., Schwiebert, E. M., Wilson, P. D., and Guggino, W. B. (1996). A role for CFTR in human autosomal dominant polycystic kidney disease. *Am. J. Physiol.* **270,** C389–C399.
Hanrahan, J. W., Gentzsch, M., and Riordan, J. R. (2003). The cystic fibrosis transmembrane conductance regulator (ABCC7). In: *ABC Proteins: From Bacteria to Man* (I. B. Holland, S. P. C. Cole, C. F. Higgins, and K. Kuchler, Eds.), pp. 589–618. Academic Press, London.
Holland, I. B., Cole, S. P. C., Kuchler, K., and Higgins, C. F. (2003). *ABC Proteins: From Bacteria to Man.* Academic Press, London.
Huang, P., Lazarowski, E., Tarran, R., Milgram, S. L., Boucher, R. C., and Stutts, M. J. (2001). Compartmentalized autocrine signaling to cystic fibrosis transmembrane conductance regulator at the apical membrane of airway epithelial cells. *Proc. Natl. Acad. Sci. USA* **98,** 14120–14125.
Hume, J. R., Duan, D., Collier, M. L., Yamazaki, J., and Horowitz, B. (2000). Anion transport in heart. *Physiol. Rev.* **80,** 31–81.
Hwang, T.-C. and Sheppard, D. N. (1999). Molecular pharmacology of the CFTR Cl$^-$ channel. *Trends Pharmacol. Sci.* **20,** 448–453.
Hwang, T.-C., Horie, M., and Gadsby, D. C. (1993). Functionally distinct phospho-forms underlie incremental activation of protein kinase-regulated Cl$^-$ conductance in mammalian heart. *J. Gen. Physiol.* **101,** 629–650.
Kidd, J. F., Ramjeesingh, M., Stratford, F., Huan, L.-J., and Bear, C. E. (2004). A heteromeric complex of the two nucleotide binding domains of cystic fibrosis transmembrane conductance regulator (CFTR) mediates ATPase activity. *J. Biol. Chem.* **279,** 41664–41669.
Ko, S. B. H., Zeng, W., Dorwart, M. R., Luo, X., Kim, K. H., Millen, L., Goto, H., Naruse, S., Soyombo, A., Thomas, P. J., and Muallem, S. (2004). Gating of CFTR by the STAS domain of SLC26 transporters. *Nat. Cell Biol.* **6,** 343–350.
Koulen, P., Cai, Y., Geng, L., Maeda, Y., Nishimura, S., Witzgall, R., Ehrlich, B. E., and Somlo, S. (2002). Polycystin-2 is an intracellular calcium channel. *Nat. Cell Biol.* **4,** 191–197.
Lakkis, M. and Zhou, J. (2003). Molecular complexes formed with polycystins. *Nephron Exp. Nephrol.* **93,** e3–e8.
Lewis, H. A., Buchanan, S. G., Burley, S. K., Conners, K., Dickey, M., Dorwart, M., Fowler, R., Gao, X., Guggino, W. B., Hendrickson, W. A., Hunt, J. F., Kearins, M. C., *et al.* (2004). Structure of nucleotide-binding domain 1 of the cystic fibrosis transmembrane conductance regulator. *EMBO J.* **23,** 282–293.
Lewis, H. A., Zhao, X., Wang, C., Sauder, J. M., Rooney, I., Noland, B. W., Lorimer, D., Kearins, M. C., Conners, K., Condon, B., Maloney, P. C., Guggino, W. B., *et al.* (2005). Impact of the ΔF508 mutation in first nucleotide-binding domain of human cystic fibrosis transmembrane conductance regulator on domain folding and structure. *J. Biol. Chem.* **280,** 1346–1353.
Li, C., Ramjeesingh, M., Wang, W., Garami, E., Hewryk, M., Lee, D., Rommens, J. M., Galley, K., and Bear, C. E. (1996). ATPase activity of the cystic fibrosis transmembrane conductance regulator. *J. Biol. Chem.* **271,** 28463–28468.
Li, H., Findlay, I. A., and Sheppard, D. N. (2004). The relationship between cell proliferation, Cl$^-$ secretion, and renal cyst growth: A study using CFTR inhibitors. *Kidney Int.* **66,** 1926–1938.
Linsdell, P. (2005). Location of a common inhibitor binding site in the cytoplasmic vestibule of the cystic fibrosis transmembrane conductance regulator chloride channel pore. *J. Biol. Chem.* **280,** 8945–8950.
Linsdell, P. (2006). Mechanism of chloride permeation in the cystic fibrosis transmembrane conductance regulator chloride channel. *Exp. Physiol.* **91,** 123–129.

Linsdell, P. and Hanrahan, J. W. (1998). Adenosine triphosphate-dependent asymmetry of anion permeation in the cystic fibrosis transmembrane conductance regulator chloride channel. *J. Gen. Physiol.* **111**, 601–614.

Linsdell, P., Tabcharani, J. A., and Hanrahan, J. W. (1997a). Multi-ion mechanism for ion permeation and block in the cystic fibrosis transmembrane conductance regulator chloride channel. *J. Gen. Physiol.* **110**, 365–377.

Linsdell, P., Tabcharani, J. A., Rommens, J. M., Hou, Y.-X., Chang, X.-B., Tsui, L.-C., Riordan, J. R., and Hanrahan, J. W. (1997b). Permeability of wild-type and mutant cystic fibrosis transmembrane conductance regulator chloride channels to polyatomic anions. *J. Gen. Physiol.* **110**, 355–364.

Liu, X., Smith, S. S., and Dawson, D. C. (2003). CFTR: What's it like inside the pore? *J. Exp. Zool.* **300A**, 69–75.

Locher, K. P., Lee, A. T., and Rees, D. C. (2002). The *E. coli* BtuCD structure: A framework for ABC transporter architecture and mechanism. *Science* **296**, 1091–1098.

Loo, T. W., Bartlett, M. C., and Clarke, D. M. (2003). Simultaneous binding of two different drugs in the binding pocket of the human multidrug resistance P-glycoprotein. *J. Biol. Chem.* **278**, 39706–39710.

Ma, T., Thiagarajah, J. R., Yang, H., Sonawane, N. D., Folli, C., Galietta, L. J. V., and Verkman, A. S. (2002a). Thiazolidinone CFTR inhibitor identified by high-throughput screening blocks cholera toxin-induced intestinal fluid secretion. *J. Clin. Invest.* **110**, 1651–1658.

Ma, T., Vetrivel, L., Yang, H., Pedemonte, N., Zegarra-Moran, O., Galietta, L. J. V., and Verkman, A. S. (2002b). High-affinity activators of cystic fibrosis transmembrane conductance regulator (CFTR) chloride conductance identified by high-throughput screening. *J. Biol. Chem.* **277**, 37235–37241.

Maertens, C., Wei, L., Tytgat, J., Droogmans, G., and Nilius, B. (2000). Chlorotoxin does not inhibit volume-regulated, calcium-activated and cyclic AMP-activated chloride channels. *Br. J. Pharmacol.* **129**, 791–801.

Matsui, H., Grubb, B. R., Tarran, R., Randell, S. H., Gatzy, J. T., Davis, C. W., and Boucher, R. C. (1998). Evidence for periciliary liquid layer depletion, not abnormal ion composition, in the pathogenesis of cystic fibrosis airways disease. *Cell* **95**, 1005–1015.

McCarty, N. A. (2000). Permeation through the CFTR chloride channel. *J. Exp. Biol.* **203**, 1947–1962.

McDonough, S., Davidson, N., Lester, H. A., and McCarty, N. A. (1994). Novel pore-lining residues in CFTR that govern permeation and open-channel block. *Neuron* **13**, 623–634.

Melin, P., Thoreau, V., Norez, C., Bilan, F., Kitzis, A., and Becq, F. (2004). The cystic fibrosis mutation G1349D within the signature motif LSHGH of NBD2 abolishes the activation of CFTR chloride channels by genistein. *Biochem. Pharmacol.* **67**, 2187–2196.

Moran, O., Galietta, L. J. V., and Zegarra-Moran, O. (2005). Binding site of activators of the cystic fibrosis transmembrane conductance regulator in the nucleotide binding domains. *Cell. Mol. Life Sci.* **62**, 446–460.

Muanprasat, C., Sonawane, N. D., Salinas, D., Taddei, A., Galietta, L. J. V., and Verkman, A. S. (2004). Discovery of glycine hydrazide pore-occluding CFTR inhibitors: Mechanism, structure-activity analysis and *in vivo* efficacy. *J. Gen. Physiol.* **124**, 125–137.

Naren, A. P., Cobb, B., Li, C., Roy, K., Nelson, D., Heda, G. D., Liao, J., Kirk, K. L., Sorscher, E. J., Hanrahan, J., and Clancy, J. P. (2003). A macromolecular complex of β2 adrenergic receptor, CFTR, and ezrin/radixin/moesin-binding phosphoprotein 50 is regulated by PKA. *Proc. Natl. Acad. Sci. USA* **100**, 342–346.

Nauli, S. M., Alenghat, F. J., Luo, Y., Williams, E., Vassilev, P., Li, X., Elia, A. E. H., Lu, W., Brown, E. M., Quinn, S. J., Ingber, D. E., and Zhou, J. (2003). Polycystins 1 and 2 mediate mechanosensation in the primary cilium of kidney cells. *Nat. Genet.* **33**, 129–137.

Novak, I. and Greger, R. (1988). Properties of the luminal membrane of isolated perfused rat pancreatic ducts: Effect of cyclic AMP and blockers of chloride transport. *Pflügers Arch.* **411**, 546–553.

Ostedgaard, L. S., Baldursson, O., Vermeer, D. W., Welsh, M. J., and Robertson, A. D. (2000). A functional R domain from cystic fibrosis transmembrane conductance regulator is predominantly unstructured in solution. *Proc. Natl. Acad. Sci. USA* **97**, 5657–5662.

Ostedgaard, L. S., Baldursson, O., and Welsh, M. J. (2001). Regulation of the cystic fibrosis transmembrane conductance regulator Cl$^-$ channel by its R domain. *J. Biol. Chem.* **276**, 7689–7692.

Pedemonte, N., Sonawane, N. D., Taddei, A., Hu, J., Zegarra-Moran, O., Suen, Y. F., Robins, L. I., Dicus, C. W., Willenbring, D., Nantz, M. H., Kurth, M. J., Galietta, L. J. V., et al. (2005). Phenylglycine and sulfonamide correctors of defective ΔF508 and G551D cystic fibrosis transmembrane conductance regulator chloride-channel gating. *Mol. Pharmacol.* **67**, 1797–1807.

Qu, Z. and Hartzell, H. C. (2001). Functional geometry of the permeation pathway of Ca^{2+}-activated Cl^- channels inferred from analysis of voltage-dependent block. *J. Biol. Chem.* **276**, 18423–18429.

Quinton, P. M. (1983). Chloride impermeability in cystic fibrosis. *Nature* **301**, 421–422.

Reddy, M. M. and Quinton, P. M. (1992). cAMP activation of CF-affected Cl^- conductance in both cell membranes of an absorptive epithelium. *J. Membr. Biol.* **130**, 49–62.

Reddy, M. M. and Quinton, P. M. (1996). Deactivation of CFTR-Cl conductance by endogenous phosphatases in the native sweat duct. *Am. J. Physiol.* **270**, C474–C480.

Rich, D. P., Berger, H. A., Cheng, S. H., Travis, S. M., Saxena, M., Smith, A. E., and Welsh, M. J. (1993). Regulation of the cystic fibrosis transmembrane conductance regulator Cl^- channel by negative charge in the R domain. *J. Biol. Chem.* **268**, 20259–20267.

Riordan, J. R. (2005). Assembly of functional CFTR chloride channels. *Annu. Rev. Physiol.* **67**, 701–718.

Riordan, J. R., Rommens, J. M., Kerem, B.-S., Alon, N., Rozmahel, R., Grzelczak, Z., Zielenski, J., Lok, S., Plavsic, N., Chou, J.-L., Drumm, M. L., Iannuzzi, M. C., et al. (1989). Identification of the cystic fibrosis gene: Cloning and characterization of complementary DNA. *Science* **245**, 1066–1073.

Rosenberg, M. F., Kamis, A. B., Aleksandrov, L. A., Ford, R. C., and Riordan, J. R. (2004). Purification and crystallization of the cystic fibrosis transmembrane conductance regulator (CFTR). *J. Biol. Chem.* **279**, 39051–39057.

Rosenberg, M. F., Callaghan, R., Modok, S., Higgins, C. F., and Ford, R. C. (2005). Three-dimensional structure of P-glycoprotein: The transmembrane regions adopt an asymmetric configuration in the nucleotide-bound state. *J. Biol. Chem.* **280**, 2857–2862.

Rubenstein, R. C. (2006). New pharmacological approaches for treatment of cystic fibrosis. In: *Cystic Fibrosis in the 21st Century* (A. Bush, E. W. F. W. Alton, J. C. Davies, U. Griesenbach, and A. Jaffe, Eds.), pp. 212–220. Karger, Basel, Switzerland.

Schultz, B. D., Singh, A. K., Devor, D. C., and Bridges, R. J. (1999). Pharmacology of CFTR chloride channel activity. *Physiol. Rev.* **79**, S109–S144.

Scott-Ward, T. S., Li, H., Schmidt, A., Cai, Z., and Sheppard, D. N. (2004). Direct block of the cystic fibrosis transmembrane conductance regulator Cl^- channel by niflumic acid. *Mol. Membr. Biol.* **21**, 27–38.

Sheppard, D. N. and Welsh, M. J. (1992). Effect of ATP-sensitive K^+ channel regulators on cystic fibrosis transmembrane conductance regulator chloride currents. *J. Gen. Physiol.* **100**, 573–591.

Sheppard, D. N. and Welsh, M. J. (1999). Structure and function of the cystic fibrosis transmembrane conductance regulator chloride channel. *Physiol. Rev.* **79**, S23–S45.

Sheppard, D. N., Ostedgaard, L. S., Winter, M. C., and Welsh, M. J. (1995). Mechanism of dysfunction of two nucleotide binding domain mutations in cystic fibrosis transmembrane conductance regulator that are associated with pancreatic sufficiency. *EMBO J.* **14**, 876–883.

Short, D. B., Trotter, K. W., Reczek, D., Kreda, S. M., Bretscher, A., Boucher, R. C., Stutts, M. J., and Milgram, S. L. (1998). An apical PDZ protein anchors the cystic fibrosis transmembrane conductance regulator to the cytoskeleton. *J. Biol. Chem.* **273**, 19797–19801.

Singh, A. K., Venglarik, C. J., and Bridges, R. J. (1995). Development of chloride channel modulators. *Kidney Int.* **48**, 985–993.

Singh, A. K., Devor, D. C., Illek, B., Schultz, B. D., and Bridges, R. J. (1996). Does ORCC contribute to transepithelial Cl^- secretion? *Pediatr. Pulmonol.* Suppl. 13, 237.

Smith, J. J., Travis, S. M., Greenberg, E. P., and Welsh, M. J. (1996). Cystic fibrosis airway epithelia fail to kill bacteria because of abnormal airway surface fluid. *Cell* **85**, 229–236.

Sonawane, N. D., Hu, J., Muanprasat, C., and Verkman, A. S. (2006). Luminally active, nonabsorbable CFTR inhibitors as potential therapy to reduce intestinal fluid loss in cholera. *FASEB J.* **20**, 130–132.

Springsteel, M. F., Galietta, L. J. V., Ma, T., By, K., Berger, G. O., Yang, H., Dicus, C. W., Choung, W., Quan, C., Shelat, A. A., Guy, R. K., Verkman, A. S., et al. (2003). Benzoflavone activators of the cystic

fibrosis transmembrane conductance regulator: Towards a pharmacophore model for the nucleotide-binding domain. *Bioorg. Med. Chem.* **11**, 4113–4120.
Sullivan, L. P., Wallace, D. P., and Grantham, J. J. (1998). Chloride and fluid secretion in polycystic kidney disease. *J. Am. Soc. Nephrol.* **9**, 903–916.
Tabcharani, J. A., Low, W., Elie, D., and Hanrahan, J. W. (1990). Low-conductance chloride channel activated by cAMP in the epithelial cell line T_{84}. *FEBS Lett.* **270**, 157–164.
Tabcharani, J. A., Rommens, J. M., Hou, Y.-X., Chang, X.-B., Tsui, L.-C., Riordan, J. R., and Hanrahan, J. W. (1993). Multi-ion pore behaviour in the CFTR chloride channel. *Nature* **366**, 79–82.
Thiagarajah, J. R. and Verkman, A. S. (2003). CFTR pharmacology and its role in intestinal fluid secretion. *Curr. Opin. Pharmacol.* **3**, 594–599.
Torres, V. E., Wang, X., Qian, Q., Somlo, S., Harris, P. C., and Gattone, V. H., II (2004). Effective treatment of an orthologous model of autosomal dominant polycystic kidney disease. *Nat. Med.* **10**, 363–364.
Travis, S. M., Berger, H. A., and Welsh, M. J. (1997). Protein phosphatase 2C dephosphorylates and inactivates cystic fibrosis transmembrane conductance regulator. *Proc. Natl. Acad. Sci. USA* **94**, 11055–11060.
Van Goor, F., Straley, K. S., Cao, D., González, J., Hadida, S., Hazlewood, A., Joubran, J., Knapp, T., Makings, L. R., Miller, M., Neuberger, T., Olson, E., *et al.* (2006). Rescue of ΔF508-CFTR trafficking and gating in human cystic fibrosis airway primary cultures by small molecules. *Am. J. Physiol.* **290**, L1117–L1130.
Vastiau, A., Cao, L., Jaspers, M., Owsianik, G., Janssens, V., Cuppens, H., Goris, J., Nilius, B., and Cassiman, J.-J. (2005). Interaction of the protein phosphatase 2A with the regulatory domain of the cystic fibrosis transmembrane conductance regulator channel. *FEBS Lett.* **579**, 3392–3396.
Vergani, P., Nairn, A. C., and Gadsby, D. C. (2003). On the mechanism of MgATP-dependent gating of CFTR Cl⁻ channels. *J. Gen. Physiol.* **120**, 17–36.
Vergani, P., Lockless, S. W., Nairn, A. C., and Gadsby, D. C. (2005). CFTR channel opening by ATP-driven tight dimerization of its nucleotide-binding domains. *Nature* **433**, 876–880.
Wang, F., Zeltwanger, S., Yang, I. C. H., Nairn, A. C., and Hwang, T.-C. (1998). Actions of genistein on cystic fibrosis transmembrane conductance regulator channel gating: Evidence for two binding sites with opposite effects. *J. Gen. Physiol.* **111**, 477–490.
Wang, F., Zeltwanger, S., Hu, S., and Hwang, T.-C. (2000). Deletion of phenylalanine 508 causes attenuated phosphorylation-dependent activation of CFTR chloride channels. *J. Physiol.* **524**, 637–648.
Welsh, M. J. and Smith, A. E. (1993). Molecular mechanisms of CFTR chloride channel dysfunction in cystic fibrosis. *Cell* **73**, 1251–1254.
Welsh, M. J., Ramsey, B. W., Accurso, F., and Cutting, G. R. (2001). Cystic fibrosis. In: *The Metabolic and Molecular Basis of Inherited Disease* (C. R. Scriver, A. L. Beaudet, W. S. Sly, and D. Valle, Eds.), pp. 5121–5188. McGraw-Hill Inc., New York.
Wilkinson, D. J., Strong, T. V., Mansoura, M. K., Wood, D. L., Smith, S. S., Collins, F. S., and Dawson, D. C. (1997). CFTR activation: Additive effects of stimulatory and inhibitory phosphorylation sites in the R domain. *Am. J. Physiol.* **273**, L127–L133.
Wilson, P. D. (2004). Polycystic kidney disease. *N. Engl. J. Med.* **350**, 151–164.
Wine, J. J. (1999). The genesis of cystic fibrosis lung disease. *J. Clin. Invest.* **103**, 309–312.
Winter, M. C. and Welsh, M. J. (1997). Stimulation of CFTR activity by its phosphorylated R domain. *Nature* **389**, 294–296.
Woodhull, A. M. (1973). Ionic blockage of sodium channels in nerve. *J. Gen. Physiol.* **61**, 687–708.
Yang, H., Shelat, A. A., Guy, R. K., Gopinath, V. S., Ma, T., Du, K., Lukacs, G. L., Taddei, A., Folli, C., Pedemonte, N., Galietta, L. J. V., and Verkman, A. S. (2003). Nanomolar affinity small molecule correctors of defective ΔF508-CFTR chloride channel gating. *J. Biol. Chem.* **278**, 35079–35085.
Zhang, Z.-R., Zeltwanger, S., and McCarty, N. A. (2000). Direct comparison of NPPB and DPC as probes of CFTR expressed in *Xenopus* oocytes. *J. Membr. Biol.* **175**, 35–52.
Zhu, T., Dahan, D., Evagelidis, A., Zheng, S.-X., Luo, J., and Hanrahan, J. W. (1999). Association of cystic fibrosis transmembrane conductance regulator and protein phosphatase 2C. *J. Biol. Chem.* **274**, 29102–29107.

Chapter 6

Gating of Cystic Fibrosis Transmembrane Conductance Regulator Chloride Channel

Zhen Zhou and Tzyh-Chang Hwang

Department of Medical Pharmacology and Physiology, Dalton Cardiovascular Research Center, University of Missouri-Columbia, Columbia, Missouri 65211

- I. CFTR Overview
- II. Regulation of CFTR by Phosphorylation/Dephosphorylation
- III. CFTR Is an ATPase
- IV. Structural Biology of ABC Transporters
  - A. Crystal Structures of ABC Transporters
  - B. The Two NBDs of ABC Proteins Form a Dimer
- V. Methods Used to Study CFTR Gating
  - A. Expression Systems
  - B. Electrophysiological Recording Systems and Technical Hurdles
  - C. Nonhydrolyzable and Hydrolyzable Nucleotide Analogues
  - D. Mutagenesis of the Two NBDs
- VI. CFTR Gating by ATP
  - A. The Role of ATP Hydrolysis in Gating
  - B. The Role of ATP Binding in Gating
  - C. Working Model of CFTR Gating
- VII. Unsettled Issues and Future Directions
  - References

I. CFTR OVERVIEW

Cystic Fibrosis Transmembrane conductance Regulator (CFTR), the protein critical in secretion and absorption of water and electrolytes across epithelia, was cloned in 1989 (Riordan *et al.*, 1989; Rommens *et al.*, 1989). Malfunction of CFTR, due to mutations of the gene coding for CFTR, results in cystic fibrosis (CF), the most common lethal hereditary disease in Caucasians (Riordan *et al.*, 1989). Based on its primary sequence and topology, CFTR is placed in the ATP-binding cassette (ABC) superfamily and is classified as ABCC7 (Riordan *et al.*, 1989; Dean *et al.*, 2001; Higgins

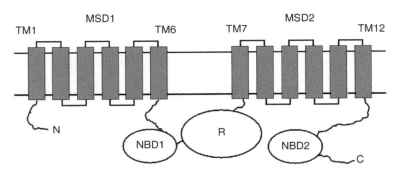

Figure 1. Topology of the CFTR channel. The postulated domain structure of a CFTR Cl⁻ channel consists of cytoplasmic N- and C-termini, two nucleotide-binding domains (NBD1, NBD2) and a regulatory (R) domain, and two predicted membrane-spanning domains (MSD1 and MSD2), each containing six α-helices (TM1–TM6 in MSD1 and TM7–TM12 in MSD2).

and Linton, 2004). Members in the ABC family include P-glycoprotein that is involved in multidrug resistance to cancer treatment presumably by extruding chemotherapeutical reagents from cells in an ATP-dependent manner, and sulfonylurea receptors that play an important role in regulating K_{ATP} channel activity. It is generally believed that most members of the ABC family actively transport substances across the cell membrane or membranes of intracellular organelles of prokaryotic and eukaryotic cells, using ATP hydrolysis as the energy source.

CFTR is composed of two membrane-spanning domains (MSDs), each followed by a nucleotide-binding domain (NBD1 and NBD2, respectively), characteristic topological features that define the ABC transporter family (Fig. 1). The regulatory (R) domain, following immediately the N-terminal NBD (i.e., NBD1), is unique for CFTR and essential for phosphorylation-dependent regulation of the CFTR channel (Section II). CFTR has six transmembrane segments in each of its MSD, although members of the ABC family can have different numbers of transmembrane segments in each MSD (e.g., 10 in each MSD of BtuCD in Locher *et al.*, 2002). The MSDs in different members of the family share little sequence identity and therefore may account for the different substrates different ABC proteins transport (Higgins and Linton, 2004). As seen in all NBDs of the ABC proteins, each NBD of CFTR contains three highly conserved motifs: Walker A motif, Walker B motif, and a signature sequence (LSGGQ). Walker A and Walker B motifs are important for ATP binding/hydrolysis (Zou and Hwang, 2001; Higgins and Linton, 2004; Lewis *et al.*, 2004). Specifically, Walker A lysines (i.e., K464 in NBD1 and K1250 in NBD2 of human CFTR) coordinate β- and γ-phosphates of the bound ATP; Walker B aspartates (D571 in NBD1 and D1370 in NBD2) coordinate Mg^{2+}, a cofactor for ATP hydrolysis; a glutamate residue immediately following the Walker B aspartate (E1371 in NBD2) serves as a catalytic base for ATP hydrolysis (Higgins and Linton, 2004; note: CFTR's NBD1 lacks this glutamate) (Fig. 2). The signature sequence is unique to the ABC transporter family. The physiological importance of the signature sequence has yet to be elucidated, but the fact that the

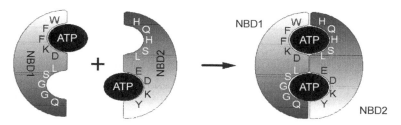

Figure 2. ATP binding at NBDs catalyzes the formation of a head-to-tail dimer. After ATP binds to the Walker A and Walker B motif of each NBD, the two NBDs are believed to form a head-to-tail dimer with two ATP molecules sandwiched in between. The head of each NBD is represented by a lighter shade, while the tail is represented by a darker shade. We define the NBD1 ATP-binding site (or the NBD1 site) as the binding pocket containing Walker A and Walker B motifs in the NBD1 sequence and the signature sequence of the NBD2. An equivalent definition is applied to the NBD2 site. This definition is based on the fact that in all crystal structures of NBD monomers, ATP is found associated with the Walker A and B motifs, suggesting that the signature sequence may not have an appreciable intrinsic binding capability to ATP. One-letter symbols depict those residues that are important for ATP binding based on previous research and our unpublished results. Specifically, Walker A and Walker B residues discussed in this chapter include those in NBD1: W401, F409, F430, K464, and D571, and those in NBD2: Y1219, K1250, D1370, and E1371. Residues in the signature sequences are labeled by white letters.

ATP-binding pocket of an ABC protein consists of the Walker A and Walker B motifs in one NBD and the signature sequence from the partner NBD suggests an important functional role of the signature sequence. Furthermore, many disease-associated mutations are located in the signature sequences of both NBDs of CFTR (CF Consortium, http://www.genet.sickkids.on.ca/cftr/).

As implied by its name, CFTR has been shown to regulate functions of other channels such as Na^+ channels (Stutts *et al.*, 1995, 1997), outwardly rectifying Cl^- channels (Egan *et al.*, 1992; Gabriel *et al.*, 1993), and K^+ channels (for review see Schwiebert *et al.*, 1999). It is, however, unclear whether such modulation is mediated through a direct interaction of CFTR with other channels (Ando-Akatsuka *et al.*, 2002) or indirectly, perhaps, through a protein–protein interaction cascade mediated by CFTR's C-terminal PDZ (postsynaptic density protein)-binding sequence (for review see Guggino, 2004). Interestingly, an autocrine mechanism involving CFTR-mediated ATP release was proposed to explain the relationship between CFTR and outwardly rectifying Cl^- channels (Schwiebert *et al.*, 1995).

In addition to its function as a channel regulator, CFTR is, without any doubt, a Cl^- channel itself. It is the only member in the ABC transporter family that functions as an ion channel. The most compelling evidence for CFTR being an ion channel comes from the reconstitution experiments elegantly carried out by Bear *et al.* (1992). When recombinant CFTR proteins, purified from infected insect cells, were incorporated into the planar lipid bilayer, they formed protein kinase A (PKA) and ATP-regulated Cl^- channels that were nonrectifying and had a single-channel conductance of ~ 10 pS under symmetrical Cl^- concentration, the same biophysical characteristics seen in native CFTR-containing epithelia or in cell lines heterologously expressing CFTR (for review see Gadsby and Nairn, 1999). The notion that the CFTR protein itself contains the Cl^- permeation pathway is further supported by the findings that mutations in the putative transmembrane segments, thought to form the channel pore for Cl^- conduction, affect the anion selectivity and conduction (for review see Dawson et al., 1999). Although the mechanism of CFTR channel permeation is not covered in this chapter due to limited space, it is no doubt another important issue in the CFTR field. Readers interested in this issue are referred to some recent review articles (Dawson et al., 1999; McCarty, 2000; Linsdell, 2006).

CFTR plays important physiological roles in multiple systems, including airways, intestine, pancreas, and sweat duct (for review see Quinton, 1999). More than 1000 different mutations of the CFTR gene have been identified as disease associated. These mutations affect CFTR function one way or the other (for review, see Welsh and Smith, 1993). For instance, one type of mutation, such as deletion of the phenylalanine 508 (ΔF508), leads to a maturation defect in that the mutant CFTR protein cannot be processed appropriately so that it is trapped in the endoplasmic reticulum and subsequently degraded (Cheng et al., 1990; Denning et al., 1992a,b). In this specific category, the ΔF508 mutation is the most common cause of CF and accounts for ~70% of all the CF-associated mutations. Surprisingly, a recent high-resolution crystal structure of NBD1 containing the ΔF508 mutation indicates that the overall structure of NBD1 is little affected by the ΔF508 mutation (Lewis et al., 2005). It is postulated that the local environment is altered such that interactions between NBD1 and the transmembrane segments may be disrupted by the deletion of F508. Another type of mutant protein, such as G551D and G1349D (both mutations are in the signature sequences), can reach the cell membrane; however, their channel activity is greatly decreased (Gregory et al., 1991; Logan et al., 1994). Although the mechanism by which these mutations affect CFTR channel function remains unknown, it was proposed that the mutations may affect dimerization of the NBDs that are essential for channel gating (Gross et al., 2006).

II. REGULATION OF CFTR BY PHOSPHORYLATION/ DEPHOSPHORYLATION

Kinase-mediated phosphorylation is an essential first step for CFTR to function as a Cl^- channel. A unique structural feature of CFTR, lacking in other ABC transporters, is the presence of an R domain, which follows NBD1 immediately (Fig. 1). The R domain contains multiple consensus sequences for PKA and protein kinase C (PKC) (Riordan et al., 1989). Previous studies indicate that phosphorylation of the R domain by PKA or PKC is a prerequisite for CFTR channel activity (Anderson et al., 1991; Tabcharani et al., 1991; Nagel et al., 1992; for reviews see Gadsby and Nairn, 1999; Sheppard and Welsh, 1999; Hanrahan and Wioland, 2005). After the channel is phosphorylated by protein kinase and ATP, the opening and closing (i.e., gating) is controlled by ATP.

Since the physiological intracellular [ATP] is in the millimolar range, a saturating concentration for the ATP-dependent gating process (Zeltwanger et al., 1999; Vergani et al., 2003), phosphorylation thus becomes a critical step to regulate CFTR activity in an intact cell. However, the molecular mechanism underlying this regulation is still poorly understood. It was thought previously that unphosphorylated R domain inhibits channel activity; phosphorylation results in dissociation of the R domain from an inhibitory site (Ma et al., 1997; Winter and Welsh, 1997). This idea is further supported by the observation that removal of the R domain partially (Rich et al., 1993) or completely (Csanády et al., 2000; Bompadre et al., 2005a) renders the channel phosphorylation-independent. However, a recent study shows that phosphorylation of the R domain may actually promote the interaction of the R domain with other domains (Chappe et al., 2005).

Mutagenesis studies indicate that phosphorylation of individual serines may not play an equal role in regulating channel activity. CFTR contains multiple PKA consensus sequences (R/K-R/K-X-S/T) in the R domain; eight of them are shown to be phosphorylated either *in vivo* or *in vitro* (Cheng et al., 1991; Picciotto et al., 1992; Neville et al., 1997). It is quite a challenge to dissect the distinct function of phosphorylation of individual sites on channel activity given the large number of consensus phosphorylation sites present in CFTR. Although different degree of phosphorylation may yield different level of channel activity (Hwang et al., 1993), the effect of phosphorylation is not simply additive. It has been suggested that there is functional redundancy among them since none of the consensus sites is essential for channel function and there is a graded decrease of channel activity with an increased number of serine residues mutated (Cheng et al., 1991; Rich et al., 1993). To make matter even more complicated, phosphorylation of two serines, S737 and S768, were found to have an inhibitory effect on channel activity (Wilkinson et al., 1997; Csanády et al., 2005b). Using mass spectrometric analysis, Csanády et al. (2005b) demonstrated that S768 is the first serine in the R domain to be phosphorylated. Detailed single-channel kinetic analysis showed that phosphorylation of S768 exerts its inhibitory effect by shortening the burst duration. How these inhibitory phosphorylation sites modulate CFTR function in Cl^- secretion or absorption in intact epithelia remains unclear.

An unsolved issue regarding phosphorylation-dependent activation of CFTR is whether all physiologically relevant phosphorylation sites are located in the R domain. CFTR channels with partial or entire R domain deleted (ΔR-CFTR) are constitutively active in the presence of ATP without prior phosphorylation (Rich et al., 1993; Csanády et al., 2000; Bompadre et al., 2005a) and are not responsive further to PKA stimulation (Rich et al., 1993; Bompadre et al., 2005a; cf. Csanády et al., 2000), supporting the notion that all phosphorylation sites that are involved in regulating channel activity are located in the R domain. However, when all 10 PKA consensus sites (9 in the R domain and 1, S422, in NBD1) were mutated to alanines (10SA), the channel still showed significant PKA-stimulated activity (Chang et al., 1993), suggesting sites other than the classic dibasic sites in the R domain or sites outside of the R domain are also involved in channel regulation (Csanády et al., 2000). The finding that a mutant CFTR with 15 serines (monobasic and dibasic), all in the R domain, mutated to alanines completely abolishes the effect of PKA seems to support the former possibility (Seibert et al., 1999). However, it should be noted that S422, 1 of the 10 classical dibasic PKA consensus sites, is actually located in NBD1. According to recently available high-resolution structures of the mouse and human CFTR NBD1, S422 is located in a region called regulatory insertion, a structurally flexible segment that is lacking in other ABC proteins (Lewis et al., 2004, 2005; Section IV.A). Phosphorylation of S422 is suggested to contribute to the regulatory role of the insertion segment for CFTR function (Chang et al., 1993), but this idea remains to be substantiated. In addition, it has been reported that a synthetic peptide containing a series of residues near the C-terminus of CFTR can be phosphorylated by PKA *in vitro* (Gadsby and Nairn, 1999), though the functional consequence remains to be elucidated. Furthermore, direct biochemical evidence of *in vivo* phosphorylation of the sites beyond the R domain is needed to validate the functional role of these sites.

PKC can also phosphorylate CFTR (Picciotto et al., 1992), but how PKC phosphorylation regulates CFTR channel function remains unsettled. Some studies showed that PKC can activate CFTR channels directly by phosphorylating consensus PKC sites, such as T604, S686, and S790, although the degree of activation is lower than that with PKA (Tabcharani et al., 1991; Berger et al., 1993; Chappe et al., 2003, 2004). On the other hand, PKC also potentiates CFTR activity elicited by PKA (Tabcharani et al., 1991; Jia et al., 1997; Chappe et al., 2003). As high as eightfold potentiation of PKA-dependent activity by PKC was observed for the *Xenopus laevis* isoform of CFTR (XCFTR) expressed in *Xenopus* oocytes. More recently, a permissive role of PKC phosphorylation was proposed by Jia et al. (1997). This idea is based on the observation that the decline in CFTR's responsiveness to PKA in excised patches can be prevented by the presence of PKC. However, if PKC indeed has the permissive role, it is puzzling why PKA and ATP alone can activate CFTR channels after the CFTR protein being purified and reconstituted into a pure lipid bilayer (Bear et al., 1992). In addition, the observation that mutant CFTR, with altered PKC consensus sequences, is still responsive to PKC leads to the possibility that PKC's effect on CFTR channel activity could be indirect, probably through phosphorylating ancillary proteins (Yamazaki et al., 1999; Button et al., 2001). Although this latest idea nicely explains the nonessential role of PKC consensus sites in CFTR, it is still puzzling why PKC is not required to activate purified CFTR in bilayers. To complicate the matter even further, recent studies show that some of the PKC consensus sites, such as S641 and T682, may play an inhibitory role when phosphorylated (Chappe et al., 2004). More detailed studies are needed to elucidate the molecular mechanism underlying PKC modulation of CFTR function.

The level of phosphorylation of the CFTR protein, a dynamic process controlled by both the activities of protein kinases and protein phosphatases, affects the channel activity (Cheng et al., 1991; Hwang et al., 1994; Mathews et al., 1998; Wang et al., 1998; Szellas and Nagel, 2003). Therefore, understanding how phosphatases affect CFTR channel activity is as important. Previous studies, using exogenous phosphatases or specific phosphatase inhibitors show that type 2A protein phosphatase (PP2A) and type 2C protein phosphatase (PP2C) are the likely candidates that regulate CFTR channel activity physiologically (Berger et al., 1993; Hwang et al., 1993; Travis et al., 1997; Yang et al., 1997; Luo et al., 1998; Zhu et al., 1999). Biochemical and functional studies demonstrate that PP2A and PP2C can interact directly with the CFTR protein, dephosphorylate the channel, and down-regulate channel activity (Zhu et al., 1999; Thelin et al., 2005). It has been observed consistently that CFTR currents show a significant run-down in isolated membrane patches excised from either native epithelial cells or cell lines expressing recombinant CFTR (Tabcharani et al., 1991; Becq et al., 1994; Haws et al., 1994; Luo et al., 1998; Weinreich et al., 1999; Vergani et al., 2003), suggesting the existence of membrane-bound phosphatases. Since the deactivation time course by exogenous PP2C is similar to that of current run-down, it was suggested that PP2C may be the major player responsible for the observed current run-down (Luo et al., 1998). The current run-down due to membrane-bound phosphatases presents serious technical difficulties in obtaining stationary single-channel recordings for kinetic studies of CFTR gating. For example, oocyte membranes are shown to contain very high level of phosphatase activities (Weinreich et al., 1999). After macroscopic CFTR current is activated by PKA and ATP in excised inside-out patches from oocytes, removal of PKA results in a drastic reduction of the current within seconds (Weinreich et al., 1999; Szellas and Nagel, 2003). This robust membrane-bound phosphatase activity poses a colossal challenge for studies of ATP-dependent gating since the level of phosphorylation has an immense impact on channel kinetics (Cheng et al., 1991; Hwang et al., 1994; Mathews et al., 1998; Wang et al., 1998; Szellas and Nagel, 2003).

Other phosphatases, such as PP1 and alkaline phosphatase, are not likely involved in regulating CFTR channels physiologically. Whether PP2B regulates CFTR channels remains controversial. PP2B can inactivate CFTR channels in some expression systems, such as NIH-3T3 cells, both intact and cell-attached mode (Fischer et al., 1998), but not in others, such as intact Calu-3 cells and intact T84 cells (Berger et al., 1993; Travis et al., 1997; Fischer et al., 1998). It is safe to conclude that multiple phosphatases are involved in regulating CFTR channel activity (Hwang et al., 1993). We are yet to learn how different phosphatases dephosphorylate different phosphoserine residues, how these phosphatases target CFTR in the membrane, and how protein kinases and phosphatases work in concert to regulate CFTR channel function under physiological conditions.

CFTR's activity is also regulated by other intracellular components. For instance, AMP-activated kinase (AMPK), a physiologically important sensor and regulator in cellular processes, can decrease the open probability of CFTR through interaction with the C-terminus of CFTR (for review see Hallows, 2005). On the other hand, the N-terminus of CFTR can interact directly with proteins such as SNAP-23 and syntaxin 1A, proteins involved in membrane fusion, resulting in inhibition of CFTR channel activity (Naren et al., 1997, 1998; Chang et al., 2002; Cormet-Boyaka et al., 2002). Other phosphorylation-independent regulation of CFTR channel activity has been proposed recently. Himmel and Nagel (2004) show that phosphatidylinositol 4,5-bisphosphate (PIP_2) is able to activate CFTR channels in the presence of ATP without prior PKA-dependent phosphorylation. It has been documented previously that PIP_2 regulates ion channels, such as inwardly rectifying K channels (Huang et al., 1998) and the epithelial Na channel (ENaC) (Yue et al., 2002), and transporters, such as Na^+/Ca^{2+} exchangers (Hilgemann and Ball, 1996), through a direct interaction (Hilgemann et al., 2001). It would be interesting to see whether PIP_2 interacts with CFTR directly and if it does so, with which region of CFTR. Interestingly, intracellular glutamate (Reddy and Quinton, 2003) and deoxy-ATP (dATP) (Aleksandrov et al., 2002a) can also activate CFTR channels bypassing the requirement of phosphorylation. These results suggest the existence of alternative pathways to activate CFTR independent of protein phosphorylation.

III. CFTR IS AN ATPase

Although phosphorylation is a prerequisite for CFTR function, after the channel is phosphorylated, it is ATP that plays a critical role in controlling the opening and closing of the phosphorylated CFTR. Gating patterns of CFTR channels are characterized by the appearance of long bursts of opening interrupted by short and fast flickery closures. Figure 3 shows a representative single channel current trace of wild-type CFTR (WT-CFTR) in the presence of 2.75-mM ATP. Closed time histograms can be well fitted by a double-exponential function, indicating that there are at least

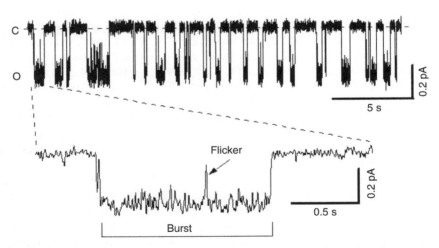

Figure 3. Phosphorylated CFTR channel is gated by ATP. Representative single-channel current trace of WT-CFTR in response to 2.75-mM ATP after prior activation by PKA and 1-mM ATP. Channel openings (O) (downward deflection) are separated by long closures (C) (top). Note that in the expanded trace (below) there are very brief closures (i.e., flicker) within the long opening (i.e., burst). Different analytical methods have been used to extract kinetic parameters of ATP-dependent gating events by excluding flickery closures. In this chapter, the open time is referred to the bursting opening; whereas the closed time depicts the interburst duration.

two closed states with different dwell times (Gunderson and Kopito, 1994). The fast one, with a time constant of ~10 ms (i.e., flickers), is ATP-independent and is presumably due to voltage-dependent block by unknown intracellular blockers (Tabcharani et al., 1991; Zhou et al., 2002). The slow one, corresponding to the long interburst event, shows apparent ATP dependence. To understand how ATP gates CFTR, it is essential to isolate ATP-dependent events, with as less contamination of flickers as possible, for kinetic analysis (see Powe et al., 2002b for details). The voltage-dependent component of CFTR gating has been studied extensively by Cai et al. (2003). In this chapter, however, we will focus on the molecular mechanism underlying the ATP-dependent gating.

It is generally accepted that ATP not only serves as a ligand for CFTR gating, during the gating cycle, ATP is also hydrolyzed by CFTR (Gadsby and Nairn, 1999). Biochemical studies show that, as other members of the ABC transporter family, the intact CFTR protein, as well as CFTR's NBD2 can hydrolyze ATP (Li et al., 1996; Randak et al., 1997). On the contrary, recent biochemical and crystallographic studies provide compelling evidence that NBD1 does not hydrolyze ATP presumably because a glutamate residue that is critical for ATP hydrolysis at NBD2 is replaced by a serine at the equivalent position of NBD1 (Lewis et al., 2004, 2005). Although details of CFTR-gating mechanisms differ in different reports, it is generally accepted that the energy from ATP hydrolysis is harvested to drive the gating cycle (Gunderson and Kopito, 1994; Hwang et al., 1994; Carson et al., 1995; Zeltwanger et al., 1999; Vergani et al., 2003; Berger et al., 2005). Studies from many research groups using NBD mutations and nonhydrolyzable ATP analogues demonstrate that CFTR gating involves not only ATP binding but also its hydrolysis. This interesting feature of CFTR gating distinguishes it from other ligand-gated channels whose gating involves only the binding and unbinding of the ligand. However, this input of free energy into gating transitions makes the study of the gating mechanism of CFTR more challenging since it is often difficult to separate the functional roles of ATP binding and ATP hydrolysis in CFTR gating. The recent breakthrough in solving the crystal structures of the mouse and human CFTR NBD1 (Lewis et al., 2004, 2005) has allowed investigators to take advantage of structure-guided mutagenesis approaches to further our understanding on how CFTR's NBDs control gating motions. In this chapter, we will provide a historical background of gating studies; more importantly, we will emphasize the structural mechanism of CFTR gating that starts to emerge in the post-crystallography era.

IV. STRUCTURAL BIOLOGY OF ABC TRANSPORTERS

A. Crystal Structures of ABC Transporters

High-resolution crystallographic structures of ion channel proteins have allowed electrophysiologists to ask detailed mechanistic questions at a submolecular level. Even though most of the K^+ channel structures are obtained from bacterial channels (Doyle et al., 1998; Jiang et al., 2002, 2003; cf. Long et al., 2005a,b), the rich information unveiled by the structure is truly remarkable. Unfortunately, scientists in the CFTR field cannot enjoy this luxury since CFTR does not have an orthologue in bacteria and CFTR is the only member in the ABC transporter family that works as an ion channel. Although the crystal structure of other ABC family members has limited values in shedding light on how CFTR catalyzes Cl^- movement, the structure information for their NBDs can be very informative for CFTR structure–function studies since increasing evidence reveals that the NBDs of all ABC family members share similar basic architectures (Higgins and Linton, 2004). To date, high-resolution structures of several ABC proteins, including BtuCD, an ABC transporter mediating vitamin B_{12} uptake (Locher et al., 2002), and MsbA, an ABC transporter that transports Lipid A across the membrane (Chang and Roth, 2001; Chang, 2003; Dong et al., 2005; Reyes and Chang, 2005). NBDs of many ABC proteins, including HisP, the ATP-binding subunit of the histidine permease (Hung et al., 1999), MalK, the ATPase subunit of the maltose transporter from *Escherichia coli* (Diederichs et al., 2000), and MJ0796 and MJ1267, NBDs of two bacterial ABC transporters (Moody et al., 2002; Smith et al., 2002), have also been solved. More recently, crystal structures of NBD1 of the mouse and human CFTR became available (Lewis et al., 2004, 2005). Not surprisingly, they show a high structural homology with known NBD structures.

A distinct and surprising difference between the mouse and human NBD1 is the way the adenine ring of the nucleotide is coordinated. In the mouse CFTR NBD1, three residues, tryptophan 401 (W401), leucine 409 (L409, an F409 in human CFTR NBD1), and F430, are in close contact of the adenine ring. Interestingly, only one of these three positions, that is, F430, was identified previously by simple sequence alignment as one of the important residues that interact with ATP (Berger and Welsh, 2000). Three (F429, F433, and F446) out of five aromatic residues in NBD1 identified by sequence alignment to be important for ATP binding turn out to be far away from the ATP-binding pockets. W401 and L409 were never suspected to be close to the adenine ring of the bound ATP until the mouse CFTR NBD1 structure was solved. On the other hand, in the human CFTR NBD1 structure, W401 is the only residue stacking with the adenine ring of the bound nucleotide in a way seen in most NBD structures (Schmitt and Tampe, 2002; Lewis et al., 2005). The reason for this discrepancy is unclear. Is it simply due to species difference? Alternatively, numerous artificial mutations introduced into human CFTR's NBD1 for optimal crystallization may cause distortion of the structure. A more interesting possibility is that these minor differences may reflect different NBD1 structures in different functional states. Since NBD1 exhibits a higher affinity to 5′-adenylylimidodiphosphate (AMP-PNP) than NBD2 (Aleksandrov et al., 2001), it is also interesting to test if multiple aromatic amino acids present in human CFTR's NBD1 are responsible for a tight binding of nucleotide at NBD1. We are currently tackling the functional role of these aromatic amino acids in CFTR gating. Some of the preliminary results will be discussed.

A unique structural feature of CFTR NBD1 is the presence of a regulatory insertion region (amino acids 404–435) and a regulatory extension at the C-terminus of NBD1 (amino acids 639–670), which are lacking in NBDs of other ABC transporters (Lewis et al., 2004, 2005). These two segments contain several consensus serine residues for

PKA-dependent phosphorylation (S422, S660, and S670) (Cheng et al., 1991, 1993; Wilkinson et al., 1997; Winter and Welsh, 1997; Lewis et al., 2004). Although recent studies suggest they may not play a critical role since CFTR channels with part of these two segments deleted can still form functional channels (Chan et al., 2000; Csanády et al., 2005a), detailed single-channel kinetic analysis revealed that the deletion of part of the regulatory insertion (amino acids 415–432) of NBD1 shortens the open time by ~30%, suggesting that the regulatory insertion may be involved in stabilizing the open state (Csanády et al., 2005a). However, it is not clear at this moment whether this shortening of the mean open time is due to removal of the consensus PKA phosphorylation site S422 or due to deletion of other residues in this segment. It is noteworthy that the two aromatic residues, F409 and F430, which are revealed in the crystal structure to be close to the adenine ring of ATP, are located in the regulatory insertion (Lewis et al., 2004, 2005). Our preliminary studies suggest that mutations of these residues can affect the stability of the open state (Zhou et al., in press). These results raise the interesting possibility that ATP binding at NBD1 may play a role in determining the stability of the open state (Section VI.B).

Another piece of important information revealed by the CFTR NBD1 structures is the inability of NBD1 to hydrolyze ATP. It was thought previously that ATP hydrolysis at NBD1 opens the channel (for review see Gadsby and Nairn, 1999). However, recent biochemical studies demonstrate that CFTR's NBD1 does not hydrolyze ATP at an appreciable rate (Szabo et al., 1999; Aleksandrov et al., 2002b; Basso et al., 2003). These biochemical results are further confirmed by the crystal structure of CFTR's NBD1 (Lewis et al., 2004, 2005). When the mouse NBD1 was crystallized in the presence of ATP, the γ-phosphate of Mg-ATP remained intact. In addition, the ADP-bound and the ATP-bound NBD1 have the same conformation. Furthermore, the structure of CFTR's NBD1 reveals that two residues (a glutamate and a histidine), important in interactions with the phosphate group of ATP for ATP hydrolysis, are replaced by two serines, S573 and S605, respectively. This structural asymmetry of the catalytic site for CFTR's two NBDs is not limited to the ABCC subfamily to which CFTR belongs; a few members of the ABC proteins in other subfamilies also exhibit this similar asymmetry (Jones and George, 1999). Does this structural asymmetry reflect a functional asymmetry? Crystal structures of CFTR's NBDs surely provide an excellent opportunity to examine this intriguing possibility.

Although a high-resolution structure of the whole CFTR protein is still lacking due to the difficulty to obtain sufficient amount of the CFTR protein and the low solubility of the protein, the first low-resolution (~20 Å) structure of CFTR from two-dimensional crystals in the presence of AMP-PNP, a nonhydrolyzable ATP analogue, was solved recently (Rosenberg et al., 2004; Awayn et al., 2005). The overall structure of CFTR is similar to that of P-glycoprotein (Rosenberg et al., 2001, 2003). Two different structures of CFTR were observed. It was speculated that one of the structure may represent the open channel conformation since AMP-PNP was included in the crystallization solution. It should be noted, however, that the opening events observed in the presence of AMP-PNP alone (in the complete absence of ATP) are short-lived (Vergani et al., 2003; Cho and Hwang, unpublished data), indicating that

AMP-PNP-opened channels exhibit unstable open conformations. It is, therefore, somewhat uncertain at this point whether the observed low-resolution structures of CFTR truly represent the open state of the channel.

B. The Two NBDs of ABC Proteins Form a Dimer

Although early reports suggest that the NBDs of the ABC transporters exist as monomers (Karpowich et al., 2001; Yuan et al., 2001), there is growing evidence for dimerization of the two NBDs (Hung et al., 1999; Moody et al., 2002; Smith et al., 2002; Chen et al., 2003; Verdon et al., 2003; Zaitseva et al., 2005). Hung et al. (1999) proposed the first dimeric structure of the NBDs for a member of the ABC transporter family. The two NBDs of HisP, the nucleotide-binding subunit of the histidine permease, adopt a back-to-back dimer conformation. However, this dimer structure is probably not physiological for several reasons. First, the bound ATP is unusually exposed to the solvent, unlike other ATPase with active sites embedded in the protein. Second, the dimer interface in HisP is not very extensive, raising the possibility that the back-to-back dimeric configuration is due to crystallization artifacts. Third, the back-to-back dimer places two ATP-binding sites far away from each other. It is thus difficult to explain biochemical data suggesting the presence of cooperativity between the two active sites (Davidson et al., 1996; Fetsch and Davidson, 2003). The first head-to-tail dimer structure was revealed in Rad50CD, a DNA repair enzyme that shares homology with the NBDs of ABC family members (Hopfner et al., 2000). The NBD of MJ0796 is the first NBD in the ABC family that was shown to form a head-to-tail homodimer (Smith et al., 2002). Many newly resolved structures of NBD dimers (Chen et al., 2003) also show a head-to-tail configuration (Fig. 2). In all head-to-tail dimers of NBDs, the two ATP-binding sites are located at the dimer interface. In those dimeric structures with bound nucleotides, it was shown that ATP molecules interact extensively with amino acids from each NBD. This new dimer structure not only explains biochemical data (Fetsch and Davidson, 2003), but also places the signature motif (LSGGQ) that defines the ABC transporter family at the dimer interface participating in interactions with ATP. Most importantly, this head-to-tail configuration of NBD dimers corroborates with the holoenzyme structures of the *E. coli* proteins BtuCD (Locher et al., 2002) and MsbA (Dong et al., 2005; Reyes and Chang, 2005).

The evidence for NBD dimerization goes beyond crystallographic data. Using an analytic gel filtration assay, Moody et al. (2002) showed that MJ0796 and MJ1267, two bacterial ABC transporters' NBDs, can form dimers in the presence of ATP, but not with ADP or AMP-PNP. Since the dimeric configuration was only seen in mutants whose ATPase activity is abolished, it was hypothesized that ATP hydrolysis by wild-type NBD dimers provides the energy to break the stable dimer formation (Moody et al., 2002; Smith et al., 2002). Later, Chen et al. (2003) showed that MalK can assume two different dimeric structures depending on whether the structures contain ATP. In the absence of bound ATP, the structure shows a wide gap between the two NBDs. On the other hand, the gap is closed in the structure with bound ATP molecules.

From these snap shots of NBD dimers, Chen et al. (2003) proposed a tweezers-like motion on ATP binding to NBDs.

It should be noted that in all the head-to-tail dimer structures published so far, there are extensive hydrogen bonds and van der Waals interactions between ATP and Walker A and Walker B sequences in one subunit, and between ATP and the LSGGQ motif in the other subunit. These extensive binding forces suggest that the dimeric configuration is an energetically stable state. Thus, it seems reasonable to propose that ATP binding elicits a large degree of molecular motion that involves closing of the dimer interface and that only hydrolysis of bound ATP could provide sufficient energy for fast dimer dissociation (Hopfner et al., 2000; Smith et al., 2002). Although there is so far no direct structural evidence of NBD dimerization for CFTR, Vergani et al. (2005) recently performed elegant functional studies to demonstrate for the first time that the dimerization of the two NBDs of CFTR is associated with the open state of the channel. By using mutant cycle analysis, they showed that R555 and T1246, residues located in NBD1 and NBD2 respectively, are energetically coupled when the channel is in the open state, whereas there is no such coupling when the channel is closed. If the two NBDs of CFTR do form a dimer, then each binding pocket in a dimeric structure consists of molecular components from both NBDs. Therefore, for the sake of clarity, we define the NBD1 ATP-binding site (or the NBD1 site) as the binding pocket containing Walker A and Walker B motifs in the NBD1 sequence and the signature sequence of the NBD2. An equivalent definition is applied to the NBD2 site (Fig. 2).

V. METHODS USED TO STUDY CFTR GATING

Gating of CFTR is different from that of other ligand-gated channels because it uses the free energy from ATP hydrolysis to drive the gating cycle. Therefore, it is important to distinguish the effect of ATP binding from that of ATP hydrolysis on channel gating, a somewhat difficult task. In addition, previous research demonstrates that gating of CFTR may involve binding of two ATP molecules at two separate sites (Gunderson and Kopito, 1994; Hwang et al., 1994). Gating mechanisms involving two ligand-binding sites are, in principle, more complicated than that with only a single binding site. To complicate things even further, ATP is also required for PKA-dependent phosphorylation, a step prior to ATP-dependent gating of the channel. The potential presence of active protein kinases in the membrane patches (unpublished data) makes it a daunting task to dissect effects of ATP on the multiple processes involved in controlling CFTR function. Fortunately, over the years, many methods and tools have been developed to meet this challenge. We will briefly summarize these methods in this section (details can be found in Powe et al., 2002b).

A. Expression Systems

Native epithelial cell lines with endogenous CFTR, such as T84, a human colon carcinoma cell line, and Calu-3, a human subbronchial gland cell line, are especially useful to study CFTR channel regulation, and effects of pharmacological reagents in a more physiological setting (Tabcharani et al., 1990; Al-Nakkash and Hwang, 1999; Devor et al., 1999; Marcet et al., 2004). Readers interested in the pharmacology of CFTR channels are referred to the review article by Dr. Sheppard in this volume (Chapter 5; see also Hwang and Sheppard, 1999). On the other hand, gating studies have benefited greatly from studying exogenously expressed, recombinant CFTR in cells naturally lacking its expression. The commonly used cells include Chinese Hamster Ovary (CHO) cells (Tabcharani et al., 1991; Bompadre et al., 2005a,b; Zhou et al., 2005a), NIH-3T3 cells (a mouse fibroblast cell line) (Zeltwanger et al., 1999; Powe et al., 2002a; Zhou et al., 2002), baby hamster kidney (BHK) cells (Luo et al., 1998; Chappe et al., 2004, 2005), Sf9 insect cells (Kartner et al., 1991; Travis et al., 1993; Szabo et al., 1999), and *Xenopus* oocytes (Szellas and Nagel, 2003; Vergani et al., 2003; Chen et al., 2004). Different types of cells have their distinct advantages and often time disadvantages. For instance, patches excised from BHK cells have relatively slower and less degree of current run-down, a characteristic that is useful in studying the effect of exogenous phosphatases (Luo et al., 1998). On the other hand, for most overexpression systems, since the channel density is very high, it is difficult to obtain patches with a limited number of channels for single-channel kinetic studies.

B. Electrophysiological Recording Systems and Technical Hurdles

An ideal way to study the function of an ion channel without the interference of other intracellular proteins is to purify the channel protein and reconstitute it into lipid bilayers. However, this method is technically challenging and only very few laboratories in the CFTR field are equipped to carry out this type of experiment (Bear et al., 1992). In addition, when studies involve multiple CFTR constructs, it becomes practically impossible to employ this costly method. Therefore, most CFTR researchers perform electrophysiological recordings either on lipid bilayers incorporated with CFTR proteins from membrane vesicles (Gunderson and Kopito, 1995; Ma et al., 1997; Aleksandrov and Riordan, 1998) or on excised membrane patches from cells expressing CFTR channels (Anderson et al., 1991; Tabcharani et al., 1991). As a result of membrane-bound phosphatases remaining after patch excision, the level of CFTR phosphorylation can change throughout the course of the experiment, consequently altering channel-gating kinetics. This dephosphorylation of CFTR in excised patches is especially problematic in the *Xenopus* oocyte expression system (Weinreich et al., 1999; Szellas and Nagel, 2003; Section II). This reduction of current can be recovered by reapplication of PKA, indicating a robust dephosphorylation of CFTR by membrane-bound phosphatases. The dephosphorylation issue can be especially detrimental when one intends to quantify mutants with defective (i.e., less) surface expression (e.g., K464 mutations in Cheng et al., 1990) since the fraction of channels dephosphorylated throughout the recording could be relatively high. Difficulty in maintaining CFTR in fully phosphorylated states may partly account for the wide variation of the closed time constant reported in the literature (Vergani et al., 2003;

Csanády et al., 2005a). Thus, to examine CFTR gating more rigorously, a proper bracketing procedure—applying a control condition for activity before and after the test condition—is necessary to reduce the influence of current run-down in excised patches (Zeltwanger et al., 1999; Powe et al., 2002a,b; Vergani et al., 2003). It should be noted, however, bracketing by itself does not solve the problem of CFTR dephosphorylation and subsequent functional consequences. Careful kinetic analyses of CFTR gating rely critically on investigators' awareness of this issue and subsequent scrutiny of the raw data. An apparent time-dependent decay of the channel activity should serve as a warning sign that precise kinetic analysis is in question. In searching for solution to this problem, various mutant CFTR constructs with an R domain deletion were made to bypass the requirement of phosphorylation. Earlier studies showed that deletion of part of the R domain (deletion of amino acids 708–835) does not alter the anion selectivity sequence and the single-channel conductance (Anderson et al., 1991; Rich et al., 1991). Most importantly, it has constitutive ATP-dependent Cl^- channel activity in the absence of PKA stimulation. Recently, Csanády et al. (2000) made a CFTR construct with the R domain (amino acids 634–836) completely deleted (ΔR-CFTR). This channel, when expressed in CHO cells, shows similar ATP-dependent gating kinetics to that of WT-CFTR (Bompadre et al., 2005a). ΔR-CFTR channels can be gated by ATP alone without prior treatment of PKA (Csanády et al., 2000; Bompadre et al., 2005a). In fact, addition of PKA does not increase the activity of ΔR-CFTR (Bompadre et al., 2005a; cf. Csanády et al., 2000). Thus, this construct allows researchers to bypass the phosphorylation-dependent steps, avoid channel run-down by dephosphorylation, and isolate ATP-dependent gating process. In addition, the relatively low expression level of ΔR-CFTR channels makes it much easier to obtain recordings from patches containing a single channel, which is somewhat more difficult for WT-CFTR channels (Bompadre et al., 2005a).

An alternative way to keep a steady level of phosphorylation of CFTR proteins was provided recently by Szellas and Nagel (2003). These investigators employed a very clever strategy to engineer a membrane-bound PKA. Specifically, the soluble catalytic subunit of cAMP-dependent protein kinase was fused with bacteriorhodopsin (bR-PKA), an integral membrane protein. After coexpression of bR-PKA and CFTR in oocytes, large CFTR currents could be elicited by ATP alone without prior exposure to PKA in excised inside-out patches. Most importantly, the rate of current run-down due to membrane-bound phosphatases was decreased when the channel was phosphorylated by bR-PKA. In addition, the apparent affinity of CFTR to ATP obtained from Michaelis–Menten fit to the ATP dose–response relationship was increased by twofold when PKA was engineered to be associated with the membrane. It is likely that the continuous presence of bR-PKA in the excised patch can keep a higher level of phosphorylation of CFTR channels. However, due to the constant phosphorylation (by bR-PKA) and dephosphorylation (by membrane-bound phosphatases) of CFTR, this method, in theory, cannot avoid blending of the phosphorylation and dephosphorylation steps with the ATP-dependent gating processes. It would be interesting to see if the kinetic parameters for CFTR in this modified oocyte system are significantly different from those obtained more traditionally.

C. Nonhydrolyzable and Hydrolyzable Nucleotide Analogues

Since both ATP binding and hydrolysis are involved in CFTR gating, studies of CFTR gating have been greatly aided by various nucleotide analogues. Depending on whether they can be hydrolyzed, these analogues can be categorized into two major groups: nonhydrolyzable (or poorly hydrolyzable) nucleotide analogues and hydrolyzable nucleotide analogues. Most commonly used nonhydrolyzable nucleotide analogues include AMP-PNP, adenosine 5′(β,γ-methylene) triphosphate (AMP-PCP), and ATPγS. These nonhydrolyzable nucleotide analogues are useful to explore the functional role of ATP hydrolysis in CFTR gating (Gunderson and Kopito, 1994; Hwang et al., 1994). On the other hand, hydrolyzable nucleotide analogues, such as GTP, ITP, UTP, CTP, ADP, 2′-dATP and 3′-dATP, and N^6-(2-phenylethyl)-ATP (P-ATP), are valuable to study the functional role of ATP binding in CFTR gating (Anderson et al., 1991; Aleksandrov et al., 2002a; Zhou et al., 2005a).

D. Mutagenesis of the Two NBDs

For obvious reasons, key residues critical for ATP hydrolysis that have been exploited by site-directed mutagenesis approaches include Walker A lysines (K464 and K1250A), Walker B aspartates (D571 and D1370), and the catalytic base E1371 in NBD2 (Anderson and Welsh, 1992; Gunderson and Kopito, 1995; Ramjeesingh et al., 1999; Zeltwanger et al., 1999; Powe et al., 2002a; Vergani et al., 2003; Bompadre et al., 2005b). While this classical approach has produced tremendous amounts of data that enrich our understanding of how ATP hydrolysis controls CFTR gating, the apparent effect of mutations at these residues on ATP hydrolysis, especially at NBD2, makes it difficult to assess the role of ATP binding by using the same mutants. On the other hand, the high-resolution structures of CFTR's NBD1 reveal exactly which amino acid residue(s) interact with the bound ATP (Lewis et al., 2004, 2005). Thus, this opens the door for designing strategies that affect ATP binding with minimal effect on ATP hydrolysis (Section VI.B). As mentioned above, many CF-associated mutations are located in the signature sequence of NBDs, for example, G551D. Although it has been demonstrated that the channel open probability (P_o) is greatly reduced by these mutations, surprisingly, little is known about the mechanism responsible for their defective function (Gregory et al., 1991; Logan et al., 1994). Nevertheless, combining structural biology and bioinformatics with real-time functional assays, research on the structural basis of CFTR gating is expected to blossom in the near future (Vergani et al., 2005).

VI. CFTR GATING BY ATP

A. The Role of ATP Hydrolysis in Gating

The importance of ATP hydrolysis in CFTR gating is implicated from earlier studies showing that hydrolyzable nucleotide triphosphates, such as ATP, GTP, ITP,

CTP, and UTP, are required to open phosphorylated CFTR channels (Anderson et al., 1991). Nonhydrolyzable ATP analogues, such as AMP-PNP, AMP-PCP, and ATPγS, fail to open the channel (Anderson et al., 1991; Nagel et al., 1992; Carson and Welsh, 1993; Gunderson and Kopito, 1994; Hwang et al., 1994; Schultz et al., 1995). In addition, Mg^{2+}, a cofactor for ATP hydrolysis, is required for ATP to efficiently open the channel (Anderson et al., 1991; Li et al., 1996; Dousmanis et al., 2002). As a member of the ABC transporter family, CFTR has consensus sequences in the two NBDs that are characteristic for ATPase activity and biochemical studies provide direct evidence that intact CFTR can act as an ATPase (Li et al., 1996). Purified CFTR proteins exhibit intrinsic ATPase activity with a hydrolysis rate compatible to the rate of gating.

The functional role of ATP hydrolysis in CFTR gating has also been examined by studying the effect of mutations of critical residues involved in ATP hydrolysis and by utilizing nonhydrolyzable ATP analogues. When the NBD1 Walker A lysine is mutated to an alanine (K464A), the opening rate decreases (Gunderson and Kopito, 1995; Vergani et al., 2003; cf. Sugita et al., 1998; Ramjeesingh et al., 1999; Powe et al., 2002a) while the closing rate is somewhat smaller than that of WT-CFTR (Carson et al., 1995; Ramjeesingh et al., 1999; Powe et al., 2002a; cf. Vergani et al., 2003). Since ATPase activity is dramatically impaired in the K464A mutant, this result is taken as evidence that ATP hydrolysis at NBD1 opens the channel. On the other hand, when the NBD2 Walker A lysine is mutated to an alanine (K1250A), the closing rate decreases dramatically, resulting in so-called "locked open" events that can last for minutes (Gunderson and Kopito, 1995; Ramjeesingh et al., 1999; Powe et al., 2002a; Vergani et al., 2003). Since biochemical studies show that K1250A abolishes ATPase activity of CFTR (Ramjeesingh et al., 1999), the prolongation of the open time seen in K1250A indicates that ATP hydrolysis is critical for channel closing. The notion that ATP hydrolysis precedes channel closing is further supported by the results using nonhydrolyzable ATP analogues. In sweat gland and T84 intestinal epithelia, CFTR channel activity can be enhanced in the presence of both ATP and AMP-PNP (Quinton and Reddy, 1992; Bell and Quinton, 1993). The same effect of AMP-PNP was also reported in guinea pig cardiac myocytes, in C127 cells stably transfected with recombinant CFTR, and in lipid bilayer reconstituted with CFTR (Gunderson and Kopito, 1994; Hwang et al., 1994; Carson et al., 1995). In addition, other nonhydrolyzable ATP analogues, such as ATPγS and polyphosphates, can also lock the channel open in the presence of ATP (Gunderson and Kopito, 1994; Carson et al., 1995). The prolongation of the channel open time by nonhydrolyzable ATP analogues in WT-CFTR channels is reminiscent of that observed in the K1250A mutant, suggesting that they likely bind to the NBD2 site to exert this prolongation effect. Based on the above results, it has been proposed that ATP hydrolysis at NBD1 opens the channel (Hwang et al., 1994) or primes the channel for opening (Gunderson and Kopito, 1994), and that ATP hydrolysis at NBD2 closes the channel. The idea that it is ATP hydrolysis at the NBD1 site that opens the channel may explain why ATP-independent spontaneous openings were not realized until very recently (Bompadre et al., 2005a; also see below).

This once popular gating model can explain many early experimental results; however, some later evidence casts serious doubt on this early model of CFTR gating. Electrophysiological recordings showed that when applied at millimolar concentration, nonhydrolyzable ATP analogues (e.g., AMP-PNP, ATPγS, and AMP-PCP) can by themselves activate CFTR channels, although with a much slower opening rate (Vergani et al., 2003; cf. Aleksandrov and Riordan, 1998). Thus, contrary to previous belief (Gunderson and Kopito, 1994; Hwang et al., 1994), ATP hydrolysis is not required to open the channel. Furthermore, although some earlier studies show that CFTR's NBD1 can hydrolyze ATP (Ko and Pedersen, 1995), more recent biochemical studies using photolabeling experiments with 8-azido-ATP, a hydrolyzable ATP analogue that can gate CFTR channels effectively, show convincing evidence that CFTR's NBD1 does not hydrolyze ATP at an appreciable rate; whereas the NBD2 site can readily hydrolyze ATP (Szabo et al., 1999; Aleksandrov et al., 2002b; Basso et al., 2003). Structural basis for the inability of CFTR's NBD1 to hydrolyze ATP is revealed by the crystal structures of the human and mouse CFTR NBD1. It was shown that the consensus catalytic glutamate is replaced by serine in CFTR's NBD1 (Lewis et al., 2004, 2005). Although it is unlikely that NBD1 can hydrolyze ATP, the fact that the effect of AMP-PNP mimics that of hydrolysis-deficient mutations at the NBD2 site (e.g., K1250A or E1371S) strongly supports the idea that hydrolysis of ATP at the NBD2 site controls the closing of the channel.

What is the structural role of ATP hydrolysis in NBD dimerization? With more structural and biochemical data becoming available, the structural role of ATP hydrolysis in ABC proteins is emerging. As discussed above, it is now accepted that the two NBDs of ABC proteins form a head-to-tail dimer on ATP binding. However, formation of a stable dimer is very difficult to be observed. Although a stable dimer can form in MalK, it is due to the presence of an additional C-terminal domain, absent in most NBDs, which anchors the dimer (Chen et al., 2003). In most cases, a stable dimer is only seen when ATP hydrolysis is diminished by mutagenesis (Moody et al., 2002; Janas et al., 2003; Verdon et al., 2003; Tombline et al., 2004). Even the observed dimer structure of Malk is probably due to the fact that the crystal is formed in the absence of Mg. Nevertheless, these results are taken as evidence to suggest that ATP hydrolysis is the driving force for fast dimer dissociation. Perhaps because of the formation of energetically stable dimers, it is necessary for the ABC transporters to use ATP hydrolysis as the energy source to break the dimer apart quickly. This speedy dissociation of NBDs is essential for an effective transport cycle. Taking these biochemical and structural data of NBD dimers as a hint, Powe et al. (2002a) proposed that CFTR's open state is associated with a dimer of NBDs (also see Hunt, 2002). It was shown that mutating the Walker A lysine at the NBD1 site (i.e., K464A) decreases slightly the open time constant of WT-CFTR, but drastically shortens the locked open time of K1250A mutant CFTR (i.e., K464A/K1250A double mutant), whose ATP hydrolysis at the NBD2 site is diminished. The fact that a mutation at the NBD1 site can affect the phenotype of a mutation at the NBD2 site suggests an interaction between these two ATP-binding sites. Although the physical nature of this interaction was unclear, our recent studies suggest an energetic coupling between these two sites (Bompadre et al., 2005b; Zhou et al., 2005a; Section VI.B).

B. The Role of ATP Binding in Gating

As other ligand-gated ion channels, CFTR channels can open spontaneously, though very rarely, in the absence of ATP. Due to the requirement of ATP for PKA phosphorylation prior to channel opening by ATP, it has always been a concern if these "ATP-independent" gating may result from some residual ATP remained after ATP washout. To circumvent this difficulty, Bompadre et al. (2005a) took advantage of ΔR-CFTR channels that are shown to be phosphorylation independent. When inside-out patches excised from CHO cells transiently expressing ΔR-CFTR were exposed to an ATP-free solution, the spontaneous openings were observed for several minutes, a time long enough for residual ATP to diffuse away from the membrane patch. The calculated spontaneous opening rate for ΔR-CFTR channels is $\sim 0.006\ s^{-1}$. The presence of spontaneous ATP-independent openings indicates that the role of ATP binding in CFTR gating is catalytic rather than permissive. In the presence of ATP, the opening rate of CFTR channels can increase as many as three orders of magnitude presumably because the activation energy for channel opening is lowered by ATP binding to the NBDs of CFTR. It is, however, interesting that the mean open time of spontaneous openings is somewhat longer than that of ATP-dependent openings (Bompadre et al., 2005b). This observation perhaps can explain why [ATP]-dependent open time is not routinely observed (Zeltwanger et al., 1999; Zhou et al., 2005a; cf. Gunderson and Kopito, 1994; Vergani et al., 2003), since, at low micromolar [ATP], inevitable inclusion of a relatively large amount of spontaneous opening events in kinetic analysis could result in an overestimation of the true mean open time for ATP-induced openings.

Where does ATP bind to catalyze channel opening? Are both NBDs involved? Or alternatively binding of one site is sufficient? As a member of the ABC family, CFTR's NBDs consist of consensus sequences, such as Walker A motif, Walker B motif, and signature sequence for ATP binding and hydrolysis. Indeed, biochemical studies have shown that both NBD1 and NBD2 can interact directly with ATP and its analogues (Thomas et al., 1991; Hartman et al., 1992; Travis et al., 1993; Ko et al., 1994). Previous research suggests that these two ATP-binding sites, however, have different affinities for nucleotide with the NBD1 site being the one with a higher affinity (Szabo et al., 1999; Aleksandrov et al., 2001, 2002b; Basso et al., 2003; Zhou et al., 2005a). However, functional studies of the role of each NBD show considerable controversy, despite lots of efforts have been devoted to tackle this fundamental question in the CFTR field.

It has been proposed that ATP binding to both NBDs is required for channel opening (Vergani et al., 2003; Berger et al., 2005). Vergani et al. (2003) reach this conclusion based on the observation that mutations of Walker A lysine (K464 and K1250) or Walker B aspartate in NBD2 (D1370) change the apparent affinity ($K_{1/2}$) of

ATP (Anderson and Welsh, 1992; Vergani et al., 2003). Detailed kinetic analysis suggested that the opening rate is affected by these mutations (Vergani et al., 2003). Assuming these mutations affect ATP-binding affinity, these results suggest that binding at either NBD can be the rate-limiting step preceding channel opening. Berger et al. (2005) also propose that ATP binding at both sites is required for normal channel gating. They show that the opening rate is diminished by introducing bulky entities into either ATP-binding pockets.

Photolabeling with 8-N_3-ATP is decreased as well by these maneuvers. However, it is unclear if the decrease of the opening rate could be rectified by increasing [ATP]. It is important to note that altering the gating process, but not necessarily the binding step, can affect apparent ligand binding (Colquhoun, 1998). Therefore, more careful single-channel kinetic analysis over a wide range of [ATP] is needed. Powe et al. (2002a) characterized both Walker A lysine mutants, K464A and K1250A, and found that the ATP dose–response relationship for K464A is nearly identical to that of WT. Because of the prolonged open time of K1250A (~120 s, Powe et al., 2002a), technically it is difficult to accurately assess the ATP dose–response relationship.

A caveat of mutating conserved Walker A and Walker B residues, particularly at the NBD2 site, is that these mutations likely affect ATP hydrolysis. To circumvent this problem, our laboratory recently focuses on residues that are in the ATP-binding pocket but do not interact directly with the phosphate groups of ATP. In the human NBD1 structure, an aromatic residue, W401, interacts directly with the adenine ring of ATP by a ring–ring stacking mechanism (Lewis et al., 2005). We made homology model of CFTR's NBD2 based on the human CFTR's NBD1 structure and identified Y1219 in NBD2 as the counterpart of W401 in NBD1. When Y1219 is mutated to a glycine (Y1219G), the ATP dose–response relationship shows a dramatic rightward shift with a $K_{1/2}$ of 4.72 ± 1.12 mM, greater than 50-fold higher than that of WT. A more conservative mutation (Y1219W), however, did not change the $K_{1/2}$ value significantly. The ATP dose–response relationships of Y1219F and Y1219I mutants lie between those of WT and Y1219G, suggesting a correlation between changes of the ATP sensitivity and the chemical natures of the side chain at this position. Single-channel kinetic analysis indicates that the shifts of the ATP dose–response relationships in Y1219G and Y1219I mutants are mainly due to the change of the opening rate (Zhou et al., 2004, 2005b, Zhou et al., manuscript in press). In the case of Y1219I, although a slower opening rate is observed at low [ATP] compared to that of WT, the maximal opening rate of Y1219I is similar to that of WT. These results are consistent with the idea that Y1219 mutations change the ATP-binding affinity at the NBD2 site and strongly suggest that ATP binding at the NBD2 site plays a critical role in catalyzing channel opening. Since at 1-mM ATP, when the NBD1 site is expected to be fully occupied, the opening rate of the Y1219G mutant is still extremely small, it is questionable if ATP binding at the NBD1 site alone increases the opening rate.

Although ATP binding at the NBD1 site may not catalyze channel opening, is ATP binding at the NBD1 site required for channel opening by ATP binding at the NBD2 site? The mere existence of spontaneous openings already argues against the absolute requirement of nucleotide binding at either NBD for channel opening.

However, conflicting results are reported regarding whether the K464A mutation, which decreases ATP-binding affinity at the NBD1 site, changes the relationship between [ATP] and the opening rate (Powe et al., 2002a; Vergani et al., 2003). Although the exact reason for this discrepancy is still unknown, different expression systems used by different investigators could be partially responsible. It is known that the level of phosphorylation can affect CFTR channel function, including the opening rate and the ATP sensitivity. It has been suggested that the opening rate, compared to the closing rate, is more sensitive to the degree of phosphorylation of the CFTR channel (Zeltwanger et al., 1999; Powe et al., 2002a; Vergani et al., 2003). In other words, the estimation of the opening rate is more prone to error if CFTR proteins are not fully phosphorylated. Since various expression systems have been used to study the mutational effects of K464A on CFTR gating, it is difficult to evaluate the individual result without knowing the activity of membrane-bound phosphatases in each expression system. For this reason, the result from recordings of purified K464A-CFTR reconstituted into the lipid bilayers where membrane-bound phosphatases are no longer an issue, can serve as a crucial reference. Ramjeesingh et al. (1999) demonstrated that the opening rate of purified K464A-CFTR in the presence of 1-mM ATP is not different from that of WT-CFTR (cf. Vergani et al., 2003).

To look further into the issue of the functional role of ATP binding at the NBD1 site in CFTR gating, we have started to characterize mutations for residues that interact with the adenine ring of ATP at the NBD1 site based on the crystal structures of NBD1 of CFTR (Lewis et al., 2004, 2005). These include W401, F409, and F430 (Lewis et al., 2004). When mutated to a glycine one at a time, none of the three mutants (i.e., W401G, F409G, or F430G) altered the ATP dose–response relationship significantly. Even when all three aromatic residues are mutated to glycines (W401G/F409G/F430G), the ATP dose–response and the opening rate are little affected by this mutation (Zhou et al., 2004, 2005b). These results cast some doubt on the requirement of ATP binding at the NBD1 site for catalyzing channel opening.

If ATP binding at the NBD1 site does not play a role as important as that at the NBD2 site in channel opening, then is ATP binding at the NBD1 site involved in channel closing? Some evidence in the literature already suggests that ATP binding at the NBD1 site may stabilize the open state. First, single-channel kinetic analysis of K464A-CFTR in several different expression systems indicates that the mean open time of this mutant channel is shortened, although some groups report a slight change (Gunderson and Kopito, 1995; Sugita et al., 1998) while others show significant changes (Carson et al., 1995; Powe et al., 2002a; cf. Vergani et al., 2003). Interestingly, it is difficult to lock open K464A-CFTR with AMP-PNP in the presence of ATP. Instead of ~60-s locked open time for WT, the locked open time for K464A is much shorter (~9 s) (Powe et al., 2002a). When the K464A mutation is introduced into K1250A or E1371S, where ATP hydrolysis at NBD2 is abolished, the locked open time is much shorter than that of K1250A or E1371S (Powe et al., 2002a; Vergani et al., 2003; Bompadre et al., 2005a). Second, CFTR channels have multiple open states and the distribution of the open time constants is dependent on [ATP] (Bompadre et al., 2005b). This result was taken as evidence that open channel conformations have different lifetimes depending on whether one or both ATP-binding sites are occupied (Bompadre *et al.*, 2005b). Third, when we examined the detailed single-channel kinetics of inhibition by ADP, an inhibitor that competitively inhibits ATP-induced current by binding to the same site as ATP (Anderson and Welsh, 1992; Schultz *et al.*, 1995; Bompadre *et al.*, 2005a; cf. Randak and Welsh, 2005), we found that ADP not only induces a new closed state (consistent with the idea that ADP and ATP compete for a site for channel opening), but also reduces the mean open time of the channel (Weinreich *et al.*, 1999; Bompadre *et al.*, 2005a; cf. Gunderson and Kopito, 1994; Winter *et al.*, 1994). This shortening of the open time by ADP is more evident in D1370N or E1371S, where the mean open time is prolonged due to impaired ATP hydrolysis (Bompadre *et al.*, 2005b). Fourth, the macroscopic current relaxation time (i.e., locked open time) of E1371S on withdrawal of ATP is shortened by NBD1 mutations (Zhou *et al.*, 2005b). Detailed single-channel kinetic analysis reveals that W401G and W401G/F409G/F430G mutations, although do not affect the closed time, shorten the channel open time by more than 40% at saturating [ATP] (Zhou *et al.*, manuscript in press). Finally, recently we have reported that P-ATP, an ATP analogue greater than 50-fold more potent than ATP, can prolong the open time of WT or the locked open time of E1371S (Zhou *et al.*, 2005a). The locked open state of WT-CFTR is more stable in the presence of P-ATP plus AMP-PNP than that with ATP plus AMP-PNP. If we assume that AMP-PNP binds to the NBD2 site to lock the channel open (Gunderson and Kopito, 1994; Hwang *et al.*, 1994), it is then likely that P-ATP prolongs the open time by binding to the NBD1 site. The mechanistic/structural interpretation of the role of ATP binding at the NBD1 site in stabilizing the open state is discussed in the Section VI.C.

C. Working Model of CFTR Gating

There is little doubt that gating of the phosphorylated CFTR channel is controlled by ATP binding and hydrolysis. Although many gating models for CFTR have been proposed (Hwang *et al.*, 1994; Carson *et al.*, 1995; Gunderson and Kopito, 1995; Zeltwanger *et al.*, 1999; Ikuma and Welsh, 2000; Vergani *et al.*, 2003; Berger *et al.*, 2005), some fundamental issues remain unsettled. As new data emerge, it becomes evident that those simplified models are inadequate. Furthermore, recent biochemical and crystallographic studies provide structural information that grants us an unprecedented opportunity to interpret gating kinetics in the context of molecular/structural events. In this section, we will synthesize a structural/kinetic model based on the premise that dimerization of CFTR's NBDs opens the channel (Powe *et al.*, 2002a; Vergani *et al.*, 2003, 2005).

Figure 4 presents a working model of CFTR gating that accommodates most of the experimental results and the available structural information of the NBDs and NBD dimers. As shown in this model, there are multiple open states and closed states. C_1 is the closed state where the two NBDs are dissociated and neither ATP-binding site is occupied. Spontaneous dimerization of the two NBDs by thermal

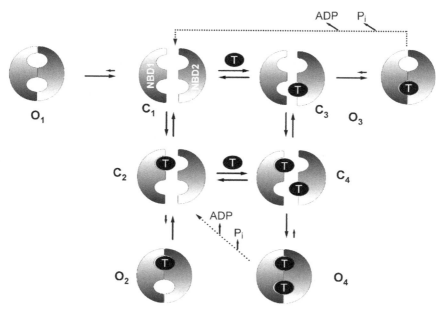

Figure 4. Working model of CFTR gating. In this model, we propose that CFTR channel can open in the absence of ATP (O_1) (i.e., spontaneous opening). ATP (T) binding at the NBD1 site (C_2) or the NBD2 site (C_3) alone can also open the channel (to O_2 and O_3, respectively). When both NBDs are occupied by ATP molecules (C_4), a stable dimer (O_4) is formed. The lifetime of each open state depends on whether the NBD2 site is vacant (O_1 and O_2) or occupied by ATP (O_3 and O_4). For O_3 and O_4, the channel closes through the dominant, fast hydrolytic pathway (dashed lines). When ATP hydrolysis is eliminated by mutations, O_3 and O_4 states close through a much slower nonhydrolytic pathway. In either hydrolytic or nonhydrolytic closing, the closing rate is faster for single-ligated O_3 than that of double-ligated O_4. For O_1 and O_2, the dissociation of the dimer is simply caused by thermo motion since the NBD2 site is not occupied by ATP. Both O_1 and O_2 are short-lived open states indicating that these two dimeric configurations are relatively unstable. Thus, ATP occupancy of the NBD2 site is required for the formation of a stable dimer (see Section VI.C. for details). The length of the solid arrow reflects qualitatively the rate of individual transition.

agitation can lead to ATP-independent channel openings in the complete absence of ATP (i.e., spontaneous opening) (O_1). This ATP-independent opening rate is fairly small (~ 0.006 s^{-1} in Bompadre *et al.*, 2005b). In the presence of ATP, ATP can bind to either the NBD1 site or the NBD2 site (C_2 and C_3, respectively), which will then induce dimerization of the two NBDs leading to open states O_2 and O_3, respectively. When both NBDs are occupied by ATP (C_4), a stable dimer, corresponding to the open state O_4, is formed.

Because of the involvement of ATP hydrolysis in channel closing (or dimer dissociation), the closing pathway (or rate) depends on whether the NBD2 site is occupied by ATP or not. Closing of the channel from open states with the NBD2 site occupied, for example, O_3 and O_4, takes place mostly through the hydrolytic pathway (dashed lines in Fig. 4). We speculate that the fast rate of dimer dissociation results from the large free energy (>30 kJ/mol) released from ATP hydrolysis (Rosing and Slater, 1972). If the hydrolysis is eliminated by mutations or by occupancy of the NBD2 site by AMP-PNP, the closing rate is slowed dramatically because now it is the thermal energy ($\sim$RT, or $\sim$2.8 kJ/mol) that breaks the dimer apart (i.e., nonhydrolytic pathway). Interestingly, for the open states O_1 and O_2, where the NBD2 site is vacant, it is somewhat surprising that their lifetimes are not much longer than that of O_3 and O_4. Since only the NBD2 site hydrolyzes ATP, closing from O_1 and O_2 can be considered nonhydrolytic. Then, why is the lifetime of states O_1 and O_2 not tens or hundreds of seconds as seen in hydrolysis-deficient mutants (e.g., K1250A or E1371S)? One has to propose that ATP binding to the NBD2 site is required to trigger conformational changes that lead to stable dimer formation. However, the mean open time of spontaneous openings (O_1) is actually slightly longer than that through hydrolysis-driven closing from O_3 or O_4 (Bompadre et al., 2005b). If the dimer hypothesis is correct, this result further supports the notion that even though ATP binding at the dimer interface stabilizes the dimer, ATP hydrolysis provides the energy to rapidly break the energetically stable dimer apart. Therefore, a dimer without ATP (i.e., O_1) dissociates somewhat more slowly than that with ATP binding at the NBD2 site (i.e., O_3 and O_4).

In our working model, there are two single-ligated closed states, C_2 and C_3. Based on our recent results, these two closed states likely have different opening rates. Since the opening rate of the Y1219G mutant is $\sim$10-fold lower than that of WT at 1-mM ATP, a concentration at which the NBD1 site is likely always occupied, the C_2 to O_2 transition must be extremely slow (perhaps as slow as spontaneous ATP-independent opening). If mutations at the W401 and K464 residues indeed dramatically alter the binding affinity of ATP at the NBD1 site as the crystal structures indicate, the fact that K464A and W401G mutations have little effect on the opening rate suggest that the opening rate for C_3, a single-ligated state, is not different from the double-ligated closed state (C_4). On the other hand, mutations of the NBD2 site affect the channel opening rate greatly, indicating the opening rate of the other single-ligated state, C3, is much faster than that of C2. In other words, it is the NBD2 site that catalyzes channel opening. This asymmetrical role of ATP binding in channel opening implies an asymmetrical motion of the NBD during dimerization (or channel opening).

Among the four open states, O_4 is the one that is most stable with the two ATP molecules sandwiched at the dimer interface. Thus, when ATP hydrolysis is abolished, the lifetime of this state is in the range of hundreds of seconds. From the energetic point of view, once ATP is sandwiched at the dimer interface, ATP molecules become part of the dimer conformation. Thus, it is reasonable to hypothesize that the binding energy of ATP contributes to the overall energetics of the dimer. The functional consequences of altering ligand-binding energy should depend on whether the channel closes through the hydrolytic or nonhydrolytic pathway. Because of the large difference between the hydrolysis energy and the thermo energy, changes of nucleotide-binding energy by using ligands with different binding affinity or by mutations of the binding pockets will affect the closing rate to different extent depending on which closing pathway the channel undertakes (see detailed discussion in Bompadre et al., 2005b). Our recent data (Zhou et al., 2005a; Bompadre et al., 2005b) indeed support this notion (Section VI.B).

An important message we try to convey through this model is that the energy from ATP at each NBD is utilized differently during channel gating. At the NBD1 site, ATP cannot be hydrolyzed and ATP-binding energy contributes to the total energy of the open state. At the NBD2 site, energy from ATP is released during ATP hydrolysis and used to drive the dimer apart. Elucidating the molecular mechanism of this energetic asymmetry as well as the kinetic asymmetry is a challenging task we are embarking on. Finally, it should be noted that this model is by no means a complete and final model. The model, nevertheless, can serve as a guide for future experiments designed to study CFTR-gating mechanism.

VII. UNSETTLED ISSUES AND FUTURE DIRECTIONS

As described above, even a complicated model like the one shown in Fig. 4 fails to explain all the experimental data. There are still many unsettled issues regarding CFTR-gating mechanism. For instance, in our working model we propose that ATP binding at the NBD2 site catalyzes channel opening based on the results with the Y1219 mutations. The role of ATP binding at the NBD1 site in channel opening remains unclear. Although NBD1 mutations, such as K464A, W401G, and the triple mutant W401G/F409G/F430G, have little effect on the relationships between [ATP] and the opening rate, this result by itself cannot rule out a possible assistant role of NBD1 in channel opening. Furthermore, conflicting results regarding the effect of the K464A mutation on channel opening have been reported (Section VI.B). Although the relationship between [ATP] and the opening rate may be used to estimate the apparent binding affinity of ATP at the NBD2 site, it is challenging to gauge quantitatively ATP–NBD1 interactions because none of the NBD1 mutations alters the ATP dose–response relationship. A functional assay that allows an estimation of the fractional occupancy of ATP at the NBD1 site is required to definitively assess the role of ATP binding at the NBD1 site in channel opening.

Another unsettled issue is whether ATP is occluded at the NBD1 site for minutes during normal hydrolysis-driven gating. Biochemical studies using 8-N_3-ATP photolabeling show that the nucleotide can be occluded at the NBD1 site, but not the NBD2 site, with a time constant as long as 14 min (Szabo et al., 1999; Aleksandrov et al., 2002b; Basso et al., 2003). Based on these results, it has been proposed that following ATP hydrolysis at NBD2 the channel closes to the ATP-bound closed state where ATP is occluded in the NBD1 site. Then on binding of ATP at the NBD2 site again, the channel reopens (Vergani et al., 2003, 2005). It should be noted that most of the occlusion experiments were carried out at ∼0°C, a condition likely to decrease ATP hydrolysis drastically (Szabo et al., 1999; Aleksandrov et al., 2002b; Basso et al., 2003). Therefore, there is little doubt that once ATP hydrolysis is diminished, there exists a state where the NBD1 site is still occupied while the NBD2 site is vacated. It seems important at this juncture to verify that this occlusion of ATP at the NBD1 site takes place in conditions that allow rapid ATP hydrolysis (Basso et al., 2003). This issue is important because, as elaborated in the Section VI.C, the amount of free energy involved in hydrolytic closing is very different from that for nonhydrolytic closing. Although the closing step is depicted as a single step in Fig. 4, some intermediate states must exist. Even if the molecular motion during hydrolytic and nonhydrolytic channel closings may follow exactly the same path, it does not mean the stability of each intermediate state will be the same. The fact that the closing rate, measured electrophysiologically, differs by nearly three orders of magnitude between hydrolytic and nonhydrolytic closings indicating that the kinetics of intermediate states must be different in these two different types of closings.

The mechanism underlying gating of CFTR by nonhydrolyzable ATP analogues, such as AMP-PNP, constitutes another intriguing, yet puzzling, issue. AMP-PNP can by itself increase the opening rate of WT-CFTR, but this effect is miniscule compared to that with ATP (Vergani et al., 2003; cf. Aleksandrov et al., 2000). Interestingly, however, the open time of AMP-PNP-opened channels is also brief ($\sim$1 s, Cho and Hwang, unpublished data) even though the closing of AMP-PNP-opened channels has to proceed through the nonhydrolytic closing. If AMP-PNP also induces NBD dimerization, these results suggest that AMP-PNP not only falls short to efficiently promote dimer formation, but also fails to stabilize the dimer. Since the structure of AMP-PNP is very similar to that of ATP, these results cannot be explained simply by assuming different binding energies for these two ligands. Most likely, a stable dimer formation involves additional conformational changes after the formation of the initial dimer (i.e., O_4); these conformational changes may be different on binding of ATP or AMP-PNP. With this in mind, it appears that once ATP binding triggers dimer formation (channel opening), it takes at least one more step to stabilize the dimer. One thus would expect that even during ATP-dependent gating, an unstable dimer should be present before the eventual stable dimer. Our preliminary single-channel data support this idea (Bompadre et al., 2005b), but pre-steady-state experiments are needed to substantiate this hypothesis.

Assuming AMP-PNP as a poor agonist still cannot explain an early observation that AMP-PNP applied in the presence of ATP greatly stabilizes the open state (Gunderson and Kopito, 1994; Hwang et al., 1994). It was proposed that after ATP opens the channel, AMP-PNP binds to the open channel conformation to exert this effect (Hwang et al., 1994). However, if the open state represents an NBD dimer, this old hypothesis becomes questionable since the ATP-binding sites should not be accessible to AMP-PNP in the open state (i.e., a dimeric configuration). For AMP-PNP to act on a binding site, more likely the dimer needs to be separated (partially or completely). We are then left with an alternative hypothesis that AMP-PNP acts on a closed state. Since AMP-PNP by itself cannot form a stable open state, we must assume that there exists a different closed state where AMP-PNP can bind and actually reopen the channel to a stable open state. As described above, the closing process likely involves multiple steps (thus multiple closed states). Do different closed states respond to AMP-PNP differently? Does the closing mechanism itself, hydrolytic versus nonhydrolytic, determine the effect of AMP-PNP? One observation that is puzzling in the field is that although AMP-PNP readily stabilizes the open channel conformation at room temperature (Gunderson and Kopito, 1994; Hwang et al., 1994), this effect is rarely seen at higher temperatures (34–37 °C in Schultz et al., 1995), suggesting that how the channel closes does influence the outcome of AMP-PNP. We are in the process of solving these mysteries.

In the past several years, with recently available structural and functional information of CFTR and other ABC proteins, a dynamic picture of CFTR channel gating starts to emerge, which has addressed some critical issues in the CFTR field and laid the new foundation for future studies. But at the same time, it also presents us with more questions that have not been raised before. One obvious question is how ATP binding catalyzes dimer formation. For instance, after ATP binds to the Walker A and Walker B of one NBD, how is the signature sequence of the partner NBD recruited? What is the role of the signature sequence in the dimerization process? So far, most of the studies on CFTR gating focus on the binding areas composed of the Walker A and Walker B motifs. Studying the functional role of the partner site (i.e., signature sequences) should provide complementary information to better our understanding of the role of each NBD. As described in the Section II, phosphorylation is a prerequisite for CFTR gating by ATP. Furthermore, different degrees of phosphorylation can affect gating kinetics. Therefore, one area for future research is to tackle how phosphorylation or degree of phosphorylation affects NBD dimerization.

In spite of significant progress made toward understanding how the gating machinery of CFTR works, we know little about how the information from NBDs is transduced to MSDs where the gate of the channel is located. Two-way communications between MSDs and NBDs have been studied in some details in other ABC transporters (Higgins and Linton, 2004). It has been suggested that substrate binding to MSDs can promote ATP binding at NBDs and the subsequent dimer formation. On the other hand, ATP hydrolysis at NBDs leads to changes of the binding affinity of substrate at MSDs and the eventual release of the substrate. A recent paper proposes a model of how energy from ATP binding and hydrolysis at NBDs is transduced to MSDs to transport substrates in MsbA (Dong et al., 2005). Although it has been suggested that CFTR can act as a transporter for glutathione flux (Kogan et al., 2003), there is little doubt that the main physiological role of CFTR is an ion channel instead of a transporter. Nevertheless, information about the underlying mechanisms of how other ABC proteins work is of great value to CFTR researchers. Of note, a recent report by Wright et al. (2004) suggests that gating of CFTR can be affected by extracellular anions. If this result is confirmed, perhaps the similarities between CFTR and other ABC transporters are more than previously thought. Indeed, the dimerization concept in CFTR was adopted from other ABC proteins (Hunt, 2002; Powe et al., 2002a). Based on the low resolution structure of CFTR, it has been suggested that the overall architecture of CFTR is similar to that of P-glycoprotein (Rosenberg et al., 2004). Interestingly, the two NBDs of P-glycoprotein hydrolyze ATP alternately between transport cycles (Senior and Gadsby, 1997). In other words, although both NBDs of P-glycoprotein are capable of hydrolyzing ATP, only one ATP molecule is hydrolyzed in each transport cycle. This alternating hydrolyzing model is supported by the structural data of MsbA (Reyes and Chang, 2005). Interestingly, in each CFTR gating cycle, only one ATP is hydrolyzed as well. Therefore, the molecular mechanism regarding how information from the NBDs is transduced to control the gate for CFTR may be also used by ABC transporters. Although CFTR, as an ion channel, possesses only one gate whereas ABC transporters demand two alternating gates for their function as pumps, the gating mechanism of CFTR may at least shed light on half of the transport cycle of ABC transporters. By combining electrophysiological, biochemical, and structural tools, we believe that our understanding about CFTR would be greatly advanced in the near future. This fundamental understanding of how CFTR works will for sure take us one step closer toward the ultimate goal of curing this devastating disease.

ACKNOWLEDGMENTS

We are indebted to our colleagues for their insightful discussions and constant encouragement. Special thanks to Drs. Silvia Bompadre, Min Li, Haoyang Liu, Yoshiro Sohma, and Xiaoqin Zou for sharing their thoughts. We are grateful to Shenghui Hu for technical assistance. This work is supported by NIHR01DK55835 (T.-C.H.) and NIHR01HL53445 (T.-C.H.). Z.Z. was supported by the Postdoctoral Fellowship from the Cystic Fibrosis Foundation.

REFERENCES

Aleksandrov, A. A. and Riordan, J. R. (1998). Regulation of CFTR ion channel gating by MgATP. *FEBS Lett.* **431**, 97–101.
Aleksandrov, A. A., Chang, X.-B., Aleksandrov, L., and Riordan, J. R. (2000). The non-hydrolytic pathway of cystic fibrosis transmembrane conductance regulator ion channel gating. *J. Physiol.* **528**, 259–265.
Aleksandrov, L., Mengos, A., Chang, X.-B., Aleksandrov, A., and Riordan, J. R. (2001). Differential interactions of nucleotides at the two nucleotide binding domains of CFTR. *J. Biol. Chem.* **276**, 12918–12923.
Aleksandrov, A. A., Aleksandrov, L., and Riordan, J. R. (2002a). Nucleoside triphosphate pentose ring impact on CFTR gating and hydrolysis. *FEBS Lett.* **518**, 183–188.
Aleksandrov, L., Aleksandrov, A. A., Chang, X.-B., and Riordan, J. R. (2002b). The first nucleotide binding domain of cystic fibrosis transmembrane conductance regulator is a site of stable nucleotide interaction, whereas the second is a site of rapid turnover. *J. Biol. Chem.* **277**, 15419–15425.
Al-Nakkash, L. and Hwang, T.-C. (1999). Activation of wild-type and ΔF508-CFTR by phosphodiesterase inhibitors through cAMP-dependent and -independent mechanisms. *Pflügers Arch.* **437**, 553–561.
Anderson, M. P. and Welsh, M. J. (1992). Regulation by ATP and ADP of CFTR chloride channels that contain mutant nucleotide-binding domains. *Science* **257**, 1701–1704.
Anderson, M. P., Berger, H. A., Rich, D. P., Gregory, R. J., Smith, A. E., and Welsh, M. J. (1991). Nucleoside triphosphates are required to open the CFTR chloride channel. *Cell* **67**, 775–784.
Ando-Akatsuka, Y., Abdullaev, I. F., Lee, E. L., Okada, Y., and Sabirov, R. Z. (2002). Down-regulation of volume-sensitive Cl^- channels by CFTR is mediated by the second nucleotide-binding domain. *Pflügers Arch.* **445**, 177–186.

Awayn, N. H., Rosenberg, M. F., Kamis, A. B., Aleksandrov, L. A., Riordan, J. R., and Ford, R. C. (2005). Crystallographic and single-particle analyses of native- and nucleotide-bound forms of the cystic fibrosis transmembrane conductance regulator (CFTR) protein. *Biochem. Soc. Trans.* **33**, 996–999.

Basso, C., Vergani, P., Nairn, A. C., and Gadsby, D. C. (2003). Prolonged nonhydrolytic interaction of nucleotide when CFTR's NH_2-terminal nucleotide binding domain and its role in channel gating. *J. Gen. Physiol.* **122**, 333–348.

Bear, C. E., Li, C., Kartner, N., Bridges, R. J., Jensen, T. J., Ramjeesingh, M., and Riordan, J. R. (1992). Purification and functional reconstitution of the cystic fibrosis transmembrane conductance regulator (CFTR). *Cell* **68**, 809–818.

Becq, F., Jensen, T. J., Chang, X.-B., Savoia, A., Rommens, J. M., Tsui, L.-C., Buchwald, M., Riordan, J. R., and Hanrahan, J. W. (1994). Phosphatase inhibitors activate normal and defective CFTR chloride channels. *Proc. Natl. Acad. Sci. USA* **91**, 9160–9164.

Bell, C. L. and Quinton, P. M. (1993). Regulation of CFTR Cl^- conductance in secretion by cellular energy levels. *Am. J. Physiol.* **264**, C925–C931.

Berger, A. L. and Welsh, M. J. (2000). Differences between cystic fibrosis transmembrane conductance regulator and HisP in the interaction with the adenine ring of ATP. *J. Biol. Chem.* **275**, 29407–29412.

Berger, A. L., Ikuma, M., and Welsh, M. J. (2005). Normal gating of CFTR requires ATP binding to both nucleotide-binding domains and hydrolysis at the second nucleotide-binding domain. *Proc. Natl. Acad. Sci. USA* **102**, 455–460.

Berger, H. A., Travis, S. M., and Welsh, M. J. (1993). Regulation of the cystic fibrosis transmembrane conductance regulator Cl^- channel by specific protein kinases and protein phosphatases. *J. Biol. Chem.* **268**, 2037–2047.

Bompadre, S. G., Ai, T., Cho, J. H., Wang, X., Sohma, Y., Li, M., and Hwang, T.-C. (2005a). CFTR gating I: Characterization of the ATP-dependent gating of a phosphorylation-independent CFTR channel (ΔR-CFTR). *J. Gen. Physiol.* **125**, 361–375.

Bompadre, S. G., Cho, J. H., Wang, X., Zou, X., Sohma, Y., Li, M., and Hwang, T.-C. (2005b). CFTR gating II: Effects of nucleotide binding on the stability of open states. *J. Gen. Physiol.* **125**, 377–394.

Button, B., Reuss, L., and Altenberg, G. A. (2001). PKC-mediated stimulation of amphibian CFTR depends on a single phosphorylation consensus site. Insertion of this site confers PKC sensitivity to human CFTR. *J. Gen. Physiol.* **117**, 457–467.

Cai, Z., Scott-Ward, T. S., and Sheppard, D. N. (2003). Voltage-dependent gating of the cystic fibrosis transmembrane conductance regulator Cl^- channel. *J. Gen. Physiol.* **122**, 605–620.

Carson, M. R. and Welsh, M. J. (1993). 5'-Adenylylimidodiphosphate does not activate CFTR chloride channels in cell-free patches of membrane. *Am. J. Physiol.* **265**, L27–L32.

Carson, M. R., Travis, S. M., and Welsh, M. J. (1995). The two nucleotide-binding domains of cystic fibrosis transmembrane conductance regulator (CFTR) have distinct functions in controlling channel activity. *J. Biol. Chem.* **270**, 1711–1717.

Chan, K. W., Csanády, L., Seto-Young, D., Nairn, A. C., and Gadsby, D. C. (2000). Severed molecules functionally define the boundaries of the cystic fibrosis transmembrane conductance regulator's NH_2-terminal nucleotide binding domain. *J. Gen. Physiol.* **116**, 163–180.

Chang, G. (2003). Structure of MsbA from *Vibrio cholera*: A multidrug resistance ABC transporter homolog in a closed conformation. *J. Mol. Biol.* **330**, 419–430.

Chang, G. and Roth, C. B. (2001). Structure of MsbA from *E. Coli*: A homolog of the multidrug resistance ATP binding cassette (ABC) transporters. *Science* **293**, 1793–1800.

Chang, S. Y., Di, A., Naren, A. P., Palfrey, H. C., Kirk, K. L., and Nelson, D. J. (2002). Mechanisms of CFTR regulation by syntaxin 1A and PKA. *J. Cell Sci.* **115**, 783–791.

Chang, X.-B., Tabcharani, J. A., Hou, Y.-X., Jensen, T. J., Kartner, N., Alon, N., Hanrahan, J. W., and Riordan, J. R. (1993). Protein kinase A (PKA) still activates CFTR chloride channel after mutagenesis of all 10 PKA consensus phosphorylation sites. *J. Biol. Chem.* **268**, 11304–11311.

Chappe, V., Hinkson, D. A., Zhu, T., Chang, X.-B., Riordan, J. R., and Hanrahan, J. W. (2003). Phosphorylation of protein kinase C sites in NBD1 and the R domain control CFTR channel activation by PKA. *J. Physiol.* **548,** 39–52.

Chappe, V., Hinkson, D. A., Howell, L. D., Evagelidis, A., Liao, J., Chang, X.-B., Riordan, J. R., and Hanrahan, J. W. (2004). Stimulatory and inhibitory protein kinase C consensus sequences regulate the cystic fibrosis transmembrane conductance regulator. *Proc. Natl. Acad. Sci. USA* **101,** 390–395.

Chappe, V., Irvine, T., Liao, J., Evagelidis, A., and Hanrahan, J. W. (2005). Phosphorylation of CFTR by PKA promotes binding of the regulatory domain. *EMBO J.* **24,** 2730–2740.

Chen, J., Lu, G., Lin, J., Davidson, A. L., and Quiocho, F. A. (2003). A tweezers-like motion of the ATP-binding cassette dimer in an ABC transport cycle. *Mol. Cell* **12,** 651–661.

Chen, Y., Button, B., Altenberg, G. A., and Reuss, L. (2004). Potentiation of effect of PKA stimulation of *Xenopus* CFTR by activation of PKC: Role of NBD2. *Am. J. Physiol.* **287,** XC1436–XC1444.

Cheng, S. H., Gregory, R. J., Marshall, J., Paul, S., Souza, D. W., White, G. A., O'Riordan, C. R., and Smith, A. E. (1990). Defective intracellular transport and processing of CFTR is the molecular basis of most cystic fibrosis. *Cell* **63,** 827–834.

Cheng, S. H., Rich, D. P., Marshall, J., Gregory, R. J., Welsh, M. J., and Smith, A. E. (1991). Phosphorylation of the R domain by cAMP-dependent protein kinase regulates the CFTR chloride channel. *Cell* **66,** 1027–1036.

Colquhoun, D. (1998). Binding, gating, affinity and efficacy: The interpretation of structure-activity relationships for agonists and of the effects of mutating receptors. *Br. J. Pharmacol.* **125,** 924–947.

Cormet-Boyaka, E., Di, A., Chang, S. Y., Naren, A. P., Tousson, A., Nelson, D. J., and Kirk, K. L. (2002). CFTR chloride channels are regulated by a SNAP-23/syntaxin 1A complex. *Proc. Natl. Acad. Sci. USA* **99,** 12477–12482.

Csanády, L., Chan, K. W., Seto-Young, D., Kopsco, D. C., Nairn, A. C., and Gadsby, D. C. (2000). Severed channels probe regulation of gating of cystic fibrosis transmembrane conductance regulator by its cytoplasmic domains. *J. Gen. Physiol.* **116,** 477–500.

Csanády, L., Chan, K. W., Nairn, A. C., and Gadsby, D. C. (2005a). Functional roles of nonconserved structural segments in CFTR's NH_2-terminal nucleotide binding domain. *J. Gen. Physiol.* **125,** 43–55.

Csanády, L., Seto-Young, D., Chan, K. W., Cenciarelli, C., Angel, B. B., Qin, J., McLachlin, D. T., Krutchinsky, A. N., Chait, B. T., Nairn, A. C., and Gadsby, D. C. (2005b). Preferential phosphorylation of R-domain serine 768 dampens activation of CFTR channels by PKA. *J. Gen. Physiol.* **125,** 171–186.

Davidson, A. L., Laghaeian, S. S., and Mannering, D. E. (1996). The maltose transport system of *Escherichia coli* displays positive cooperativity in ATP hydrolysis. *J. Biol. Chem.* **271,** 4858–4863.

Dawson, D. C., Smith, S. S., and Mansoura, M. K. (1999). CFTR: Mechanisms of anion conduction. *Physiol. Rev.* **79,** S47–S75.

Dean, M., Hamon, Y., and Chimini, G. (2001). The human ATP-binding cassette (ABC) transporter superfamily. *J. Lipid Res.* **42,** 1007–1017.

Denning, G. M., Ostedgaard, L. S., and Welsh, M. J. (1992a). Abnormal localization of cystic fibrosis transmembrane conductance regulator in primary cultures of cystic fibrosis airway epithelia. *J. Cell Biol.* **118,** 551–559.

Denning, G. M., Anderson, M. P., Amara, J. F., Marshall, J., Smith, A. E., and Welsh, M. J. (1992b). Processing of mutant cystic fibrosis transmembrane conductance regulator is temperature-sensitive. *Nature* **358,** 761–764.

Devor, D. C., Singh, A. K., Lambert, L. C., DeLuca, A., Frizzell, R. A., and Bridges, R. J. (1999). Bicarbonate and chloride secretion in Calu-3 human airway epithelial cells. *J. Gen. Physiol.* **113,** 743–760.

Diederichs, K., Diez, J., Greller, G., Müller, C., Breed, J., Schnell, C., Vonrhein, C., Boos, W., and Welte, W. (2000). Crystal structure of MalK, the ATPase subunit of the trehalose/maltose ABC transporter of the archaeon Thermococcus litoralis. *EMBO J.* **19,** 5951–5961.

Dong, J., Yang, G., and Mchaourad, H. S. (2005). Structural basis of energy transduction in the transport cycle of MsbA. *Science* **308,** 1023–1028.

Dousmanis, A. G., Nairn, A. C., and Gadsby, D. C. (2002). Distinct Mg^{2+}-dependent steps rate limit opening and closing of a single CFTR Cl^- channel. *J. Gen. Physiol.* **119,** 545–559.

Doyle, D. A., Cabral, J. M., Pfuetzner, R. A., Kuo, A., Gulbis, J. M., Cohen, S. L., Chait, B. T., and MacKinnon, R. (1998). The structure of the potassium channel: Molecular basis of K^+ conduction and selectivity. *Science* **280**, 69–77.

Egan, M., Flotte, T., Afione, S., Solow, R., Zeitlin, P. L., Carter, B. J., and Guggino, W. B. (1992). Defective regulation of outwardly rectifying Cl^- channels by protein kinase A corrected by insertion of CFTR. *Nature* **358**, 581–584.

Fetsch, E. E. and Davidson, A. L. (2003). Maltose transport through the inner membrane of *E. coli*. *Front. Biosci.* **8**, d652–d660.

Fischer, H., Illek, B., and Machen, T. E. (1998). Regulation of CFTR by protein phosphatase 2B and protein kinase C. *Pflügers Arch.* **436**, 175–181.

Gabriel, S. E., Clarke, L. L., Boucher, R. C., and Stutts, M. J. (1993). CFTR and outward rectifying chloride channels are distinct proteins with a regulatory relationship. *Nature* **363**, 263–268.

Gadsby, D. C. and Nairn, A. C. (1999). Control of CFTR channel gating by phosphorylation and nucleotide hydrolysis. *Physiol. Rev.* **79**, S77–S107.

Gregory, R. J., Rich, D. P., Cheng, S. H., Souza, D. W., Paul, S., Manavalan, P., Anderson, M. P., Welsh, M. J., and Smith, A. E. (1991). Maturation and function of cystic fibrosis transmembrane conductance regulator variants bearing mutations in putative nucleotide-binding domains 1 and 2. *Mol. Cell. Biol.* **11**, 3886–3893.

Gross, C. H., Abdul-Manan, N., Fulghum, J., Lippke, J., Liu, X., Prabhaker, P., Brennan, D., Swope, W. M., Faerman, C., Connelly, P., Raybuck, S., and Moore, J. (2006). Nucleotide binding domains of cystic fibrosis transmembrane regulator, an ABC-transporter, catalyze adenylate kinase activity but not ATP hydrolysis. *J. Biol. Chem.* **281**, 4058–4068.

Guggino, W. B. (2004). The cystic fibrosis transmembrane regulator forms macromolecular complexes with PDZ domain scaffold proteins. *Proc. Am. Thorac. Soc.* **1**, 28–32.

Gunderson, K. L. and Kopito, R. R. (1994). Effects of pyrophosphate and nucleotide analogs suggest a role for ATP hydrolysis in cystic fibrosis transmembrane regulator channel gating. *J. Biol. Chem.* **269**, 19349–19353.

Gunderson, K. L. and Kopito, R. R. (1995). Conformational states of CFTR associated with channel gating: The role ATP binding and hydrolysis. *Cell* **82**, 231–239.

Hallows, K. R. (2005). Emerging role of AMP-activated protein kinase in coupling membrane transport to cellular metabolism. *Curr. Opin. Nephrol. Hypertens.* **14**, 464–471.

Hanrahan, J. W. and Wioland, M.-A. (2005). Revisiting cystic fibrosis transmembrane conductance regulator structure and function. *Proc. Am. Thorac. Soc.* **1**, 17–21.

Hartman, J., Huang, Z., Rado, T. A., Peng, S., Jilling, T., Muccio, D. D., and Sorscher, E. J. (1992). Recombinant synthesis, purification, and nucleotide binding characteristics of the first nucleotide binding domain of the cystic fibrosis gene product. *J. Biol. Chem.* **267**, 6455–6458.

Haws, C. M., Finkbeiner, W. E., Widdicombe, J. H., and Wine, J. J. (1994). CFTR in Calu-3 human airway cells: Channel properties and role in cAMP-activated Cl^- conductance. *Am. J. Physiol.* **266**, L502–L512.

Higgins, C. F. and Linton, K. J. (2004). The ATP switch model for ABC transporters. *Nat. Struct. Mol. Biol.* **11**, 918–926.

Hilgemann, D. W. and Ball, R. (1996). Regulation of cardiac Na^+,Ca^{2+} exchange and K_{ATP} potassium channels by PIP_2. *Science* **273**, 956–959.

Hilgemann, D. W., Feng, S., and Nasuhoglu, C. (2001). The complex and intriguing lives of PIP2 with ion channels and transporters. *Sci. STKE* **111**, RE19.

Himmel, B. and Nagel, G. (2004). Protein kinase-independent activation of CFTR by phosphatidylinositol phosphates. *EMBO Rep.* **5**, 85–90.

Hopfner, K. P., Karcher, A., Shin, D. S., Craig, L., Arthur, L. M., Carney, J. P., and Tainer, J. A. (2000). Structural biology of Rad50 ATPase: ATP-driven conformational control in DNA double-strand break repair and the ABC-ATPase superfamily. *Cell* **101**, 789–800.

Huang, C. L., Feng, S., and Hilgemann, D. W. (1998). Direct activation of inward rectifier potassium channels by PIP_2 and its stabilization by $G\beta\gamma$. *Nature* **391**, 803–806.

Hung, L.-W., Wang, I. X., Nikaido, K., Liu, P.-Q., Ames, G. F. L., and Kim, S.-H. (1999). Crystal structure of the ATP-binding subunit of an ABC transporter. *Nature* **396**, 703–707.

Hunt, J. F. (2002). The perplexing challenges of a pump turned channel. *J. Physiol.* **539**, 331.

Hwang, T.-C. and Sheppard, D. N. (1999). Molecular pharmacology of the CFTR Cl⁻ channel. *Trends Pharmacol. Sci.* **20**, 448–453.

Hwang, T.-C., Horie, M., and Gadsby, D. C. (1993). Functionally distinct phospho-forms underlie incremental activation of protein kinase-regulated Cl⁻ conductance in mammalian heart. *Am. J. Physiol.* **101**, 629–650.

Hwang, T.-C., Nagel, G., Nairn, A. C., and Gadsby, D. C. (1994). Regulation of the gating of cystic fibrosis transmembrane conductance regulator Cl⁻ channels by phosphorylation and ATP hydrolysis. *Proc. Natl. Acad. Sci. USA* **91**, 4698–4702.

Ikuma, M. and Welsh, M. J. (2000). Regulation of CFTR Cl⁻ channel gating by ATP binding and hydrolysis. *Proc. Natl. Acad. Sci. USA* **97**, 8675–8680.

Janas, E., Hofacker, M., Chem, M., Gompf, S., van der Does, C., and Tampe, R. (2003). The ATP hydrolysis cycle of the nucleotide-binding domain of the mitochondrial ATP-binding cassette transporter Mdl1p. *J. Biol. Chem.* **278**, 26862–26869.

Jia, Y., Mathews, C. J., and Hanrahan, J. W. (1997). Phosphorylation by protein kinase C is required for acute activation of cystic fibrosis transmembrane conductance regulator by protein kinase A. *J. Biol. Chem.* **272**, 4978–4984.

Jiang, Y., Lee, A., Chen, J., Cadene, M., Chait, B. T., and MacKinnon, R. (2002). Crystal structure and mechanism of a calcium-gated potassium channel. *Nature* **417**, 515–522.

Jiang, Y., Lee, A., Chen, J., Ruta, V., Cadene, M., Chait, B. T., and MacKinnon, R. (2003). X-ray structure of a voltage-dependent K⁺ channel. *Nature* **423**, 33–41.

Jones, P. M. and George, A. M. (1999). Subunit interactions in ABC transporters: Towards a functional architecture. *FEMS Microbiol. Lett.* **179**, 187–202.

Kartner, N., Hanrahan, J. W., Jensen, T. J., Naismith, A. L., Sun, S., Ackerley, C. A., Reyes, E. F., Tsui, L.-C., Rommens, J. M., Bear, C. E., and Riordan, J. R. (1991). Expression of the cystic fibrosis gene in non-epithelial invertebrate cells produces a regulated anion conductance. *Cell* **64**, 681–691.

Karpowich, N., Martsinkevich, O., Millen, L., Yuan, Y.-R., Dai, P. L., MacVey, K., Thomas, P. J., and Hunt, J. F. (2001). Crystal structures of the MJ1267 ATP binding cassette reveal an induced-fit effect at the ATPase active site of an ABC transporter. *Structure* **9**, 571–586.

Ko, Y. H. and Pedersen, P. L. (1995). The first nucleotide binding fold of the cystic fibrosis transmembrane conductance regulator can function as an active ATPase. *J. Biol. Chem.* **270**, 22093–22096.

Ko, Y. H., Thomas, P. J., and Pederson, P. L. (1994). The cystic fibrosis transmembrane conductance regulator. Nucleotide binding to a synthetic peptide segment from the second predicted nucleotide binding fold. *J. Biol. Chem.* **269**, 14584–14588.

Kogan, I., Ramjeesingh, M., Li, C., Kidd, J. F., Wang, Y., Leslie, E. M., Cole, S. P. C., and Bear, C. E. (2003). CFTR directly mediates nucleotide-regulated glutathione flux. *EMBO J.* **22**, 1981–1989.

Lewis, H. A., Buchanan, S. G., Burley, S. K., Conners, K., Dickey, M., Dorwart, M., Fowler, R., Gao, X., Guggino, W. B., Hendrickson, W. A., Hunt, J. F., Kearins, M. C., et al. (2004). Structure of nucleotide-binding domain 1 of the cystic fibrosis transmembrane conductance regulator. *EMBO J.* **23**, 282–293.

Lewis, H. A., Zhao, X., Wang, C., Sauder, J. M., Rooney, I., Noland, B. W., Lorimer, D., Kearins, M. C., Conners, K., Condon, B., Maloney, P. C., Guggino, W. B., et al. (2005). Impact of the ΔF508 mutation in first nucleotide-binding domain of human cystic fibrosis transmembrane conductance regulator on domain folding and structure. *J. Biol. Chem.* **280**, 1346–1353.

Li, C., Ramjeesingh, M., Wang, W., Garami, E., Hewryk, M., Lee, D., Rommens, J. M., Galley, K., and Bear, C. E. (1996). ATPase activity of the cystic fibrosis transmembrane conductance regulator. *J. Biol. Chem.* **271**, 28463–28468.

Linsdell, P. (2006). Mechanism of chloride permeation in the cystic fibrosis transmembrane conductance regulator chloride channel. *Exp. Physiol.* **91**, 123–129.

Locher, K. P., Lee, A. T., and Rees, D. C. (2002). The *E. coli* BtuCD structure: A framework for ABC transporter architecture and mechanism. *Science* **296**, 1091–1098.

Logan, J., Hiestand, D., Daram, P., Huang, Z., Muccio, D. D., Hartman, J., Haley, B., Cook, W. J., and Sorscher, E. J. (1994). Cystic fibrosis transmembrane conductance regulator mutations that disrupt nucleotide binding. *J. Clin. Invest.* **94**, 228–236.

Long, S. B., Campbell, E. B., and MacKinnon, R. (2005a). Crystal structure of a mammalian voltage-dependent Shaker family K^+ channel. *Science* **309**, 897–903.

Long, S. B., Campbell, E. B., and MacKinnon, R. (2005b). Voltage sensor of Kv1.2: Structural basis of electromechanical coupling. *Science* **309**, 903–908.

Luo, J., Pato, M. D., Riordan, J. R., and Hanrahan, J. W. (1998). Differential regulation of single CFTR channels by PP2C, PP2A, and other phosphatases. *Am. J. Physiol.* **274**, C1397–C1410.

Ma, J., Zhao, J., Drumm, M. L., Xie, J., and Davis, P. B. (1997). Function of the R domain in the cystic fibrosis transmembrane conductance regulator chloride channel. *J. Biol. Chem.* **272**, 28133–28141.

Marcet, B., Becq, F., Norez, C., Delmas, P., and Verrier, B. (2004). General anesthetic octanol and related compounds activate wild-type and ΔF508 cystic fibrosis chloride channels. *Br. J. Pharmacol.* **141**, 905–914.

Mathews, C. J., Tabcharani, J. A., Chang, X.-B., Jensen, T. J., Riordan, J. R., and Hanrahan, J. W. (1998). Dibasic protein kinase A sites regulate bursting rate and nucleotide sensitivity of the cystic fibrosis transmembrane conductance regulator chloride channel. *J. Physiol.* **508**, 365–377.

McCarty, N. A. (2000). Permeation through the CFTR chloride channel. *J. Exp. Biol.* **203**, 1947–1962.

Moody, J. E., Millen, L., Binns, D., Hunt, J. F., and Thomas, P. J. (2002). Cooperative, ATP-dependent association of the nucleotide binding cassettes during the catalytic cycle of ATP-binding cassette transporters. *J. Biol. Chem.* **277**, 21111–21114.

Nagel, G., Hwang, T.-C., Nastiuk, K. L., Nairn, A. C., and Gadsby, D. C. (1992). The protein kinase A-regulated cardiac Cl^- channel resembles the cystic fibrosis transmembrane conductance regulator. *Nature* **360**, 81–84.

Naren, A. P., Nelson, D. J., Xie, W., Jovov, B., Pevsner, J., Bennett, M. K., Benos, D. J., Quick, M. W., and Kirk, K. L. (1997). Regulation of CFTR chloride channels by syntaxin and Munc18 isoforms. *Nature* **390**, 302–305.

Naren, A. P., Quick, M. W., Collawn, J. F., Nelson, D. J., and Kirk, K. L. (1998). Syntaxin 1A inhibits CFTR chloride channels by means of domain-specific protein-protein interactions. *Proc. Natl. Acad. Sci. USA* **95**, 10972–10977.

Neville, D. C., Rozanas, C. R., Price, E. M., Gruis, D. B., Verkman, A. S., and Townsend, R. R. (1997). Evidence for phosphorylation of serine 753 in CFTR using a novel metal-ion affinity resin and matrix-assisted laser desorption mass spectrometry. *Protein Sci.* **6**, 2436–2445.

Picciotto, M. R., Cohn, J. A., Bertuzzi, G., Greengard, P., and Nairn, A. C. (1992). Phosphorylation of the cystic fibrosis transmembrane conductance regulator. *J. Biol. Chem.* **267**, 12742–12752.

Powe, A. C., Jr., Al-Nakkash, L., Li, M., and Hwang, T.-C. (2002a). Mutation of Walker-A lysine 464 in cystic fibrosis transmembrane conductance regulator reveals functional interaction between its nucleotide-binding domains. *J. Physiol.* **539**, 333–346.

Powe, A. C., Jr., Zhou, Z., Hwang, T.-C., and Nagel, G. (2002b). Quantitative analysis of ATP-dependent gating of CFTR. *Methods Mol. Med.* **70**, 67–98.

Quinton, P. M. (1999). Physiological basis of cystic fibrosis: A historical perspective. *Physiol. Rev.* **79**, S3–S22.

Quinton, P. M. and Reddy, M. M. (1992). Control of CFTR chloride conductance by ATP levels through non-hydrolytic binding. *Nature* **360**, 79–81.

Ramjeesingh, M., Li, C., Garami, E., Huan, L.-J., Galley, K., Wang, Y., and Bear, C. E. (1999). Walker mutations reveal loose relationship between catalytic and channel-gating activities of purified CFTR (cystic fibrosis transmembrane conductance regulator). *Biochemistry* **38**, 1463–1468.

Randak, C., Neth, P., Auerswald, E. A., Eckerskorn, C., Assfalg-Machleidt, I., and Machleidt, W. (1997). A recombinant polypeptide model of the second nucleotide-binding fold of the cystic fibrosis trans-membrane conductance regulator functions as an active ATPase, GTPase and adenylate kinase. *FEBS Lett.* **410**, 180–186.

Randak, C. O. and Welsh, M. J. (2005). ADP inhibits function of the ABC transporter cystic fibrosis transmembrane conductance regulator via its adenylate kinase activity. *Proc. Natl. Acad. Sci. USA* **102**, 2216–2220.

Reddy, M. M. and Quinton, P. M. (2003). Control of dynamic CFTR selectivity by glutamate and ATP in epithelial cells. *Nature* **423**, 756–760.

Reyes, C. and Chang, G. (2005). Structure of the ABC transporter MsbA in complex with ADP.vanadate and lipopolysaccharide. *Science* **308**, 1028–1031.

Rich, D. P., Gregory, R. J., Anderson, H. A., Manavalan, P., Smith, A. E., and Welsh, M. J. (1991). Effect of deleting the R domain on CFTR-generated chloride channels. *Science* **253**, 205–207.

Rich, D. P., Berger, H. A., Cheng, S. H., Travis, S. M., Saxena, M., Smith, A. E., and Welsh, M. J. (1993). Regulation of the cystic fibrosis transmembrane conductance regulator Cl$^-$ channel by negative charge in the R domain. *J. Biol. Chem.* **268**, 20259–20267.

Riordan, J. R., Rommens, J. M., Kerem, B.-S., Alon, N., Rozmahel, R., Grzelczak, Z., Zielenski, J., Lok, S., Plavsic, N., Chou, J.-L., Drumm, M. L., Iannuzzi, M. C., et al. (1989). Identification of the cystic fibrosis gene: Cloning and characterization of complementary DNA. *Science* **245**, 1066–1072.

Rommens, J. M., Iannuzzi, M. C., Kerem, B.-S., Drumm, M. L., Melmer, G., Dean, M., Rozmahel, R., Cole, J. L., Kennedy, D., Hidaka, N., Zsiga, M., Buchwald, M., et al. (1989). Identification of the cystic fibrosis gene: Chromosome walking and jumping. *Science* **245**, 1059–1065.

Rosenberg, M. F., Velarde, G., Ford, R. C., Martin, C., Berridge, G., Kerr, I. D., Callagham, R., Schmidlin, A., Wooding, C., Linton, K. J., and Higgins, C. F. (2001). Repacking of the transmembrane domains of P-glycoprotein during the transport ATPase cycle. *EMBO J.* **20**, 5616–5625.

Rosenberg, M. F., Kamis, A. B., Callagham, R., Higgins, C. F., and Ford, R. C. (2003). Three-dimensional structures of the mammalian multidrug resistance P-glycoprotein demonstrate major conformational changes in the transmembrane domains upon nucleotide binding. *J. Biol. Chem.* **278**, 8294–8299.

Rosenberg, M. F., Kamis, A. B., Aleksandrov, L., Ford, R. C., and Riordan, J. R. (2004). Purification and crystallization of the cystic fibrosis transmembrane conductance regulator (CFTR). *J. Biol. Chem.* **279**, 39051–39057.

Rosing, J. and Slater, E. C. (1972). The value of G degrees for the hydrolysis of ATP. *Biochim. Biophys. Acta* **267**, 275–290.

Schmitt, L. and Tampe, R. (2002). Structure and mechanism of ABC transporters. *Curr. Opin. Struct. Biol.* **12**, 754–760.

Schultz, B. D., Venglarik, C. J., Bridges, R. J., and Frizzell, R. A. (1995). Regulation of CFTR Cl$^-$ channel gating by ADP and ATP analogues. *J. Gen. Physiol.* **105**, 329–361.

Schwiebert, E. M., Egan, M. E., Hwang, T.-H., Fulmer, S. B., Allen, S. S., Cutting, G. R., and Guggino, W. B. (1995). CFTR regulates outwardly rectifying chloride channels through an autocrine mechanism involving ATP. *Cell* **81**, 1063–1073.

Schwiebert, E. M., Benos, D. J., Egan, M. E., Stutts, M. J., and Guggino, W. B. (1999). CFTR is a conductance regulator as well as a chloride channel. *Physiol. Rev.* **79**, S145–S163.

Seibert, F. S., Chang, X. B., Aleksandrov, A. A., Clarke, D. M., Hanrahan, J. W., and Riordan, J. R. (1999). Influence of phosphorylation by protein kinase A on CFTR at the cell surface and endoplasmic reticulum. *Biochim. Biophys. Acta* **1461**, 275–283.

Senior, A. E. and Gadsby, D. C. (1997). ATP hydrolysis cycles and mechanism in P-glycoprotein and CFTR. *Semin. Cancer Biol.* **8**, 143–150.

Sheppard, D. N. and Welsh, M. J. (1999). Structure and function of the CFTR chloride channel. *Physiol. Rev.* **79**, S23–S45.

Smith, P. C., Karpowith, N., Millen, L., Moody, J. E., Rosen, J., Thomas, P. J., and Hunt, J. F. (2002). ATP binding to the motor domain from an ABC transporter drives formation of a nucleotide sandwich dimer. *Mol. Cell* **10**, 139–149.

Stutts, M. J., Canessa, C. M., Olsen, J. C., Hamrick, M., Cohn, J. A., Rossier, B. C., and Boucher, R. C. (1995). CFTR as a cAMP-dependent regulator of sodium channels. *Science* **269**, 847–850.

Stutts, M. J., Rossier, B. C., and Boucher, R. C. (1997). Cystic fibrosis transmembrane conductance regulator inverts protein kinase A-mediated regulation of epithelial sodium channel single channel kinetics. *J. Biol. Chem.* **272**, 14037–14040.

Sugita, M., Yue, Y., and Foskett, J. K. (1998). CFTR Cl$^-$ channel and CFTR-associated ATP channel: Distinct pores regulated by common gates. *EMBO J.* **17**, 898–908.

Szabo, K., Szakacs, G., Hegedus, T., and Sarkadi, B. (1999). Nucleotide occlusion in the human cystic fibrosis transmembrane conductance regulator: Different patterns in the two nucleotide binding domains. *J. Biol. Chem.* **274**, 12209–12212.

Szellas, T. and Nagel, G. (2003). Apparent affinity of CFTR for ATP is increased by continuous kinase activity. *FEBS Lett.* **535**, 141–146.

Tabcharani, J. A., Low, W., Edie, D., and Hanrahan, J. W. (1990). Low-conductance chloride channel activated by cAMP in the epithelial cell line T84. *FEBS Lett.* **270**, 157–164.

Tabcharani, J. A., Chang, X.-B., Riordan, J. R., and Hartshorn, M. J. (1991). Phosphorylation-regulated Cl$^-$ channel in CHO cells stably expressing the cystic fibrosis gene. *Nature* **352**, 628–631.

Thelin, W. R., Kesimer, M., Tarran, R., Kreda, S. M., Grubb, B. R., Sheehan, J. K., Stutts, M. J., and Milgram, S. L. (2005). The cystic fibrosis transmembrane conductance regulator is regulated by a direct interaction with the protein phosphatase 2A. *J. Biol. Chem.* **280**, 41512–41520.

Thomas, P. J., Shenbagamurthi, P., Ysern, X., and Pedersen, P. L. (1991). Cystic fibrosis transmembrane conductance regulator: Nucleotide binding to a synthetic peptide. *Science* **251**, 555–557.

Tombline, G., Bartholomew, L. A., Tyndall, G. A., Gimi, K., Urbatsch, I. L., and Senior, A. E. (2004). Properties of P-glycoprotein with mutations in the "catalytic carboxylate" glutamate residues. *J. Biol. Chem.* **279**, 46518–46526.

Travis, S. M., Carson, M. R., Ries, D. R., and Welsh, M. J. (1993). Interaction of nucleotides with membrane-associated cystic fibrosis transmembrane conductance regulator. *J. Biol. Chem.* **268**, 15336–15339.

Travis, S. M., Berger, H. A., and Welsh, M. J. (1997). Protein phosphatase 2C dephosphorylates and inactivates cystic fibrosis transmembrane conductance regulator. *Proc. Natl. Acad. Sci. USA* **94**, 11055–11060.

Verdon, G., Albers, S. V., van Oosterwijk, N., Dijkstra, B. W., Driessen, A. J., and Thunnissen, A. M. (2003). Formation of the productive ATP-Mg^{2+}-bound dimer of GlcV, an ABC-ATPase from Sulfolobus solfataricus. *J. Mol. Biol.* **334**, 255–267.

Vergani, P., Nairn, A. C., and Gadsby, D. C. (2003). On the mechanism of MgATP-dependent gating of CFTR Cl$^-$ channels. *J. Gen. Physiol.* **120**, 17–36.

Vergani, P., Lockless, S. W., Nairn, A. C., and Gadsby, D. C. (2005). CFTR channel opening by ATP-driven tight dimerization of its nucleotide-binding domains. *Nature* **433**, 876–880.

Wang, F., Zeltwanger, S., Yang, I. C. H., Nairn, A. C., and Hwang, T.-C. (1998). Actions of genistein on cystic fibrosis transmembrane conductance regulator channel gating: Evidence of two binding sites with opposite effects. *J. Gen. Physiol.* **111**, 477–490.

Weinreich, F., Riordan, J. R., and Nagel, G. (1999). Dual effects of ADP and adenylylimidodiphosphate on CFTR channel kinetics show binding to two different nucleotide binding sites. *J. Gen. Physiol.* **114**, 55–70.

Welsh, M. J. and Smith, A. E. (1993). Molecular mechanisms of CFTR chloride channel dysfunction in cystic fibrosis. *Cell* **73**, 1251–1254.

Wilkinson, D. J., Strong, T. V., Mansoura, M. K., Wood, D. L., Smith, S. S., Collins, F. S., and Dawson, D. C. (1997). CFTR activation: Additive effects of stimulatory and inhibitory phosphorylation sites in the R domain. *Am. J. Physiol.* **273**, L127–L133.

Winter, M. C. and Welsh, M. J. (1997). Stimulation of CFTR activity by its phosphorylated R domain. *Nature* **389**, 294–296.

Winter, M. C., Sheppard, D. N., Carson, M. R., and Welsh, M. J. (1994). Effect of ATP concentration on CFTR Cl$^-$ channels: A kinetic analysis of channel regulation. *Biophys. J.* **66**, 1398–1403.

Wright, A. M., Gong, X., Verdon, B., Linsdell, P., Mehta, A., Riordan, J. R., Argent, B. E., and Gray, M. A. (2004). Novel regulation of cystic fibrosis transmembrane conductance regulator (CFTR) channel gating by external chloride. *J. Biol. Chem.* **279**, 41658–41663.

Yamazaki, J., Britton, F., Collier, M. L., Horowitz, B., and Hume, J. R. (1999). Regulation of recombinant cardiac cystic fibrosis transmembrane conductance regulator chloride channels by protein kinase C. *Biophys. J.* **76**, 1972–1987.

Yang, I. C. H., Cheng, T.-H., Wang, F., Price, E. M., and Hwang, T.-C. (1997). Modulation of CFTR chloride channels by calyculin A and genistein. *Am. J. Physiol.* **272**, C142–C155.

Yuan, Y.-R., Blecker, S., Martsinkevich, O., Millen, L., and Thomas, P. J. (2001). The crystal structure of the MJ0796 ATP-binding cassette. Implications for the structural consequences of ATP hydrolysis in the active site of an ABC transporter. *J. Biol. Chem.* **276**, 32313–32321.

Yue, G., Malik, B., Yue, G., and Eaton, D. C. (2002). Phosphatidylinositol 4,5-bisphosphate (PIP_2) stimulates epithelial sodium channel activity in A6 cells. *J. Biol. Chem.* **277**, 11965–11969.

Zaitseva, J., Jenewein, S., Wiedenmann, A., Benabdelhak, H., Holland, I. B., and Schmitt, L. (2005). Functional characterization and ATP-induced dimerization of the isolated ABC-domain of the haemolysin B transporter. *Biochemistry* **44**, 9680–9690.

Zeltwanger, S., Wang, F., Wang, G.-T., Gillis, K. D., and Hwang, T.-C. (1999). Gating of cystic fibrosis transmembrane conductance regulator chloride channels by adenosine triphosphate hydrolysis: Quantitative analysis of a cyclic gating scheme. *J. Gen. Physiol.* **113**, 541–554.

Zhou, Z., Hu, S., and Hwang, T.-C. (2002). Probing an open CFTR pore with organic anion blockers. *J. Gen. Physiol.* **120**, 647–662.

Zhou, Z., Wang, X., Liu, H., Li, M., Zou, X., and Hwang, T.-C. (2004). ATP opens and closes the CFTR channel via NBD2. (Abstract) *Pediatr. Pulmonol. Suppl.* **24**, 197.

Zhou, Z., Wang, X., Li, M., Sohma, Y., Zou, X., and Hwang, T.-C. (2005a). High affinity ATP/ADP analogues as new tools for studying CFTR gating. *J. Physiol.* **569**, 447–457.

Zhou, Z., Wang, X., Liu, H., Li, M., Zou, X., and Hwang, T.-C. (2005b). Distinct functions of individual nucleotide binding domains in CFTR gating revealed by mutations in ATP binding pockets (Abstract). *Biophys. J.* **88**, 546a.

Zhu, T., Dahan, D., Evagelidis, A., Zheng, S.-X., Luo, J., and Hanrahan, J. W. (1999). Association of cystic fibrosis transmembrane conductance regulator and protein phosphatase 2C. *J. Biol. Chem.* **274**, 29102–29107.

Zou, X. and Hwang, T.-C. (2001). ATP-hydrolysis-coupled gating of CFTR chloride channels: Structure and function. *Biochemistry* **40**, 5579–5586.

Chapter 7
Functional Properties of Ca^{2+}-Dependent Cl^- Channels and Bestrophins: Do They Correlate?

Jorge Arreola[1] and Patricia Pérez-Cornejo[2]

[1]*Institute of Physics, University of San Luis Potosí, San Luis Potosí, SLP 78000, México*
[2]*School of Medicine, University of San Luis Potosí, San Luis Potosí, SLP 78000, México*

I. Summary
II. Introduction
III. Functional Properties of CaCCs
    A. Activation Kinetics
    B. Single Channel Conductance
    C. Anion Permeation and Selectivity
    D. Pharmacology
    E. Regulation
IV. Molecular Candidates of CaCCs
    A. The Bestrophin Family
V. Functional Properties of Bestrophins
    A. Activation Kinetics
    B. Anion Permeation and Selectivity
    C. Pharmacology
    D. Regulation
VI. Concluding Remarks
    References

I. SUMMARY

Calcium-dependent chloride channels (CaCCs) play an important role in numerous physiological processes including electrolyte and water secretion, sensory transduction, and regulation of neuronal and cardiac excitability, as well as vascular tone. In spite of this, the molecular domains that bestow sensitivity to intracellular calcium ($[Ca^{2+}]_i$), voltage (Vm) and second messengers, as well as domains that participate in CaCC gating are still unknown. Characterization of these domains awaits the discovery of a gene (or genes) that when expressed in heterologous systems will recapitulate key functional properties of CaCCs observed in native cells. Thus, findings that point out the bestrophin family as molecular candidates for CaCCs have been received with interest. However, there are many properties of CaCCs that have been studied in endogenous channels that are not yet characterized in bestrophins. In this chapter, we will review the functional properties (activation, kinetics, permeation, ion selectivity, pharmacology, and regulation) of CaCCs and bestrophins in an attempt to highlight the similarities and differences between these two families of chloride (Cl^-) channels.

II. INTRODUCTION

Calcium-dependent chloride channels (CaCCs) are expressed in different cell types including neurons, various epithelial cells, olfactory and photoreceptors, cardiac, smooth, and skeletal muscles, Sertoli cells, mast cells, neutrophils, lymphocytes, uterine muscles, brown fat adipocytes, hepatocytes, insulin-secreting beta cells, mammary and sweat glands, and *Vicia faba* guard cells (reviewed by Hartzell *et al.*, 2005a). When intracellular Ca^{2+} concentrations ($[Ca^{2+}]_i$) rise to 0.2–5 µM, CaCCs become activated resulting in Cl^- exit and subsequent depolarization of the cell membrane. This Cl^- efflux has a critical role in epithelial secretion (Wagner *et al.*, 1991; Boucher, 1994; Grubb and Gabriel, 1997), membrane excitability in cardiac muscle and neurons (Zygmunt, 1994; Kawano *et al.*, 1995; Frings *et al.*, 2000), olfactory transduction (Frings *et al.*, 2000), regulation of vascular tone (Large and Wang, 1996), and modulation of photoreceptor light responses, while membrane depolarization in *Xenopus* oocytes somehow prevents the fusion of additional sperm during fertilization. Despite the well-established physiological relevance of CaCC, the molecular identification of the gene (or genes) coding for this channel still remains elusive. The search for gene candidates has led to the cloning of a family of proteins known as bestrophins that, when expressed in several heterologous systems, generate a Cl^- current that is sensitive to $[Ca^{2+}]_i$ (Sun *et al.*, 2002; Hartzell *et al.*, 2005a). Although expression of bestrophins do not reproduce all functional CaCCs properties, ablation of bestrophin expression with interfering RNA has provided strong support to the idea that these proteins are critical components of endogenous CaCCs (Chien *et al.*, 2006).

Several excellent reviews have been published where the functional properties of CaCCs in a variety of tissues are discussed in detail (Large and Wang, 1996; Begenisich and Melvin, 1998; Kotlikoff and Wang, 1998; Frings *et al.*, 2000; Kidd and Thorn, 2000a; Fuller, 2002; Hartzell *et al.*, 2005a; Melvin *et al.*, 2005). In addition, the identity of several molecular candidates for CaCCs has been reviewed elsewhere (Jentsch *et al.*, 2002; Eggermont, 2004).

III. FUNCTIONAL PROPERTIES OF CaCCs

A. Activation Kinetics

CaCCs are activated by an increase in intracellular Ca^{2+}. Sources contributing to this increase are Ca^{2+} influx and Ca^{2+} release from intracellular stores. In many instances, Ca^{2+}

influx results from activation of Ca^{2+} channels and Ca^{2+} release results from activation of Ca^{2+}-permeable inositol triphosphate (IP_3) or ryanodine-sensitive receptors located in intracellular membranes. These two sources of Ca^{2+} usually increase the $[Ca^{2+}]_i$ into the micromolar range, which is sufficient to activate CaCCs.

Ca^{2+} can activate CaCCs by at least two known mechanisms: (1) direct gating and (2) indirectly via Ca^{2+}-binding proteins or Ca^{2+}-dependent enzymes. These two mechanisms seem to operate in different cell types, but may not be exclusive. The first mechanism appears to operate in salivary gland acinar cells (Martin, 1993; Arreola et al., 1996a), pulmonary endothelial cells (Nilius et al., 1997), ventricular myocytes (Collier et al., 1996), hepatocytes (Koumi et al., 1994), and glomerular mesangial cells (Ling et al., 1993). Thus, in acinar cells isolated from pancreas and parotid glands, CaCCs can be activated by photoreleasing Ca^{2+} (Park et al., 2001; Giovannucci et al., 2002) or by direct application of Ca^{2+} in excised patches (hepatocytes and Xenopus oocytes) while showing little or no rundown (Kuruma and Hartzell, 2000). In addition, CaCC activity can be recorded in the absence of intracellular ATP indicating that phosphorylation is not involved. Furthermore, CaCC activation is not prevented by preexposure of parotid acinar cells to KN-62 or peptide inhibitors of calmodulin and of calcium/calmodulin-dependent protein kinase II (CaMKII) (Arreola et al., 1998). These experiments support the direct gating mechanism.

In contrast, gating of CaCCs from human colonic tumor cell line, T84 (Worrell and Frizzell, 1991; Chan et al., 1994; Kaetzel et al., 1994a; Xie et al., 1996, 1998), airway epithelia (Wagner et al., 1991), T lymphocytes and neutrophils (Nishimoto et al., 1991), human macrophages (Holevinsky et al., 1994), biliary epithelial cells (Schlenker and Fitz, 1996), and cystic fibrosis-derived pancreatic epithelial cells (Chao et al., 1995) is mediated by phosphorylation via CaMKII. The participation of CaMKII has been demonstrated in experiments where the cell is dialyzed with either purified enzyme or pharmacological inhibitors of calmodulin or CaMKII such as KN62 and the autocamtide inhibitory peptide (Nishimoto et al., 1991; Wagner et al., 1991; Worrell and Frizzell, 1991).

In general, CaCC currents activate slowly with depolarization, exhibit a linear instantaneous I–V relationship and an outwardly rectifying steady state I–V. CaCCs activate slowly reaching a steady state in $\sim$2 s when $[Ca^{2+}]_i$ is less than $\sim$1 µM, however, activation is accelerated if the $[Ca^{2+}]_i$ is further increased. The activation time course nearly follows a single exponential function with Vm-independent time constants. By contrast, deactivation follows a single exponential time course with a time constant that is Vm dependent (Arreola et al., 1996a; Kuruma and Hartzell, 2000; Hartzell et al., 2005a).

The apparent open probability (P_o) of CaCC increases as $[Ca^{2+}]_i$ increases. Hill coefficients of 2–5 have been estimated from dose–response curves, which suggest the presence of multiple Ca^{2+}-binding sites in the channel protein (Arreola et al., 1996a; Kuruma and Hartzell, 2000). In addition, a Vm dependence of P_o has been reported (EC_{50} decreased at positive Vm), which may explain why at low $[Ca^{2+}]_i$ (<500 nM) a strong outward rectification is observed but it disappears by raising $[Ca^{2+}]_i$ to >1 µM (Evans and Marty, 1986; Arreola et al., 1996a; Kuruma and Hartzell, 2000).

An interesting characteristic of CaCCs is the dependence of gating kinetics on external anions (Greenwood and Large, 1999; Perez-Cornejo et al., 2004). Anions with higher permeability than Cl⁻ are able to accelerate activation and slow down deactivation at the same time. Thus, activation and deactivation kinetics are changed by a factor that nearly matches the permeability ratios without changing Ca^{2+} sensitivity (Greenwood and Large, 1999; Perez-Cornejo et al., 2004). Although CaCCs from *Xenopus* oocytes are an exception because replacing internal Cl⁻ with SCN⁻ produces a slight increase in the Ca^{2+} affinity of these channels (Kuruma and Hartzell, 2000). Taken together, the data suggest a possible coupling between channel gating and permeation analogous to what has been demonstrated for CLC channels (Perez-Cornejo et al., 2004).

B. Single Channel Conductance

Whole cell currents through CaCCs from various cell types have similar kinetics, permeation, and pharmacological properties. However, at the single channel level at least five distinct conductance values have been reported. Thus, in cardiac myocytes (Collier et al., 1996), arterial smooth muscle (Klockner, 1993; Hirakawa et al., 1999; Piper and Large, 2003), A6 kidney cells (Marunaka and Eaton, 1990), endocrine cells (Taleb et al., 1988), and *Xenopus* oocytes (Takahashi et al., 1987) the channels described have a conductance of 1–3 pS. Channels in endothelial cells (Nilius et al., 1997) and hepatocytes (Koumi et al., 1994) have a reported conductance of 8 pS, while the channels in colon (Morris and Frizzell, 1993) and a biliary cell line (Schlenker and Fitz, 1996) have a 15-pS conductance. Larger conductances are reported for channels described in Jurkat T cells (Nishimoto et al., 1991), vascular smooth muscle cells (Piper and Large, 2003), and airway epithelial cells (Frizzell et al., 1986) with 40–50 pS and 310 pS for maxi-CaCC found in *Xenopus* spinal neurons (Hussy, 1992). Some of these channels are activated by CaMKII.

This apparent contradiction between single channel and whole cell data suggests that there might be more than one type of protein underlying this diversity of channel behavior. In line with this idea, several gene candidates have been cloned and are proposed to be responsible for CaCC expression (Sun et al., 2002; Eggermont, 2004; Hartzell et al., 2005a).

C. Anion Permeation and Selectivity

Although CaCCs are relatively nonselective channels, their selectivity sequence has been characterized in different cells (Qu and Hartzell, 2000; Perez-Cornejo et al., 2004). Most CaCCs studied including those from *Xenopus* oocytes, rat parotid and lachrymal glands display the following selectivity sequence: $SCN^- > NO_3^- > I^- > Br^- > Cl^- > F^-$ (Evans and Marty, 1986; Large and Wang, 1996; Nilius et al., 1997; Qu and Hartzell, 2000; Kidd and Thorn, 2000b; Perez-Cornejo and Arreola, 2004). In addition, CaCCs from *Xenopus* oocytes also exhibit a relatively high Na^+ permeability with a P_{Na}/P_{Cl^-} of 0.1 (Qu and Hartzell, 2000).

The relative permeability of different anions through the pore of CaCC is related to the hydration energy of the anion (Qu and Hartzell, 2000; Perez-Cornejo et al., 2004). Thus, larger ions that have lower hydration energies are relatively more permeant than smaller ions. Particularly, hydrophobic anions with small hydration energies, like SCN^- and $C(CN)_3^-$, bind tightly to the pore and block CaCC conductance (Qu and Hartzell, 2000). The EC_{50} at 0 mV determined for those anions is 1.7 and ~0.4 mM, respectively. Interestingly, the ability of certain permeant ions to block Cl^- conductance is commonly observed in other Cl^- channels (Fahlke et al., 1997; Rychkov et al., 1998; Dawson et al., 1999) and it has been used to characterize CaCC molecular candidates such as bestrophins.

D. Pharmacology

The most common blockers for native CaCCs are niflumic acid (NFA) and flufenamic acid, which block CaCCs in *Xenopus* oocytes at concentrations in the 10 μM-range (Qu and Hartzell, 2001). NFA is often considered a specific blocker and even used in different tissues to identify anion currents as CaCCs. However, NFA also acts on other conductances, for example, it enhances CaCCs in smooth muscle at negative Vms and blocks volume-regulated anion channels, K^+ channels, and Ca^{2+} currents (Reinsprecht et al., 1995; Wang et al., 1997; Xu et al., 1997; Doughty et al., 1998; Piper et al., 2002).

Less effective blockers of CaCCs include tamoxifen, 4,4'-diisothiocyanato-2,2'-stilbenedisulfonic acid (DIDS), 4-acetamido-4'-isothiocyanostilbene-2,2'-disulphonic acid (SITS), 5-nitro-2-(3-phenylpropylamino)-benzoic acid (NPPB), 9-anthracene-carboxylic acid (9-AC), diphenylamine-2-carboxylic acid (DPC), fluoxetine, and mefloquine (Frings et al., 2000; Maertens et al., 2000b). The small peptide chlorotoxin appears to specifically block CaCCs from rat astrocytoma cells (Dalton et al., 2003), but results ineffective when used on CaCCs expressed in secretory epithelia (Maertens et al., 2000a).

Interestingly, some CaCCs blockers exert their effect in a Vm-dependent manner (Qu and Hartzell, 2001; Qu et al., 2003) and thus, have been used to partially deduce the pore's architecture. The accessibility of each blocker was assessed estimating the distance that the blocker has to travel within the electrical field to reach its binding site. This is indicative of how far the blocker can travel within the pore before it gets stuck. From these measurements and considering the blockers dimension it was proposed that the CaCC pore has an elliptical cone shape with a large opening (0.6 × 0.94 nm) that faces the extracellular space (Qu and Hartzell, 2001).

E. Regulation

Several regulatory mechanisms of CaCCs have been described, including those with participation of Ca^{2+}-dependent enzymes (CaMKII, protein phosphatases, annexins), inositol 3,4,5,6-tetrakisphosphate (IP_4), pH, cGMP, G-proteins, and other regulatory proteins such as cystic fibrosis transmembrane conductance regulator (CFTR). The mechanisms that underlie CaCC regulation are still waiting to be elucidated but because CaCCs participate in a variety of physiological functions understanding those regulatory mechanisms would of major importance.

As mentioned before CaMKII can activate CaCCs, however, in arterial and tracheal smooth muscle CaMKII inhibits CaCCs (Greenwood et al., 2001; Ledoux et al., 2003; Leblanc et al., 2005). Other Ca^{2+}-dependent enzymes, such as calcineurin and alkaline phosphatase, act as positive regulators of CaCCs (Marunaka and Eaton, 1990; Wang and Kotlikoff, 1997). In CFPAC-1 and T84 colonic carcinoma cells, the CaMKII effect is modulated by IP_4 (Vajanaphanich et al., 1994; Ismailov et al., 1996; Xie et al., 1996; Nilius et al., 1998; Xie et al., 1998) via channel dephosphorylation since the IP_4 effect is prevented by inhibiting phosphatase activity. It has been proposed that this regulation could serve to tune CaCC activation during electrical activity.

Annexins, a class of phospholipids and Ca^{2+}-binding proteins that are concentrated in the apical membrane of secretory epithelia where CaCCs have been described, inhibited CaCCs (Chan et al., 1994; Kaetzel et al., 1994a,b; Jorgensen et al., 1997). CaCCs from Xenopus oocytes are inhibited quite potently (IC_{50} near 50 nM) by annexins II, III, and V isolated from Ehrlich ascites cells but not by porcine and bovine annexins types II and V. Annexins block is potentiated by IP_4 (Xie et al., 1996).

Second messengers like cGMP also regulate CaCCs. In rat mesenteric artery smooth muscle cells, CaCC is activated by Ca^{2+} in a cGMP-dependent manner (Piper and Large, 2003; Matchkov et al., 2004). It is thought that cGMP acts via phosphorylation by a cGMP-dependent protein kinase (Matchkov et al., 2004). Interestingly, single channel recording of cGMP-dependent CaCCs show substate conductances of 15, 35, and 55 pS. These numbers are larger than the CaCC conductances determined in rabbit pulmonary artery smooth muscle cells (Piper and Large, 2003) or salivary glands (Martin, 1993) which may imply that this channel is different from the CaCCs previously described in other tissues. This idea is supported by recent evidence showing pharmacological differences between cGMP-dependent CaCC (blocked by Zn^{2+} but not by NFA) and classical CaCC (blocked by NFA but not by Zn^{2+}) (Matchkov et al., 2004; Hartzell et al., 2005a). In addition, cGMP-dependent CaCC is potentiated by CaM but unaffected by CaMKII blockade (Piper and Large, 2004).

Direct regulation of CaCCs by G-proteins has been reported only in inside-out patches isolated from submandibular acinar cells where application of GTPγS induced the appearance of small conductance CaCCs (Martin, 1993). However, indirect regulation of CaCCs via IP_3 production by a PLC β3-like enzyme following G-proteins stimulation with ginsenosides or GTPγS have been reported in Xenopus oocytes and HTC hepatoma cells (Choi et al., 2001; Kaibara et al., 2001; Kilic and Fitz, 2002). Further experiments are needed to clarify these regulatory mechanisms.

The CFTR, a Cl^- channel defective in cystic fibrosis (CF) disease (see Chapters 5 and 6), seems to be able to regulate various ion channels, among those CaCCs (Kunzelmann et al., 2000, 2001; Tarran et al., 2002). Expression of CFTR in either calf pulmonary artery endothelial (CPAE) cells (Wei et al., 1999) or Xenopus oocytes (Kunzelmann et al., 1997) reduces the current magnitude through endogenous CaCCs.

Similarly, the CaCC current obtained from mouse parotid acinar cells isolated from $CFTR^{-/-}$ mice is larger in amplitude than CaCC current from wild-type (WT) mice (Perez-Cornejo and Arreola, 2004). A possible underlying mechanism involved in this type of regulation might involve an interaction between the C-terminal part of the CFTR's R domain and CaCCs (Wei et al., 2001).

Nearly all anion channels are sensitive to changes in internal pH, external pH, or both. Intracellular acidification inhibits CaCCs in acinar cells from lachrymal and parotid glands and T84 cells (Arreola et al., 1995; Park and Brown, 1995). In contrast, CaCCs from *Xenopus* oocytes are less sensitive to intracellular acidification but intracellular alkalinization blocks inward current (Qu and Hartzell, 2000). The underlying mechanism of regulation of CaCCs by intracellular pH is unknown but it has been proposed that internal H^+ ions may compete for Ca^{2+}-binding sites on the channel. Extracellular alkalinization decreased CaCC currents at positive Vms (Qu and Hartzell, 2000).

IV. MOLECULAR CANDIDATES OF CaCCs

Discovering the molecular identity of CaCCs would be determinant in elucidating the role of CaCCs in normal physiology as well as in disease. The search has been hindered by lack of high-affinity, selective blockers. Several molecular candidates have been identified as Ca^{2+}-activated Cl^- channels. However, none recapitulate all functional properties of CaCCs observed in native cells and some are not even expressed in tissues where CaCCs have been described.

The human genes *hTTHY2* and *hTTYH3*, known as *tweety* in *Drosophila*, have been proposed to be Ca^{2+}-regulated maxi-Cl^- channels with a conductance of ~260 pS (Suzuki and Mizuno, 2004). These gene products might correspond to the maxi-Cl^- channel found in spinal neurons (Hussy, 1992) and skeletal muscle (Fahmi et al., 1995). However, *hTTYH1* encodes a Ca^{2+}-insensitive channel and *hTTYH3*, although being regulated by Ca^{2+}, is not expressed in salivary glands. Hence, it is unlikely that these genes play any role in CaCCs found in acinar cells of secretory glands. A ClC-3 isoform, a member of the CLC chloride channel family (see Chapters 2–4), was reported to be regulated by Ca^{2+} and was proposed to represent a Ca^{2+}-dependent Cl^- channel activated by CaMKII (Huang et al., 2001; Robinson et al., 2004), since it has been shown that the CaMKII-activated conductance is absent in $ClC-3^{-/-}$ mice (Robinson et al., 2004). In contrast, secretory CaCCs are still present in these mice and their properties are unaffected (Arreola et al., 2002), which suggests that the CaMKII-activated and the Ca^{2+}-activated Cl^- conductance are different from each other.

The CLCA family members represent another class of molecular CaCC candidates (Cunningham et al., 1995; Jentsch et al., 2002; Eggermont, 2004). This family was originally cloned from a bovine tracheal cDNA expression library screened with an antibody generated against a purified protein that induced a Ca^{2+}-dependent conductance in artificial lipid bilayer (Cunningham et al., 1995). Although Ca^{2+}-dependent currents were generated in various cell types transfected with cDNA encoding various CLCA proteins (Elble et al., 1997; Gandhi et al., 1998; Gruber *et al.*, 1998a,b; Pauli *et al.*, 2000), the properties of the resulting currents were different from those of native CaCCs. These differences included discrepancies in Ca^{2+} sensitivity, kinetics, Vm sensitivity, and pharmacology of the channel. To this date, there are no structure–function studies to help determine permeation and Ca^{2+} sensitivity of these proteins. The role of CLCA as CaCCs is further clouded by learning that some cell types that express native CaCCs do not express CLCA proteins (Papassotiriou *et al.*, 2001). To make matters more complex, CLCA proteins have very high homology to known cell adhesion proteins, with some of them being actually secreted as soluble proteins (Gruber and Pauli, 1998; Pauli *et al.*, 2000). These observations cast doubt on the suitability of CLCA proteins to form CaCCs. Instead, it has been suggested that CLCA proteins may not be a channel but a modulator of endogenous Cl^- channels (Loewen *et al.*, 2002, 2003, 2004) or a subunit of CaCCs.

All the molecular candidates described in preceding paragraph seem to be missing some auxiliary subunits that may be necessary for these proteins (tweety, ClC–3, and CLCA) to function as Ca^{2+}-dependent Cl^- channels able to recapitulate key functional properties of endogenous CaCCs.

A. The Bestrophin Family

Bestrophins comprise a new family of Cl^- channel proteins (Sun *et al.*, 2002; Hartzell *et al.*, 2005b) that in mammals is composed of four members. Mutations in human bestrophin-1 (hBest-1) produce vitelliform macular dystrophy, an early onset form of macular degeneration (Petrukhin *et al.*, 1998). hBest-1 has been localized to the basolateral membrane of retinal pigment epithelial cells (Marmorstein *et al.*, 2000; Sun *et al.*, 2002). When expressed heterologously, bestrophins function as Ca^{2+}-activated Cl^- channels (Sun *et al.*, 2002). Mutations introduced in the transmembrane domain 2 of mouse bestrophin-2 (mBest-2) (Qu and Hartzell, 2004; Qu *et al.*, 2004) altered the anionic selectivity and conductance of the channel. Expression of these mutants inhibited WT currents by a dominant negative effect suggesting that Best-2 forms a multimeric protein permeable to anions. Although multimers are formed in the plasma membrane, a fraction of the bestrophins remains in intracellular organelles, which suggests that bestrophins could be intracellular Cl^- channels (Qu and Hartzell, 2003; Tsunenari *et al.*, 2003, Hartzell *et al.*, 2005a,b). Table I shows a brief summary of some properties of bestrophins and compares them to those of CaCCs. As can be readily seen (see following sections), not all CaCC properties are replicated by bestrophins.

V. FUNCTIONAL PROPERTIES OF BESTROPHINS

A. Activation Kinetics

Nathan's group was the first to clone hBest-1 from ocular tissue using PCR (Sun *et al.*, 2002). Subsequent expression of hBest-1 in HEK 293 cells resulted in Cl^- currents that activated by increasing intracellular Ca^{2+} (Sun *et al.*, 2002). Since then, other

Table I
Functional and Pharmacological Properties of CaCCs and Bestrophins[a]

Parameter	CaCC	WT bestrophins (hBest-1, hBest-2, xBest-1a, xBest-1b, and mBest-2)
Kinetics	Time dependent at $[Ca^{2+}]_i = EC_{50}$, time independent at high $[Ca^{2+}]_i$	Time independent at any $[Ca^{2+}]_i$
I–V relationship	Outward rectifying	Linear
Ca^{2+} sensitivity (EC_{50}; Hill coefficient; electrical distance)	0.06–0.9; 2–3; 0.13	0.21–0.23, Vm independent
Anion selectivity	$C(CN)_3 > SCN^- > N(CN)_2 > ClO_4 > I^- > N_3 > Br^- > Cl^- >$ formate $> HCO_3 >$ acetate $= F >$ gluconate	$SCN^- > I > Br^- > Cl^- > F$
Blockade (EC_{50}; electrical distance)	NFA (10.1; 0.1) > A9C (18.3; 0.6) > DIDS (48; 0.3) > DPC (111;0.3) Vm independent	DIDS (3.1), NFA Vm dependent
Blockade by SCN^- (EC_{50})	~2000	~12,000
Sensitivity to cell volume	No	Yes

[a]This table is not an exhaustive comparison since not all parameters studied in endogenous CaCCs have been determined or analyzed in bestrophins. Ca sensitivity and blockade parameters are those obtained at positive Vm. EC_{50} values are given in μM, Hill coefficient and electrical distance values are dimensionless.

members of the bestrophin family have been cloned and electrophysiologically analyzed. xBest-2a and xBest-2b cloned from *Xenopus* oocytes and expressed in HEK 293 cells are activated by intracellular Ca^{2+} with EC_{50} of 210 and 228 nM, respectively (Qu and Hartzell, 2003). Similarly, mBest-2 is activated with an EC_{50} of 230 nM (Qu et al., 2004). Thus, Ca^{2+} sensitivity of bestrophin proteins is well in agreement with that of CaCCs from different cells, including neuronal, *Xenopus* oocytes, hepatocytes, lachrymal and salivary glands, medullary duct cells, and pancreatic acinar cells (Melvin et al., 2002; Hartzell et al., 2005a). However, there is an important difference between bestrophin-generated currents and those recorded from CaCCs. Currents through CaCCs exhibit time- and Ca^{2+}-dependent activation, Vm-dependent kinetics, and outward rectification that are not displayed by bestrophins (Sun et al., 2002; Qu and Hartzell, 2003; Qu et al., 2004). Currents from hBest-1 and hBest-2 are time independent and Vm independent, while Best-1 from *Caenorhabditis elegans* or *Drosophila* displays inward and outward rectification, respectively (Sun et al., 2002). Furthermore, hBest-3 displays slow activation kinetics and strong inward rectification (Tsunenari et al., 2003). Since these electrophysiological characteristics are not matched by endogenous CaCCs, this would suggest two things: not all bestrophins form Ca^{2+}-dependent Cl^- channels or native CaCCs have auxiliary subunits not present in the systems used to express bestrophins. Independently of the final explanation to these observations, it has become clear that bestrophins are essential part of endogenous CaCCs as has been shown using siRNA (Chien et al., 2006).

B. Anion Permeation and Selectivity

Both CaCCs and bestrophin channels exhibit a generic lyotropic anion selectivity sequence. The anionic selectivity of the current induced by hBest-1, hBest-2, xBest-2a, xBest-2b, and mBest-2 expression in HEK 293 cells has been estimated from changes in the reversal potential (Sun et al., 2002; Qu and Hartzell, 2003, 2004; Qu et al., 2004). In general, these currents exhibited an anionic permeability sequence of $SCN^- > NO_3^- > I^- > Br^- > Cl^- > F^-$, which is identical to that of CaCCs in several tissues (Melvin et al., 2002; Hartzell et al., 2005a). Nonetheless, there are some significant differences in permeability ratios. For example, $P_{NO_3^-}/P_{Cl^-}$ for hBest-1 and hBest-2 are 5.8 and 2.7, respectively (Sun et al., 2002). The selectivity sequence determined for xBest-2a is $I^->Br^->Cl^->$aspartate with the following approximates P_{X^-}/P_{Cl^-} values: 2.6, 1.65, 1, <0.2, respectively (Qu and Hartzell, 2003). Anion selectivity has been studied in more detail for mBest-2 with a sequence $SCN^- > NO_3^- > I^- > Br^- > Cl^- > F^-$, which correspond to P_{X^-}/P_{Cl^-} values of 8.2 ± 1.3, 2.1 ± 0.2, 1.9 ± 0.1, 1.4 ± 0.1, 1 ± 0, and 0.5 ± 0.1, respectively (Qu and Hartzell, 2004; Qu et al., 2004).

As discussed before, highly permeant anions, like SCN^-, block in a Vm-dependent manner CaCCs from *Xenopus* oocytes (Qu and Hartzell, 2000). Similarly, mBest-2 currents are inhibited by SCN^- with an EC_{50} of 12 mM at +100 mV or 9.8 mM at +/− 50 mV (Qu et al., 2004). This EC_{50} value is larger than the one observed for *Xenopus* oocyte CaCC, that shows an EC_{50} of about 2 mM at the same potential (Qu and Hartzell, 2000). Anions like I^- and Br^- increase whole cell conductance. Thus, the blockade of mBest-2 and *Xenopus* oocytes CaCC by SCN^- is Vm independent. However, CaCC conductance in parotid acinar cells is not decreased by extracellular SCN^- (Perez-Cornejo and Arreola, 2004). In these acinar cells, the P_{SCN^-}/P_{Cl^-} ratio determined is 4.3 ± 0.4, much smaller than that determined for both mBest-2 and CaCC from *Xenopus* oocytes. Moreover, the permeation process appears to be coupled to the gating mechanism (Perez-Cornejo and Arreola, 2004).

S79C mutant mBest-2 channels display a fivefold enhancement in conductance and a drastically reduced P_{SCN^-}/P_{Cl^-} when compared to WT channels. In addition, the S79C mutation decreased the EC_{50} for Vm-independent block of SCN^- and DIDS by about two- and fivefold, respectively (Qu and Hartzell, 2004; Qu et al., 2004). However, currents through F80R mBest-2 mutants are blocked by DIDS in a Vm-dependent manner and display a decrease in P_{SCN^-}/P_{Cl^-} from 8.2 to 3.3. These mutant channels have a fivefold larger conductance in the presence of external SCN^- but in the presence of NO_3^- the conductance decreases nearly fourfold (Qu and Hartzell, 2004). Therefore characterization of these pore properties lend support to the hypothesis that mBest-2 is a Ca^{2+}-dependent Cl^- channel.

C. Pharmacology

CaCCs are sensitive to many nonspecific inhibitors (Hartzell et al., 2005a). It is generally accepted that NFA blocks CaCCs in a mildly Vm-dependent manner. So far, block by NFA in bestrophins has not been characterized in detail but 100 μM NFA seems to block most of the mBest-2 currents (Criss Hartzell, personal communication). In addition, currents through mBest-2 as well as human bestrophins can be blocked in a Vm-independent manner by DIDS with an EC_{50} (at +100 mV) of 3.1 ± 0.3 μM (Sun et al., 2002; Qu et al., 2004). In contrast, CaCC currents from *Xenopus* oocytes, medullary collecting ducts, and parotid acinar cells are blocked by DIDS in a Vm-dependent manner with an EC_{50} (at +100 mV) of ≥ 40 μM (Qu and Hartzell, 2001; Melvin et al., 2002; Qu et al., 2003). Interestingly, mutations in the second transmembrane segment of mBest-2 alter the pharmacology of these channels. Thus, sensitivity to DIDS is decreased in S79C mutant channels, which have an EC_{50} of 14.6 ± 3.6 μM (Qu et al., 2004). In addition, the Vm sensitivity of DIDS blockade is changed in F80R mutant channels, where the block becomes Vm dependent just like in CaCCs (Qu and Hartzell, 2004). Tamoxifen, at 20 μM, a concentration that blocks volume-sensitive Cl^- channels, appeared not to block xBest-2 currents (Qu and Hartzell, 2003).

D. Regulation

Little is known about bestrophin regulation. It has been demonstrated that Ca^{2+}-dependent currents generated by bestrophin expression are regulated by changes in cell volume (Fischmeister and Hartzell, 2005). hBest-1 and mBest-2 were expressed in HEK 293, HeLa, and ARPE-19 cells and then subjected to a 20% increase in extracellular osmolarity. This maneuver caused cell shrinkage and an approximately 70–80% reduction in bestrophin current. However, a more typical manipulation used to determine regulation of channels by changes in cell volume consists in exposing cells to a hypotonic media. When cells expressing bestrophins were exposed to this protocol, a slight increase in bestrophin current was observed which was masked by activation of the endogenous volume-sensitive Cl^- channels. At present, it is not known if bestrophin expression up-regulates volume-sensitive Cl^- channels. Interestingly, when exposed to a hyperosmotic media, mouse retinal pigment epithelium (RPE) cells show a response similar to that of hBest-1 (Fischmeister and Hartzell, 2005). Regulation of hBest-1 by cell volume is a puzzling finding since generally the current through volume-sensitive, but not CaCC, channels is abolished when CaCC-expressing cells are exposed to a hypertonic media (Arreola et al., 1996b).

As noted in preceding paragraph, CaCCs are strongly regulated by both internal and external pH, as well as CaMKII and annexins. However, none of those modulators of CaCCs have tested in bestrophin currents.

VI. CONCLUDING REMARKS

Functional and pharmacological analyses suggest that CaCCs represent a channel family composed of many different members. Therefore, more than one gene product would be expected to underlie the molecular identity of these channels. Nonetheless, each candidate gene when expressed in heterologous systems will be expected to display some key characteristics for activation that include Ca^{2+} sensitivity, time dependence, and Vm dependence and outward rectification.

To date, the most promising molecular candidates for CaCCs belong to the bestrophin family, a set of genes recently cloned from human, mice, fruit fly, *C. elegans*, and *Xenopus laevis*. Although bestrophins have been expressed in heterologous systems and their functional activity studied in isolation, there are many functional properties that have not yet been determined (Table I). This is also true for *tweety* and ClC-3. For example, no single channel data are available and little is known about their regulation by second messengers or other proteins. Thus, a complete characterization of the functional properties of bestrophins and their regulation is essential to best correlate CaCCs with bestrophin proteins.

A key property that defines an ion channel pore is the ability to select among different ions. Consequently, for a protein to be considered a molecular candidate of the native channel it must replicate this property. As mentioned before, bestrophins display Ca^{2+} sensitivity and exhibit an anionic permeability sequence identical to that of CaCCs in several tissues. Mutational analysis of mBest-2 has shown that mBest-2 forms a channel pore that is selective for anions and sensitive to SCN^-. The ability of SCN^- to block the Cl^- conductance is a distinctive feature commonly observed in CaCCs. Pharmacological tools are indispensable to assay the physiological role of ion channels and to dissect their contribution to whole cell current. Unfortunately, few CaCC blockers are available and require high concentrations to completely block Cl^- currents, which may result in many side effects. Nonetheless, it seems that bestrophins share with CaCCs their sensitivity to NFA, which is often considered a specific blocker and has been used to identify CaCCs in different tissues.

A more thorough functional characterization of several bestrophins is obviously needed to classify them as a family or subfamily of Ca^{2+}-dependent Cl^- channels. Thus, further studies, such as single channel recordings, block by pharmacological agents, and regulation by different molecules and second messengers, are needed. It has been shown that small interference RNA (siRNA) against bestrophin 1 or bestrophin 2 inhibits expression of a 1 pS endogenous CaCC present in *Drosophila* S2 cells (Chien et al., 2006). This result provides the strongest evidence in favor of a role for bestrophins in forming CaCCs. Further experiments using knock out mice are critical to also establish the role of bestrophins as essential molecular components of mammalian CaCCs.

ACKNOWLEDGMENTS

The authors thank Criss Hartzell for sharing NFA data. This work was supported in part by NIH grants DE-09692 and DE13539, Fogarty International Center grant R03TW006429 (J. E. Melvin), P01-HL18208 (R. Waugh), and by CONACyT-Mexico grants 42561 (J.A.) and 45895 (P.P.-C.).

REFERENCES

Arreola, J., Melvin, J. E., and Begenisich, T. (1995). Inhibition of Ca^{2+}-dependent Cl^- channels from secretory epithelial cells by low internal pH. *J. Membr. Biol.* **147**, 95–104.

Arreola, J., Melvin, J., and Begenisich, T. (1996a). Activation of calcium dependent chloride channels in rat parotid acinar cells. *J. Gen. Physiol.* **108**, 35–47.

Arreola, J., Park, K., Melvin, J. E., and Begenisich, T. (1996b). Three distinct chloride channels control anion movements in rat parotid acinar cells. *J. Physiol.* **490**, 351–362.

Arreola, J., Melvin, J. E., and Begenisich, T. (1998). Differences in regulation of Ca^{2+}-activated Cl^- channels in colonic and parotid secretory cells. *Am. J. Physiol.* **274**, C161–C166.

Arreola, J., Begenisich, T., Nehrke, K., Nguyen, H. V., Park, K., Richardson, L., Yang, B. L., Schutte, B. C., Lamb, F. S., and Melvin, J. E. (2002). Secretion and cell volume regulation by salivary acinar cells from mice lacking expression of the Clcn3 Cl^- channel gene. *J. Physiol.* **545**, 207–216.

Begenisich, T. and Melvin, J. E. (1998). Regulation of chloride channels in secretory epithelia. *J. Membr. Biol.* **163**, 77–85.

Boucher, R. C. (1994). Human airway ion transport. *Am. J. Respir. Crit. Care Med.* **150**, 271–281.

Chan, H. C., Kaetzel, M. A., Gotter, A. L., Dedman, J. R., and Nelson, D. J. (1994). Annexin IV inhibits calmodulin-dependent protein kinase II-activated chloride conductance: A novel mechanism for ion channel regulation. *J. Biol. Chem.* **269**, 32464–32468.

Chao, A. C., Kouyama, K., Heist, E. K., Dong, Y. J., and Gardner, P. (1995). Calcium- and CaMKII-dependent chloride secretion induced by the microsomal Ca^{2+}-ATPase inhibitor 2,5-di-(tert-butyl)-1, 4-hydroquinone in cystic fibrosis pancreatic epithelial cells. *J. Clin. Invest.* **96**, 1794–1801.

Chien, L. T., Zhang, Z. R., and Hartzell, H. C. (2006). Single Cl^- channels activated by Ca^{2+} in *Drosophila* S2 cells are mediated by bestrophins. *J. Gen. Physiol.* **128**, 247–259.

Choi, S., Rho, S. H., Jung, S. Y., Kim, S. C., Park, C. S., and Nah, S. Y. (2001). A novel activation of Ca^{2+}-activated Cl^- channel in *Xenopus* oocytes by ginseng saponins: Evidence for the involvement of phospholipase C and intracellular Ca^{2+} mobilization. *Br. J. Pharmacol.* **132**, 641–648.

Collier, M. L., Levesque, P. C., Kenyon, J. L., and Hume, J. R. (1996). Unitary Cl^- channels activated by cytoplasmic Ca^{2+} in canine ventricular myocytes. *Circ. Res.* **78**, 936–944.

Cunningham, S. A., Awayda, M. S., Bubien, J. K., Ismailov, I. I., Arrate, M. P., Berdiev, B. K., Benos, D. J., and Fuller, C. M. (1995). Cloning of an epithelial chloride channel from bovine trachea. *J. Biol. Chem.* **270**, 31016–31026.

Dalton, S., Gerzanich, V., Chen, M., Dong, Y., Shuba, Y., and Simard, J. M. (2003). Chlorotoxin-sensitive Ca^{2+}-activated Cl^- channel in type R2 reactive astrocytes from adult rat brain. *Glia* **42**, 325–339.

Dawson, D. C., Smith, S. S., and Mansoura, M. K. (1999). CFTR: Mechanism of anion conduction. *Physiol. Rev.* **79**(Suppl. 1), S47–S75.

Doughty, J. M., Miller, A. L., and Langton, P. D. (1998). Non-specificity of chloride channel blockers in rat cerebral arteries: Block of the L-type calcium channel. *J. Physiol.* **507**, 433–439.

Eggermont, J. (2004). Calcium-activated chloride channels (Un)known, (Un)loved? *Proc. Am. Thorac. Soc.* **1**, 22–27.

Elble, R. C., Widom, J., Gruber, A. D., Abdel-Ghany, M., Levine, R., Goodwin, A., Cheng, H. C., and Pauli, B. U. (1997). Cloning and characterization of lung-endothelial cell adhesion molecule-1 suggest it is an endothelial chloride channel. *J. Biol. Chem.* **272**, 27853–27861.

Evans, M. G. and Marty, A. (1986). Calcium-dependent chloride currents in isolated cells from rat lacrimal glands. *J. Physiol.* **378**, 437–460.

Fahlke, C., Dürr, C., and George, A. L., Jr. (1997). Mechanism of ion permeation in skeletal muscle chloride channels. *J. Gen. Physiol.* **110**, 551–564.

Fahmi, M., Garcia, L., Taupignon, A., Dufy, B., and Sartor, P. (1995). Recording of a large-conductance chloride channel in normal rat lactotrophs. *Am. J. Physiol.* **269**, E969–E976.

Fischmeister, R. and Hartzell, C. (2005). Volume-sensitivity of the bestrophin family of chloride channels. *J. Physiol.* **562**, 477–491.

Frings, S., Reuter, D., and Kleene, S. J. (2000). Neuronal Ca^{2+}-activated Cl^- channels—homing in on an elusive channel species. *Prog. Neurobiol.* **60,** 247–289.

Frizzell, R. A., Rechkemmer, G., and Shoemaker, R. L. (1986). Altered regulation of airway epithelial cell chloride channels in cystic fibrosis. *Science* **233,** 558–560.

Fuller, C. M. (2002). *Calcium*-Activated Chloride Channels, Vol. 53. Academic Press, San Diego.

Gandhi, R., Elble, R. C., Gruber, A. D., Schreur, K. D., Ji, H. L., Fuller, C. M., and Pauli, B. U. (1998). Molecular and functional characterization of a calcium-sensitive chloride channel from mouse lung. *J. Biol. Chem.* **273,** 32096–32101.

Giovannucci, D. R., Bruce, J. I., Straub, S. V., Arreola, J., Sneyd, J., Shuttleworth, T. J., and Yule, D. I. (2002). Cytosolic Ca^{2+} and Ca^{2+}-activated Cl^- current dynamics: Insights from two functionally distinct mouse exocrine cells. *J. Physiol.* **540,** 469–484.

Greenwood, I. A. and Large, W. A. (1999). Modulation of the decay of Ca^{2+}-activated Cl^- currents in rabbit portal vein smooth muscle cells by external anions. *J. Physiol.* **516,** 365–376.

Greenwood, I. A., Ledoux, J., and Leblanc, N. (2001). Differential regulation of Ca^{2+}-activated Cl^- currents in rabbit arterial and portal vein smooth muscle cells by Ca^{2+}-calmodulin-dependent kinase. *J. Physiol.* **534,** 395–408.

Grubb, B. R. and Gabriel, S. E. (1997). Intestinal physiology and pathology in gene-targeted mouse models of cystic fibrosis. *Am. J. Physiol.* **273,** G258–G266.

Gruber, A. D. and Pauli, B. U. (1998). Molecular cloning and biochemical characterization of a truncated secreted member of the human family of Ca^{2+}-activated Cl^- channels. *Biochim. Biophys. Acta* **1444,** 418–423.

Gruber, A. D., Elble, R. C., Ji, H.-L., Schreur, K. D., Fuller, C. M., and Pauli, B. U. (1998a). Genomic cloning, molecular characterization, and functional analysis of human CLCA1, the first human member of the family of Ca^{2+}-activated Cl^- channel proteins. *Genomics* **54,** 200–214.

Gruber, A. D., Gandhi, R., and Pauli, B. U. (1998b). The murine calcium-sensitive chloride channel (mCaCC) is widely expressed in secretory epithelia and in other select tissues. *Histochem. Cell Biol.* **110,** 43–49.

Hartzell, C., Putzier, I., and Arreola, J. (2005a). Calcium-activated chloride channels. *Ann. Rev. Physiol.* **67,** 719–758.

Hartzell, C., Qu, Z., Putzier, I., Artinian, L., Chien, L. T., and Cui, Y. (2005b). Looking chloride channels straight in the eye: Bestrophins, lipofuscinosis, and retinal degeneration. *Physiology* **20,** 292–302.

Holevinsky, K. O., Jow, F., and Nelson, D. J. (1994). Elevation in intracellular calcium activates both chloride and proton currents in human macrophages. *J. Membr. Biol.* **140,** 13–30.

Huang, P., Liu, J., Di, A., Robinson, N. C., Musch, M. W., Kaetzel, M. A., and Nelson, D. J. (2001). Regulation of human CLC-3 channels by multifunctional Ca^{2+}/calmodulin-dependent protein kinase. *J. Biol. Chem.* **276,** 20093–20100.

Hussy, N. (1992). Calcium-activated chloride channels in cultures embryonic *Xenopus* spinal neurons. *J. Neurophysiol.* **68,** 2042–2050.

Ismailov, I. I., Fuller, C. M., Berdiev, B. K., Shlyonsky, V. G., Benos, D. J., and Barrett, K. E. (1996). A biologic function for an "orphan" messenger: D-myo-inositol 3,4,5,6-tetrakisphosphate selectively blocks epithelial calcium-activated chloride channels. *Proc. Natl. Acad. Sci. USA* **93,** 10505–10509.

Jentsch, T. J., Stein, V., Weinreich, F., and Zdebik, A. A. (2002). Molecular structure and physiological function of chloride channels. *Physiol. Rev.* **82,** 503–568.

Jorgensen, A. J., Bennekou, P., Eskesen, K., and Kristensen, B. I. (1997). Annexins from Ehrlich ascites inhibit the calcium-activated chloride current in *Xenopus laevis* oocytes. *Eur. J. Physiol.* **434,** 261–266.

Kaetzel, M. A., Chan, H. C., Dubinsky, W. P., Dedman, J. R., and Nelson, D. J. (1994a). A role for annexin IV in epithelial cell function. Inhibition of calcium-activated chloride conductance. *J. Biol. Chem.* **269,** 5297–5302.

Kaetzel, M. A., Pula, G., Campos, B., Uhrin, P., Horseman, N., and Dedman, J. R. (1994b). Annexin VI isoforms are differentially expressed in mammalian tissues. *Biochim. Biophys. Acta* **1223,** 368–374.

Kaibara, M., Nagase, Y., Murasaki, O., Uezono, Y., Doi, Y., and Taniyama, K. (2001). GTPγS-induced Ca^{2+} activated Cl^- currents: Its stable induction by G_q alpha overexpression in *Xenopus* oocytes. *Jpn. J. Pharmacol.* **86,** 244–247.

Kawano, S., Hirayama, Y., and Hiraoka, M. (1995). Activation mechanism of Ca^{2+} sensitive transient outward current in rabbit ventricular myocytes. *J. Physiol.* **486**, 593–604.

Kidd, J. F. and Thorn, P. (2000a). Intracellular Ca^{2+} and Cl^- channel activation in secretory cells. *Annu. Rev. Physiol.* **62**, 493–513.

Kidd, J. F. and Thorn, P. (2000b). The properties of the secretagogue-evoked chloride current in mouse pancreatic acinar cells. *Pflügers Arch.* **441**, 489–497.

Kilic, G. and Fitz, J. G. (2002). Heterotrimeric G-proteins activate Cl^- channels through stimulation of a cyclooxygenase-dependent pathway in a model liver cell line. *J. Biol. Chem.* **277**, 11721–11727.

Klockner, U. (1993). Intracellular calcium ions activate a low-conductance chloride channel in smooth-muscle cells isolated from human mesenteric artery. *Pflügers Arch.* **424**, 231–237.

Kotlikoff, M. I. and Wang, Y. X. (1998). Calcium release and calcium-activated chloride channels in airway smooth muscle cells. *Am. J. Resp. Crit. Care Med.* **158**, S109–S114.

Koumi, S., Sato, R., and Aramaki, T. (1994). Characterization of the calcium-activated chloride channel in isolated guinea-pig hepatocytes. *J. Gen. Physiol.* **104**, 357–373.

Kunzelmann, K. (2001). CFTR: Interacting with everything? *News Physiol. Sci.* **16**, 167–170.

Kunzelmann, K., Mall, M., Briel, M., Hipper, A., Nitschke, R., Ricken, S., and Greger, R. (1997). The cystic fibrosis transmembrane conductance regulator attenuates the endogenous Ca^{2+} activated Cl^- conductance of *Xenopus* oocytes. *Eur. J. Physiol.* **435**, 178–181.

Kunzelmann, K., Schreiber, R., Nitschke, R., and Mall, M. (2000). Control of epithelial Na^+ conductance by the cystic fibrosis transmembrane conductance regulator. *Pflügers Arch.* **440**, 193–201.

Kuruma, A. and Hartzell, H. C. (2000). Bimodal control of a Ca^{2+}-activated Cl^- channel by different Ca^{2+} signals. *J. Gen. Physiol.* **115**, 59–80.

Large, W. A. and Wang, Q. (1996). Characteristics and physiological role of the Ca^{2+}-activate Cl^- conductance in smooth muscle. *Am. J. Physiol.* **271**, C435–C454.

Leblanc, N., Ledoux, J., Saleh, S., Sanguinetti, A., Angermann, J., O'Driscoll, K., Britton, F., Perrino, B. A., and Greenwood, I. A. (2005). Regulation of calcium-activated chloride channels in smooth muscle cells: A complex picture is emerging. *Can. J. Physiol. Pharmacol.* **83**, 541–556.

Ledoux, J., Greenwood, I., Villeneuve, L. R., and Leblanc, N. (2003). Modulation of Ca^{2+}-dependent Cl^- channels by calcineurin in rabbit coronary arterial myocytes. *J. Physiol.* **552**, 701–714.

Ling, B. N., Seal, E. E., and Eaton, D. C. (1993). Regulation of mesangial cell ion channels by insulin and angiotensin II. Possible role in diabetic glomerular hyperfiltration. *J. Clin. Invest.* **92**, 2141–2151.

Loewen, M. E., Bekar, L. K., Gabriel, S. E., Walz, W., and Forsyth, G. W. (2002). pCLCA1 becomes a cAMP-dependent chloride conductance mediator in Caco-2 cells. *Biochem. Biophys. Res. Commun.* **298**, 531–536.

Loewen, M. E., Smith, N. K., Hamilton, D. L., Grahn, B. H., and Forsyth, G. W. (2003). CLCA protein and chloride transport in canine retinal pigment epithelium. *Am. J. Physiol.* **285**, C1314–C1321.

Loewen, M. E., Bekar, L. K., Walz, W., Forsyth, G. W., and Gabriel, S. E. (2004). pCLCA1 lacks inherent chloride channel activity in an epithelial colon carcinoma cell line. *Am. J. Physiol.* **287**, G33–G41.

Maertens, C., Wei, L., Tytgat, J., Droogmans, G., and Nilius, B. (2000a). Chlorotoxin does not inhibit volume-regulated, calcium-activated and cyclic AMP-activated chloride channels. *Br. J. Pharmacol.* **129**, 791–801.

Maertens, C., Wei, L., Droogmans, G., and Nilius, B. (2000b). Inhibition of volume-regulated and calcium-activated chloride channels by the antimalarial mefloquine. *J. Pharm. Exp. Ther.* **295**, 29–36.

Marmorstein, A. D., Mormorstein, L. Y., Rayborn, M., Wang, X., Hollyfield, J. G., and Petrukhin, K. (2000). Bestrophin, the product of the best vitelliform macular dystrophy gene (VMD2), localizes to the basolateral membrane of the retinal pigment epithelium. *Proc. Natl. Acad. Sci. USA* **97**, 12758–12763.

Martin, D. K. (1993). Small conductance chloride channels in acinar cells from the rat mandibular salivary gland are directly controlled by a G-protein. *Biochem. Biophys. Res. Commun.* **192**, 1266–1273.

Marunaka, Y. and Eaton, D. C. (1990). Effects of insulin and phosphatase on a Ca^{2+}-dependent Cl^- channel in a distal nephron cell line (A6). *J. Gen. Physiol.* **95**, 773–789.

Matchkov, V. V., Aalkjaer, C., and Nilsson, H. (2004). A cyclic GMP-dependent calcium-activated chloride current in smooth-muscle cells from rat mesenteric resistance arteries. *J. Gen. Physiol.* **123**, 121–134.

Melvin, J. E., Arreola, J., Nehrke, K., and Begenisich, T. (2002). Ca^{2+}-activated Cl^- currents in salivary and lacrimal glands. In: *Calcium*-Activated Chloride Channels (C. M. Fuller, Ed.). Academic Press, New York.

Melvin, J. E., Yule, D., Shuttleworth, T. J., and Begenisich, T. (2005). Regulation of fluid and electrolyte secretion in salivary gland cells. *Ann. Rev. Physiol.* **67**, 445–469.

Morris, A. P. and Frizzell, R. A. (1993). Ca^{2+}-dependent Cl^- channels in undifferentiated human colonic cells (HT-29). II. Regulation and rundown. *Am. J. Physiol.* **264**, C977–C985.

Nilius, B., Prenen, J., Szucs, G., Wei, L., Tanzi, F., Voets, T., and Droogmans, G. (1997). Calcium-activated chloride channels in bovine pulmonary artery endothelial cells. *J. Physiol.* **498**, 381–396.

Nilius, B., Prenen, J., Voets, T., Eggermont, J., Bruzik, K. S., Shears, S. B., and Droogmans, G. (1998). Inhibition by inositoltetrakisphosphates of calcium- and volume-activated Cl^- currents in macrovascular endothelial cells. *Pflügers Arch.* **435**, 637–644.

Nishimoto, I., Wagner, J., Schulman, H., and Gardner, P. (1991). Regulation of Cl^- channels by multifunctional CaM kinase. *Neuron* **6**, 547–555.

Papassotiriou, J., Eggermont, J., Droogmans, G., and Nilius, B. (2001). Ca^{2+}-activated Cl^- channels in Ehrlich ascites tumor cells are distinct from mCLCA1, 2 and 3. *Pflügers Arch.* **442**, 273–279.

Park, K. and Brown, P. D. (1995). Intracellular pH modulates the activity of chloride channels in isolated lacrimal gland acinar cells. *Am. J. Physiol.* **268**, C647–C650.

Park, M. K., Lomax, R. B., Tepikin, A. V., and Petersen, O. H. (2001). Local uncaging of caged Ca^{2+}-activated Cl^- channels in pancreatic acinar cells. *Proc. Natl. Acad. Sci. USA* **98**, 10948–10953.

Pauli, B. U., Abdel-Ghany, M., Cheng, H. C., Gruber, A. D., Archibald, H. A., and Elble, R. C. (2000). Molecular characteristics and functional diversity of CLCA family members. *Clin. Exp. Pharmacol. Physiol.* **27**, 901–905.

Perez-Cornejo, P. and Arreola, J. (2004). Regulation of Ca^{2+}-activated chloride channels by cAMP and CFTR in parotid acinar cells. *Biochem. Biophys. Res. Commun.* **316**, 612–617.

Perez-Cornejo, P., De Santiago, J. A., and Arreola, J. (2004). Permeant anions control gating of calcium-dependent chloride channels. *J. Membr. Biol.* **198**, 125–133.

Petrukhin, K., Koisti, M. J., Bakall, B., Li, W., Xie, G., Marknell, T., Sandgren, O., Forsman, K., Holmgren, G., Andreasson, S., Vujic, M., Bergen, A. A. B., *et al.* (1998). Identification of the gene responsible for Best macular dystrophy. *Nat. Genet.* **19**, 241–247.

Piper, A. S. and Large, W. A. (2003). Multiple conductance states of single Ca^{2+}-activated Cl^- channels in rabbit pulmonary artery smooth muscle cells. *J. Physiol.* **547**, 181–196.

Piper, A. S. and Large, W. A. (2004). Single cGMP-activated Ca^{2+}-dependent Cl^- channels in rat mesenteric artery smooth muscle cells. *J. Physiol.* **555**, 397–408.

Piper, A. S., Greenwood, I. A., and Large, W. A. (2002). Dual effect of blocking agents on Ca^{2+}-activated Cl^- currents in rabbit pulmonary artery smooth muscle cells. *J. Physiol.* **539**, 119–131.

Qu, Z. and Hartzell, H. C. (2000). Anion permeation in Ca^{2+}-activated Cl^- channels. *J. Gen. Physiol.* **116**, 825–844.

Qu, Z. and Hartzell, H. C. (2001). Functional geometry of the permeation pathway of Ca^{2+} activated Cl^- channels inferred from analysis of voltage-dependent block. *J. Biol. Chem.* **276**, 18423–18429.

Qu, Z. and Hartzell, H. C. (2003). Two bestrophins cloned from *Xenopus laevis* oocytes express Ca-activated Cl currents. *J. Biol. Chem.* **278**, 49563–49572.

Qu, Z. and Hartzell, H. C. (2004). Determinants of anion permeation in the second transmembrane domain of the mouse bestrophin-2 chloride channel. *J. Gen. Physiol.* **124**, 371–382.

Qu, Z., Wei, R. W., and Hartzell, H. C. (2003). Characterization of Ca^{2+}-activated Cl^- currents in mouse kidney inner medullary collecting duct cells. *Am. J. Physiol.* **285**, F326–F335.

Qu, Z., Fischmeister, R., and Hartzell, H. C. (2004). Mouse bestrophin-2 is a bona fide Cl^- channel: Identification of a residue important in anion binding and conduction. *J. Gen. Physiol.* **123**, 327–340.

Reinsprecht, M., Rohn, M. H., Spadinger, R. J., Pecht, I., Schindler, H., and Romanin, C. (1995). Blockade of capacitive Ca^{2+} influx by Cl^- channel blockers inhibits secretion from rat mucosal-type mast cells. *Mol. Pharmacol.* **47**, 1014–1020.

Robinson, N. C., Huang, P., Kaetzel, M. A., Lamb, F. S., and Nelson, D. J. (2004). Identification of an N-terminal amino acid of the CLC-3 chloride channel critical in phosphorylation-dependent activation of a CaMKII-activated chloride current. *J. Physiol.* **556**, 353–368.

Rychkov, G. Y., Pusch, M., Roberts, M. L., Jentsch, T. J., and Bretag, A. H. (1998). Permeation and block of the skeletal muscle chloride channel ClC-1 by foreign anions. *J. Gen. Physiol.* **111**, 653–665.

Schlenker, T. and Fitz, J. G. (1996). Ca^{2+}-activated Cl^- channels in human biliary cell line: Regulation by Ca^{2+}/calmodulin-dependent protein kinase. *Am. J. Physiol.* **271**, G304–G310.

Sun, H., Tsunenari, T., Yau, K.-W., and Nathans, J. (2002). The vitelliform macular dystrophy protein defines a new family of chloride channels. *Proc. Natl. Acad. Sci. USA* **99**, 4008–4013.

Suzuki, M. and Mizuno, A. (2004). A novel human Cl^- channel family related to Drosophila flightless locus. *J. Biol. Chem.* **279**, 22461–22468.

Takahashi, T., Neher, E., and Sakmann, B. (1987). Rat brain serotonin receptors in *Xenopus* oocytes are coupled by intracellular calcium to endogenous channels. *Proc. Natl. Acad. Sci. USA* **84**, 5063–5067.

Taleb, O., Feltz, P., Bossu, J.-L., and Felta, A. (1988). Small-conductance chloride channels activated by calcium on cultured endocrine cells from mammalian pars intermedia. *Pflügers Arch.* **412**, 641–646.

Tarran, R., Loewen, M. E., Paradiso, A. M., Olsen, J. C., Gray, M. A., Argent, B. E., Boucher, R. C., and Gabriel, S. E. (2002). Regulation of murine airway surface liquid volume by CFTR and Ca-activated Cl conductances. *J. Gen. Physiol.* **120**, 407–418.

Tsunenari, T., Sun, H., Williams, J., Cahill, H., Smallwood, P., Yau, K. W., and Nathans, J. (2003). Structure-function analysis of the bestrophin family of anion channels. *J. Biol. Chem.* **278**, 41114–41125.

Vajanaphanich, M., Schultz, C., Rudolf, M. T., Wasserman, M., Enyedi, P., Craxton, A., Shears, S. B., Tsien, R. Y., Barrett, K. E., and Traynor-Kaplan, A. (1994). Long-term uncoupling of chloride secretion from intracellular calcium levels by Ins(3,4,5,6)P4. *Nature* **371**, 711–714.

Wagner, J. A., Cozens, A. L., Schulman, H., Gruenert, D. C., Stryer, L., and Gardner, P. (1991). Activation of chloride channels in normal and cystic fibrosis airway epithelial cells by multifunctional calcium/calmodulin-dependent protein kinase. *Nature* **3**, 793–796.

Wang, H. S., Dixon, J. E., and McKinnon, D. (1997). Unexpected and differential effects of Cl^- channel blockers on the Kv4.3 and Kv4.2 K+ channels. Implications for the study of the I(to2) current. *Circ. Res.* **81**, 711–718.

Wang, Y. X. and Kotlikoff, M. I. (1997). Inactivation of calcium-activated chloride channels in smooth muscle by calcium/calmodulin-dependent protein kinase. *Proc. Natl. Acad. Sci. USA* **94**, 14918–14923.

Wei, L., Vankeerberghen, A., Cuppens, H., Eggermont, J., Cassiman, J. J., Droogmans, G., and Nilius, B. (1999). Interaction between calcium-activated chloride channels and the cystic fibrosis transmembrane conductance regulator. *Pflügers Arch.* **438**, 635–641.

Wei, L., Vankeerberghen, A., Cuppens, H., Cassiman, J. J., Droogmans, G., and Nilius, B. (2001). The C-terminal part of the R-domain, but not the PDZ binding motif, of CFTR is involved in interaction with Ca^{2+}-activated Cl^- channels. *Pflügers Arch.* **442**, 280–285.

Worrell, R. T. and Frizzell, R. A. (1991). CaMKII mediates stimulation of chloride conductance by calcium in T84 cells. *Am. J. Physiol.* **260**, C877–C882.

Xie, W., Kaetzel, M. A., Bruzik, K. S., Dedman, J. R., Shears, S. B., and Nelson, D. J. (1996). Inositol 3,4,5,6-tetrakisphosphate inhibits the calmodulin-dependent protein kinase II-activated chloride conductance in T84 colonic epithelial cells. *J. Biol. Chem.* **271**, 14092–14097.

Xie, W., Solomons, K. R., Freeman, S., Kaetzel, M. A., Bruzik, K. S., Nelson, D. J., and Shears, S. B. (1998). Regulation of Ca^{2+}-dependent Cl^- conductance in a human colonic epithelial cell line (T_{84}): Cross-talk between Ins(3,4,5,6)P_4 and protein phosphates. *J. Physiol.* **510**(Pt. 3), 661–673.

Xu, W. X., Kim, S. J., So, I., Kang, T. M., Rhee, J. C., and Kim, K. W. (1997). Volume-sensitive chloride current activated by hyposmotic swelling in antral gastric myocytes of the guinea-pig. *Pflügers Arch.* **435**, 9–19.

Zygmunt, A. C. (1994). Intracellular calcium activates a chloride current in canine ventricular myocytes. *Am. J. Physiol.* **267**, H1984–H1995.

Chapter 8

Cell Volume Homeostasis: The Role of Volume-Sensitive Chloride Channels

Alessandro Sardini

MRC Clinical Sciences Centre, Faculty of Medicine, Imperial College, London W12 0NN, United Kingdom

I. Introduction
II. Cell Volume Homeostasis
   A. RVD and RVI: How Cells Regulate Their Volume
III. Volume-Sensitive Cl$^-$ Channels
   A. $I_{Cl,swell}$ and VSOR
   B. Volume-Sensitive Cl$^-$ Channels and Candidates for VSOC
   C. Identification of the Molecular Identity of the VSOR: How Many Arrows Are Left in Our Quiver
IV. Concluding Remarks
   References

I. INTRODUCTION

Although we may not have experienced it, we are all aware of and have probably witnessed the fight-or-flight response that human beings, in common with all vertebrates, display when confronted with a threatening or very stressful situation. This is an evolutionary selected behavior since it increases the chance of survival of the organism. Now, we may think of the cell as an open system, namely a system that allows exchange of material and energy with the environment, at the steady state. Changes in the extracellular environment perturb this balance and consequently the cell reacts by activating the following distinct responses: cellular homeostasis and cellular stress. Both these responses are designed, just as the fight-or-flight response mentioned above, to increase the cell's chances of survival, but differ in the way this is achieved. The cellular stress response stabilizes the cell to a different steady state by operating on a well-defined set of cellular functions such as the cellular metabolism, chromatin repair, protein repair/disposal, and cell cycle control (Kultz, 2005). This is quite well exemplified by the simple fact that a common stress response is growth arrest: a clearly different steady state from a cell progressing through the cell cycle. On the other hand, cellular homeostasis counteracts the effects of environmental changes on the parameters that define the steady state of a cell (such as cell volume, temperature, ion composition, and pH) by restoring them to their original values. Therefore, cellular homeostasis is a resistance to change. In this chapter, I will focus on this property and in particular I will describe how a cell counteracts perturbations of its volume by activation of cell volume-activated Cl^- channels.

II. CELL VOLUME HOMEOSTASIS

From an evolutionary point of view, cell volume regulation is considered to be a fundamental requirement for the emergence of life. Living organisms have evolved from a water environment, an ancient ocean, where the earliest cells had to develop osmoregulatory mechanisms to preserve their intracellular milieu from the high salinity of the extracellular environment (Kultz, 2001). The observation that virtually all the cells have conserved throughout evolution the property of regulating their volume reinforces the importance of this function. And yet, this is somehow surprising when considering mammalian cells in the context of the whole organism. Mammals have evolved very sophisticated organs whose function is to minimize changes in the extracellular environment. For example, the kidney, under hormonal control from the hippocampus, maintains plasma osmolarity within a narrow range. Indeed, other than cells of the kidney medulla, which are exposed to a very hyperosmotic environment, and intestinal cells, which face anisosmotic luminal fluids, mammalian cells do not experience, under physiological conditions, changes in osmolarity of their extracellular environment. It is then valid to ask why such a property has been maintained in mammalian cells or whether it is merely an evolutionary vestige. Changes in cell volume can occur without changes in the osmolarity of the extracellular milieu by alterations of the intracellular solute content. The intracellular solute content is affected by changes in substrate transport and/or metabolic rates. For example, cells experience changes in their metabolic rates during cell cycle progression (Davies *et al.*, 2004), hepatocyte metabolism is regulated in function of levels of insulin and glucagon (Lang *et al.*, 1998), and transporting epithelia show changes in their apical and basolateral transport rates with consequent changes of their osmolytes' concentrations (Lang *et al.*, 1998). Therefore, two types of volume changes can be envisaged: anisosmotic volumes changes, due to changes in extracellular osmolarity, and isosmotic changes, which are due to change in concentration of intracellular osmotically active substances without changes in extracellular osmolarity. Since the latter is probably experienced by the majority of cells, this explains the fact that the function of the regulation of cell volume has been preserved as well as in mammalian cells. But why do cells regulate their volume? Cell membranes are highly permeable to water and unable to withstand hydrostatic pressure. Following an influx or efflux of water, as consequence of an alteration of its thermodynamic equilibrium, cells tend then to swell or shrink. This strains cellular structures, alters optimized intracellular concentrations of essential elements for the cellular metabolism as well as induces significant disturbance of the cell surface to volume ratio with severe consequences for signaling and membrane transport. Cells therefore react to the change of the values of all these parameters which define their steady state. Plant cells and bacteria have a relatively rigid wall that partially opposes swelling; animal cells instead have to rely entirely on their molecular mechanisms of cell volume regulation. After the change in cell volume is sensed, an immediate homeostatic response follows by activation of membrane transports due either to activity of channels or transporters that tend to restore the original cell volume. A long-term response is also elicited resulting in alteration of gene expression, although this response should be considered a stress response more than a homeostatic response. How cells sense their size, the volume-sensing mechanism, is a fundamental but as yet unresolved issue. In particular, in animal cells this mechanism is poorly understood and a variety of hypothesis has been brought forward (Lang et al., 1998; Kultz, 2001). Recently, important advances to our understanding of the molecular mechanism of osmosensing in bacteria (Poolman et al., 2002) as well as in *Caenorhabditis elegans* (Strange et al., 2006) have been described. These findings should soon start to be translated to mammalian cells.

A. RVD and RVI: How Cells Regulate Their Volume

Cellular volume changes and regulation of cellular volume may be investigated by the use of fluorescent dyes and exposure to anisosmotic solutions. The description of the following experiment will exemplify the method. After loading with the fluorescent dye Calcein, cells have been imaged by confocal microscopy. The intensity of the fluorescence signal derived from a small volume within the cell, the sample volume, was analyzed in function of the time; for a detailed explanation of the technique see Alvarez-Leefmans et al. (1995) and Sardini et al. (2003). Since Calcein is membrane impermeant, any change of cellular volume will alter its concentration and consequently the intensity of the fluorescent signal collected from the sample volume (Fig. 1, panel A). The fluorescence signal is calibrated by brief exposure to small changes in extracellular osmolarity ($\pm 15\%$). The linear relationship between the reciprocal of the relative steady state fluorescence (F_0/F_t) and the reciprocal of the relative osmotic pressure of the extracellular solution (π_0/π_t), for the tested osmolarity range, demonstrates that the changes in the measured fluorescence intensity account for changes in cellular volume (Fig. 1, panel B). The relative change in volume (V_t/V_0) following exposure to a hypotonic solution (-40%) could be then computed from the F_0/F_t ratio and it is shown in panel C of Fig. 1. We observe a fast volume increase on hypotonic solution exposure followed by a decrease of the cellular volume when the cell is still exposed to hypotonic solution, such a phenomenon is defined recovery volume decrease (RVD). If the bathing solution is brought back to the original value, the cells shrinks to a smaller volume than the original and then we assist to a recovery of the

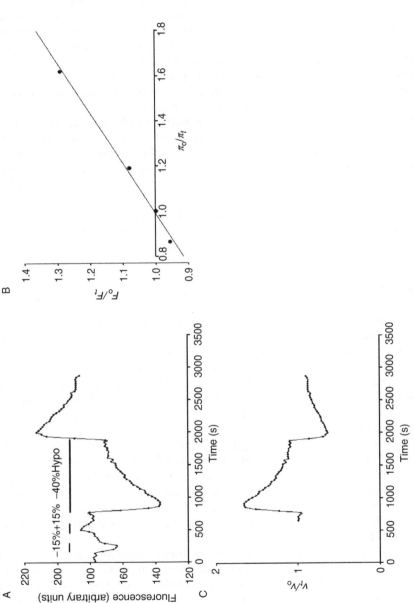

Figure 1. Measurement of volume change in a single HEK 293. Cells were imaged by confocal microscopy as described previously (Weylandt *et al.*, 2001). Panel A: average fluorescence signal of a cellular area plotted against the time from a confocal optical section. Data points were acquired every 30 s. To calibrate the fluorescence signal, brief exposure to 15% hypotonic and 15% hypertonic solutions were used. On exposure of the cell to 40% hypotonic solution, the fluorescence signal decreased suddenly and recovered its original intensity following a different kinetic. Panel B: relationship between the

volume to the initial original value: recovery volume increase (RVI). Both the described phenomena, RVD and RVI, are due to the rapid activation of preexisting membrane transport mechanisms which allow movements of osmolytes followed by the thermodynamically obliged movement of water. RVD, in the majority of animal cells, is mainly due to extrusion of KCl following the activation of separated K^+ and Cl^- channels or to the activation of a K^+/Cl^--cotransporter (KCC) as for example in erythrocytes, where it constitutes the principal pathway for RVD (Lauf et al., 1992; Lauf and Adragna, 2000). The contribution to RVD by KCCs in cells other than red blood cells should not be underestimated (O'Neill, 1999). Indeed, disruption of KCC3 in mice impaired RVD, as assessed in renal tubular proximal cells and hippocampal pyramidal cells (Boettger et al., 2003). In contrast to RVD, RVI is due to the influx of KCl and NaCl by the activation of Na^+/H^+ and Cl^-/HCO_3^- exchangers or the $Na^+/K^+/2Cl^-$-cotransporter. The movement of these electrolytes is also accompanied by the membrane movement of organic osmolytes such as the amino acid taurine (Strange, 2004). The advantage is that changes in concentrations of such osmolytes, differently from electrolytes such as K^+ and Cl^-, do not influence ion concentrations essential for membrane potential and metabolism. Moreover, changes in the synthesis rate of organic osmolytes have been also observed, following volume changes. This is a slow process and does not play a role in the process of RVD described in Fig. 1, although it may play an important role in long-term response and adaptation to change in extracellular osmolarity.

We will now consider the Cl^- channels that are activated following cell swelling.

III. VOLUME-SENSITIVE Cl^- CHANNELS

A. $I_{Cl,swell}$ and VSOR

The current activated by cell swelling is mainly known as $I_{Cl,swell}$, although the definition $I_{Cl,vol}$ has also been used in the literature to identify the same current. Furthermore, the channel responsible for this current is known with a variety of acronyms, such as volume-regulated anion channel (VRAC), volume-sensitive outwardly rectifying Cl^- channel (VSOR), and volume-sensitive organic osmolyte/anion channel (VSOAC). The explanation for such acronyms will appear clear in the following description of the current but clearly, the inconsistent channel nomenclature reflects the absence of an established molecular candidate for the volume-sensitive Cl^- channel.

---

reciprocal of the steady state changes in fluorescence (F_0/F_t) and the reciprocal of the external osmotic pressure (π_0/π_t) for the cell shown in panel A. The slope of the linear fitting is 0.447 (correlation coefficient 0.997) giving a background fluorescence $F_{bkg} = 0.55$. Panel C: the fluorescence signal shown in panel A was analyzed and converted to relative volume measurement (Alvarez-Leefmans, 1995). RVD is evident on exposure to 40% hypotonic solution. [Reprinted from Sardini et al. (2003), with permission from Elsevier].

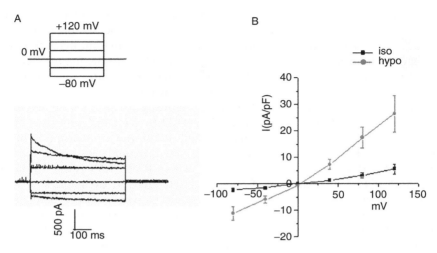

Figure 2. Measurement of $I_{Cl,Swell}$ properties in HEK 293. Cl^- currents were measured in the whole cell recording mode of the patch-clamp technique as described previously (Bond et al., 1998a,b). Panel A: a family of Cl^- currents after exposure to 30% hypo-osmotic solution is shown (bottom). HEK 293 cell was stimulated with square voltage pulses from -80 to $+120$ mV in 40 mV steps from a 0 mV holding potential (as shown at the top of the panel). Panel B: I/V relationship in presence of isotonic and hypotonic solutions. The elicited peak currents are computed against the pulse voltage. The currents recorded in hypotonic solution showed the typical outwardly rectifying behavior. [Reprinted from Sardini et al. (2003), with permission from Elsevier].

I will adopt the definition $I_{Cl,swell}$ for the swelling-activated current and VSOR for the channel supporting it.

$I_{Cl,swell}$ can be summarily described as a current that exhibits outward rectification (Fig. 2, panel B), fast voltage-dependent activation, time- and voltage-dependent inactivation at potentials more positive than $+40$ mV (Fig. 2, panel A), and with a slow rate of activation in response to increase in cell volume (cell swelling), reaching a maximum over a period of minutes (Nobles et al., 2004). $I_{Cl,swell}$ shows the following relative anion selectivity $SNC^- > I^- > Br^- > Cl^- > F^- >$ gluconate (Okada, 1997), referred as "Eisenman sequence I," and a rather complex pharmacological profile of inhibitors; for a list of inhibitors and their IC_{50} see d'Anglemont de Tassigny et al. (2003). Among the inhibitors widely employed are the stilbene derivative DIDS, the anti-inflammatory drug niflumic acid, the carboxylate analogue DPC, and the diterpene DDFSK. All these inhibitors are poorly selective, and indeed this constitutes a major problem for the study of $I_{Cl,swell}$. The anti-oestrogens compounds such as tamoxifen, clomifen, and nafoxidine show an inhibition of $I_{Cl,swell}$ with an IC_{50} in the order of 1 µM (Valverde et al., 1993; Maertens et al., 2001), and constitute the class of inhibitors with the highest affinity.

1. $I_{Cl,swell}$ and Taurine Efflux

The volume-sensitive channel responsible for $I_{Cl,swell}$ has also been proposed to constitute a pathway for the efflux of the amino acid taurine, the main organic osmolyte of mammalian cells (reviewed in Strange et al., 1996). This phenomenon was first observed in rat C6 glioma cells (Jackson and Strange, 1993) and Madin–Darby canine kidney cells (Banderali and Roy, 1992). A following study disputed such hypothesis and on the basis of differences in anion sensitivity, time course of activation, and sensitivity to DIDS observed in HeLa cells, the authors proposed that the cell swelling-activated permeability pathways for taurine and Cl^- are separate (Stutzin et al., 1999). Since the onset of taurine efflux, on cell swelling, is delayed in comparison to the onset of the efflux of Cl^-, two distinct permeability pathways play a role in RVD: an anion-selective channel early activated during RVD (the VSOR) and a permeability (either a channel or a transporter) allowing taurine permeation of later activation.

2. $I_{Cl,swell}$ and Ca^{2+}

Following cell swelling, in many cell types a raise in intracellular Ca^{2+} concentrations is observed (McCarty and O'Neil, 1992). Similarly to the established Ca^{2+} activation of K^+ channels involved in RDV, a Ca^{2+} activation of VSOR would be then foreseeable. However, $I_{Cl,swell}$ is, in the majority of tested cell types, insensitive to changes in intracellular Ca^{2+} concentration (reviewed by Okada, 1997). Moreover, changes in extracellular Ca^{2+} concentration are also ineffective (Lemonnier et al., 2002). These considerations exclude then that VSOR is Ca^{2+} activated. It is noteworthy that in prostate cancer epithelial cells LNCaP, Ca^{2+} entering the cells through plasma membrane store-operated Ca^{2+} channels actually inhibits $I_{Cl,swell}$ (Lemonnier et al., 2002). The authors suggested an inhibition of VSOR by Ca^{2+} raise in a submembrane microdomain that is not accessible to changes in global intracellular Ca^{2+} concentrations, implying then a close localization of VSOR and store-operated Ca^{2+} channels in LNCaP cells.

3. VSOR Is Not a Stretch-Activated Cl^- Channel

VSOR does not sense decrease of osmolarity (anisosmotic stimulus) since it is activated as well as by isosmotic cell swelling (for example, following active solute uptake) but it can be considered as sensing cell volume increase. Although VSOR senses cell volume increases, it is not a stretch-activated channel as opposite to the well-known MscL channel in *Escherichia coli*. MscL is recognized as component of the osmolyte efflux system of *E. coli*, which is activated on hypo-osmotic exposure and gated by tension applied through the surrounding lipid bilayer (Sukharev and Anishkin, 2004). VSOR-channel activity cannot be recorded in cell-attached mode by applying a negative pressure through the patch pipette and therefore stretching the patch of membrane; it can only be recorded in cell-attached mode on cells that have been previously swollen. These lines of evidence rule out the activation of VSOR by membrane stretching. Besides, cell swelling may not increase membrane tension (membrane stretching) but can be accomplished by distension of membrane folds (Okada, 1997). The "infolding-associated protein–protein interaction hypothesis" has been proposed to explain the activation of VSOR by cell swelling (Okada, 1997). Briefly, VSOR gating could be regulated by interaction with the submembrane F-actin network which forms a scaffold to the membrane folds. On cell swelling, membrane unfolding occurs with a concomitant alteration of the F-actin network and VSOR activation follows as a consequence of release from interaction with F-actin proteins.

B. Volume-Sensitive Cl^- Channels and Candidates for VSOC

The molecular identity of the volume-activated channel has been sought for a long period of time. Many laboratories have invested considerable resources in this scientific venture, resulting in the proposal of several candidate proteins for the intense scrutiny by the scientific community. P-glycoprotein (P-gp) and pIcln were both proposed as candidates for $I_{Cl,swell}$ more than a decade ago (reviewed in Sardini et al., 2003). Briefly, P-gp, a member of the ATP-binding cassette superfamily, was shown to be associated to volume-regulated Cl^- channels and suggested as candidate for $I_{Cl,swell}$ or a component thereof (Gill et al., 1992; Valverde et al., 1992). Since then, it has been shown that P-gp is not itself a Cl^- channel but instead a regulator of $I_{Cl,swell}$. P-gp expression does not influence the magnitude of $I_{Cl,swell}$ but increases the sensitivity of an endogenous swelling-activated Cl^- channel to hypo-osmotic exposure and its rate of activation (Bond et al., 1998a). pIcln was identified as Cl^- channel by expression cloning (Paulmichl et al., 1992), its association with $I_{Cl,swell}$ has been controversial (Eggermont et al., 1998; Strange, 1998). To date, its relationship to $I_{Cl,swell}$ has still to be proven.

I will focus on the two other classes of Cl^- channels proposed as sensitive to cell swelling: bestrophins and the CLC proteins family.

1. Bestrophins

Bestrophins have been proposed to be a new family of Cl^- channels (Qu et al., 2004). Expression of human bestrophin-1 (hBest-1) as well as of mouse bestrophin-2 (mBest-2) gave rise to Ca^{2+}-activated Cl^- current, sensitive to changes in cell volume (Fischmeister and Hartzell, 2005). Bestophins are not the VSOR since the current elicited by their expression has an linear I/V relationship, is time independent and Ca^{2+} activated, although they may play a role in volume regulation in addition to VSOR.

2. Cl⁻ Channels and Transporters of the CLC Family

Members of the CLC protein family are widely distributed from prokaryotes to mammals. Nine mammalian members have been identified, covering a vast array of functions which have been identified in the last recent years by the use of mouse models as well as by the study of genetic human diseases (Jentsch et al., 2005a,b) (see Chapter 2 of the present volume). The CLC family has been considered until recently as a protein family of Cl⁻ channels. The discovery that the prokaryotic protein ClC-e1 is a Cl⁻/H⁺ exchanger (Accardi and Miller, 2004) has prompted studies on the mammalian ClC-4 and ClC-5. Both proteins were also found to behave as Cl⁻/H⁺ exchangers (Picollo and Pusch, 2005; Scheel et al., 2005). Since ClC-3 is highly homologous to ClC-4 and ClC-5, it is likely to behave also as an exchanger, although a formal proof is still lacking. Here we will consider two mammalian CLC proteins, ClC-2 and ClC-3, which have been suggested to play a role in cell volume regulation.

ClC-2 is an inwardly rectifying Cl⁻ channel, with an anion selectivity sequence Cl⁻ ≥ Br⁻ > I⁻, activated by hyperpolarization and cell swelling (Thiemann et al., 1992); its N-terminal cytoplasmic tail has been shown to play a role in volume-sensing gating (Gründer et al., 1992). Although ClC-2 has properties distinct from the properties of VSOR, it has nevertheless been implicated in cell volume regulation. Expression of ClC-2 in Sf9 insect cells enhanced the rate of RVD (Xiong et al., 1999) as well as in *Xenopus* oocytes (Furukawa et al., 1998). Moreover, the delivery of a specific antibody against an essential domain of ClC-2 to rat HTC hepatoma cells that express endogenously ClC-2 delayed RVD (Roman et al., 2001). Furthermore, infection of erythrocytes by *Plasmodium falciparium* induced swelling-activated ClC-2 currents. Inhibition of ClC-2 resulted in a cell volume increase in parasitized red blood cells, suggesting a functional role of ClC-2 in the volume maintenance (Huber et al., 2004). However, it must be noted that $Clcn2^{-/-}$ mice did not display severe histological alterations in organs exposed to change in osmolarity (Bösl et al., 2001) and parotid acinar cells from $Clcn2^{-/-}$ mice displayed similar swelling-activated currents and rate of RVD to wild-type mice (Nehrke et al., 2002). In conclusion, although ClC-2 is not the VSOR, it may play a role in volume regulation. RVD may well be supported by more than one type of swelling-activated Cl⁻ channels.

ClC-3 was proposed as the volume-activated Cl⁻ channel in 1997 (Duan et al., 1997). Duan et al. cloned ClC-3 from guinea-pig heart and when expressed in murine fibroblasts it generated a current strongly modulated by cell volume, sharing many properties with the ubiquitous $I_{Cl,swell}$. Crucially, mutation of asparagine to lysine at position 579 changed the current's characteristics and seemed to confirm that ClC-3 is responsible for the ubiquitous $I_{Cl,swell}$. Since then, a vast and often contradictory literature regarding this hypothesis has been produced. It is impossible to reconcile all these publications that have appeared in a period spanning almost a decade, although we need to notice that a large amount of these publications does not support the hypothesis of ClC-3 as volume-activated Cl⁻ channel. Surely, evaluating the number of publications against or pro a scientific hypothesis is not the right criterion for reaching a conclusion but an analysis of the lines of evidence is the way forward.

Such an account has already been provided (Sardini et al., 2003) and here more recent published studies are taken in consideration. We initially tested the hypothesis of ClC-3 as the volume-activated Cl⁻ channel by creating cell lines permanently expressing human ClC-3. The magnitude of $I_{Cl,swell}$ and the rate of RVD were not affected by expression of ClC-3 and we concluded that ClC-3 is not responsible for $I_{Cl,swell}$ (Weylandt et al., 2001). The function of a gene can be identified through the analysis of the appearance of new properties in the recipient system where the gene has been transfected. In the case of candidate genes for VSOR, the difficulty is posed by the fact that $I_{Cl,swell}$ and RVD are ubiquitous properties of cells. The effect of transfection of ClC-3 therefore had to be evaluated as changes of parameters of properties already present: the magnitude of $I_{Cl,swell}$ and the rate of RVD. Alternatively, the role of a protein can be tested by impairing its function with the use of antibodies against the protein itself or by suppressing the expression of the protein, for example by using antisense oligonucleotides. Both methods were actually employed to test the functional role of ClC-3. Specific antibodies against ClC-3 were developed and used to inactivate ClC-3 in cardiac and smooth muscle cells (Wang et al., 2003); while expression of ClC-3 in *Xenopus* oocytes and HeLa cells was suppressed by the use of antisense oligonucleotides (Hermoso et al., 2002). In both cases these maneuver reduced significantly $I_{Cl,swell}$, supporting the role of ClC-3 as the volume-activated Cl⁻ channel. The role of ClC-3 has been scrutinized also with a further different approach, by disruption of the *Clcn3* gene: three independent ClC-3 KO mice have been generated and cells derived from them analyzed for volume-activated Cl⁻ currents. Hepatocytes and pancreatic acinar cells isolated from the first created ClC-3 KO mouse were found to have $I_{Cl,swell}$ currents identical to $I_{Cl,swell}$ currents recorded from cells isolated from wild-type mice (Stobrawa et al., 2001). This finding was also confirmed in parotid salivary acinar cells (Arreola et al., 2002) derived from a second independently generated ClC-3 KO mouse (Dickerson et al., 2002). Furthermore, cardiac atrial myocytes and pulmonary arterial smooth muscle cells isolated from the second KO mouse also showed $I_{Cl,swell}$ currents remarkably similar to the currents recorded in cells derived from wild-type mice although possessed altered properties. For example, activation of endogenous protein kinase C (PKC) inhibited $I_{Cl,swell}$ in wild-type mouse-derived cells but this inhibition was lost in ClC-3 KO mouse-derived cells (Yamamoto-Mizuma et al., 2004). This is agreement with the inhibition of the activity of ClC-3 following PKC activation (Duan et al., 1997, 1999). The authors suggested that following *Clnc3* gene deletion, compensatory changes of protein expression can give rise to similar $I_{Cl,swell}$ with different properties. A third ClC-3 KO mouse was generated (Yoshikawa et al., 2002) and its ventricular cardiomyocytes tested for $I_{Cl,swell}$. The magnitude as well as the biophysical and pharmacological properties of $I_{Cl,swell}$ were identical in the ClC-3 KO and wild-type cells. Furthermore, irrespective of whether ClC-3 was expressed, activation of PKC increased significantly $I_{Cl,swell}$ providing an ulterior proof against the hypothesis that expression of ClC-3 is required for supporting $I_{Cl,swell}$ (Gong et al., 2004). The study previously mentioned was conducted by recording whole cell currents (in whole cell patch-clamp configuration) and the possibility that expression of ClC-3 could modulate volume-activated unitary currents then remains. This possibility has been recently tested in a study that utilized cells derived from the third ClC-3 KO mouse. It was shown that volume-activated unitary currents in ventricular cardiomyocytes were identical independently of expression of ClC-3 (Wang et al., 2006), ruling out that ClC-3 serves as a volume-activated channel or regulator thereof. The authors also argued that since the macroscopic currents recorded were indistinguishable between $Clcn3^{-/-}$ and $Clcn3^{+/+}$ ventricular myocytes (Gong et al., 2004) the number of functional channels expressed in both types of myocytes must be identical, excluding that ClC-3 could influence the expression of volume-activated Cl⁻ channels.

C. Identification of the Molecular Identity of the VSOR: How Many Arrows Are Left in Our Quiver

It is quite evident then that although $I_{Cl,swell}$ has been extensively characterized and several candidates for VSOR have been proposed, we have not yet discovered the molecular identity of VSOR. Since the properties of $I_{Cl,swell}$ have been extensively tested in different cell types and tissues and found to be very similar, the search for a single protein, that is, only a single type of VSOR, responsible for $I_{Cl,swell}$ is justified. However, this does not exclude that more than one type of swelling-activated channel could participate to the RVD phenomenon. Furthermore, from the study of $I_{Cl,swell}$ it is known that the activity of VSOR is modulated and protein–protein interaction has been also suggested as a possible mechanism of modulation (Okada, 1997; Davies et al., 2004). Identification of proteins that modulate and may interact directly with VSOR, such as cytoskeleton proteins or P-gp, could offer an important tool for the identification of the molecular nature of VSOR. Such proteins could be used in immunoprecipitation studies and the coprecipitated proteins could be identified, for example, by mass spectroscopy. Alternatively, the use of the modified yeast two-hybrid system for detection of interaction of membrane proteins, "split-ubiquitin-system," could be employed (Zhu et al., 2003). The candidate proteins for VSOR, identified by the approaches described above, have to be functionally tested by expression of their genes in an adequate system. We would be faced again with the problems that have hindered the research in this field: since $I_{Cl,swell}$ is an ubiquitous current, the current generated by our candidate protein would need to be detectable over and above a significant background; moreover, highly specific inhibitors of $I_{Cl,swell}$ are not available. A different approach to the problem is the use of model organism, such as C. elegans, that has been extensively used for investigating channel and transport biology (Strange, 2003). Model organisms are relatively less complex systems, genomically well defined and easy to manipulate, therefore, they are ideal for functional genomics. For example, it is easy to generate transgenic C. elegans worms as well as silencing their genes by just soaking them in dsRNA solutions so that genome-wide RNA interference (RNAi) screen can be conducted. This approach is useful for gathering cues but the difficulties in translating results obtained in animal models to mammalian cells remain.

1. An Iconoclastic Idea for VSOR

TRK proteins are K^+ transporters expressed in plants, fungi, and bacteria where they are responsible for the accumulation of K^+. ScTrk1s and ScTrk2p, present on the membrane of *Saccharomyces cerevisiae*, support a large accessory Cl^- current (Kuroda *et al.*, 2004). They are structurally related to K^+ channels and have been proposed to oligomerize into dimers or tetramers in order to accommodate the large fractions of hydrophilic residues that otherwise would face the lipid environment of the membrane (Durell and Guy, 1999). The SpTrk1p from *Schizosaccharomyces pombe* has been modeled into a tetrameric structure: such arrangement gives rise to a central pore lined with positive charges which could create a well for mobile anions, the Cl^- channel. This has been referred as the central pore hypothesis and could explain the Cl^- current observed for TRK proteins and accommodate at the same time their behavior of K^+ transporters (Rivetta *et al.*, 2005). In a similar way, the channel-like property of neurotransmitter transporters has been explained. For example, four separate serotonin transporter molecules (SERT) have been modeled in an oligomeric structure forming a common pore delimited by homologous transmembrane domains. Through this pore Na^+ and serotonin could permeate in single-line diffusion and accounting then for the channel mode of conduction (De Felice and Adams, 2001). Glutamate transporters exhibit a glutamate-activated Cl^- current in addition to their role of secondary active transporters of glutamate. It has been shown that the human neuron glutamate transporter (EAAT3) forms a pentameric assembly with a central pore opening to the external membrane surface. It has been suggested that the central pore could be the pathway for the glutamate-activated Cl^- current (Eskandari *et al.*, 2000). We may envisage then that a Cl^- channel could be generated by oligomerization of membrane proteins as a consequence of cell swelling; the channel so formed could constitute a pathway for $I_{Cl,swell}$. If this were the case, VSOR would not exist as a molecular entity *per se* but it would be simply an emerging property of oligomeric assembly of proteins. Since $I_{Cl,swell}$ characteristics are very similar in different cell type tested, VSOR could be created by oligomerization, perhaps mediated by a volume-sensitive protein, of ubiquitous membrane proteins. Certainly, such a scenario would make our cloning efforts futile.

IV. CONCLUDING REMARKS

A large amount of information has been gathered on cell volume regulation and on $I_{Cl,swell}$ but the molecular identity of VSOR has not yet been identified. The limitations of our current knowledge and investigation strategies have been discussed and suggestions have been provided for the search of the long sought VSOR. Identification of the molecular nature of VSOR would be not only an important breakthrough in cell physiology but also it would offer a useful therapeutic target for several pathological conditions encountered routinely by the clinicians.

REFERENCES

Accardi, A. and Miller, C. (2004). Secondary active transport mediated by a prokaryotic homologue of ClC Cl⁻ channels. *Nature* **427**, 803–807.
Alvarez-Leefmans, F. J., Altamirano, J., and Crowe, W. E. (1995). Use of ion-selective microelectrodes and fluorescent probes to measure cell volume. *Methods Neurosci.* **27**, 361–391.
Arreola, J., Begenisich, T., Nehrke, K., Nguyen, H. V., Park, K., Richardson, L., Yang, B., Schutte, B. C., Lamb, F. S., and Melvin, J. E. (2002). Secretion and cell volume regulation by salivary acinar cells from mice lacking expression of the Clcn3 Cl(−) channel gene. *J. Physiol.* **545**, 207–216.
Banderali, U. and Roy, G. (1992). Anion channels for amino acids in MDCK cells. *Am. J. Physiol.* **263**, C1200–C1207.
Boettger, T., Rust, M. B., Maier, H., Seidenbecher, T., Schweizer, M., Keating, D. J., Faulhaber, J., Ehmke, H., Pfeffer, C., Scheel, O., Lemcke, B., Horst, J., *et al.* (2003). Loss of K-Cl co-transporter KCC3 causes deafness, neurodegeneration and reduced seizure threshold. *EMBO J.* **22**, 5422–5434.
Bond, T. D., Higgins, C. F., and Valverde, M. A. (1998a). P-glycoprotein and swelling-activated chloride channels. *Methods Enzymol.* **292**, 359–370.
Bond, T. D., Valverde, M. A., and Higgins, C. F. (1998b). Protein kinase C phosphorylation disengages human and mouse-1a P-glycoproteins from influencing the rate of activation of swelling-activated chloride currents. *J. Physiol.* **508**, 333–340.
Bösl, M. R., Stein, V., Hübner, C., Zdebik, A. A., Jordt, S. E., Mukhopadhyay, A. K., Davidoff, M. S., Holstein, A. F., and Jentsch, T. J. (2001). Male germ cells and photoreceptors, both dependent on close cell-cell interactions, degenerate upon ClC-2 Cl(−) channel disruption. *EMBO J.* **20**, 1289–1299.
d'Anglemont de Tassigny, A., Souktani, R., Ghaleh, B., Henry, P., and Berdeaux, A. (2003). Structure and pharmacology of swelling-sensitive chloride channels, I(Cl,swell). *Fundam. Clin. Pharmacol.* **17**, 539–553.
Davies, A. R. L., Belsey, M. J., and Kozlowski, R. Z. (2004). Volume-sensitive organic osmolyte/anion channels in cancer. Novel approaches to studying channel modulation employing proteonomic technologies. *Ann. N. Y. Acad. Sci.* **1028**, 38–55.
De Felice, L. J. and Adams, S. V. (2001). Serotonin and norepinephrine transporters: Possible relationship between oligomeric structure and channel modes of conduction. *Mol. Membr. Biol.* **18**, 45–51.
Dickerson, L. W., Bonthius, D. J., Schutte, B. C., Yang, B., Barna, T. J., Bailey, M. C., Nehrke, K., Williamson, R. A., and Lamb, F. S. (2002). Altered GABAergic function accompanies hippocampal degeneration in mice lacking ClC-3 voltage-gated chloride channels. *Brain Res.* **958**, 227–250.
Duan, D., Winter, C., Cowley, S., Hume, J. R., and Horowitz, B. (1997). Molecular identification of a volume-regulated chloride channel. *Nature* **390**, 417–421.
Duan, D., Cowley, S., Horowitz, B., and Hume, J. R. (1999). A serine residue in ClC-3 links phosphorylation-dephosphorylation to chloride channel regulation by cell volume. *J. Gen. Physiol.* **113**, 57–70.
Durell, S. R. and Guy, H. R. (1999). Structural models of the KtrB, TrkH, and Trk1,2 symporters based on the structure of the KcsA K(+) channel. *Biophys. J.* **77**, 789–807.
Eggermont, J., Buyse, G., Voets, T., Tytgat, J., Droogmans, G., and Nilius, B. (1998). Is there a link between protein pICln and volume-regulated anion channels? *Biochem. J.* **331**, 347–349.
Eskandari, S., Kreman, M., Kavanaugh, M. P., Wright, E. M., and Zampighi, G. A. (2000). Pentameric assembly of a neuronal glutamate transporter. *Proc. Natl. Acad. Sci. USA* **97**, 8641–8646.
Fischmeister, R. and Hartzell, H. C. (2005). Volume sensitivity of the bestrophin family of chloride channels. *J. Physiol.* **562**, 477–491.
Furukawa, T., Ogura, T., Katayama, Y., and Hiraoka, M. (1998). Characteristics of rabbit ClC-2 current expressed in *Xenopus* oocytes and its contribution to volume regulation. *Am. J. Physiol.* **274**, C500–C512.
Gill, D. R., Hyde, S. C., Higgins, C. F., Valverde, M. A., Mintenig, G. M., and Sepulveda, F. V. (1992). Separation of drug transport and chloride channel functions of the human multidrug resistance P-glycoprotein. *Cell* **71**, 23–32.

Gong, W., Xu, H., Shimizu, T., Morishima, S., Tanabe, S., Tachibe, T., Uchida, S., Sasaki, S., and Okada, Y. (2004). ClC-3-independent, PKC-dependent activity of volume-sensitive Cl channel in mouse ventricular cardiomyocytes. *Cell. Physiol. Biochem.* **14**, 213–224.

Gründer, S., Thiemann, A., Pusch, M., and Jentsch, T. J. (1992). Regions involved in the opening of ClC-2 chloride channel by voltage and cell volume. *Nature* **360**, 759–762.

Hermoso, M., Satterwhite, C. M., Andrade, Y. N., Hidalgo, J., Wilson, S. M., Horowitz, B., and Hume, J. R. (2002). ClC-3 is a fundamental molecular component of volume-sensitive outwardly rectifying Cl$^-$ channels and volume regulation in HeLa cells and *Xenopus laevis* oocytes. *J. Biol. Chem.* **277**, 40066–40074.

Huber, S. M., Duranton, C., Henke, G., Van De Sand, C., Heussler, V., Shumilina, E., Sandu, C. D., Tanneur, V., Brand, V., Kasinathan, R. S., Lang, K. S., Kremsner, P. G., et al. (2004). Plasmodium induces swelling-activated ClC-2 anion channels in the host erythrocyte. *J. Biol. Chem.* **279**, 41444–41452.

Jackson, P. S. and Strange, K. (1993). Volume-sensitive anion channels mediate swelling-activated inositol and taurine efflux. *Am. J. Physiol.* **265**, C1489–C1500.

Jentsch, T. J., Maritzen, T., and Zdebik, A. A. (2005a). Chloride channel diseases resulting from impaired transepithelial transport or vesicular function. *J. Clin. Invest.* **115**, 2039–2046.

Jentsch, T. J., Neagoe, I., and Scheel, O. (2005b). CLC chloride channels and transporters. *Curr. Opin. Neurobiol.* **15**, 319–325.

Kultz, D. (2001). Cellular osmoregulation: Beyond ion transport and cell volume. *Zoology (Jena)* **104**, 198–208.

Kultz, D. (2005). Molecular and evolutionary basis of the cellular stress response. *Annu. Rev. Physiol.* **67**, 225–257.

Kuroda, T., Bihler, H., Bashi, E., Slayman, C. L., and Rivetta, A. (2004). Chloride channel function in the yeast TRK-potassium transporters. *J. Membr. Biol.* **198**, 177–192.

Lang, F., Busch, G. L., Ritter, M., Volkl, H., Waldegger, S., Gulbins, E., and Haussinger, D. (1998). Functional significance of cell volume regulatory mechanisms. *Physiol. Rev.* **78**, 247–306.

Lauf, P. K. and Adragna, N. C. (2000). K-Cl cotransport: Properties and molecular mechanism. *Cell. Physiol. Biochem.* **10**, 341–354.

Lauf, P. K., Bauer, J., Adragna, N. C., Fujise, H., Zade-Oppen, A. M., Ryu, K. H., and Delpire, E. (1992). Erythrocyte K-Cl cotransport: Properties and regulation. *Am. J. Physiol.* **263**, C917–C932.

Lemonnier, L., Prevarskaya, N., Shuba, Y., Vanden Abeele, F., Nilius, B., Mazurier, J., and Skryma, R. (2002). Ca^{2+} modulation of volume-regulated anion channels: Evidence for colocalization with store-operated channels. *FASEB J.* **16**, 222–224.

Maertens, C., Droogmans, G., Chakraborty, P., and Nilius, B. (2001). Inhibition of volume-regulated anion channels in cultured endothelial cells by the anti-oestrogens clomiphene and nafoxidine. *Br. J. Pharmacol.* **132**, 135–142.

McCarty, N. A. and O'Neil, R. G. (1992). Calcium signaling in cell volume regulation. *Physiol. Rev.* **72**, 1037–1061.

Nehrke, K., Arreola, J., Nguyen, H. V., Pilato, J., Richardson, L., Okunade, G., Baggs, R., Shull, G. E., and Melvin, J. E. (2002). Loss of hyperpolarization-activated Cl($-$) current in salivary acinar cells from Clcn2 knockout mice. *J. Biol. Chem.* **277**, 23604–23611.

Nobles, M., Higgins, C. F., and Sardini, A. (2004). Extracellular acidification elicits a chloride current that shares characteristics with ICl(swell). *Am. J. Physiol. Cell Physiol.* **287**, C1426–C1435.

Okada, Y. (1997). Volume expansion-sensing outward-rectifier Cl$^-$ channel: Fresh start to the molecular identity and volume sensor. *Am. J. Physiol.* **273**, C755–C789.

O'Neill, W. C. (1999). Physiological significance of volume-regulatory transporters. *Am. J. Physiol.* **276**, C995–C1011.

Paulmichl, M., Li, Y., Wickman, K., Ackerman, M., Peralta, E., and Clapham, D. (1992). New mammalian chloride channel identified by expression cloning. *Nature* **356**, 238–241.

Picollo, A. and Pusch, M. (2005). Chloride/proton antiporter activity of mammalian CLC proteins ClC-4 and ClC-5. *Nature* **436**, 420–423.

Poolman, B., Blount, P., Folgering, J. H., Friesen, R. H., Moe, P. C., and van der Heide, T. (2002). How do membrane proteins sense water stress? *Mol. Microbiol.* **44,** 889–902.

Qu, Z., Fischmeister, R., and Hartzell, C. (2004). Mouse bestrophin-2 is a bona fide Cl(−) channel: Identification of a residue important in anion binding and conduction. *J. Gen. Physiol.* **123,** 327–340.

Rivetta, A., Slayman, C., and Kuroda, T. (2005). Quantitative modeling of chloride conductance in yeast TRK potassium transporters. *Biophys. J.* **89,** 2412–2426.

Roman, R. M., Smith, R. L., Feranchak, A. P., Clayton, G. H., Doctor, R. B., and Fitz, J. G. (2001). ClC-2 chloride channels contribute to HTC cell volume homeostasis. *Am. J. Physiol. Gastrointest. Liver Physiol.* **280,** G344–G353.

Sardini, A., Amey, J. S., Weylandt, K. H., Nobles, M., Valverde, M. A., and Higgins, C. F. (2003). Cell volume regulation and swelling-activated chloride channels. *Biochim. Biophys. Acta* **1618,** 153–162.

Scheel, O., Zdebik, A. A., Lourdel, S., and Jentsch, T. J. (2005). Voltage-dependent electrogenic chloride/proton exchange by endosomal CLC proteins. *Nature* **436,** 424–427.

Stobrawa, S. M., Breiderhoff, T., Takamori, S., Engel, D., Schweizer, M., Zdebik, A. A., Bösl, M. R., Ruether, K., Jahn, H., Draguhn, A., Jahn, R., and Jentsch, T. J. (2001). Disruption of ClC-3, a chloride channel expressed on synaptic vesicles, leads to a loss of the hippocampus. *Neuron* **29,** 185–196.

Strange, K. (1998). Molecular identity of the outwardly rectifying, swelling-activated anion channel: Time to reevaluate pICln. *J. Gen. Physiol.* **111,** 617–622.

Strange, K. (2003). From genes to integrative physiology: Ion channel and transporter biology in *Caenorhabditis elegans. Physiol. Rev.* **83,** 377–415.

Strange, K. (2004). Cellular volume homeostasis. *Adv. Physiol. Educ.* **28,** 155–159.

Strange, K., Emma, F., and Jackson, P. S. (1996). Cellular and molecular physiology of volume-sensitive anion channels. *Am. J. Physiol.* **270,** C711–C730.

Strange, K., Denton, J., and Nehrke, K. (2006). Ste20-type kinases: Evolutionarily conserved regulators of ion transport and cell volume. *Physiology (Bethesda)* **21,** 61–68.

Stutzin, A., Torres, R., Oporto, M., Pacheco, P., Eguiguren, A. L., Cid, L. P., and Sepulveda, F. V. (1999). Separate taurine and chloride efflux pathways activated during regulatory volume decrease. *Am. J. Physiol.* **277,** C392–C402.

Sukharev, S. and Anishkin, A. (2004). Mechanosensitive channels: What can we learn from 'simple' model systems? *Trends Neurosci.* **27,** 345–351.

Thiemann, A., Gründer, S., Pusch, M., and Jentsch, T. J. (1992). A chloride channel widely expressed in epithelial and non-epithelial cells. *Nature* **356,** 57–60.

Valverde, M. A., Diaz, M., Sepulveda, F. V., Gill, D. R., Hyde, S. C., and Higgins, C. F. (1992). Volume-regulated chloride channels associated with the human multidrug-resistance P-glycoprotein. *Nature* **355,** 830–833.

Valverde, M. A., Mintenig, G. M., and Sepulveda, F. V. (1993). Differential effects of tamoxifen and I⁻ on three distinguishable chloride currents activated in T84 intestinal cells. *Pflügers Arch.* **425,** 552–554.

Wang, G. X., Hatton, W. J., Wang, G. L., Zhong, J., Yamboliev, I., Duan, D., and Hume, J. R. (2003). Functional effects of novel anti-ClC-3 antibodies on native volume-sensitive osmolyte and anion channels in cardiac and smooth muscle cells. *Am. J. Physiol. Heart Circ. Physiol.* **285,** H1453–H1463.

Wang, J., Xu, H., Morishima, S., Tanabe, S., Jishage, K., Uchida, S., Sasaki, S., Okada, Y., and Shimizu, T. (2006). Single-channel properties of volume-sensitive Cl(−) channel in ClC-3-deficient cardiomyocytes. *Jpn. J. Physiol.* Advance Publication by J-STAGE; DOI: 10.2170/jjphysiol.S655.

Weylandt, K. H., Valverde, M. A., Nobles, M., Raguz, S., Amey, J. S., Diaz, M., Nastrucci, C., Higgins, C. F., and Sardini, A. (2001). Human ClC-3 is not the swelling-activated chloride channel involved in cell volume regulation. *J. Biol. Chem.* **276,** 17461–17467.

Xiong, H., Li, C., Garami, E., Wang, Y., Ramjeesingh, M., Galley, K., and Bear, C. E. (1999). ClC-2 activation modulates regulatory volume decrease. *J. Membr. Biol.* **167,** 215–221.

Yamamoto-Mizuma, S., Wang, G. X., Liu, L. L., Schegg, K., Hatton, W. J., Duan, D., Horowitz, T. L., Lamb, F. S., and Hume, J. R. (2004). Altered properties of volume-sensitive osmolyte and anion channels (VSOACs) and membrane protein expression in cardiac and smooth muscle myocytes from Clcn3−/− mice. *J. Physiol.* **557,** 439–456.

Yoshikawa, M., Uchida, S., Ezaki, J., Rai, T., Hayama, A., Kobayashi, K., Kida, Y., Noda, M., Koike, M., Uchiyama, Y., Marumo, F., Kominami, E., *et al.* (2002). CLC-3 deficiency leads to phenotypes similar to human neuronal ceroid lipofuscinosis. *Genes Cells* **7,** 597–605.

Zhu, H., Bilgin, M., and Snyder, M. (2003). Proteomics. *Annu. Rev. Biochem.* **72,** 783–812.

Chapter 9
GABAergic Synaptic Transmission

Andreas Draguhn and Kristin Hartmann

Institut für Physiologie und Pathophysiologie, University of Heidelberg, Germany

I. Postsynaptic GABA Responses
   A. Changes in Membrane Potential
   B. Other Postsynaptic Changes
II. Molecular Composition and Function of GABAergic/Glycinergic Synapses
   A. Vesicular GABA Transporters
   B. GABA Metabolism
   C. GABA Uptake
   D. Further Elements of GABAergic Synapses
   E. Phasic Versus Tonic Inhibition
III. Modulation of Synaptic Functions Through Endogenous Substances and Drugs
   A. Benzodiazepines
   B. Other Modulators
IV. Perspective: Specificity of GABAergic and Glycinergic Signaling in Defined Networks
   References

Ionotropic GABA receptors are essentially ligand-gated Cl^- channels. They are clustered at postsynaptic membranes underneath the respective release sites of GABAergic neurons but are also abundant at extrasynaptic locations. In this chapter, we will first summarize the postsynaptic effects of activation of these channels. We will then consider GABAergic signaling in a broader context of synaptic transmission, taking into account all major molecular and cellular constituents of a GABAergic/glycinergic synapse. Specific attention will be devoted to receptor modulation, heterogeneity, and its possible implications for the development of highly specific drugs. Most of the principles of GABAergic synaptic function do also apply to glycinergic synapses which will be addressed more briefly in this chapter.

I. POSTSYNAPTIC GABA RESPONSES

Ionotropic GABA receptors (iGABARs) are ligand-gated ion channels with selectivity for anions, preferentially for Cl^-. In addition, they can conduct HCO_3^- which has been estimated to contribute about 20% to total permeability (Bormann et al., 1987). Opening of iGABAR will cause several effects:

(local) membrane potential will shift toward the Cl^- equilibrium;
(local) membrane conductance will increase;
HCO_3^- will flow out of the cell;
ion fluxes may change the osmotic load of the cell.

Most attention is usually given to the immediate effects on postsynaptic potential and conductance which will also be the focus of this chapter. However, the effects on pH and cell volume should not be neglected and may be of great importance, although less well studied.

A. Changes in Membrane Potential

The best known reaction of a cell to a GABAergic synaptic input is hyperpolarization. Indeed, a shift toward more negative potentials is often taken as synonymous with an inhibitory postsynaptic potential (IPSP). This notion can be justified in certain situations since hyperpolarizing potentials do shift the membrane potential away from action potential threshold and can, thereby, be inhibitory. However, the equation "inhibition = hyperpolarization" is misleading for a variety of reasons and should be abolished altogether. A more general definition of "inhibition" may be that inhibition is a reduction of the probability of action potential discharges. With increasing knowledge about inhibitory synapses, it has become more and more difficult to describe the precise conditions for inhibition by general mechanisms such as "hyperpolarization." The membrane potential change following activation of GABARs or GlyRs depends on the reversal potential for anionic currents through the channels which is, therefore, a key parameter for the prediction of the effect of GABAergic/glycinergic synapses. In most situations, the Cl^- gradient across the membrane will be the dominating determinant of the reversal potential. However, the contribution of $[HCO_3^-]$ must not be forgotten: its contribution to net current flow is not negligible and may increase on prolonged activation of iGABAR when the membrane potential approaches $E_{chloride}$ (Staley and Proctor, 1999). In addition, it can cause considerable changes in intra- and extracellular pH (Kaila and Voipio, 1987; Kaila et al., 1993; Smirnov et al., 1999). In most situations, the relative contributions of Cl^- and HCO_3^- to the total current are not precisely known. We will, therefore, refer to the mixed equilibrium potential as E_{GABA} or $E_{glycine}$, respectively.

Imagine a GABAergic synapse where E_{GABA} is more negative than the resting membrane potential. Activation of iGABAR will cause a net influx of Cl^- ions and hyperpolarize the membrane under these conditions. Intuition would lead us to believe that the postsynaptic cell will be inhibited by such a synaptic potential—which indeed is frequently the case. However, at a longer timescale, GABA-mediated hyperpolarization may also have different effects and sometimes even constitutes a genuine pacemaker mechanism. A famous example is the strong GABAergic projection from neurons of the reticulate thalamic nucleus (RTN) to thalamic projection neurons. Input from the RTN activates a hyperpolarization-activated cation current called I_h. This, in turn, depolarizes the cell and activates T-type (low-threshold) Ca channels which then trigger a series of classical Na-K action potentials (Destexhe et al., 1996, 1998). This mechanism underlies the slow rhythmic discharges in the corticothalamic network and is relevant for physiological sleep as well as for absence epilepsy (Steriade, 2005). It should be noted, however, that the strong hyperpolarization of thalamic projection cells is not only due to the activation of Cl^- channels but does also involve $GABA_B$ receptor ($GABA_BR$)-mediated activation of K channels (i.e., metabotropic GABA responses).

Another way in which hyperpolarizing IPSPs may trigger action potentials is by rebound activation, that is, by decreasing action potential threshold through deinactivation of Na channels. Rebound activation of cells after IPSPs may occur in many brain regions (Bevan et al., 2002). Even if discharges are not directly triggered by rebound activation, their probability will usually be increased at the end of a hyperpolarizing IPSP. This time-dependent variation in discharge probability is an important mechanism for neuronal network oscillations where rhythmic inhibitory potentials from synchronized assemblies of interneurons may entrain multiple target cells into periodical changes between low and high discharge probability (Whittington et al., 2000; Whittington and Traub, 2003; Traub et al., 2004). While IPSPs may be inhibitory at the time of occurrence, the resulting spatiotemporal patterns of activity is much more complex than a simple reduction in probability of action potentials. These comments shall not contradict the notion that hyperpolarization does, often, inhibit the generation of action potentials. They should, however, emphasize the complexity of signaling in neuronal circuits.

If E_{GABA} is more positive than resting membrane potential, opening of iGABAR will invariably depolarize the membrane, at least locally. Such depolarizing GABA responses are characteristic for specific inhibitory synapses in the mammalian spinal cord, which were actually among the earliest well-studied GABAergic synapses (Eccles et al., 1962, 1963; Eccles, 1964). They are located on axons and axon terminals of excitatory afferents to spinal motoneurons, forming axo-axonic synapses (Gray, 1962). Once activated, they give rise to a local depolarization which can be measured with extracellular field electrodes and has been termed primary afferent depolarization (PAD). This depolarization of afferent axons is caused by their high intracellular Cl^- concentration, probably due to the expression of the inward Cl^- transporter NKCC (Plotkin et al., 1997; Vardi et al., 2000; Jang et al., 2001). Surprisingly, presynaptic inhibition of primary afferents to motoneurons is truly "inhibitory," that is, action potentials and subsequent excitatory transmission to the target neurons are blocked. The precise mechanisms for this apparently paradoxical inhibition by depolarization are still under debate and may involve either of two mechanisms: inactivation of voltage-gated Na^+ or Ca^{2+} channels or shunting of the propagating action potential at the site of increased membrane conductance (Rudomin and Schmidt, 1999). Another famous example for depolarizing GABA responses has been discovered in the developing rodent hippocampus. Activation of GABAergic afferents to CA3 pyramidal cells leads to a pronounced depolarization which is, again, due to the high Cl^- load of these cells (Ben-Ari et al., 1989, 1990). In the immature hippocampus, such depolarizations occur spontaneously and are mediated by the synchronous activation of multiple GABAergic neurons, causing spontaneous "giant" depolarizing potentials, GDPs. After maturation of Cl^- extrusion mechanisms, $E_{chloride}$ shifts to more negative values and the GABA-mediated depolarizations disappear. Meanwhile, it is clear that depolarizing GABA- and glycine-responses are typical for most, if not all, immature "inhibitory" synapses in the mammalian CNS (Ben-Ari, 2002). It is feasible that future inhibitory transmitters act as trophic factors during early development and that their excitatory action is needed to activate neuronal networks at early stages, preventing pruning of synapses or neuronal death (Owens and Kriegstein, 2002; Represa and Ben-Ari, 2005). Later on, excitatory innervation and activation by sensory input are able to replace this depolarizing action of GABA or glycine while, at the same time, efficient inhibition is needed to keep the balance between excitation and inhibition. Recent evidence shows that high postsynaptic intracellular Cl^- levels and depolarizing (and truly excitatory) postsynaptic potentials can also occur at certain mature GABAergic synapses, for example at axon initial segments of pyramidal cells (Szabadics et al., 2006). In addition, "normal" inhibitory signaling in the mature CNS may often go along with depolarization: cortical neurons in brain slices *in vitro* have resting membrane potentials clearly negative from E_{GABA} (Gulledge and Stuart, 2003). The functional outcome of such depolarizing GABAergic potentials is complex and depends on the precise value of $E_{GABA}/E_{glycine}$, the resting membrane potential of the cell, the amplitude and duration of the Cl^- currents, the concomitant synaptic events, and the subcellular localization of the respective synapse.

Cl^- equilibrium potentials (the most important determinants of E_{GABA} and $E_{glycine}$) are frequently discussed with relationship to Cl^- transporters (Fig. 1). Indeed, most IPSPs are hyperpolarizing, indicating that Cl^- is being transported outwardly beyond equilibrium (which would result in a passive distribution of Cl^- following resting membrane potential, provided the cell membrane has sufficient Cl^- conductance). Even if reversal potentials of GABA- or glycine-mediated synaptic currents equal E_{rest}, activation of Cl^- channels can, in principle, block action potential generation (see below in this paragraph). However, in most neurons there is an ontogenetic shift of Cl^- reversal potential toward values negative from E_{rest}, driven by the functional expression of Cl^- outward transporters, mainly the K^+/Cl^--cotransporter KCC-2 (Rivera et al., 1999). Importantly, the common transport of Cl^- and K couples the efficacy of Cl^- outward transport to the transmembraneous gradient of K^+ (Jarolimek et al., 1999). Thus, elevated extracellular K concentration can be accompanied by an increase in intracellular Cl^-, reducing the efficacy of GABA- or glycine-mediated inhibition. This mechanism might be of pathophysiological importance in situations like epilepsy where local extracellular K^+ concentrations can reach up to 12 mM (Lux and Heinemann, 1978). KCC-2 expression increases during the first postnatal weeks in rodents, following a caudal-to-rostral gradient (Stein et al., 2004). Interestingly, the mere presence of KCC-2 in the plasma membrane does not suffice to change $E_{chloride}$. Apparently, the molecule has to undergo

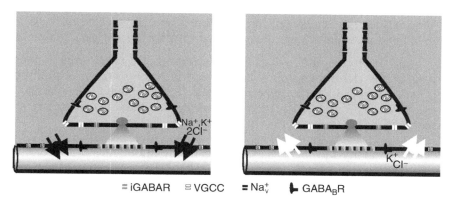

Figure 1. Simplified scheme of GABAergic synapses with different postsynaptic Cl^- equilibrium potentials. Presynaptic axon contains voltage-dependent Na and Ca channels. Further important elements are GABA-filled vesicles and pre- and postsynaptic ionotropic and metabotropic GABAR. Left panel shows a postsynaptic dendritic segment with inward Cl^- transport, leading to high $[Cl^-]_i$ (immature synapses, pathophysiological conditions). Right panel shows KCC-2-mediated Cl^- outward transport, leading to low intracellular Cl^- concentration.

phosphorylation by tyrosine kinases in order to become activated (Kelsch et al., 2001) while other modifications suppress activity (Rivera et al., 2002). It has been suggested that the depolarizing action of GABA itself promotes the expression of KCC-2 and the transition to hyperpolarizing responses (Ganguly et al., 2001), although this suggestion is subject to debate (Titz et al., 2003). Opposite mechanisms of Cl^- distribution have been found in neurons with depolarizing GABA- or glycince-responses, including dorsal root ganglions: here, Cl^- is transported inwardly by the secondarily active $Na^+/K^+/Cl^-$-cotransporter NKCC (Sung et al., 2000). This increases the intracellular Cl^- concentration above equilibrium and, thereby, causes depolarizing Cl^- currents. Recent evidence indicates that increased expression of NKCC may mediate transitions from hyperpolarizing to depolarizing inhibition in chronic pain (Morales-Aza et al., 2004), epilepsy (Okabe et al., 2002), and ischemia (Pond et al., 2006). As mentioned above, the normal resting membrane potential of cortical neurons may be more negative than previously assumed, also giving rise to depolarizing GABAergic potentials (Gulledge and Stuart, 2003). Such depolarizing IPSPs result in complex interactions between different synaptic inputs where GABAergic activity can be both, excitatory or inhibitory, depending on timing and location of the respective synapses. In discussing equilibrium potentials for GABA- or glycine-mediated synaptic potentials the contribution of HCO_3^- should always be considered. Generally, HCO_3^- will shift the effective equilibrium from Cl^- reversal toward more positive values. The relative contribution of HCO_3^- remains, however, often unknown and may vary in different situations, especially during increased GABAergic activity (Kaila et al., 1993; Staley et al., 1995; Kaila et al., 1997; Staley and Proctor, 1999).

In the following, we will outline some possible scenarios which shall illustrate the interplay between the various mechanisms involved and the complexity of synaptic

GABAergic/glycinergic signaling (for simplicity, we will mostly refer to E_{GABA} keeping in mind that similar considerations are valid for glycinergic signaling).

1. E_{GABA} is more negative than E_{rest}. Activation of a GABAergic synapse will hyperpolarize the cell membrane. Usually, this will decrease the probability of discharge. After the end of the hyperpolarizing IPSP, rebound spikes may occur. In addition, the IPSP may activate hyperpolarization-induced inward currents which can depolarize the postsynaptic cell.

2. $E_{GABA} = E_{rest}$. No potential fluctuation will be visible on activation of iGABAR (this logical possibility is probably a rare case in reality). Local membrane conductance will increase and propagation of other potentials along the membrane will be hampered. This attenuates the propagation of distally generated excitatory postsynaptic potentials (EPSPs) and IPSPs from dendrites to the soma and the backpropagation of action potentials and synaptic potentials to dendrites beyond the GABAergic synapse. It may also shunt action potentials at afferent fibers (like in PAD).

3. E_{GABA} is positive from E_{rest} but negative from action potential threshold. Essentially, this situation causes shunt inhibition similar to the scenario in (2). The depolarization can, however, boost EPSPs which arrive at sites proximal from the GABAergic synapse where the shunting conductance has no blocking effect. On the other hand, voltage-gated ion channels may be inactivated, further impeding cellular activation.

4. E_{GABA} is positive from E_{rest} as well as from action potential threshold. In this situation, GABAergic potentials can become truly excitatory and trigger action potentials. In the mature brain, directly excitatory GABA responses are supposedly an exception.

It should be noted that GABAergic synapses can induce very large conductance increases, especially when multiple synapses are activated. In this situation, Cl^- concentration underneath the synapse may be changed and the transport-generated Cl^- gradient dissipates. It has been suggested that large biphasic hyperpolarizing–depolarizing responses in CA1 on stimulation of the Schaffer collateral are generated in this way. Here, the subsequent depolarization would result from the remaining conductance for HCO_3^- which has a reversal potential around -20 mV. However, other mechanisms for the transition to depolarizing potentials have also been suggested, including extracellular accumulation of K^+ in hyperactivated networks (Smirnov et al., 1999).

B. Other Postsynaptic Changes

As mentioned at the beginning of this chapter, activation of subsynaptic or extrasynaptic ionotropic iGABARs or GlyRs may have several further effects. The outward flow of HCO_3^- can lower intracellular pH (Luckermann et al., 1997; Rivera et al., 2005). For technical reasons, this reaction is mostly measured through the corresponding extracellular alkaline shift (Chen and Chesler, 1991, 1992; Kaila et al., 1992). The functional sequelae of such activity-dependent changes in intra- and extracellular pH have not been sufficiently elucidated. It is important to note that the activity of intra- and extracellular carboanhydrases may alter these pH transients in a localized, cell-specific, and developmentally regulated manner (Ruusuvuori et al., 2004; Rivera et al., 2005).

A further effect of large Cl^- ion fluxes is the accompanying change in osmotic pressure. The Cl^- influx by a hyperpolarizing IPSP will increase intracellular ion concentration and, subsequently, water content. As a general rule, concentration changes caused by membrane ion fluxes during electrical signaling between neurons are rather small. However, synaptic and extrasynaptic Cl^- conductance through GABARs or GlyRs can be massive and may induce significant changes in osmolarity, especially within restricted microenvironments (e.g., small dendritic branches). GABA-induced swelling can, for example, induce transient elevations in intracellular Ca levels in cerebellar neurons (Chavas et al., 2004). Such actions, while difficult to study experimentally, should be kept in mind.

II. MOLECULAR COMPOSITION AND FUNCTION OF GABAergic/GLYCINERGIC SYNAPSES

In Section I we have described the immediate electrophysiological consequences of ion fluxes through GABARs or GlyRs. In order to put this knowledge into context, we will now describe a prototypic GABAergic synapse, including cellular and molecular constituents beyond the postsynaptic receptors (Fig. 2). Most of the described mechanisms can also be ascribed to glycinergic synapses. It should be kept in mind, however, that the specific properties of synapses may vary strongly between different regions of the CNS (and even between synapses at different subcellular locations of the same postsynaptic neuron).

Ionotropic GABAR and GlyR are composed of five subunits which are encoded by large families of phylogenetically related genes. This diversity allows for the assembly of an immense variety of different functional receptors which share basic properties but may vary significantly in their biophysical and pharmacological properties. Important differences have been identified in affinity toward endogenous ligands, gating, desensitization, single-channel conductance, interaction with intracellular scaffolding proteins, modulation by protein kinases and by several endogenous and pharmacological agents including dis-inhibitory toxins, benzodiazepines (bzds), and Zn^{2+}. Alterations of receptor subunit composition during development and during pathophysiological states like epilepsy have important consequences for the function of inhibitory signaling and may be exploited for selective pharmacological interventions (see Section III). An extensive review of the vast amount of data on receptor heterogeneity is beyond the scope of this article—therefore, we refer to several excellent respective reviews (Möhler et al., 1990; Luddens and Korpi, 1995; Möhler et al., 1995; Hevers and Luddens, 1998; Korpi et al., 2002; Lynch, 2004; Mody and Pearce, 2004; Rudolph and Möhler, 2004, 2006) (see also Table I).

Chemical synapses work by rapid, Ca-dependent exocytosis of transmitter from the presynaptic terminal which, after fast diffusion through the synaptic cleft, activates ligand-gated ion channels at the postsynaptic membrane. In order to secure efficient

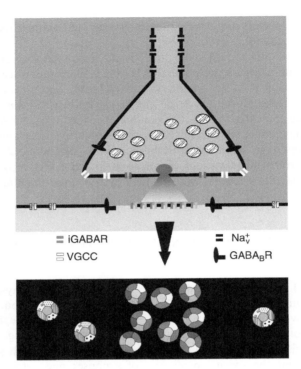

Figure 2. Sub- and extrasynaptic GABA$_A$R. Note different subunit composition of the extrasynaptic receptors which mediate tonic inhibition and are more sensitive to GABA.

transmission at a rapid (sub)-millisecond timescale, GABAR must cluster at the postsynaptic membrane opposite to the release sites. Whereas postsynaptic clustering of glycine receptors is mediated by the scaffolding protein gephyrin (Kirsch et al., 1993), the organization of the postsynaptic site at GABAergic synapses is less well understood. Although gephyrin is present at GABAergic synapses (Sassoe-Pognetto et al., 1995; Giustetto et al., 1998), it may not be directly linked to iGABAR. Rather, a variety of interacting proteins appears to act together in order to assemble and fix GABAR at the postsynaptic site (Kittler et al., 2002). Clustering of GABAR does also depend on their subunit composition. Posytsynaptic clusters of GABA$_A$ receptors (GABA$_A$Rs) do only form in the presence of a γ-subunit (mostly γ2) which appears to carry the respective sorting and anchoring signal (Alldred et al., 2005). Mice with a targeted deletion of γ2 lack postsynaptic clusters of GABA subunits (Schweizer et al., 2003).

A. Vesicular GABA Transporters

In order to secure rapid regulated release of high concentrations of GABA or glycine from presynaptic vesicles, the respective transmitter molecules have to be

Table I
Important Molecular Determinants of Inhibitory Synaptic Functions[a]

Function	Most important isoforms	Selected references
$GABA_A R$ subunits	$\alpha 1-\alpha 6$	Hevers and Luddens (1998),
	$\beta 1-\beta 3$	Mody and Pearce (2004),
	$\gamma 1-\gamma 3$	Möhler et al. (1995)
	$\delta, \varepsilon, \pi, \tau$	
$GABA_C R$ subunits	$\rho 1-\rho 3$	Bormann (2000)
GlyR subunits	$\alpha 1-\alpha 4$	Lynch (2004)
	$\beta 1$	
Vesicular uptake	VIAAT = VGAT	Sagne et al. (1997)
GABA synthesis	GAD65, GAD67	Pinal and Tobin (1998),
		Soghomonian and Martin (1998)
GABA degradation	GABA-T	Sarup et al. (2003)
GABA-uptake	GAT-1, GAT-3	Liu et al. (1993), Sarup et al. (2003)
Glutamate-uptake	EAAC-1 = EAAT-3	Conti et al. (1998),
		Mathews and Diamond (2003),
		Sepkuty et al. (2002)
Glycine-uptake	GlyT1, GlyT2	Gomeza et al. (2003a,b)
K^+/Cl^- cotransporters	KCC-2	Rivera et al. (1999)
NKCC cotransporters	NKCC1	Plotkin et al. (1997), Vardi et al. (2000)
Scaffolding proteins	Gephyrin, other?	Kirsch et al. (1993), Kittler et al. (2002)

[a]Middle column lists the most important isoforms of the respective molecules for GABAergic or glycinergic signaling. Note that in special regions or at early developmental stages, further isoforms can be present. For details, see chapter or key references.

enriched in the presynaptic cytoplasm and sequestered into vesicles. Both functions are mediated by specialized molecules and—most importantly—both functions are subject to activity-dependent plastic changes. The vesicular membrane carries a transporter for GABA and glycine, called VGAT or VIAAT for *vesicular GABA transporter* or *vesicular inhibitory amino acid transporter*, respectively (Sagne et al., 1997). This secondarily active transporter for inhibitory amino acids depends on the pH- and the voltage-gradient across the vesicular membrane which is provided by a proton pump (Gasnier, 2000). As mentioned above, VGAT does not differentiate between GABA and glycine, opening the possibility to enrich vesicles with a mixture of both transmitters. Direct evidence for the parallel use of both transmitters has been provided by analyzing miniature inhibitory postsynaptic currents (IPSCs) in spinal cord neurons which show both, a GABAergic and a glycinergic component (Jonas et al., 1998). This rare example of a synapse with two different transmitters violates "Dale's law" (a rule first formulated by John Eccles, according to which one neuron does only synthesize and release one transmitter). It also requires the clustering and stabilization of a mixture of GABA- and glycine-receptors at the postsynaptic membrane. Recent evidence suggests that the vesicular GABA transporter contributes to adaptive (homeostatic) plasticity by up-regulating filling of GABAergic vesicles in situations of network hyperexcitability (Gasnier, 2000). The most radical (and controversial) example is activity-dependent expression of VGAT in mossy fibers, that is, glutamatergic axons of dentate granule cells which project to CA3 pyramidal cells. There is evidence that such altered mossy fibers do indeed release GABA and elicit inhibitory postsynaptic responses on strong activation, as, for example, in epileptic tissue (Lamas et al., 2001; Ramirez and Gutierrez, 2001; Gutierrez et al., 2003). While this example is exceptional and creates a spectacular violation of Dale's law, VGAT can also be up-regulated in typical inhibitory neurons on high activity in the respective networks. This up-regulation indicates that the amount of GABA (or glycine) in presynaptic vesicles is variable and can be adapted to the "needs" of the system (Kang et al., 2003).

B. GABA Metabolism

GABA is synthesized from glutamate by the enzyme glutamate decarboxylase (GAD). In the mature CNS, two isoforms are found, termed GAD65 and GAD67, according to their different molecular weight (Esclapez et al., 1993). GAD65 is directly associated with the membrane of presynaptic vesicles and interacts with VGAT (Jin et al., 2003). This opens the intriguing possibility that GABA may be synthesized locally from glutamate and may be immediately used for the filling of vesicles. Moreover, glutamate can be taken up from the extracellular space through the plasma membrane glutamate transporter EAAT-3 (also called EAAC-1) which is expressed at the terminals of interneurons (Conti et al., 1998). This mechanism may constitute a basis for homeostatic plasticity, enriching GABAergic neurons with the inhibitory transmitter when glutamate release (hence network activity) is high (Sepkuty et al., 2002; Mathews and Diamond, 2003). GAD65 knockout mice are viable and show relatively subtle functional alterations, for example a reduced seizure threshold (Stork et al., 2000; Ramirez and Gutierrez, 2001). Thus, GAD65 may be involved in the activity-dependent fine-tuning of GABA synthesis. The basal synthesis of GABA seems to be secured by GAD67, the constitutively active form of GAD. Consistently, GAD67 knockout mice die briefly after birth, indicating that this enzyme is absolutely required in order to maintain normal synaptic function at GABAergic neurons (Asada et al., 1997).

Similar to VGAT, the expression and activity of GAD are regulated in an activity-dependent manner. Esclapez and Houser (1999) have shown that following repeated temporal seizures both isoforms of the enzyme are up-regulated. A contrasting example is the down-regulation of GAD and GABA on afferent denervation of circuits in the somatosensory cortex, for example following amputations (Garraghty et al., 1991). Together, these findings indicate that the amount of GABA is regulated by network activity and that the key to this homeostatic plasticity is the regulated expression of VGAT, GAD65, and, possibly, GAD67. For completeness, the GABA-degrading enzyme GABA-transaminase (GABA-T) should be mentioned, which is present in the mitochondria of presynaptic terminals and glial cells. GABA-T submits GABA to degradation in the Krebs cycle after transaminating it into succinyl semialdehyde. It has been estimated that more than 90% of the cerebral GABA go this way, leaving only a small proportion for synaptic signaling (Waagepetersen et al., 1999).

The relative activity of the enzymes for synthesis and degradation of GABA determines the level of the transmitter in the presynaptic terminal. This balance has been exploited in a pharmacological approach, elevating intracellular GABA content by inhibition of GABA-T. The GABA analogue γ-vinyl-GABA, a suicide blocker of GABA-T, has anticonvulsant potency. Unfortunately, its clinical use is limited by severe side effects (irreversible scotoma). Therefore, the use of this drug is largely confined to the severe epileptic syndrome of infantile spasms (Nabbout, 2001; Spence and Sankar, 2001). Nevertheless, the anticonvulsant effect of the agent proves that altering presynaptic GABA concentration does indeed alter the efficacy of inhibition in a predictable manner. We have used this paradigm to study the precise consequences of altered GABA concentrations at the level of single synapses (Engel et al., 2001). Incubation of cultured rat hippocampal slices with γ-vinyl-GABA caused increased amplitudes of miniature IPSCs (mIPSCs) in CA3 pyramidal cells. This finding indicates that the transmitter content of normal GABAergic vesicles is subsaturating, leaving room for potentiation by elevated levels of GABA. Previously, the question of saturating versus nonsaturating amounts of vesicular GABA had been mainly approached by enhancing GABA affinity of the receptors with bzds. Modulating presynaptic GABA concentration provides an alternative approach to tackle this question which does probably have different answers for different synapses. In addition to increased mIPSC amplitudes we found increased GABAergic noise (pointing toward an increased tonic, nonsynaptic inhibition) and an increased frequency of mIPSCs. The latter is difficult to explain but may be caused by presynaptic autoreceptors for GABA (Axmacher and Draguhn, 2004; Axmacher et al., 2004a,b). Opposite changes could be induced by the GAD-inhibiting agent 3-mercaptopropionic acid. Thus, changing GABA concentration in presynaptic terminals is an efficient way to alter inhibitory synaptic efficacy. It can be used in pharmacological therapies but appears to be implemented as a homeostatic mechanism even in normal inhibitory circuits.

C. GABA Uptake

The classical definition of a neurotransmitter includes its elimination from the synaptic cleft. With exception of acetylcholine (which is chemically inactivated by cholinesterases), this function is provided by transmitter transporters which use ionic gradients to move transmitter molecules across the membrane. There are four members of the gene family encoding plasma membrane GABA transporters, three of which are expressed in the mammalian CNS, termed GAT-1 to GAT-3 (Fig. 3) (Liu et al., 1993). In the mature mammalian CNS, GAT-1 and GAT-3 are the most abundant isoforms. They take up GABA into adjacent glia cells (where it is metabolized into succinyl semialdehyde and serves as an energy metabolite) or into axonal endings of GABAergic neurons where it can be reused as for synaptic transmission. The quantitative contribution of this recycling pathway to the presynaptic vesicular

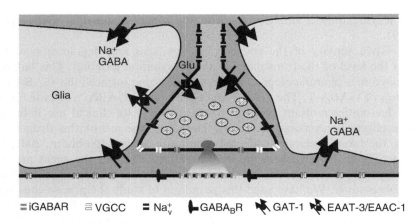

Figure 3. Transmitter transport at GABAergic synapses. Note Na-driven GABA inward transport into adjacent glia cells (mainly through GAT-3) and presynaptic terminal (mainly through GAT-1). Glutamate transporters are present at glia cells and presynaptic terminals (EAAC-1/EAAT-3). For consequences of transmitter transport for presynaptic [GABA] see text.

GABA content is unknown. If too much GABA is being "burned" in energy metabolism (especially in glia cells), this might induce a vicious circle between overactivity and reduced inhibitory strength. Such mechanisms of GABA depletion have been discussed in the case of sustained epileptic seizures, especially for *status epilepticus* and with respect to therapeutic interventions (Waagepetersen et al., 1999; Sarup et al., 2003). Besides regulating the distribution of synaptically released GABA between different cell types and compartments, GABA-uptake regulates the decay time course of IPSPs. Experimentally, application of GABA-uptake blockers leads to a massive prolongation of IPSPs or IPSCs (the currents corresponding to IPSPs when the voltage-clamp method is used for measurement). This effect has been shown in various cell types, including principal neurons in the hippocampus (Dingledine and Korn, 1985; Roepstorff and Lambert, 1994; Draguhn and Heinemann, 1996; Engel et al., 1998). It can be exploited pharmacologically by inhibiting GABA-uptake with the GABA analogue tiagabine. This substance has anticonvulsant activity and is being used in the treatment of complex partial seizures. GABA transporters may not only remove transmitter from the synaptic cleft and provide it for reuse to the presynaptic terminal but may, in contrary, also release GABA. This so-called nonvesicular transmitter release has been shown to occur from depolarized presynaptic terminals under special ionic conditions which alter the thermodynamic equilibrium of the transmitter-ion cotransport (Cammack and Schwartz, 1993; Cammack et al., 1994). It has been argued that pathophysiological situations like sustained depolarization and disturbed ion homeostasis during epileptic seizures favor this mechanism of GABA release (Belhage et al., 1993; Rossi et al., 2003; Liu et al., 2006). Nonvesicular liberation of GABA does also seem to play some role during synaptogenesis where GABA acts as a trophic factor (Takayama and Inoue, 2004; Represa and Ben-Ari, 2005).

An open question remains whether GABA transport is reduced in chronic temporal lobe epilepsy (TLE). Several authors have provided evidence for a functional or structural down-regulation of GABA transport in animal models of TLE or in human tissue. In hippocampal slices from chronically epileptic rats we found, however, normal function of GAT-1-mediated uptake of GABA (Frahm et al., 2003; Stief et al., 2005). In any case, the functional sequelae of a down-regulation of GAT would be difficult to predict, demonstrating once more the complexity of inhibitory signaling. Functional loss of the neuronal GABA transporter GAT-1 may:

decrease the pool of recycling GABA, thereby causing transmitter depletion of the presynaptic terminal;
decrease the contribution of nonvesicular transmitter release to inhibition;
prolong the time constant of IPSPs, thus boosting inhibition;
increase ambient GABA levels and tonic inhibition.

It is unclear which of these mechanisms would prevail after down-regulation of GAT-1 and therefore the pro- or anticonvulsant effect of this regulation cannot be predicted.

Similar to GAT-isoforms, recent work has revealed an important role of glycine plasma membrane transporters for the maintenance of glycinergic transmission. There are two isoforms of these secondarily active transporters, termed GlyT1 and GlyT2, respectively. Genetic ablation ("knockout") of GlyT1 shows that this transporter is essential for the regulation of ambient levels of glycine in the mouse brainstem (Gomeza et al., 2003a), whereas GlyT2 is needed to sequester glycine into inhibitory terminals of inhibitory spinal cord and brainstem neurons (Gomeza et al., 2003b). As a consequence, both mutants are lethal.

D. Further Elements of GABAergic Synapses

It is impossible to cover the full set of active molecules at GABAergic synapses in one review. It should be briefly mentioned, however, that the dynamics (time- and use-dependent function) of GABAergic synapses is strongly influenced by a variety of presynaptic receptors, most notably by the negative modulation of GABA release through metabotropic autoreceptors for GABA, the $GABA_BR$ (Misgeld et al., 1995; Bowery and Brown, 1997; Kaupmann et al., 1997). Besides autoinhibition of GABA release, these receptors mediate the late K^+-dependent component of postsynaptic IPSPs and are also present on glutamatergic terminals and somato-dendritic compartments of many neurons. Interestingly, these members of the heptahelical G-protein–coupled receptor family function as heterodimers (Kuner et al., 1999). There are several other ionotropic and metabotropic receptors at inhibitory terminals, including mGluR (Schoepp, 2001), iGABAR (Axmacher and Draguhn, 2004; Axmacher et al., 2004b; Kullmann et al., 2005), and, importantly, receptors for endocannabinoids which mediate depolarization-induced suppression of inhibition, *DSI*, the suppression of inhibition by depolarization of the postsynaptic neuron (Wilson et al., 2001; Wilson and Nicoll, 2002).

Other important regulators of the dynamics of GABA release are Ca-binding proteins which are differentially expressed in different subtypes of GABAergic neurons and are, therefore, frequently used for histochemically subtyping of interneurons (Somogyi and Klausberger, 2005). Differences in the Ca-buffering capacity of proteins in the presynaptic terminal can account for differences in the frequency-dependent facilitation or depression of transmitter release and are therefore important determinants for the function of interneurons in temporal signal processing (Blatow et al., 2003; Burnashev and Rozov, 2005).

Of course, a large variety of postsynaptic molecules and mechanisms is involved in the plasticity of GABAergic signaling, including activation of second messengers, protein kinases, transcriptional regulators, and others. This field has been much less extensively explored as compared to the plasticity of excitatory synapses but increasing evidence shows that such mechanisms do also exist at inhibitory synapses, including classical Hebbian plasticity (Gaiarsa et al., 2002; Lamsa et al., 2005).

E. Phasic Versus Tonic Inhibition

While "synaptic inhibition" is mostly discussed in the context of GABAergic and glycinergic signaling at defined synapses, it should be kept in mind that the extrasynaptic membrane domains of neurons are also covered with ionotropic GABA- and glycine-receptors at considerable densities. In CA1 pyramidal cells, about 50% of $GABA_AR$ appear to be located extrasynaptically (Banks and Pearce, 2000). In many cells, tonic GABAergic inhibition is mediated by specific subtypes of $GABA_AR$ with high affinity, probably securing efficient signaling by the low ($\sim \mu M$) ambient GABA concentrations (Mody, 2001; Stell and Mody, 2002). Recent evidence indicates that certain interneurons in the hippocampus receive a massive inhibitory input through extrasynaptic receptors which can efficiently regulate their excitability (Semyanov et al., 2003, 2004). The importance of tonic inhibition is also underlined by the compensatory changes following genetic ablation of the $\alpha 6$-subunit which is expressed in cerebellar granule cells. Tonic inhibition in these cells is mediated by $\alpha 6$-containing $GABA_AR$ which display high ligand affinity. In granule cells from mice lacking this subunit, K conductance is up-regulated, apparently restoring the disturbed balanced between excitation and inhibition (Brickley et al., 2001). Thus, point-to-point transmission through synapses, while indispensable for information processing, is accompanied by a tonic activation of extrasynaptic GABA- and glycine-receptors which are involved in the homeostasis of network excitability.

III. MODULATION OF SYNAPTIC FUNCTIONS THROUGH ENDOGENOUS SUBSTANCES AND DRUGS

GABAR can be modulated by a large variety of modifications from the intra- or extracellular space. Here, we will focus on substances which act from the external side and which are, therefore, more easily usable for pharmacological approaches.

Most pro-GABAergic drugs are used against pathophysiological states which go along with high neuronal excitability, for example, epilepsy or anxiety. As stated above, molecular targets at GABAergic synapses are not confined to the receptors itself but include the machinery for GABA metabolism and GABA-uptake. In contrast, the pharmacology of glycine receptors is less advanced and will not be discussed in detail.

A. Benzodiazepines

The classical potentiating drugs for GABAergic signaling are bzds which are widely used in different clinical conditions. These substances bind to a domain on $GABA_A R$ which is constituted by both the α- and γ-subunit. Their efficacy does, therefore, depend on the expression of a γ-subunit while functional $GABA_A R$ composed of only α- and β-subunit will not show the full reaction to bzds (Pritchett et al., 1989). Binding of bzds leads to an allosteric modulation of the pentameric ion channel complex and changes the binding affinity for the natural agonist GABA. A peculiar feature of bzd receptors (BzdR) is the bidirectionality of action: the efficacy of $GABA_A R$ can be either up- or down-regulated, depending on the binding substance. Therefore, ligands of the BzdR are classified as agonists or partial agonists (i.e., agonists with reduced efficacy) when they increase affinity toward GABA, and as full or partial inverse agonists (negative modulators) when they decrease GABA binding (Braestrup et al., 1983; Haefely, 1990). Pure bzd antagonists are substances which block the binding site without exerting any intrinsic activity (e.g., Ro 15-1788). At inhibitory synapses, BzdR agonists prolong the duration of IPSPs due to the more sustained activation of $GABA_A R$. Consequently, inhibitory efficacy can be massively increased. Likewise, tonic (extrasynaptic) inhibition can be boosted by bzds which can make a major contribution to increased net inhibition. An interesting and widely studied question is whether the amplitude of IPSPs/IPSCs can be enhanced by BzdR agonists. While this is usually the case after action-potential evoked (multivesicular) release of transmitter from presynaptic GABAergic neurons, the situation is more complex when we look at the effect of single vesicles. The postsynaptic response to the GABA content of a single vesicle can only be enhanced through bzds if the postsynaptically available receptors are normally not saturated by exocytosis of a single vesicle. Miniature IPSCs, recorded in the presence of tetrodotoxin, are believed to largely fulfill the single-vesicle criterion and are, therefore, used to study postsynaptic receptor saturation. In most synapses studied so far, a potentiating effect of bzds on mIPSC amplitude has been demonstrated, indicating a subsaturating concentration of GABA in presynaptic vesicles (Perrais and Ropert, 1999; Hajos et al., 2000). However, this experimental tool for analysis of saturation has to be applied with caution since some results appear contradictory or are, at least, difficult to explain: in dentate granule cells, diazepam and zolpidem did only exert a potentiating amplitude effect at room temperature but not at body temperature (Hajos et al., 2000). The precise mechanism behind this discrepancy is not known but this observation, together with the general variability of bzd-induced amplitude changes, underline the necessity to find alternative experimental approaches to the question of saturation. One such approach may be experimental manipulation of vesicular GABA content, as explained above (Engel et al., 2001).

Clinically, bzds are used in disorders where enhanced inhibition is desired. One obvious example is epilepsy which sometimes has been defined as a mismatch between excitation and inhibition. In this case, boosting inhibition would be a rational therapeutic strategy. However, a more complete understanding of epilepsy has to take into account that this chronic disease is characterized by repetitive seizures interrupted by phases of normal brain function. Seizures are, again, better described as pathologically synchronized brain activity rather than as a relative lack of inhibitory strength. Moreover, epilepsy is frequently a progressive disease, causing the problem of epileptogenesis rather than just that of seizure generation. The end point of this development may be an epileptic syndrome which is resistant against pharmacological treatment. Potentiating inhibitory synaptic transmission by bzds does, unfortunately, only solve a part of the clinical problems of epilepsy. Bzds, in general, cannot avoid chronic progression, that is, they are anticonvulsant, rather than antiepileptic, drugs. The brain reacts to chronic bzd treatment with a decreased expression of susceptible GABAR (down-regulation), causing tolerance and addiction. Moreover, suppressing network activity by enhanced inhibitory efficacy during normal functional states of the brain causes severe side effects like sedation, impaired vigilance, and subsequent intellectual retardation in children. Thus, the use of bzds in epilepsy is very efficient against acute seizures but has severe shortcomings on chronic administration. Further applications of bzds are antianxiety and sedation. Again, the above-mentioned caveats apply to the chronic administration.

As mentioned above, most bzds bind to the α- and γ-subunit of $GABA_AR$. Keeping in mind the enormous heterogeneity of subunits and subunit combination, this selective binding offers the possibility of a subtype-specific pharmacology, enabling the generation of more specific drugs. In the brain, $GABA_AR$ subunits are expressed in a highly regulated spatial and temporal pattern with clear differences between regions, cell types, subcellular compartments, functional state of the networks, or age. It should, therefore, be possible to design bzds which do exclusively modulate a specific subpopulation of synapses and cells and which, therefore, should have less side effects. Using classical pharmacological techniques, two different binding sites had been even distinguished long before the molecular diversity of $GABA_AR$ was known. Meanwhile, it has been shown that the different pharmacological properties of both isoforms are based on the expression of different α subunits (Pritchett et al., 1989).

Recently, transgenic approaches have been used to design mice which exhibit selective pharmacological properties in their reaction toward ligands of the BzdR. Möhler and coworkers made use of the fact that binding of bzdz depends on the presence of a highly conserved histidine residue in a specific position of the α subunits (Rudolph and Möhler, 2004, 2006). They mutated this amino acid in certain subunits and expressed it in mice by homologous recombination. The resulting mutant mice showed very specific pharmacological responses to bzds, allowing for the dissection of specific behavioral effects of bzds. For example, mice lacking the α1-mediated bzd response lose the sedative, amnesic, and anticonvulsant drug actions while anxiolytic, myorelaxant, motor-impairing, and ethanol-potentiating effects are preserved (Rudolph et al., 1999). In contrast, the α5-subunit is essential for trace-fear conditioning and development to tolerance against the sedative effect of the drug (Crestani et al., 2002; van Rijnsoever et al., 2004). These experiments show that a subunit-selective and specific symptom-orientated pharmacotherapy with BzdR ligands is possible. However, the search for subunit-specific ligands is still going on. At present, the clinical practice is still dominated by bzds with general action on the widely distributed "standard" bzd-sensitive $GABA_AR$.

B. Other Modulators

While the bzd receptor is clinically the most important modulatory binding site of $GABA_AR$, the heteromeric ion channel has several other pharmacological interactions. One important class of modulatory agents is barbiturates which are also used against epileptic seizures because they boost GABA responses. Barbiturates prolong the opening times of $GABA_AR$ channels with little subunit selectivity. The resulting potentiation of sub- and extrasynaptic GABA responses is grossly similar to the action of bzds. Accordingly, barbiturates are also used against epileptic seizures. It has to be kept in mind, however, that the therapeutic concentration range of barbiturates is much narrower than that of bzds. Dangerous toxic side effects, including respiratory arrest, can be easily induced.

Somewhat similar actions are exerted by certain steroids like allopregnanolone, tetrahydrodeoxicorticosterone, and others. These substances derive from steroid hormones and can be generated in the brain. Therefore, they have been termed neurosteroids. The specificity of binding and interaction is indicated by their strict stereoselectivity and by their very high affinity, allowing for marked effects on inhibition in the nanomolar range. In general, their action is similar to that of barbiturates, increasing efficacy of inhibition by prolonging channel opening time. It is very well feasible that the physiological importance of neurosteroids has been underestimated until now and that inhibitory transmission under "standard" conditions is subject to a steady (plastic?) modulatory activity of these endogenous agents. For review of the actions of neurosteroids see Belelli and Lambert (2005) and Belelli et al. (2006).

A negative modulatory binding site of some importance is that for Zn^{2+}. This divalent cation is abundant in the brain, especially in certain fiber tracts like the axons connecting granule cells of the dentate gyrus to postsynaptic CA3 pyramidal cells (mossy fibers). Zn^{2+} ions are released from Zn-containing fibers (e.g., hippocampal mossy fibers) on synaptic activity and, when present at micromolar concentrations, block $GABA_AR$ of certain subunit compositions. While we initially reported that those GABAR lacking a γ-subunit are selectively affected by Zn^{2+} (Draguhn et al., 1990), subsequent studies identified a number of further subunit combinations with sensitivity to Zn^{2+} (Krishek et al., 1998) and revealed the molecular positions of the binding sites (Hosie et al., 2003). The physiological role of synaptically released Zn in the CNS is unknown. However, one well-documented example from a pathophysiological condition illustrates its potential importance: overexpression of $GABA_AR$ is part of the adaptive changes of the hippocampal circuitry in chronic TLE. At the same time, the cellular pattern of GABAR subunit expression is changed (Coulter, 2001). In dentate granule cells from chronically epileptic rats, Mody and coworkers (Buhl et al., 1996) have found an enhanced sensitivity of $GABA_AR$ toward Zn^{2+}. This leads to enhanced block of the receptors by synaptically released Zn^{2+} ions and therefore will—at least in part—destroy the adaptive effect of overexpression of GABAR (Coulter, 2000).

Further blockers of ligand-gated anion channels are mostly toxic and can cause disinhibition of neuronal networks. They are, therefore, not used in human pharmacology. The competitive GABA antagonist bicuculline (a plant alkaloid) is of experimental importance because it allows distinguishing the majority of $GABA_AR$ from a subgroup of receptors which are not sensitive toward bicuculline. These ion channels are densely expressed in the retina but do also appear in other brain regions and are composed of ρ subunits encoded by the same gene family as $GABA_AR$ subunits. They have been termed $GABA_C$ receptors ($GABA_CRs$) and, in contrast to "classical" $GABA_AR$ they can form homooligomeric channels (Bormann, 2000). It is feasible, however, that mixed forms of both receptors exist, combining properties of $GABA_AR$ and $GABA_CR$ (Hartmann et al., 2004). $GABA_CR$ should be treated as one peculiar isoform of $GABA_AR$ and their different nomenclature should not be mistaken as an indication for strong molecular differences between both groups, like those between ionotropic $GABA_AR$ and G-protein–coupled $GABA_BR$.

IV. PERSPECTIVE: SPECIFICITY OF GABAergic AND GLYCINERGIC SIGNALING IN DEFINED NETWORKS

This chapter has focused on the general principles of GABAergic and—in less detail—glycinergic transmission. As a summary and outlook we will briefly describe the diversity of inhibitory transmission in the hippocampus and its consequences for network behavior. This section shall highlight the emerging understanding of synaptic heterogeneity in the framework of signal processing within defined networks in the CNS.

Inhibitory synapses in the rodent hippocampus are organized in a laminar way. As a consequence, the inhibitory input to distinct parts of pyramidal cells (soma, axon initial segment, proximal dendrites, distal dendrites) comes from different types or groups of interneurons. Recent work has revealed that these interneurons can be classified according to different histochemical markers as well as to their different intrinsic properties and circuit integration (Somogyi and Klausberger, 2005). At the network level, it has become clear that only defined types of interneurons participate in the different states of network activity, most evidently during different network oscillations (Klausberger et al., 2003, 2004). At the microphysiological level, synapses from different hippocampal interneurons elicit different postsynaptic responses most markedly distinguished by their different kinetics. GABAergic synapses to distal dendrites of CA1 pyramidal neurons cause very slow IPSPs probably due to a specific set of subunits composing the local $GABA_AR$ (Pearce, 1993; Banks et al., 1998). At the other extreme, IPSPs at basket-cell-to-basket-cell synapses in the dentate gyrus are extremely fast (Bartos et al., 2001). The duration of IPSPs is a key parameter in determining the intervals between action potentials of target cells. By this mechanism, interneurons can entrain many postsynaptic cells into a common rhythm of discharges and—especially when synchronized with other interneurons—generate coherent network oscillations at defined frequencies (Whittington et al., 2000; Bartos et al., 2002; Whittington and Traub, 2003; Traub et al., 2004). The type(s) of interneurons which participate in a certain network pattern will then determine the frequency and discharge probability of the participating neurons.

Besides IPSP kinetics (which is determined by $GABA_AR$ subunit composition, efficacy of GABA-uptake, dimensions of the synaptic cleft, and others) many other determinants of interneuron function have to be taken into account for a complete picture: frequency-dependent firing properties, frequency facilitation versus depression of synaptic release, modulation by presynaptic metabotropic or ionotropic receptors, amplitude of postsynaptic conductance change, efficacy of synapses at given subcellular locations, and so on. It becomes clear, however, that the complex individual properties of inhibitory synapses are important determinants of signaling within neuronal networks. We are now equipped with the necessary tools in molecular and cellular neurosciences to understand this functional diversity and to apply it to the analysis of defined neuronal networks.

REFERENCES

Alldred, M. J., Mulder-Rosi, J., Lingenfelter, S. E., Chen, G., and Luscher, B. (2005). Distinct gamma2 subunit domains mediate clustering and synaptic function of postsynaptic $GABA_A$ receptors and gephyrin. *J. Neurosci.* **25**, 594–603.

Asada, H., Kawamura, Y., Maruyama, K., Kume, H., Ding, R. G., Kanbara, N., Kuzume, H., Sanbo, M., Yagi, T., and Obata, K. (1997). Cleft palate and decreased brain gamma-aminobutyric acid in mice lacking the 67-kDa isoform of glutamic acid decarboxylase. *Proc. Natl. Acad. Sci. USA* **94**, 6496–6499.

Axmacher, N. and Draguhn, A. (2004). Inhibition of GABA release by presynaptic ionotropic GABA receptors in hippocampal CA3. *Neuroreport* **15**, 329–334.

Axmacher, N., Stemmler, M., Engel, D., Draguhn, A., and Ritz, R. (2004a). Transmitter metabolism as a mechanism of synaptic plasticity: A modeling study. *J. Neurophysiol.* **91**, 25–39.

Axmacher, N., Winterer, J., Stanton, P. K., Draguhn, A., and Muller, W. (2004b). Two-photon imaging of spontaneous vesicular release in acute brain slices and its modulation by presynaptic GABAA receptors. *Neuroimage* **22**, 1014–1021.

Banks, M. I. and Pearce, R. A. (2000). Kinetic differences between synaptic and extrasynaptic GABA(A) receptors in CA1 pyramidal cells. *J. Neurosci.* **20**, 937–948.

Banks, M. I., Li, T. B., and Pearce, R. A. (1998). The synaptic basis of GABAA, slow. *J. Neurosci.* **18**, 1305–1317.

Bartos, M., Vida, I., Frotscher, M., Geiger, J. R., and Jonas, P. (2001). Rapid signaling at inhibitory synapses in a dentate gyrus interneuron network. *J. Neurosci.* **21**, 2687–2698.

Bartos, M., Vida, I., Frotscher, M., Meyer, A., Monyer, H., Geiger, J. R., and Jonas, P. (2002). Fast synaptic inhibition promotes synchronized gamma oscillations in hippocampal interneuron networks. *Proc. Natl. Acad. Sci. USA* **99**, 13222–13227.
Belelli, D. and Lambert, J. J. (2005). Neurosteroids: Endogenous regulators of the GABA(A) receptor. *Nat. Rev. Neurosci.* **6**, 565–575.
Belelli, D., Herd, M. B., Mitchell, E. A., Peden, D. R., Vardy, A. W., Gentet, L., and Lambert, J. J. (2006). Neuroactive steroids and inhibitory neurotransmission: Mechanisms of action and physiological relevance. *Neuroscience* **138**, 821–829.
Belhage, B., Hansen, G. H., and Schousboe, A. (1993). Depolarization by K^+ and glutamate activates different neurotransmitter release mechanisms in GABAergic neurons: Vesicular versus non-vesicular release of GABA. *Neuroscience* **54**, 1019–1034.
Ben-Ari, Y. (2002). Excitatory actions of GABA during development: The nature of the nurture. *Nat. Rev. Neurosci.* **3**, 728–739.
Ben-Ari, Y., Cherubini, E., Corradetti, R., and Gaiarsa, J. L. (1989). Giant synaptic potentials in immature rat CA3 hippocampal neurones. *J. Physiol.* **416**, 303–325.
Ben-Ari, Y., Rovira, C., Gaiarsa, J. L., Corradetti, R., Robain, O., and Cherubini, E. (1990). GABAergic mechanisms in the CA3 hippocampal region during early postnatal life. *Prog. Brain Res.* **83**, 313–321.
Bevan, M. D., Magill, P. J., Hallworth, N. E., Bolam, J. P., and Wilson, C. J. (2002). Regulation of the timing and pattern of action potential generation in rat subthalamic neurons *in vitro* by GABA-A IPSPs. *J. Neurophysiol.* **87**, 1348–1362.
Blatow, M., Caputi, A., Burnashev, N., Monyer, H., and Rozov, A. (2003). Ca^{2+} buffer saturation underlies paired pulse facilitation in calbindin-D28k-containing terminals. *Neuron* **38**, 79–88.
Bormann, J. (2000). The 'ABC' of GABA receptors. *Trends Pharmacol. Sci.* **21**, 16–19.
Bormann, J., Hamill, O. P., and Sakmann, B. (1987). Mechanism of anion permeation through channels gated by glycine and gamma-aminobutyric acid in mouse cultured spinal neurones. *J. Physiol.* **385**, 243–286.
Bowery, N. G. and Brown, D. A. (1997). The cloning of GABA(B) receptors. *Nature* **386**, 223–224.
Braestrup, C., Nielsen, M., Honore, T., Jensen, L. H., and Petersen, E. N. (1983). Benzodiazepine receptor ligands with positive and negative efficacy. *Neuropharmacology* **22**, 1451–1457.
Brickley, S. G., Revilla, V., Cull-Candy, S. G., Wisden, W., and Farrant, M. (2001). Adaptive regulation of neuronal excitability by a voltage-independent potassium conductance. *Nature* **409**, 88–92.
Buhl, E. H., Otis, T. S., and Mody, I. (1996). Zinc-induced collapse of augmented inhibition by GABA in a temporal lobe epilepsy model. *Science* **271**, 369–373.
Burnashev, N. and Rozov, A. (2005). Presynaptic Ca^{2+} dynamics, Ca^{2+} buffers and synaptic efficacy. *Cell Calcium* **37**, 489–495.
Cammack, J. N. and Schwartz, E. A. (1993). Ions required for the electrogenic transport of GABA by horizontal cells of the catfish retina. *J. Physiol.* **472**, 81–102.
Cammack, J. N., Rakhilin, S. V., and Schwartz, E. A. (1994). A GABA transporter operates asymmetrically and with variable stoichiometry. *Neuron* **13**, 949–960.
Chavas, J., Forero, M. E., Collin, T., Llano, I., and Marty, A. (2004). Osmotic tension as a possible link between GABA(A) receptor activation and intracellular calcium elevation. *Neuron* **44**, 701–713.
Chen, J. C. and Chesler, M. (1991). Extracellular alkalinization evoked by GABA and its relationship to activity-dependent pH shifts in turtle cerebellum. *J. Physiol.* **442**, 431–446.
Chen, J. C. and Chesler, M. (1992). Modulation of extracellular pH by glutamate and GABA in rat hippocampal slices. *J. Neurophysiol.* **67**, 29–36.
Conti, F., DeBiasi, S., Minelli, A., Rothstein, J. D., and Melone, M. (1998). EAAC1, a high-affinity glutamate transporter, is localized to astrocytes and GABAergic neurons besides pyramidal cells in the rat cerebral cortex. *Cereb. Cortex* **8**, 108–116.
Coulter, D. A. (2000). Mossy fiber zinc and temporal lobe epilepsy: Pathological association with altered "epileptic" gamma-aminobutyric acid A receptors in dentate granule cells. *Epilepsia* **41**(Suppl. 6), S96–S99.

Coulter, D. A. (2001). Epilepsy-associated plasticity in gamma-aminobutyric acid receptor expression, function, and inhibitory synaptic properties. *Int. Rev. Neurobiol.* **45**, 237–252.

Crestani, F., Keist, R., Fritschy, J. M., Benke, D., Vogt, K., Prut, L., Bluthmann, H., Möhler, H., and Rudolph, U. (2002). Trace fear conditioning involves hippocampal alpha5 GABA(A) receptors. *Proc. Natl. Acad. Sci. USA* **99**, 8980–8985.

Destexhe, A., Bal, T., McCormick, D. A., and Sejnowski, T. J. (1996). Ionic mechanisms underlying synchronized oscillations and propagating waves in a model of ferret thalamic slices. *J. Neurophysiol.* **76**, 2049–2070.

Destexhe, A., Contreras, D., and Steriade, M. (1998). Mechanisms underlying the synchronizing action of corticothalamic feedback through inhibition of thalamic relay cells. *J. Neurophysiol.* **79**, 999–1016.

Dingledine, R. and Korn, S. J. (1985). Gamma-aminobutyric acid uptake and the termination of inhibitory synaptic potentials in the rat hippocampal slice. *J. Physiol.* **366**, 387–409.

Draguhn, A. and Heinemann, U. (1996). Different mechanisms regulate IPSC kinetics in early postnatal and juvenile hippocampal granule cells. *J. Neurophysiol.* **76**, 3983–3993.

Draguhn, A., Verdorn, T. A., Ewert, M., Seeburg, P. H., and Sakmann, B. (1990). Functional and molecular distinction between recombinant rat GABAA receptor subtypes by Zn^{2+}. *Neuron* **5**, 781–788.

Eccles, J. C. (1964). Presynaptic inhibition in the spinal cord. *Prog. Brain Res.* **12**, 65–91.

Eccles, J. C., Schmidt, R. F., and Willis, W. D. (1962). Presynaptic inhibition of the spinal monosynaptic reflex pathway. *J. Physiol.* **161**, 282–297.

Eccles, J. C., Schmidt, R., and Willis, W. D. (1963). Pharmacological studies on presynaptic inhibition. *J. Physiol.* **168**, 500–530.

Engel, D., Schmitz, D., Gloveli, T., Frahm, C., Heinemann, U., and Draguhn, A. (1998). Laminar difference in GABA uptake and GAT-1 expression in rat CA1. *J. Physiol.* **512**, 643–649.

Engel, D., Pahner, I., Schulze, K., Frahm, C., Jarry, H., Ahnert-Hilger, G., and Draguhn, A. (2001). Plasticity of rat central inhibitory synapses through GABA metabolism. *J. Physiol.* **535**, 473–482.

Esclapez, M. and Houser, C. R. (1999). Up-regulation of GAD65 and GAD67 in remaining hippocampal GABA neurons in a model of temporal lobe epilepsy. *J. Comp. Neurol.* **412**, 488–505.

Esclapez, M., Tillakaratne, N. J., Tobin, A. J., and Houser, C. R. (1993). Comparative localization of mRNAs encoding two forms of glutamic acid decarboxylase with nonradioactive *in situ* hybridization methods. *J. Comp. Neurol.* **331**, 339–362.

Frahm, C., Stief, F., Zuschratter, W., and Draguhn, A. (2003). Unaltered control of extracellular GABA-concentration through GAT-1 in the hippocampus of rats after pilocarpine-induced status epilepticus. *Epilepsy Res.* **52**, 243–252.

Gaiarsa, J. L., Caillard, O., and Ben-Ari, Y. (2002). Long-term plasticity at GABAergic and glycinergic synapses: Mechanisms and functional significance. *Trends Neurosci.* **25**, 564–570.

Ganguly, K., Schinder, A. F., Wong, S. T., and Poo, M. (2001). GABA itself promotes the developmental switch of neuronal GABAergic responses from excitation to inhibition. *Cell* **105**, 521–532.

Garraghty, P. E., LaChica, E. A., and Kaas, J. H. (1991). Injury-induced reorganization of somatosensory cortex is accompanied by reductions in GABA staining. *Somatosens Mot. Res.* **8**, 347–354.

Gasnier, B. (2000). The loading of neurotransmitters into synaptic vesicles. *Biochimie* **82**, 327–337.

Giustetto, M., Kirsch, J., Fritschy, J. M., Cantino, D., and Sassoe-Pognetto, M. (1998). Localization of the clustering protein gephyrin at GABAergic synapses in the main olfactory bulb of the rat. *J. Comp. Neurol.* **395**, 231–244.

Gomeza, J., Hulsmann, S., Ohno, K., Eulenburg, V., Szoke, K., Richter, D., and Betz, H. (2003a). Inactivation of the glycine transporter 1 gene discloses vital role of glial glycine uptake in glycinergic inhibition. *Neuron* **40**, 785–796.

Gomeza, J., Ohno, K., Hulsmann, S., Armsen, W., Eulenburg, V., Richter, D. W., Laube, B., and Betz, H. (2003b). Deletion of the mouse glycine transporter 2 results in a hyperekplexia phenotype and postnatal lethality. *Neuron* **40**, 797–806.

Gray, E. G. (1962). A morphological basis for pre-synaptic inhibition? *Nature* **193**, 82–83.

Gulledge, A. T. and Stuart, G. J. (2003). Excitatory actions of GABA in the cortex. *Neuron* **37**, 299–309.

Gutierrez, R., Romo-Parra, H., Maqueda, J., Vivar, C., Ramirez, M., Morales, M. A., and Lamas, M. (2003). Plasticity of the GABAergic phenotype of the "glutamatergic" granule cells of the rat dentate gyrus. *J. Neurosci.* **23**, 5594–5598.

Haefely, W. (1990). The GABA-benzodiazepine interaction fifteen years later. *Neurochem. Res.* **15**, 169–174.

Hajos, N., Nusser, Z., Rancz, E. A., Freund, T. F., and Mody, I. (2000). Cell type- and synapse-specific variability in synaptic $GABA_A$ receptor occupancy. *Eur. J. Neurosci.* **12**, 810–818.

Hartmann, K., Stief, F., Draguhn, A., and Frahm, C. (2004). Ionotropic GABA receptors with mixed pharmacological properties of $GABA_A$ and $GABA_C$ receptors. *Eur. J. Pharmacol.* **497**, 139–146.

Hevers, W. and Luddens, H. (1998). The diversity of $GABA_A$ receptors. Pharmacological and electrophysiological properties of $GABA_A$ channel subtypes. *Mol. Neurobiol.* **18**, 35–86.

Hosie, A. M., Dunne, E. L., Harvey, R. J., and Smart, T. G. (2003). Zinc-mediated inhibition of GABA(A) receptors: Discrete binding sites underlie subtype specificity. *Nat. Neurosci.* **6**, 362–369.

Jang, I. S., Jeong, H. J., and Akaike, N. (2001). Contribution of the Na-K-Cl cotransporter on GABA (A) receptor-mediated presynaptic depolarization in excitatory nerve terminals. *J. Neurosci.* **21**, 5962–5972.

Jarolimek, W., Lewen, A., and Misgeld, U. (1999). A furosemide-sensitive K^+-Cl^- cotransporter counteracts intracellular Cl^- accumulation and depletion in cultured rat midbrain neurons. *J. Neurosci.* **19**, 4695–4704.

Jin, H., Wu, H., Osterhaus, G., Wei, J., Davis, K., Sha, D., Floor, E., Hsu, C. C., Kopke, R. D., and Wu, J. Y. (2003). Demonstration of functional coupling between gamma-aminobutyric acid (GABA) synthesis and vesicular GABA transport into synaptic vesicles. *Proc. Natl. Acad. Sci. USA* **100**, 4293–4298.

Jonas, P., Bischofberger, J., and Sandkuhler, J. (1998). Corelease of two fast neurotransmitters at a central synapse. *Science* **281**, 419–424.

Kaila, K. and Voipio, J. (1987). Postsynaptic fall in intracellular pH induced by GABA-activated bicarbonate conductance. *Nature* **330**, 163–165.

Kaila, K., Paalasmaa, P., Taira, T., and Voipio, J. (1992). pH transients due to monosynaptic activation of GABAA receptors in rat hippocampal slices. *Neuroreport* **3**, 105–108.

Kaila, K., Voipio, J., Paalasmaa, P., Pasternack, M., and Deisz, R. A. (1993). The role of bicarbonate in $GABA_A$ receptor-mediated IPSPs of rat neocortical neurones. *J. Physiol.* **464**, 273–289.

Kaila, K., Lamsa, K., Smirnov, S., Taira, T., and Voipio, J. (1997). Long-lasting GABA-mediated depolarization evoked by high-frequency stimulation in pyramidal neurons of rat hippocampal slice is attributable to a network-driven, bicarbonate-dependent K^+ transient. *J. Neurosci.* **17**, 7662–7672.

Kang, T. C., An, S. J., Park, S. K., Hwang, I. K., Bae, J. C., and Won, M. H. (2003). Changed vesicular GABA transporter immunoreactivity in the gerbil hippocampus following spontaneous seizure and vigabatrin administration. *Neurosci. Lett.* **335**, 207–211.

Kaupmann, K., Huggel, K., Heid, J., Flor, P. J., Bischoff, S., Mickel, S. J., McMaster, G., Angst, C., Bittiger, H., Froestl, W., and Bettler, B. (1997). Expression cloning of GABA(B) receptors uncovers similarity to metabotropic glutamate receptors. *Nature* **386**, 239–246.

Kelsch, W., Hormuzdi, S., Straube, E., Lewen, A., Monyer, H., and Misgeld, U. (2001). Insulin-like growth factor 1 and a cytosolic tyrosine kinase activate chloride outward transport during maturation of hippocampal neurons. *J. Neurosci.* **21**, 8339–8347.

Kirsch, J., Wolters, I., Triller, A., and Betz, H. (1993). Gephyrin antisense oligonucleotides prevent glycine receptor clustering in spinal neurons. *Nature* **366**, 745–748.

Kittler, J. T., McAinsh, K., and Moss, S. J. (2002). Mechanisms of $GABA_A$ receptor assembly and trafficking: Implications for the modulation of inhibitory neurotransmission. *Mol. Neurobiol.* **26**, 251–268.

Klausberger, T., Magill, P. J., Marton, L. F., Roberts, J. D., Cobden, P. M., Buzsaki, G., and Somogyi, P. (2003). Brain-state- and cell-type-specific firing of hippocampal interneurons *in vivo*. *Nature* **421**, 844–848.

Klausberger, T., Marton, L. F., Baude, A., Roberts, J. D., Magill, P. J., and Somogyi, P. (2004). Spike timing of dendrite-targeting bistratified cells during hippocampal network oscillations *in vivo*. *Nat. Neurosci.* **7**, 41–47.

Korpi, E. R., Grunder, G., and Luddens, H. (2002). Drug interactions at GABA(A) receptors. *Prog. Neurobiol.* **67**, 113–159.

Krishek, B. J., Moss, S. J., and Smart, T. G. (1998). Interaction of H^+ and Zn^{2+} on recombinant and native rat neuronal $GABA_A$ receptors. *J. Physiol.* **507**, 639–652.

Kullmann, D. M., Ruiz, A., Rusakov, D. M., Scott, R., Semyanov, A., and Walker, M. C. (2005). Presynaptic, extrasynaptic and axonal $GABA_A$ receptors in the CNS: Where and why? *Prog. Biophys. Mol. Biol.* **87**, 33–46.

Kuner, R., Kohr, G., Grunewald, S., Eisenhardt, G., Bach, A., and Kornau, H. C. (1999). Role of heteromer formation in $GABA_B$ receptor function. *Science* **283**, 74–77.

Lamas, M., Gomez-Lira, G., and Gutierrez, R. (2001). Vesicular GABA transporter mRNA expression in the dentate gyrus and in mossy fiber synaptosomes. *Brain Res. Mol. Brain Res.* **93**, 209–214.

Lamsa, K., Heeroma, J. H., and Kullmann, D. M. (2005). Hebbian LTP in feed-forward inhibitory interneurons and the temporal fidelity of input discrimination. *Nat. Neurosci.* **8**, 916–924.

Liu, J., Tai, C., de Groat, W. C., Peng, X. M., Mata, M., and Fink, D. J. (2006). Release of GABA from sensory neurons transduced with a GAD67-expressing vector occurs by non-vesicular mechanisms. *Brain Res.* **1073–1074**, 297–304.

Liu, Q. R., Lopez-Corcuera, B., Mandiyan, S., Nelson, H., and Nelson, N. (1993). Molecular characterization of four pharmacologically distinct gamma-aminobutyric acid transporters in mouse brain [corrected]. *J. Biol. Chem.* **268**, 2106–2112.

Luckermann, M., Trapp, S., and Ballanyi, K. (1997). GABA- and glycine-mediated fall of intracellular pH in rat medullary neurons *in situ*. *J. Neurophysiol.* **77**, 1844–1852.

Luddens, H. and Korpi, E. R. (1995). Biological function of $GABA_A$/benzodiazepine receptor heterogeneity. *J. Psychiatr. Res.* **29**, 77–94.

Lux, H. D. and Heinemann, U. (1978). Ionic changes during experimentally induced seizure activity. *Electroencephalogr. Clin. Neurophysiol. Suppl.* **34**, 289–297.

Lynch, J. W. (2004). Molecular structure and function of the glycine receptor chloride channel. *Physiol. Rev.* **84**, 1051–1095.

Mathews, G. C. and Diamond, J. S. (2003). Neuronal glutamate uptake contributes to GABA synthesis and inhibitory synaptic strength. *J. Neurosci.* **23**, 2040–2048.

Misgeld, U., Bijak, M., and Jarolimek, W. (1995). A physiological role for $GABA_B$ receptors and the effects of baclofen in the mammalian central nervous system. *Prog. Neurobiol.* **46**, 423–462.

Mody, I. (2001). Distinguishing between GABA(A) receptors responsible for tonic and phasic conductances. *Neurochem. Res.* **26**, 907–913.

Mody, I. and Pearce, R. A. (2004). Diversity of inhibitory neurotransmission through GABA(A) receptors. *Trends Neurosci.* **27**, 569–575.

Möhler, H., Malherbe, P., Draguhn, A., and Richards, J. G. (1990). $GABA_A$-receptors: Structural requirements and sites of gene expression in mammalian brain. *Neurochem. Res.* **15**, 199–207.

Möhler, H., Knoflach, F., Paysan, J., Motejlek, K., Benke, D., Luscher, B., and Fritschy, J. M. (1995). Heterogeneity of $GABA_A$-receptors: Cell-specific expression, pharmacology, and regulation. *Neurochem. Res.* **20**, 631–636.

Morales-Aza, B. M., Chillingworth, N. L., Payne, J. A., and Donaldson, L. F. (2004). Inflammation alters cation chloride cotransporter expression in sensory neurons. *Neurobiol. Dis.* **17**, 62–69.

Nabbout, R. (2001). A risk-benefit assessment of treatments for infantile spasms. *Drug Saf.* **24**, 813–828.

Okabe, A., Ohno, K., Toyoda, H., Yokokura, M., Sato, K., and Fukuda, A. (2002). Amygdala kindling induces upregulation of mRNA for NKCC1, a Na(+), K(+)-2Cl(−) cotransporter, in the rat piriform cortex. *Neurosci. Res.* **44**, 225–229.

Owens, D. F. and Kriegstein, A. R. (2002). Is there more to GABA than synaptic inhibition? *Nat. Rev. Neurosci.* **3**, 715–727.

Pearce, R. A. (1993). Physiological evidence for two distinct GABAA responses in rat hippocampus. *Neuron* **10**, 189–200.

Perrais, D. and Ropert, N. (1999). Effect of zolpidem on miniature IPSCs and occupancy of postsynaptic $GABA_A$ receptors in central synapses. *J. Neurosci.* **19**, 578–588.

Pinal, C. S. and Tobin, A. J. (1998). Uniqueness and redundancy in GABA production. *Perspect. Dev. Neurobiol.* **5**, 109–118.

Plotkin, M. D., Snyder, E. Y., Hebert, S. C., and Delpire, E. (1997). Expression of the Na-K-2Cl cotransporter is developmentally regulated in postnatal rat brains: A possible mechanism underlying GABA's excitatory role in immature brain. *J. Neurobiol.* **33**, 781–795.

Pond, B. B., Berglund, K., Kuner, T., Feng, G., Augustine, G. J., and Schwartz-Bloom, R. D. (2006). The chloride transporter Na(+)-K(+)-Cl(−) cotransporter isoform-1 contributes to intracellular chloride increases after *in vitro* ischemia. *J. Neurosci.* **26**, 1396–1406.

Pritchett, D. B., Sontheimer, H., Shivers, B. D., Ymer, S., Kettenmann, H., Schofield, P. R., and Seeburg, P. H. (1989). Importance of a novel GABAA receptor subunit for benzodiazepine pharmacology. *Nature* **338**, 582–585.

Ramirez, M. and Gutierrez, R. (2001). Activity-dependent expression of GAD67 in the granule cells of the rat hippocampus. *Brain Res.* **917**, 139–146.

Represa, A. and Ben-Ari, Y. (2005). Trophic actions of GABA on neuronal development. *Trends Neurosci.* **28**, 278–283.

Rivera, C., Voipio, J., Payne, J. A., Ruusuvuori, E., Lahtinen, H., Lamsa, K., Pirvola, U., Saarma, M., and Kaila, K. (1999). The K^+/Cl^- co-transporter KCC2 renders GABA hyperpolarizing during neuronal maturation. *Nature* **397**, 251–255.

Rivera, C., Li, H., Thomas-Crusells, J., Lahtinen, H., Viitanen, T., Nanobashvili, A., Kokaia, Z., Airaksinen, M. S., Voipio, J., Kaila, K., and Saarma, M. (2002). BDNF-induced TrkB activation down-regulates the K^+-Cl^- cotransporter KCC2 and impairs neuronal Cl^- extrusion. *J. Cell Biol.* **159**, 747–752.

Rivera, C., Voipio, J., and Kaila, K. (2005). Two developmental switches in GABAergic signalling: The K^+-Cl^- cotransporter KCC2 and carbonic anhydrase CAVII. *J. Physiol.* **562**, 27–36.

Roepstorff, A. and Lambert, J. D. (1994). Factors contributing to the decay of the stimulus-evoked IPSC in rat hippocampal CA1 neurons. *J. Neurophysiol.* **72**, 2911–2926.

Rossi, D. J., Hamann, M., and Attwell, D. (2003). Multiple modes of GABAergic inhibition of rat cerebellar granule cells. *J. Physiol.* **548**, 97–110.

Rudolph, U. and Möhler, H. (2004). Analysis of GABA$_A$ receptor function and dissection of the pharmacology of benzodiazepines and general anesthetics through mouse genetics. *Annu. Rev. Pharmacol. Toxicol.* **44**, 475–498.

Rudolph, U. and Möhler, H. (2006). GABA-based therapeutic approaches: GABA$_A$ receptor subtype functions. *Curr. Opin. Pharmacol.* **6**, 18–23.

Rudolph, U., Crestani, F., Benke, D., Brunig, I., Benson, J. A., Fritschy, J. M., Martin, J. R., Bluethmann, H., and Möhler, H. (1999). Benzodiazepine actions mediated by specific gamma-aminobutyric acid(A) receptor subtypes. *Nature* **401**, 796–800.

Rudomin, P. and Schmidt, R. F. (1999). Presynaptic inhibition in the vertebrate spinal cord revisited. *Exp. Brain Res.* **129**, 1–37.

Ruusuvuori, E., Li, H., Huttu, K., Palva, J. M., Smirnov, S., Rivera, C., Kaila, K., and Voipio, J. (2004). Carbonic anhydrase isoform VII acts as a molecular switch in the development of synchronous gamma-frequency firing of hippocampal CA1 pyramidal cells. *J. Neurosci.* **24**, 2699–2707.

Sagne, C., El, M. S., Isambert, M. F., Hamon, M., Henry, J. P., Giros, B., and Gasnier, B. (1997). Cloning of a functional vesicular GABA and glycine transporter by screening of genome databases. *FEBS Lett.* **417**, 177–183.

Sarup, A., Larsson, O. M., and Schousboe, A. (2003). GABA transporters and GABA-transaminase as drug targets. *Curr. Drug Targets CNS Neurol. Disord.* **2**, 269–277.

Sassoe-Pognetto, M., Kirsch, J., Grunert, U., Greferath, U., Fritschy, J. M., Möhler, H., Betz, H., and Wassle, H. (1995). Colocalization of gephyrin and GABA$_A$-receptor subunits in the rat retina. *J. Comp. Neurol.* **357**, 1–14.

Schoepp, D. D. (2001). Unveiling the functions of presynaptic metabotropic glutamate receptors in the central nervous system. *J. Pharmacol. Exp. Ther.* **299**, 12–20.

Schweizer, C., Balsiger, S., Bluethmann, H., Mansuy, I. M., Fritschy, J. M., Möhler, H., and Luscher, B. (2003). The gamma 2 subunit of GABA(A) receptors is required for maintenance of receptors at mature synapses. *Mol. Cell. Neurosci.* **24**, 442–450.

Semyanov, A., Walker, M. C., and Kullmann, D. M. (2003). GABA uptake regulates cortical excitability via cell type-specific tonic inhibition. *Nat. Neurosci.* **6**, 484–490.

Semyanov, A., Walker, M. C., Kullmann, D. M., and Silver, R. A. (2004). Tonically active GABA A receptors: Modulating gain and maintaining the tone. *Trends Neurosci.* **27**, 262–269.

Sepkuty, J. P., Cohen, A. S., Eccles, C., Rafiq, A., Behar, K., Ganel, R., Coulter, D. A., and Rothstein, J. D. (2002). A neuronal glutamate transporter contributes to neurotransmitter GABA synthesis and epilepsy. *J. Neurosci.* **22**, 6372–6379.

Smirnov, S., Paalasmaa, P., Uusisaari, M., Voipio, J., and Kaila, K. (1999). Pharmacological isolation of the synaptic and nonsynaptic components of the GABA-mediated biphasic response in rat CA1 hippocampal pyramidal cells. *J. Neurosci.* **19**, 9252–9260.

Soghomonian, J. J. and Martin, D. L. (1998). Two isoforms of glutamate decarboxylase: Why? *Trends Pharmacol. Sci.* **19**, 500–505.

Somogyi, P. and Klausberger, T. (2005). Defined types of cortical interneurone structure space and spike timing in the hippocampus. *J. Physiol.* **562**, 9–26.

Spence, S. J. and Sankar, R. (2001). Visual field defects and other ophthalmological disturbances associated with vigabatrin. *Drug Saf.* **24**, 385–404.

Staley, K. J. and Proctor, W. R. (1999). Modulation of mammalian dendritic GABA(A) receptor function by the kinetics of Cl^- and HCO_3^- transport. *J. Physiol.* **519**, 693–712.

Staley, K. J., Soldo, B. L., and Proctor, W. R. (1995). Ionic mechanisms of neuronal excitation by inhibitory $GABA_A$ receptors. *Science* **269**, 977–981.

Stein, V., Hermans-Borgmeyer, I., Jentsch, T. J., and Hübner, C. A. (2004). Expression of the KCl cotransporter KCC2 parallels neuronal maturation and the emergence of low intracellular chloride. *J. Comp. Neurol.* **468**, 57–64.

Stell, B. M. and Mody, I. (2002). Receptors with different affinities mediate phasic and tonic GABA(A) conductances in hippocampal neurons. *J. Neurosci.* **22**, RC223.

Steriade, M. (2005). Sleep, epilepsy and thalamic reticular inhibitory neurons. *Trends Neurosci.* **28**, 317–324.

Stief, F., Piechotta, A., Gabriel, S., Schmitz, D., and Draguhn, A. (2005). Functional GABA uptake at inhibitory synapses in CA1 of chronically epileptic rats. *Epilepsy Res.* **66**, 199–202.

Stork, O., Ji, F. Y., Kaneko, K., Stork, S., Yoshinobu, Y., Moriya, T., Shibata, S., and Obata, K. (2000). Postnatal development of a GABA deficit and disturbance of neural functions in mice lacking GAD65. *Brain Res.* **865**, 45–58.

Sung, K. W., Kirby, M., McDonald, M. P., Lovinger, D. M., and Delpire, E. (2000). Abnormal $GABA_A$ receptor-mediated currents in dorsal root ganglion neurons isolated from Na-K-2Cl cotransporter null mice. *J. Neurosci.* **20**, 7531–7538.

Szabadics, J., Varga, C., Molnar, G., Olah, S., Barzo, P., and Tamas, G. (2006). Excitatory effect of GABAergic axo-axonic cells in cortical microcircuits. *Science* **311**, 233–235.

Takayama, C. and Inoue, Y. (2004). Extrasynaptic localization of GABA in the developing mouse cerebellum. *Neurosci. Res.* **50**, 447–458.

Titz, S., Hans, M., Kelsch, W., Lewen, A., Swandulla, D., and Misgeld, U. (2003). Hyperpolarizing inhibition develops without trophic support by GABA in cultured rat midbrain neurons. *J. Physiol.* **550**, 719–730.

Traub, R. D., Bibbig, A., LeBeau, F. E., Buhl, E. H., and Whittington, M. A. (2004). Cellular mechanisms of neuronal population oscillations in the hippocampus *in vitro*. *Annu. Rev. Neurosci.* **27**, 247–278.

van Rijnsoever, R. C., Tauber, M., Choulli, M. K., Keist, R., Rudolph, U., Möhler, H., Fritschy, J. M., and Crestani, F. (2004). Requirement of alpha5-$GABA_A$ receptors for the development of tolerance to the sedative action of diazepam in mice. *J. Neurosci.* **24**, 6785–6790.

Vardi, N., Zhang, L. L., Payne, J. A., and Sterling, P. (2000). Evidence that different cation chloride cotransporters in retinal neurons allow opposite responses to GABA. *J. Neurosci.* **20**, 7657–7663.

Waagepetersen, H. S., Sonnewald, U., and Schousboe, A. (1999). The GABA paradox: Multiple roles as metabolite, neurotransmitter, and neurodifferentiative agent. *J. Neurochem.* **73,** 1335–1342.

Whittington, M. A. and Traub, R. D. (2003). Interneuron diversity series: Inhibitory interneurons and network oscillations *in vitro. Trends Neurosci.* **26,** 676–682.

Whittington, M. A., Traub, R. D., Kopell, N., Ermentrout, B., and Buhl, E. H. (2000). Inhibition-based rhythms: Experimental and mathematical observations on network dynamics. *Int. J. Psychophysiol.* **38,** 315–336.

Wilson, R. I. and Nicoll, R. A. (2002). Endocannabinoid signaling in the brain. *Science* **296,** 678–682.

Wilson, R. I., Kunos, G., and Nicoll, R. A. (2001). Presynaptic specificity of endocannabinoid signaling in the hippocampus. *Neuron* **31,** 453–462.

Chapter 10
Physiology of Cation-Chloride Cotransporters

Christian A. Hübner[1,2] and Marco B. Rust[2,]*

[1]*Institut für Humangenetik, Universitätskrankenhaus Hamburg-Eppendorf, Butenfeld 42, D-22529 Hamburg, Germany*
[2]*Zentrum für Molekulare Neurobiologie, Universität Hamburg, Falltenried 94, 20246 Hamburg, Germany*

I. Introduction
II. The Cation-Chloride Cotransporter Family
   A. Na^+-Coupled Cation-Chloride Cotransporters
   B. Na^+-Independent Cation-Chloride Cotransporters
   C. CCC8/CIP (*SLC12A9*) and CCC9 (*SLC12A8*)
III. Cation-Chloride Cotransporters in the Nervous System
IV. Cation-Chloride Cotransporters in the Inner Ear
V. Cation-Chloride Cotransporters in the Kidney
VI. Cation-Chloride Cotransporters in RBCs
VII. Cation-Chloride Cotransporter and Cancer
VIII. Cation-Chloride Cotransport and Hypertension
IX. Concluding Remarks
   References

I. INTRODUCTION

Cl^- is the most abundant anion in vertebrate cells. In contrast to ions like Na^+ and K^+, the electrochemical gradient of Cl^- across the plasma membrane is generally not far from equilibrium. At a membrane voltage of -60 mV and an extracellular Cl^- concentration of 150 mM, the intracellular Cl^- concentration ($[Cl^-]_i$) is 15 mM at equilibrium. Various transporters and channels transport Cl^- across membranes, both at the cell surface and in intracellular organelles. While there is no evidence for an active Cl^--transporting ATPase in mammalian plasma membranes, several of these transporters use the energy stored in the transmembrane gradients of Na^+ and K^+

---

*Present address: Mouse Biology Programme, European Molecular Biology Laboratory, Rome/Monterotondo, Italy.

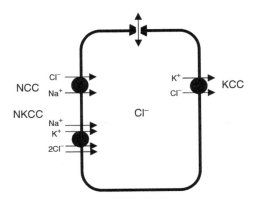

Figure 1. The Na^+ and K^+ plasma membrane gradients are generated by the Na^+/K^+-ATPase. The coupling of Cl^- to the Na^+ gradient by Na^+-Cl^- and Na^+-K^+-$2Cl^-$ cotransporters leads to the accumulation of Cl^- into the cell, whereas the K^+-Cl^- cotransporters usually mediate the efflux of Cl^- out of the cell. Depending on the particular ion gradients, the transport direction can be reversed. This may be the case in supporting cells of outer hair cells in the inner ear, where KCC4 is thought to mediate K^+ uptake which floods to supporting cells from outer hair cells, or for KCC2 during ongoing epileptic activity with a concomitant rise of the extracellular K^+ concentration.

generated by the Na^+/K^+-ATPase to transport Cl^- against its electrochemical gradient. The coupling to the Na^+ gradient by Na^+-Cl^- and Na^+-K^+-$2Cl^-$ cotransporters thus leads to an accumulation of Cl^- into the cell, whereas the K^+-Cl^- cotransporters usually mediate the efflux of Cl^- out of the cell (Fig. 1). However, the direction can be reversed depending on the existing ion gradients (Payne, 1997). As the stoichiometry of cations versus anions is 1:1, these transporters are known as electroneutral cation-chloride cotransporters. Na^+ influx and K^+ efflux is rapidly corrected by the Na^+/K^+-ATPase, as a consequence the net effect of cation-chloride cotransport activity is Cl^- movement into or out of the cell. The directional transport of Cl^- across epithelia is achieved by selectively targeting Cl^- transporters and channels into different plasma membrane domains. Thus the apical targeting of NCC and NKCC2 and the basolateral targeting of Cl^- channels results in the reabsorption of Cl^- as realized in the kidney, whereas the opposite arrangement will lead to the secretion of Cl^-.

The molecular characterization of cation-chloride cotransporters started with the identification of the thiazide-sensitive Na^+-Cl^- cotransporter from the winter flounder urinary bladder epithelium (Gamba et al., 1993) and the Na^+-K^+-$2Cl^-$ cotransporter NKCC1 from the shark rectal gland (Xu et al., 1994). Since then, other members of this gene family have been cloned and important roles in transepithelial transport, cell volume regulation, and neuronal excitability have been identified. All members are structurally closely related and form a gene family of seven functionally well-characterized members (Fig. 2) and two additional recently identified orphan members, cation-chloride cotransporter 8 (CCC8)/cation-chloride cotransporter interacting protein (CIP) and

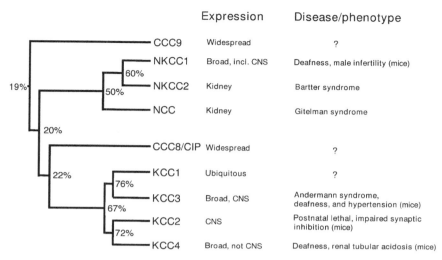

Figure 2. The cation-chloride cotransporter family. This ion transporter family can be subdivided into Na^+-dependent cotransporters comprising NKCC1, NKCC2, and NCC and the Na^+-independent K^+-Cl^- cotransporters comprising KCC1–KCC4. The dendogram represents the expression pattern of each family member as well as known human disorders linked to mutations in these genes and described phenotypes of corresponding knockout mouse models. The physiological roles of the orphan members CCC8/CIP and CCC9 are still unknown. Percentage values indicate the degree of homology at the amino acid sequence level.

cation-chloride cotransporter 9 (CCC9), with so far unknown function. The family of cation-chloride cotransporters (SLC12A) encompasses two major branches, one including the two bumetanide-sensitive Na^+-K^+-$2Cl^-$ cotransporters (NKCC1 and NKCC2) and the thiazide-sensitive Na^+-Cl^- cotransporter NCC. The second branch includes four genes encoding the K^+-Cl^- cotransporters.

After a short introduction into the various members of cation-chloride cotransporter family, this chapter will focus on recent functional aspects regarding cation-chloride cotransport in the central nervous system (CNS), the kidney, the inner ear, red blood cells (RBCs), and vascular smooth muscle cells. For more details the readers are referred to other recent reviews (Adragna *et al.*, 2004; Hebert *et al.*, 2004; Gamba, 2005a).

II. THE CATION-CHLORIDE COTRANSPORTER FAMILY

A. Na^+-Coupled Cation-Chloride Cotransporters

The branch of Na^+-coupled cation-chloride cotransporters includes the bumetanide-sensitive Na^+-K^+-$2Cl^-$ cotransporters NKCC1 (BSC2) and NKCC2 (BSC1) and the thiazide-sensitive Na^+-Cl^- cotransporter NCC (NCCT/TSC). The existence of a Na^+-Cl^- cotransport was first described in the epithelium of the winter flounder urinary bladder

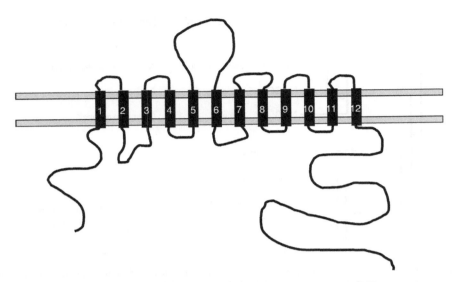

Figure 3. Hypothetical topology of K^+-Cl^- cotransporters. In contrast to K^+-Cl^- cotransporters and CCC8/CIP, which have a large extracellular loop between transmembrane spans TM5 and TM6, NKCC1, NKCC2, and NCC have a larger extracellular loop between transmembrane spans TM7 and TM8. CCC9 is the most distant member, because this protein consists of only 11 predicted transmembrane spans and the C-teminal is thought to be located outside the cell.

(Renfro, 1975), an organ that is related to the distal tubule of the mammalian kidney. The independence of NCC on transepithelial voltage revealed that this transport is electroneutral (Renfro, 1977). This and the sensitivity to thiazides (Stokes, 1984) were then exploited to clone its cDNA by a functional expression strategy in *Xenopus laevis* oocytes. The identified cDNA encoded a protein of 1023 amino acids with a molecular mass of 112 kDa. Hydropathy analysis by Kyte and Doolittle revealed a putative structure with 12 α-helical transmembrane spans flanked by a short hydrophilic N-terminal and a long C-terminal domain, both localized within the cell (Fig. 3). The cDNA encoding the basolateral Na^+-K^+-$2Cl^-$ cotransporter NKCC1 was isolated from a shark rectal gland library by screening with monoclonal antibodies to the native shark cotransporter (Xu et al., 1994), whereas the apical Na^+-K^+-$2Cl^-$ cotransporter NKCC2 was cloned by screening rabbit cortical and medullary kidney cDNA libraries with a probe derived from the cDNA coding for the human NKCC1 (Payne and Forbush, 1994). With a similar approach, the rat cDNA was identified simultaneously (Gamba et al., 1994). Hydropathy analyses of NKCC1 and NKCC2 primary structures revealed general topologies similar to that of NCC with a large intracellular hydrophilic N- and C-terminal domain on either side of a central, relatively well-conserved, hydrophobic domain with 10–12 transmembrane spans. For NKCC1, NKCC2, and NCC evidence for the formation of homodimers in the plasma membrane has been obtained (Moore-Hoon and Turner, 2000; de Jong et al., 2003). Whether heterodimers are formed is still unclear.

1. NKCC1 (*SLC12A2*)

NKCC1 is expressed in both epithelial and non-epithelial cells. In epithelial cells NKCC1 is predominantly localized to the basolateral side and thus provides Cl^- (or in the stria vascularis K^+) for secretion at the apical side. Different to this rule, NKCC1 is localized to the apical membrane in the choroid plexus (Plotkin *et al.*, 1997). In non-epithelial cells, NKCC1 was shown to play an important role for the regulation of cell volume, being activated by hypertonic cell shrinking (Lytle, 1997).

To date NKCC1 is the only family member for which the proposed topology with 12 transmembrane α-helices has been supported by experimental data (Gerelsaikhan and Turner, 2000). Structure–function relationship has also mainly been addressed for NKCC1. For this purpose, the differences with respect to ion and bumetanide affinities between human and shark NKCC1 have been exploited. Construction of chimeric proteins helped to specify regions in human or shark NKCC1 involved in the binding of Na^+, K^+, Cl^-, and its inhibitor bumetanide. Exchange of the N- and C-terminal domains between the proteins revealed that the affinities for Na^+, K^+, Cl^-, and bumetanide are determined by the central portion of the protein containing the 12 transmembrane spans (Isenring and Forbush, 1997). Further experiments showed that mutations in transmembrane domains beyond transmembrane span 7 had no effect on ion affinities, but that bumetanide affinity was affected by mutations in transmembrane spans 11 or 12. Using a similar approach, the amino acid sequence of transmembrane span 2 was shown to affect the transport affinity for cations, but not for Cl^- (Isenring *et al.*, 1998). Binding and transport of K^+ was mainly determined by transmembrane spans 2, 4, and 7, whereas Cl^- transport depended on transmembrane spans 4 and 7 (Isenring and Forbush, 2001). Experiments in RBCs have shown that Na^+ binds first to NKCC1, followed by one Cl^-, then by K^+, and finally, the second Cl^- (Lytle *et al.*, 1998).

The N-terminal domain of NKCC1 contains a RVXFXD motif that binds protein-phosphatase-1 (PP1) (Darman *et al.*, 2001) and an (R/K)FX(V/I) motif for the binding of the Ste20-related kinase SPAK (PASK/STK39) (Piechotta *et al.*, 2002). Inhibition of NKCC1 by WNK4 (Kahle *et al.*, 2004), a member of the novel family of serine-threonine kinases (*with no K* = lysine), is mediated via phosphorylation of the cotransporter by SPAK or by the closely related oxidative stress response kinase-1 (OSR1), which in turn are phosphorylated and activated by WNK4 (Moriguchi *et al.*, 2005; Vitari *et al.*, 2005; Gagnon *et al.*, 2006a,b). WNK3 is another regulator of NKCC1 cotransport. In its kinase-active form, WNK3 is a potent activator of NKCC1-mediated transport as demonstrated by coexpression studies in *X. laevis* oocytes (Kahle *et al.*, 2005). WNK3 effects are imparted via altered phosphorylation and surface expression of the cotransporter bypassing the normal requirement of altered tonicity for activation.

Though no human disease has been linked to NKCC1 so far, many interesting phenotypes have been identified in mice with a targeted or spontaneous disruption of NKCC1, including deafness and the typical *shaker/waltzer* phenotype in vestibular dysfunction, male infertility due to a defect in sperm production, reduction in saliva secretion, defects in sensory perception, and growth deficit (Delpire *et al.*, 1999; Flagella *et al.*, 1999; Evans *et al.*, 2000; Pace *et al.*, 2000; Sung *et al.*, 2000).

2. NKCC2 (*SLC12A1*)

NKCC2 expression is mainly confined to the kidney. In the mouse kidney, different isoforms resulting from alternative exon splicing have been identified. Three mutually exclusive alternative exons termed A, B, and F encode slightly different putative second transmembrane spans with the adjacent part of the following intracellular loop (Mount *et al.*, 1999a). The isoforms have a different expression along the thick ascending limb (TAL) and show differences in the affinities for the ions transported (B>A>F) (Gimenez *et al.*, 2002). The important role of NKCC2 in distal salt reabsorption in the kidney has been proven by identification of loss-of-function mutations in Bartter syndrome type I patients (Simon *et al.*, 1996a), which are hypotensive due to severe renal salt loss. Because of the role of NKCC2 for salt reabsorption, it was hypothesized that altered activity of NKCC2 may have a quantitative effect on blood pressure (Lifton *et al.*, 2001). NKCC2 expression is up-regulated either by acidic intracellular pH (Attmane-Elakeb *et al.*, 1998), vasopressin (Kim *et al.*, 1999), or glucocorticoids (Attmane-Elakeb *et al.*, 2000). In NKCC2-deficient mice, the phenotype seen in Bartter syndrome is even aggravated compared to humans (Takahashi *et al.*, 2000).

Coexpression studies in *X. laevis* oocytes revealed that kinase-active WNK3 not only activates NKCC1 but also NKCC2- and NCC-mediated transport (Rinehart *et al.*, 2005). Conversely, in its kinase-inactive state, WNK3 is a potent inhibitor of NKCC2 and NCC activity. As for NKCC1, WNK3 regulates the activity of NKCC2 and NCC by altering the expression at the plasma membrane. Wild-type WNK3 increases and kinase-inactive WNK3 decreases NKCC2 phosphorylation at T184 and T189, sites required for the vasopressin-mediated plasmalemmal translocation and activation of NKCC2 *in vivo*. Thus WNK3 may be involved in signaling downstream of vasopressin. On the other hand, NKCC2 is inhibited by WNK4, another member of this family of kinases.

3. NCC (*SLC12A3*)

Though thiazide diuretics have been used for the treatment of hypertension for many decades, their mode of action remained unclear until it was shown that thiazides inhibit apical NCC-mediated Na^+-Cl^- cotransport in the distal convoluted tubule (Ellison *et al.*, 1987). Inactivating mutations of NCC cause Gitelman syndrome (Simon *et al.*, 1996c), which is characterized by renal salt loss. Its disruption in mice not only reproduces the human phenotype but also results in a drastic reduction in the distal convoluted tubule cell number (Schultheis *et al.*, 1998). Expression of NCC is predominantly confined to the kidney, though transcripts have also been identified in an osteoblast cell line (Barry et al., 1997). Again WNK-kinases play a central role in the regulation of NCC. Inactivating mutations in WNK4 or activating mutations of WNK1 result in increased NCC-activity and hence cause pseudohypoaldosteronism type II (PHAII), a severe form of familiar hypertension due to salt retention (Wilson et al., 2001). WNK1 was shown to be an upstream regulator of WNK4 at NCC (Yang et al., 2003). In contrast to the inhibitory activity of WNK4, kinase-active WNK3 was shown to act as a potent activator of NCC-mediated transport (Rinehart et al., 2005). Conversely, in its kinase-inactive state, WNK3 inhibited NCC-mediated transport by altering the expression at the plasma membrane.

B. Na^+-Independent Cation-Chloride Cotransporters

Initially, K^+-Cl^- cotransport has been identified in low-K^+ RBCs from sheep as swelling and N-ethylmaleimide (NEM)-activated, Cl^--dependent K^+-efflux pathway (Dunham et al., 1980; Lauf and Theg, 1980). Consecutively, RBCs have remained the primary model tissue to characterize this transport mechanism, but functional and physiological evidences for the existence of similar K^+-Cl^- cotransport mechanisms were reported in several tissues and cell types including neurons, endothelial, and epithelial cells. Four K^+-Cl^- cotransporters, denominated KCC1, KCC2, KCC3, and KCC4, have been identified so far, which are encoded by the genes *SLC12A4*, *SLC12A5*, *SLC12A6*, and *SLC12A7*. KCCs differ in expression patterns and biophysical properties. KCC2, the isoform exclusively expressed in neurons, is active under isotonic conditions and cell swelling only weakly stimulates its transport activity (Song et al., 2002). In contrast KCC1, KCC3, and KCC4 are silent under isotonic conditions, but can be stimulated much stronger by hypotonic stress (Gamba, 2005a). Besides cell swelling, thiol oxidation, cellular Mg^{2+}-depletion, protein kinase inhibition, and free radicals are further stimuli, whereas cell shrinkage, protein phosphatase inhibitors, most bivalent cations, marked acidification below pH 6.5, and polyamines inhibit K^+-Cl^- cotransporters (Adragna et al., 2004). It has been shown that K^+-Cl^- cotransporters are not only part of the cell volume regulation machinery, but are also implicated in transepithelial ion transport and the regulation of the $[Cl^-]_i$ of neurons. Interestingly, all K^+-Cl^- cotransporters are capable of transporting NH_4^+ instead of K^+, with a proposed K_m for NH_4^+ similar to that of K^+ (Liu et al., 2003; Williams and Payne, 2004). Furthermore, the activity of K^+-Cl^- cotransporters can be influenced by intracellular pH (pH_i) with KCC1 and KCC3 exhibiting a lower activity at $pH_i < 7.0$ and $pH_i > 7.5$, KCC2 at $pH_i < 7.5$, and KCC4 at $pH_i > 7.5$ implicating a role of K^+-Cl^- cotransporters in pH_i regulation.

At the moment, little information exists regarding the structure–function relationship. Though KCCs exhibit only an ~20% similarity compared to NKCCs at the amino acid sequence level, the secondary structure is remarkably related to Na^+-coupled cation-chloride transporters with 12 putative transmembrane spans and an intracellular N- and C-terminal domain. A structural difference between K^+-Cl^- cotransporters and Na^+-coupled cation-chloride transporters is the location and size of the larger extracellular loop, which contains potential N-linked glycosylation sites. This loop is located between putative transmembrane spans TM7 and TM8 in NCC and NKCCs but between TM5 and TM6 in K^+-Cl^- cotransporters (Fig. 3). Whether the formation of homodimers and heterodimers occurs *in vivo* is yet unknown though a truncated mutant of KCC1 was shown to have a dominant negative effect on KCC2, KCC3, and KCC4 in *X. laevis* oocytes (Casula *et al.*, 2001).

1. KCC1 (*SLC12A4*)

For the identification of KCC1, the first K^+-Cl^- cotransporter identified at the molecular level, its homology to Na^+-coupled cation-chloride cotransporters was exploited (Gillen *et al.*, 1996). By Northern blot analysis, KCC1 is expressed ubiquitously, suggesting a role as a housekeeping gene involved in cell volume regulation. The raising or lowering of $[Cl^-]_i$ is an important mechanism by which intracellular volume is maintained in epithelial cells in response to changes in extracellular tonicity (Lang *et al.*, 1998). Functional analysis of KCC1 in different heterologous expression systems demonstrated a 2- to 20-fold increase of transport rate by hypotonic challenge (Gillen and Forbush, 1999; Mercado *et al.*, 2000). Hypotonicity-induced activation of KCC1 could be prevented by the protein phosphatase inhibitor calyculin A. WNK3 strongly inhibits Cl^- exit through KCC1 in *X. laevis* oocytes probably by regulating downstream kinases/phosphatases (Kahle *et al.*, 2005).

Truncation of eight C-terminal amino acids of the murine KCC1 protein abolished function despite expression but lacked a dominant negative phenotype (Casula *et al.*, 2001), whereas removal of 89 or 117 N-terminal amino acids abolished function and exhibited a dominant negative phenotype that required the presence of the C-terminal cytoplasmic domain. The dominant negative loss-of-function mutant could be coimmunoprecipitated with wild-type KCC1 polypeptide suggesting that KCC1 forms homodimers as has been shown for NKCCs. It also exhibited dominant negative inhibition of human KCC1 and KCC3 and, with lower potency, mouse KCC4 and rat KCC2 (Casula *et al.*, 2001).

So far, no KCC1-knockout mouse model has been published and the localization of the protein remains elusive.

2. KCC2 (*SLC12A5*)

KCC2 is specifically expressed in neurons of the CNS including the retina (Payne *et al.*, 1996; Williams *et al.*, 1999). Of the four KCCs, KCC2 is unique in mediating constitutive K^+-Cl^- cotransport under isotonic conditions, whereas the other KCCs are swelling activated with no isotonic activity (Gamba, 2005a). Using a series of chimeric and mutant cDNAs, amino acid residues 1021–1035 of KCC2 were identified as essential for conferring isotonic transport activity (Mercado *et al.*, 2006). As K^+-Cl^- cotransport is influenced by tyrosine kinases, the role of a conserved consensus tyrosine phosphorylation site located in the C-terminal domain of KCC2 at position 1087 was addressed by substitution to aspartate to mimick phosphorylation (Strange et al., 2000). This resulted in a strongly diminished basal activity and blocked further activation by cell swelling. However, substitution by phenylalanine, alanine, or isoleucine did not change cotransport activity. Hence it was conferred that the C-terminus plays an essential role in maintaining the functional conformation of K^+-Cl^- cotransporters and may be involved in essential regulatory protein–protein interactions but is not a regulative phosphorylation site. Like Na^+-coupled cation-chloride cotransporters, activity of KCC2 is regulated by WNK-kinases with inhibiting effect of both WNK3 (Kahle et al., 2005) and WNK4 (Gagnon et al., 2006b).

Disruption of KCC2 in mice results in a spastic phenotype and perinatal death due to respiratory failure (Hübner et al., 2001b).

3. KCC3 (*SLC12A6*)

KCC3 is broadly expressed (Mount et al., 1999b). Two major isoforms have been described, KCC3a and KCC3b, which are generated by transcriptional initiation 5′ of two distinct first coding exons. KCC3a is 50 amino acids longer including several potential phosphorylation sites for protein kinase C, which are not present in KCC3b (Hiki et al., 1999). KCC3a expression is more widespread than that of KCC3b (Mount et al., 1999b). Northern blot analysis of mouse tissues indicates that KCC3b expression is particularly abundant in the kidney. This was confirmed by Western blotting of mouse tissue using an exon 3-specific antibody. Though KCC3 has been shown to be localized in white matter tracts of the spinal cord (Pearson et al., 2001), this could not be reproduced with a different antiserum (Boettger et al., 2003). Recently, loss-of-function mutations were shown to give rise to the Andermann syndrome (ACCPN) (Howard et al., 2002). This rare syndrome, which is particularly common in Canada due to a founder effect, is characterized by peripheral neuropathy and a variable agenesis of the corpus callosum, mental retardation, and a progressive course. KCC3-knockout mice not only display peripheral polyneuropathy (Howard et al., 2002; Boettger et al., 2003) but also a characteristic brain pathology, deafness, and hypertension (Boettger et al., 2003; Rust et al., 2006).

4. KCC4 (*SLC12A7*)

KCC4 is expressed in multiple tissues including the inner ear, the kidney, and several epithelia. Like KCC1 and KCC3, KCC4 did not mediate K^+-Cl^- cotransport under isotonic conditions when expressed in *X. laevis* oocytes, but showed 200-fold activation by cell swelling (Mount et al., 1999b; Mercado et al., 2000). Similar to other KCCs, activation of KCC4 during cell swelling is prevented by calyculin A, but not by okadaic acid and cypermethrin, suggesting that PP1 is involved in the activation of this transporter. Interestingly, the transport activity of KCC4 also depends on pH, with maximal activity at an extracellular pH < 7.0 (Bergeron et al., 2003). This is worth noting because KCC4 is expressed in acid-secreting α-intercalated cells of the collecting duct in the kidney (Boettger et al., 2002).

C. CCC8/CIP (*SLC12A9*) and CCC9 (*SLC12A8*)

Two orphan members of the cation-chloride cotransporter family denominated as CCC8 or CIP and CCC9 have been identified (Caron et al., 2000; Gamba, 2005a). Neither for CCC8/CIP nor for CCC9 function is clear and, though the putative topology of CCC8 is closely related to the other family members, transport activity has not been shown so far. Coexpression with CCC8/CIP consistently inhibited the activity of NKCC1, but not that of KCC1 or NKCC2 and supports the idea that CCC8/CIP physically interacts with NKCC1. CCC9 is unique among cation-chloride cotransporters because of a putative membrane topology with 11 transmembrane spans thus predicting an extracellular C-terminal domain (Gamba, 2005a).

III. CATION-CHLORIDE COTRANSPORTERS IN THE NERVOUS SYSTEM

As $GABA_A$ and glycine receptors are ligand-gated Cl^- channels, the action of GABA and glycine critically depends on the Cl^- gradients across the plasma membrane and hence on the intracellular Cl^- concentration ($[Cl^-]_i$) of the particular neuron. $[Cl^-]_i$ is the result of various Cl^--extruding and Cl^--accumulating processes. Changes in the expression of cation-chloride cotransporters thus have a significant role in the ontogeny of neuronal Cl^- homeostasis (Clayton et al., 1998). In most mature neurons $[Cl^-]_i$ is below its equilibrium (Fig. 4). Then opening of $GABA_A$ receptors will result in hyperpolarization due to the influx of Cl^-. In some neurons, $[Cl^-]_i$ is above its equilibrium (Smith et al., 1995), which is a general rule in the developing nervous system (Ben-Ari et al., 1989). In this situation, opening of $GABA_A$- or glycine receptors will entail the efflux of Cl^- and hence depolarization of neurons. When the glutamatergic system is not yet fully established early in development, glutamatergic synaptic transmission is purely NMDA-receptor based and lacks functional AMPA receptors. NMDA channels are blocked by Mg^{2+} and hence glutamatergic synapses are initially "silent" at the resting membrane potential. However, when GABA and glutamatergic synapses are coactivated during physiological patterns of activity, GABA-dependent depolarization can remove the Mg^{2+} block of NMDA receptors (Ben-Ari et al., 1997). It is speculated that this GABA-dependent excitation is important for the formation of the neuronal network through an increase of $[Ca^{2+}]_i$ by activation of voltage-dependent Ca^{2+} channels. Indeed, spontaneous neural activity is a basic property of the developing brain and probably regulates key developmental processes, including migration, neural differentiation, and formation and refinement of connections (Ben-Ari, 2002).

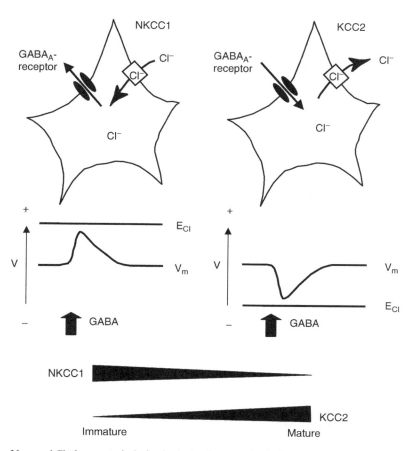

Figure 4. Neuronal Cl⁻ homeostasis during brain development. In the immature neuron the intracellular Cl⁻ concentration ($[Cl^-]_i$) is above the equilibrium, hence GABA results in Cl⁻ efflux and cell depolarization. This excitatory action is deemed to be important for the development of neuronal networks and depends in part due to the accumulation of Cl⁻ via NKCC1. With ongoing maturation, NKCC1 expression is down-regulated and that of KCC2 up-regulated. Finally, $[Cl^-]_i$ is below its equilibrium. Then opening of $GABA_A$ receptors results in the influx of Cl⁻ and the hyperpolarization of the neuron.

In early development the Na^+-coupled K^+-Cl^- cotransporter NKCC1 appears to be the primary transporter responsible for Cl⁻ accumulation in many neurons. This has been shown for pyramidal cells of the rat neocortex (Yamada *et al.*, 2004) and newly generated neurons in the adult brain (Ge *et al.*, 2006). Accumulation of Cl⁻ is maintained in adult vertebrate olfactory receptor neurons (Kaneko *et al.*, 2004). Whereas NKCC1 is broadly expressed in neurons of the developing mouse brain, its expression is shifted toward a glial pattern during postnatal maturation (Hübner *et al.*, 2001a). NKCC1 is very sensitive to changes in $[Cl^-]_i$ such that a fall in $[Cl^-]_i$ below a set point promotes direct phosphorylation and activation of the transporter, leading to restoration of $[Cl^-]_i$ (Lytle and Forbush, 1992, 1996). In some structures, NKCC1 expression is maintained into adulthood like in dorsal root ganglia, where this cotransporter participates in the modulation of GABAergic neurotransmission and sensory perception (Sung et al., 2000). NKCC1 expression is maintained as well in vertebrate olfactory receptor neurons (Kaneko et al., 2004). The odorant-triggered receptor current flows through two distinct ion channels on the sensory cilia: Ca^{2+} influx through a cyclic nucleotide-gated channel followed by Cl^- efflux through a Ca^{2+}-activated anion channel. The excitatory Cl^- efflux amplifies the response and crucially depends on a high intracellular Cl^- concentration established by NKCC1 (Bradley et al., 2005).

Recently it was suggested that NKCC1-mediated Cl^- accumulation and hence its "proexcitatory" effect may contribute to epileptic activity in newborns and it was suggested that inhibition of NKCC1 might be a therapeutic option in some subgroups of infantile seizures (Dzhala et al., 2005). However, this regimen might interfere with brain maturation, if the concept of the importance of the excitatory action of GABA for establishing neuronal networks holds true. Intriguingly, retroviral knockdown of NKCC1 in newly generated neurons resulted in the conversion of GABA-induced depolarization (excitation) into hyperpolarization (inhibition) and marked defects in their synapse formation and dendritic development (Ge et al., 2006). This study identified an essential role for GABA in the synaptic integration of newly generated neurons in the adult brain, and suggests an unexpected mechanism for activity-dependent regulation of adult neurogenesis, in which newborn neurons may sense neuronal network activity through tonic and phasic GABA activation. Though a spontaneous mouse mutant devoid of NKCC1 (Dixon et al., 1999) and several knockout mouse models for NKCC1 are around (Delpire et al., 1999; Flagella et al., 1999; Pace et al., 2000), brain morphology has not been studied in detail and subtle behavioral studies were precluded as knockout mice display a *shaker/waltzer* phenotype typical of vestibular dysfunction.

As KCC2 expression increases dramatically after the first week of postnatal life, when $[Cl^-]_i$ is shifted below its equilibrium, it was postulated that KCC2 could be a key player in this process (Clayton et al., 1998; Lu et al., 1999). Kinetic characterization indicates a much higher ion affinity than for other KCCs. If extracellular K^+ increases beyond 10–12 mM, as it happens during ongoing neuronal activity, the driving force for K^+-Cl^- cotransport will switch from efflux to influx (Payne, 1997). Expression analysis revealed that KCC2 is specifically expressed in neurons of the CNS but absent in glia cells or the peripheral nervous system (Payne et al., 1996). KCC2 expression follows neuronal differentiation and the emergence of low $[Cl^-]_i$ (Stein et al., 2004). The concurrent down-regulation of NKCC1 in many neurons (Hübner et al., 2001a) contributes to the functional GABA-switch from excitation to inhibition. In a cell culture system, the switch and the expression of KCC2 could be delayed by chronic blockade of $GABA_A$ receptors and was accelerated by increased $GABA_A$ receptor activation (Ganguly et al., 2001). However, blockade of $GABA_A$-mediated transmission with picrotoxin did not affect the expression levels of KCC2 protein in mouse hippocampal-dissociated cultures as well as in organotypic cultures

(Ludwig et al., 2003). Furthermore, chronic application of the action-potential blocker TTX or NMDA and non-NMDA antagonists APV and NBQX did not alter the developmental up-regulation of KCC2 expression. Expression of exogenous KCC2 in immature neurons effectively decreased neuronal $[Cl^-]_i$ and increased the formation of functional GABAergic synapses (Chudotvorova et al., 2005).

GABA and the $GABA_A$ receptor agonist muscimol induced brain-derived neurotrophic factor (BDNF) expression in embryonic hippocampal neurons (Berninger et al., 1995). Thus, BDNF may be involved in the trophic effect of GABA in the immature brain. Indeed, BDNF regulated spontaneous correlated activity at early developmental stages by increasing synaptogenesis and expression of KCC2 in vivo (Aguado et al., 2003). However, overexpression of BDNF did not alter the expression of GABA and glutamate ionotropic receptors.

Little is known about the transcriptional regulation of KCC2. Though the presence of a neuronal-restrictive silencing element (NRSE) 3' to exon 1 in the human and mouse KCC2 gene had been described (Karadsheh and Delpire, 2001), this motif is dispensable for neuronal expression, as KCC2 reporters with or without deletion of the NRSE in transgenic mice were expressed exclusively in neurons and in the CNS with a similar pattern and developmental up-regulation as endogenous KCC2 (Uvarov et al., 2005).

Immunolocalization studies using C-terminal antibodies against KCC2 confirmed a widespread neuronal localization of the protein (Williams et al., 1999; Vu et al., 2000). KCC2 resides in the plasma membrane of neuronal somata and dendrites but is mostly absent from axons (Szabadics et al., 2006). KCC2 partially colocalizes with inhibitory synapses as demonstrated by colocalization of KCC2 and gephyrin, a scaffolding protein of inhibitory synapses, and by electron microscopy, showing KCC2 flanking the postsynaptic density of inhibitory synapses (Hübner et al., 2001b). KCC2 was shown to be highly expressed in the vicinity of excitatory inputs in the hippocampus, perhaps in close association with extrasynaptic $GABA_A$ receptors. A high level of excitation is known to lead to a simultaneous net influx of Na^+ and Cl^-, as evidenced by dendritic swelling. Hence it was postulated that KCC2 in the same microenvironment may provide a Cl^- extrusion mechanism to deal with both ion and water homeostasis in addition to its role in setting the driving force of Cl^- currents involved in fast postsynaptic inhibition (Gulyas et al., 2001).

It was known that furosemide-sensitive Cl^- transport is important for GABAergic inhibition and due to differences in GABA effects on granule and CA3 neurons a Cl^--accumulating and a Cl^--extruding transport system was postulated (Misgeld et al., 1986). The distinct postsynaptic localization of KCC2 and its colocalization with $GABA_A$ receptors are consistent with its putative role to function as an active Cl^- extrusion pathway important in postsynaptic inhibition mediated by $GABA_A$ and glycine receptors. Whole cell recordings of spontaneous $GABA_A$ receptor-mediated postsynaptic currents in cultured neurons suggested that an outward Cl^- transport reduced dendritic $[Cl^-]_i$ if the somata of cells were loaded with Cl^- via the patch pipette (Jarolimek et al., 1999). $[Cl^-]_i$ and $[K^+]_o$ were tightly coupled by a furosemide-sensitive K^+-Cl^- cotransport. Indeed, the knockdown of KCC2 in vitro by an antisense approach shifted the reversal potential of $GABA_A$ responses in functionally mature hippocampal pyramidal neurons and supported the conclusion that KCC2 is the main Cl^- extruder to promote fast hyperpolarizing postsynaptic inhibition in the brain (Rivera et al., 1999). These results could be confirmed in vivo by its targeted disruption in mice. Complete absence of KCC2 resulted in perinatal lethality due to a severe spastic motor disorder and the inability to breathe due to absent rhythmogenesis in the respiratory centre in the brainstem (Hübner et al., 2001b). Analysis of spinal cord motoneurons at birth revealed a shift in the GABA reversal potential by +30 mV in $Kcc2^{-/-}$ mice (Hübner et al., 2001b). In another mouse model, ~5% residual KCC2 protein is produced (Woo et al., 2002). These mice survive for 2–3 weeks and finally die of an epileptic phenotype. Cortical neurons lacking KCC2 not only fail to show a developmental decrease in $[Cl^-]_i$, but are unable to regulate $[Cl^-]_i$ on Cl^- loading (Zhu et al., 2005). Mice heterozygous for a KCC2-null and a hypomorphic allele that retains 15–20% of normal KCC2 protein levels in the brain displayed increased anxiety-like behavior in several tests including elevated plus-maze and were more susceptible to pentylenetetrazole-induced seizures (Tornberg et al., 2005). Though the mice were impaired in water maze learning and showed reduced sensitivity to tactile and noxious thermal stimuli, they exhibited normal spontaneous locomotor activity and intact motor coordination.

Interestingly, the depolarizing action of GABA is recapitulated in traumatized brain tissue (Katchman et al., 1994; van den Pol et al., 1996; Inglefield and Schwartz-Bloom, 1998). A dramatic loss of KCC2 protein (60–80% of total) has been observed after either sustained interictal-like activity induced in hippocampal slices in the absence of Mg^{2+} (Rivera et al., 2004) or in an in vivo-kindling model of epilepsy (Rivera et al., 2002). Up-regulation of NKCC1 expression was observed in the cortex after focal ischemia (Yan et al., 2001). Furthermore, inhibition of NKCC1 with bumetanide significantly reduced the infarction volume in this ischemia model (Yan et al., 2003).

Modern pain-control theory predicts that a loss of inhibition in the dorsal horn of the spinal cord is a crucial substrate for chronic pain syndromes. Though no change in NKCC1 transcript or protein was observed after in vivo axonal injury (Nabekura et al., 2002), it involves a trans-synaptic reduction in the expression of KCC2, and the consequent disruption of anion homeostasis in neurons of lamina I of the superficial dorsal horn, one of the main spinal nociceptive output pathways. Local blockade or knockdown of the spinal KCC2 exporter in intact rats markedly reduced the nociceptive threshold, confirming that the reported disruption of anion homeostasis in lamina I neurons was sufficient to cause neuropathic pain (Coull et al., 2003).

The activity of cation-chloride cotransporters is known to be regulated by phosphorylation and/or dephosphorylation (Adragna et al., 2004). Although regulated in opposite directions, the Na^+-K^+-$2Cl^-$ cotransporter and the K^+-Cl^+ cotransporter seem to share common regulatory pathways, as factors activating one transporter usually inhibit the other and the other way round. The close interplay between KCC2 and NKCC1 in the regulation of neuronal $[Cl^-]_i$ was recently highlighted by the opposite effects of WNK3, a member of the WNK family of serine-threonine kinases, which colocalizes with NKCC1 and KCC2 in many neurons expressing $GABA_A$ receptors. By expression studies in *X. laevis* oocytes, kinase-active WNK3 increased Cl^- influx via NKCC1, and at the same time inhibited Cl^- exit through KCC1 and KCC2, whereas kinase-inactive WNK3 had the opposite effects (Kahle et al., 2005). WNK3's effects were imparted via altered phosphorylation and surface expression of its downstream targets. Together, these data indicate that WNK3 can modulate the level of intracellular Cl^- via opposing actions on entry and exit pathways. They suggest that WNK3 is part of the Cl^-/volume-sensing mechanism necessary for the maintenance of cell volume during osmotic stress and the dynamic modulation of GABA neurotransmission. The stress kinase SPAK may be downstream in the WNK4-signaling pathway (Gagnon et al., 2006a).

Another K^+-Cl^- cotransporter isoform that is extensively expressed in the brain is KCC3, which is encoded by *SLC12A6*. The chromosomal region 15q14 in humans encompassing *SLC12A6* has been linked to several neurological syndromes. Though no mutations in KCC3 have been found in patients with epilepsy syndromes linked to 15q14 (Steinlein et al., 2001), it has been subsequently shown that loss-of-function mutations of KCC3 result in the rare human disorder Andermann syndrome or ACCPN (Howard et al., 2002), which is characterized by a variable *a*genesis of the corpus *c*allosum and a *p*eripheral *n*europathy (Andermann et al., 1972). It is inherited as an autosomal recessive trait. Most patients are significantly mentally retarded, are prone to seizures, and show various dysmorphic features. Disruption of *Kcc3* in mice resulted in peripheral polyneuropathy and a severe locomotor deficit (Howard et al., 2002). In addition to peripheral polyneuropathy, $Kcc3^{-/-}$ mice displayed a widespread neurodegeneration of the CNS (Boettger et al., 2003). A peculiarity of the degeneration was the vacuolization and the presence of enlarged myelinated and unmyelinated axons. Synapses were found in degenerating hippocampal axons, suggesting that these neurons had established synaptic connections before degeneration. The related postsynaptic spines were normal in size and had well-developed postsynaptic densities. The observed morphological changes were not due to an impaired development since the degeneration occurred postnatal. The analysis of sciatic nerves and ventral and dorsal roots of the spinal cord indicated that axons of both motoneurons and sensory neurons degenerated. However, no alterations could be seen in the gray matter in semi-thin sections of the spinal cord. As in patients with Andermann syndrome, $Kcc3^{-/-}$ mice displayed spike–wave complexes in electrocorticograms. Furthermore, KCC3-deficient mice had a reduction of the seizure threshold.

Localization studies for KCC3 in the nervous system are controversial. KCC3 has been reported to be predominantly expressed in large fiber tracts. Immunofluorescence studies demonstrated strong signals in myelinated tracts of the spinal cord, consistent with individual myelin sheaths (Pearson et al., 2001). Brain sections not only showed white matter enhancement, but also cellular signals consistent with pyramidal neurons and Purkinje cells. With a different antibody, KCC3 could not be detected in myelin sheaths, but was mainly localized to neurons of the CNS (Boettger et al., 2003) and at low levels in astrocytes. As the gene targeting for the generation of $KCC3^{-/-}$ mice resulted in the fusion of a lacZ cassette to the Kcc3 open reading frame,

Kcc3-expression could be analyzed by lacZ-stainings of tissue sections from heterozygous mice. Using this approach, *Kcc3*-expression followed a neuronal expression pattern in the CNS (Boettger *et al.*, 2003).

The function of KCC3 in the nervous system and how this is related to neurodegeneration is not yet understood. Patch clamp analysis of P14 Purkinje cells, which strongly express KCC3, demonstrated that E_{GABA} is shifted by +10 mV compared to +30 mV in P0 motor neurons of KCC2-knockout mice, which may indicate that KCC3 plays a role in determining Purkinje-cell $[Cl^-]_i$ (Boettger *et al.*, 2003). On the contrary, KCC3 may be more important for cell volume regulation as cell swelling is known to activate KCC3 (Mount *et al.*, 1999b). Indeed, regulatory volume decrease (RVD) was almost absent in cultured primary hippocampal neurons from $Kcc3^{-/-}$ mice (Boettger *et al.*, 2003). Hence, swollen axons as a hallmark of neurodegeneration in $Kcc3^{-/-}$ mice may be related to a cell volume regulation defect.

Still open questions are whether KCCs are involved in the generation of the circadian rhythm in the suprachiasmatic nucleus of the hypothalamus or in the control of GnRH secretion. A diurnal oscillation of $[Cl^-]_i$ in SCN neurons has been shown (Wagner *et al.*, 1997). In a similar manner, steroid-mediated sexual differentiation of the mammalian brain also appears to be based on mechanisms involving excitatory versus inhibitory actions of GABA (Auger *et al.*, 2001). Although a switch to a hyperpolarizing response at puberty would require active Cl^- extrusion (Han *et al.*, 2002), most adult GnRH neurons appear to maintain a high $[Cl^-]_i$ and are excited by GABA (DeFazio *et al.*, 2002).

The role of KCC1 and KCC4 in the nervous system is still elusive. Though KCC4 mRNA was abundantly expressed in the ventricular zone where neurogenesis takes place, it is down-regulated perinatally (Li *et al.*, 2002). KCC4-knockout mice do not have an obvious brain phenotype (unpublished observations). KCC4 may be expressed in peripheral neurons, where its function remains to be determined (Karadsheh *et al.*, 2004). Though KCC1 is present in the CNS, its expression is only low, with increased expression in the olfactory bulb, hippocampus, cerebellum, and the choroid plexus (Kanaka *et al.*, 2001).

IV. CATION-CHLORIDE COTRANSPORTERS IN THE INNER EAR

Disruption of NKCC1, KCC3, and KCC4 in mice resulted in deafness (Delpire *et al.*, 1999; Pace *et al.*, 2001; Boettger *et al.*, 2002, 2003), highlighting the importance of cation-chloride cotransporters for inner ear ion homeostasis. Hearing critically depends on the specific ionic composition of the endolymph, the fluid surrounding the upper surface of hair cells. Endolymph has a high K^+ and low Na^+ concentration and is maintained at a high resting potential of around +100 mV. This ionic composition and voltage depend on active transport processes in the stria vascularis, a specialized epithelium at the lateral cochlear wall (Fig. 5). A large proportion of K^+ ions secreted by this epithelium is thought to be derived from ions entering hair cells from the endolymph and then transported back to the stria vascularis via a K^+-recycling

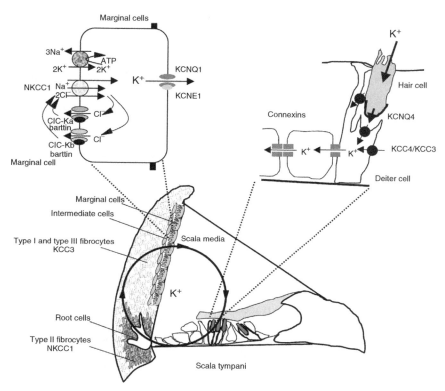

Figure 5. Ion transport pathways in the inner ear. The main diagram shows a cochlear duct, highlighting the proposed K^+ ion recycling pathways. Hearing depends on a high K^+ concentration (150 mM) in the endolymph which fills the scala media and bathes the apical membranes of sensory hair cells. During sound stimulation, K^+ enters the hair cells via apical mechanosensitive cation channels. K^+ exits from outer hair cells through KCNQ4 K^+ channels. K^+ is then taken up via the K^+-Cl^- cotransporter KCC4 by supporting Deiters' cells (shown in detail at top right). Deiters' cells are connected to root cells in the spiral ligament by an epithelial gap junction system, which provides a cytoplasmic route for K^+ diffusion. After exiting from root cells to the extracellular space, K^+ is taken up by type II fibrocytes, which accumulated K^+ through NKCC1 and the Na^+/K^+-ATPase. Type II cells express NKCC1, but not KCC3, and the reverse is true for type I and type III fibroblasts. This may create a K^+ gradient within the fibrocyte gap junction system. The exit into the space between intermediate and marginal cells of the stria occurs through Kir4.1 K^+ channels. These cells are coupled to type I and type III fibrocytes in a fibrocyte gap junction system that also includes the basal and intermediate cells of the stria vascularis. K^+ leaving intermediate cells is then taken up by marginal cells and is secreted through apical K^+ channels. K^+ passes fibrocytes of the lateral wall through gap junctions to the stria vascularis, where it is pumped into the endolymph. In strial marginal cells (magnified at top left), basolateral NKCC1 raises intracellular K^+ concentration. Parallel ClC-Ka/barttin and ClC-Kb/barttin channels recycle Cl^-. K^+ exits through apical channels formed by KCNQ1 and KCNE1 subunits. Loss of KCNQ1, KCNE1, NKCC1, or barttin causes deafness in humans or mice. Neither loss of ClC-Ka nor of ClC-Kb alone entails deafness.

pathway (Kikuchi *et al.*, 2000). The unique composition of the endolymph is needed because the apical mechanosensitive channels of sensory hair cells function as K^+ channels. The sound-induced relative movement of the basilar and tectorial membrane deflects the cilia of sensory hair cells and opens this K^+ channels. The resulting K^+ influx depolarizes the hair cell, leading to the exocytosis of synaptic vesicles. K^+ leaves outer hair cells through KCNQ4 K^+ channels at the basolateral side (Kharkovets *et al.*, 2006). After leaving outer hair cells, K^+ must be removed, partially by uptake into Deiters' cells. This uptake probably occurs through the K^+-Cl^- cotransporter KCC4, which is highly and specifically expressed in these supporting cells (Boettger *et al.*, 2002). In the mouse, this pattern of KCC4 expression in the inner ear is established during postnatal maturation, while at earlier stages KCC4 exhibits a broader distribution in the inner ear. Around postnatal day 8, before the onset of hearing, antibodies against KCC4 stained membranes in the stria vascularis and in almost all cells of the organ of Corti. This included the developing epithelial and supporting cells, as well as the membranes of outer hair cells.

In the inner ear K^+-recycling model, K^+ that has entered Deiters' cells, diffuses through a gap junction system connecting Deiters' cells to adjacent epithelial cells and back to the stria vascularis through a distinct fibrocyte gap junction system (Fig. 5). In this model, K^+ leaves Deiters' cells and enters root cells in the spiral ligament. After exiting from root cells to the extracellular space, K^+ is then taken up by type II fibrocytes (Kikuchi *et al.*, 2000). These cells express NKCC1, but not KCC3 and are coupled to type I and type III fibrocytes via another gap junction system that also includes the basal and intermediate cells of the stria vascularis. As type I and type III fibrocytes express KCC3 (Boettger *et al.*, 2003) but not NKCC1, this may create a K^+ gradient within this fibrocyte gap junction system. After having passed the fibrocyte gap junction system, K^+ leaves intermediate cells through K^+ channels. The exit step into the space between intermediate and marginal cells of the stria occurs through the K^+ channel Kir4.1 (KCNJ10) and generates the endocochlear potential (EP) (Ando and Takeuchi, 1999). It is then taken up by marginal cells via the Na^+-K^+-$2Cl^-$ cotransporter NKCC1 and the Na^+/K^+-ATPase and is secreted through apical KCNQ1/KCNE1 K^+ channels, as the potential of marginal cells is even higher than the EP (Offner *et al.*, 1987). The identification of mutations of many genes encoding components of the ion transport machinery in deaf humans (Hübner and Jentsch, 2002) support this model as well as several knockout mouse models. Mice either lacking KCNQ1 (Lee *et al.*, 2000) or NKCC1 (Delpire *et al.*, 1999; Pace *et al.*, 2001) display a collapse of the scala media due to defective endolymph production. However, so far no mutations have been identified in the human genes encoding NKCC1, KCC3, and KCC4 in patients suffering from deafness.

Intriguingly, the hearing loss develops much more rapidly with a loss of KCC4 than KCC3, although its expression is more restricted (Boettger *et al.*, 2003). Even more surprising, the supporting cells (in particular Deiters' cells), the only cells of the adult cochlea expressing KCC4, also express KCC3. Thus, a loss of KCC3 in Deiters' cells is better tolerated than a loss of KCC4. It is unclear whether this was due to a higher expression of KCC4 or different properties of these isoforms. Whereas the cellular distribution of KCC3 is compatible with K^+ recycling, the transporter is dispensable for strial K^+ secretion because young $Kcc3^{-/-}$ mice hear normally and the EP and the endolymph K^+ concentration ($[K^+]_{el}$) were unchanged even in old animals (Boettger et al., 2003). Hence, KCC3 is either not crucial for K^+ recycling, or the stria does not critically depend on recycling for its supply of K^+. The latter conclusion was supported by mice with a selective knockout of connexin 26 in the epithelial gap junction system (Cohen-Salmon et al., 2002). Similar to mice with a disruption of KCC4, $Kcc3^{-/-}$ mice had normal hearing thresholds at P14. Hence neither K^+-Cl^- cotransporter *per se* is essential for hearing. Deafness apparently resulted from a secondary degeneration of sensory hair cells, which express neither KCC3 nor KCC4. As postulated for $Kcc4^{-/-}$ mice and mice lacking connexin26, hair cells may degenerate because of a changed ionic environment. The loss of type I and type III fibrocytes in $Kcc3^{-/-}$ mice, in contrast, may occur in a cell autonomous manner, possibly related to a defect in volume regulation, as they frequently disappeared earlier than type II fibrocytes, which are devoid of KCC3 (Boettger et al., 2003).

The importance of cation-chloride cotransporters for inner ear ion homeostasis and endolymph secretion may explain the well-known ototoxic effect of loop diuretics (Ikeda et al., 1997). Inhibition of this transport system may relate to the ototoxic potential of these compounds due to the resulting changes in the ionic composition and fluid volume within the endolymph. Edema within the stria vascularis and a decrease in EP have been described with the administration of these drugs (Arnold et al., 1981; Rybak, 1993).

V. CATION-CHLORIDE COTRANSPORTERS IN THE KIDNEY

Primary urine is formed by glomerular filtration. Then ions, water, and organic substances are reabsorbed by tubular epithelial cells. Although the bulk of solutes and water is reabsorbed in the proximal tubule and the loop of Henle, the fine-tuning takes place in the distal nephron, where Na^+, K^+, and acid excretion occurs under the control of hormones such as aldosterone and vasopressin (antidiuretic hormone). For this purpose, the different nephron segments are equipped with various sets of channels and transporters, including cation-chloride cotransporters (Mount and Gamba, 2001). The diuretic-sensitive cotransport of Na^+ and/or K^+ with Cl^- (cation-chloride cotransport) can be separated into three transport subtypes with divergent but related properties: Na^+-K^+-$2Cl^-$ cotransport mediated by NKCCs, Na^+-Cl^- cotransport by NCC, and K^+-Cl^- cotransport by KCCs. Electroneutral Na^+-K^+-$2Cl^-$ cotransport by NKCC2 is sensitive to loop diuretics like furosemide or bumetanide. The thiazide diuretics specifically inhibit Cl^--independent Na^+-Cl^- cotransport by NCC. Finally, K^+-Cl^- cotransport mediated by KCCs is only weakly sensitive to loop diuretics. Soon after the identification of K^+-Cl^- cotransport in erythrocytes, this type of transport was also found at the basolateral side of the *Necturus* gallbladder epithelium (Reuss, 1983). This pathway has long been thought to provide an exit pathway for Cl^- in absorptive epithelia with a small or absent basolateral Cl^- conductance. Though other Cl^- exit mechanisms in the basolateral membrane of the proximal tubule are known (Seki et al., 1993), basolateral membrane vesicles from renal cortex exhibit K^+-Cl^- cotransport activity (Eveloff and Warnock, 1987). Furthermore, the apical Na^+-glucose transport in proximal tubule cells strongly activates a barium-resistant K^+ efflux that is inhibited by furosemide (Avison et al., 1988), and RVD in barium-blocked proximal tubules under hypotonic conditions is furosemide sensitive (Welling and Linshaw, 1988). Both KCC3 and KCC4 are abundantly expressed at the basolateral site of proximal tubule epithelial cells (Boettger et al., 2002, 2003). As proximal tubules face large osmotic challenges in vivo, it was suggested that K^+-Cl^- cotransport may be important for the regulation of cell volume of proximal tubule epithelial cells. Indeed, disrupting KCC3 severely impaired RVD of intact proximal tubules (Boettger et al., 2003). Disruption of KCC4 had a smaller effect (Boettger et al., 2003), although KCC4 was more drastically activated by cell swelling than KCC3 in heterologous expression systems (Race et al., 1999; Mercado et al., 2000). KCC3 and KCC4 may thus protect these cells from swelling on an increased urinary load of solutes that are reabsorbed in the proximal tubule. Neither in $Kcc3^{-/-}$ nor in $Kcc4^{-/-}$ mice histological changes of the proximal tubules were observed. No changes of electrolyte concentrations were found in urine samples from $Kcc3^{-/-}$ or $Kcc4^{-/-}$ mice, in line with a minor role of K^+-Cl^- cotransport in transepithelial transport of the proximal tubule. Further insights into the role of K^+-Cl^- cotransport in the proximal tubule may arise from $Kcc3^{-/-}/Kcc4^{-/-}$ double-knockout mice in the future.

Immunofluorescence revealed that KCC4 is expressed in α-intercalated cells in the collecting duct (Boettger et al., 2002), which are specialized for the secretion of protons into the urine. Deafness in $Kcc4^{-/-}$ mice was associated with renal tubular acidosis. The urine of knockout mice was more alkaline than in wild-type littermates, whereas concentrations of Na^+, K^+, and Cl^- were not changed. Consistent with a defect in urinary acidification, blood gas analysis indicated a compensated metabolic acidosis with significantly decreased base excess. The apical proton ATPase of α-intercalated cells secretes H^+ into the lumen of the distal nephron. At the basolateral membrane, acid equivalents are transported by the anion exchanger AE1 (Karet et al., 1998). As basolateral HCO_3^- efflux is coupled to Cl^- uptake, KCC4 may be required for basolateral Cl^- extrusion. Energy-dispersive X-ray microanalysis was used to compare $[Cl^-]_i$ in renal cells of wild-type and knockout mice. Indeed, $[Cl^-]_i$ was increased in proximal tubules and particularly in α-intercalated cells of $Kcc4^{-/-}$ mice. Considering the prominent Cl^-/HCO_3^- exchange activity in α-intercalated cells, the rise in $[Cl^-]_i$ predicts a more alkaline intracellular pH in the knockout mouse. This will decrease apical H^+ secretion by increasing the electrochemical gradient against which pumping has to occur. Thus, KCC4 joins the H^+-ATPase and the AE1 anion exchange as the third transport protein of α-intercalated cells whose mutation entails renal tubular acidosis. However, so far no mutations have been identified in humans with renal tubular acidosis and hearing loss. In the rabbit, KCC4 was strongly expressed at the basolateral site of the distal convoluted tubule and the connecting tubule (Velazquez and Silva, 2003). The same authors described at least two different variants of KCC4, a full-length form and a truncated form with a C-terminal domain that is 92 amino acids shorter, the first 92 amino acids being highly divergent from KCC4 from other species. The significance of these variants is still unknown.

Several studies suggest the possibility of an apical K^+-Cl^- cotransport in the distal convoluted tubule, as the reduction of luminal Cl^- markedly increased K^+ secretion (Velazquez et al., 1982; Ellison et al., 1986; Wingo, 1989). Whether KCC1 is the apical secretory K^+-Cl^- cotransporter that has been proposed in the rat distal convoluted tubule (Velazquez et al., 1992) remains to be determined. The role of KCC1 in the kidney is not yet clear though it has been described that KCC1 is expressed throughout the distal and proximal renal tubular epithelium as well as in glomerular mesangial cells and endothelial cells of the renal blood vessels (Liapis et al., 1998).

Different from KCC1, KCC3, and KCC4, NKCC2 is almost exclusively expressed in the kidney. NKCC2 was localized to the apical membrane and the subapical cytosol of TAL epithelial cells (Mount et al., 1999a). Evidence for a loop diuretic-sensitive Cl^- absorption in the mammalian TAL was first described in 1973 (Rocha and Kokko, 1973) and it took more than 10 years until the molecular identity of this transporter was unravelled (Gamba et al., 1993). As already mentioned, loss-of-function mutations in NKCC2 result in Bartter syndrome (Simon et al., 1996a), a genetically heterogeneous disease characterized by severe renal salt loss with volume depletion and hypotension. Given that arterial hypotension is a feature of Bartter syndrome, altered activity of NKCC2 may have a quantitative effect on blood pressure (Lifton et al., 2001). Targeted disruption of $Nkcc2$ in mice reproduced the human disorder: Homozygous $Nkcc2^{-/-}$ pups were born in expected numbers and appeared normal. However, by day 1 they showed signs of extracellular volume depletion and subsequently failed to thrive. By day 7, they were small and markedly dehydrated and exhibited renal insufficiency, high plasma K^+, metabolic acidosis, hydronephrosis of varying severity, and high plasma renin concentrations. None survived to weaning. Treatment of knockout pups with indomethacin from day 1 prevented growth retardation and 10% treated for 3 weeks survived, although as adults they exhibited severe polyuria, extreme hydronephrosis, low plasma K^+, high blood pH, hypercalciuria, and proteinuria (Takahashi et al., 2000). Concerted efforts of geneticists and physiologists have identified the other components of the salt transport machinery in the TAL (Fig. 5). Again their loss of function entails different types of Bartter syndrome, which can be either caused by the impairment of ROMK-mediated apical K^+ recycling (Simon et al., 1996b), by mutations of the basolateral Cl^- channel ClC-Kb (Simon et al., 1997) or its β-subunit barttin (Birkenhäger et al., 2001; Estévez et al., 2001). As Cl^- recycling at the basolateral membrane of the marginal cells of the stria vascularis is also mediated via ClC-Ka and ClC-Kb/barttin Cl^- channels (Fig. 4), Bartter syndrome caused by barttin mutations also includes deafness due to an endolymph secretion defect (Estévez et al., 2001).

Adjacent to the TAL is the distal convoluted tubule, where NCC mediates salt uptake. NCC expression is limited to the kidney. Loss of function mutations of NCC cause Gitelman syndrome, an autosomal recessive disorder which is characterized by volume depletion with a reduction of blood pressure, hypocalciuria, hypomagnesemia, and hypokalemic acidosis. The phenotype of $Ncc^{-/-}$ mice reflects the human disorder and revealed that the number of distal convoluted tubular cells is drastically reduced (Schultheis et al., 1998). PHAII or Gordon syndrome, which is associated with high blood K^+ and hypertension that probably results from increased renal Na^+ absorption, is caused by mutations in two genes encoding the protein kinases WNK1 and WNK4 (Wilson et al., 2001). Immunohistochemical analysis revealed that WNK1 and WNK4 are predominantly expressed in the distal convoluted tubule and collecting duct. Further studies have shown that WNK4 down-regulates the activity of ion transport pathways expressed in these nephron segments, such as the apical thiazide-sensitive Na^+-Cl^- cotransporter and apical secretory K^+ channel ROMK. At the same time, it up-regulates paracellular Cl^- transport and phosphorylation of tight junction proteins such as claudins (Gamba, 2005b). In addition, WNK4 down-regulates other Cl^- influx pathways such as the basolateral Na^+-K^+-$2Cl^-$ cotransporter NKCC1 and Cl^-/HCO_3^- exchanger. WNK4 coexpression inhibits the trafficking of NCC to the plasma membrane, whereas WNK1 prevents WNK4-mediated inhibition of NCC (Wilson et al., 2003; Yang et al., 2003) (Fig. 6). Mutations of WNK4 that are found in individuals with PHAII abolish the inhibition of NCC trafficking, implying a pathologically increased NaCl absorption in the distal convoluted tubule. Remarkably, wild-type WNK4 also reduced the plasma membrane expression of ROMK. In contrast to the mechanism proposed for NCC, however, this reduced expression of ROMK involved a stimulation of endocytosis (Kahle et al., 2003). Notably, the specific WNK4 mutations found in PHAII further stimulated retrieval of ROMK from the plasma membrane in a gain-of-function mechanism. The predicted decrease of ROMK in the apical membrane *in vivo* should lead to decreased renal K^+ secretion and hence to the increased serum K^+ as observed in PHAII. Coexpression studies in *X. laevis* oocytes revealed that kinase-active WNK3 is a potent activator of NKCC2-mediated transport (Rinehart et al., 2005). On the other hand, in its kinase-inactive state, WNK3 was a potent inhibitor of NKCC2 and NCC activity. WNK3 regulates the activity of NKCC2 and NCC by altering their expression at the plasma membrane. Wild-type WNK3 increased and kinase-inactive WNK3 decreased NKCC2 phosphorylation at T184 and T189, sites required for the vasopressin-mediated plasmalemmal translocation and activation of NKCC2 *in vivo*. Thus WNK3 may be involved in signaling downstream of vasopressin.

VI. CATION-CHLORIDE COTRANSPORTERS IN RBCs

As already mentioned, K^+-Cl^- cotransport was originally described in low-K^+ sheep RBCs as a swelling- and NEM-activated, Cl^--dependent K^+-efflux pathway (Dunham et al., 1980; Lauf and Valet, 1980) and the majority of functional characterization of K^+-Cl^- cotransport was performed in erythrocytes (Lauf et al., 1992; Lauf and Adragna, 2000).

The extrusion of K^+ and Cl^-, osmotic relevant molecules, generates an osmotic force across the cell membrane, which is followed by an obligatory exit of water

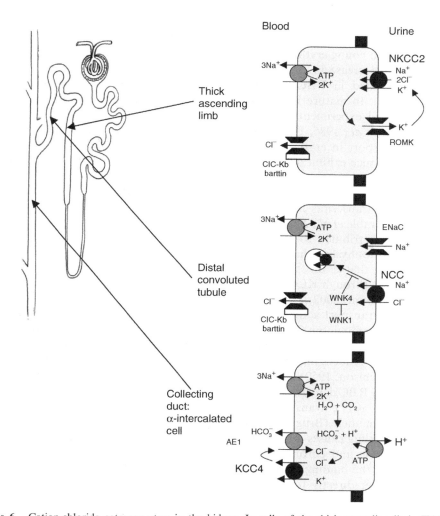

Figure 6. Cation-chloride cotransporters in the kidney. In cells of the thick ascending limb (TAL) of Henle's loop, apical NKCC2 cotransporters drive Cl^- uptake, K^+ being recycled by the K^+ channel ROMK. Cl^- exits through basolateral channels formed by ClC-Kb and barttin. Mutations in all four genes can cause Bartter syndrome. Salt uptake in the distal convoluted tubule is mediated by NCC. An increase of NaCl uptake through NCC contributes to pseudohypoaldosteronism type II (PHAII), which is caused by mutations of WNK1 or WNK4. Both kinases differentially affect the apical insertion of NCC Na^+-Cl^- cotransporters. In the cortical collecting duct, WNK4 stimulates the endocytosis of ROMK. PHAII-specific WNK4 mutations further increase this effect. The predicted decrease of ROMK in the apical membrane leads to decreased K^+ secretion. KCC4 is expressed in basolateral membranes of α-intercalated cells. The apical H^+-ATPase of α-intercalated cells secretes H^+ into the lumen of the distal nephron. At the basolateral membrane, acid equivalents are transported by the anion exchanger AE1. Basolateral HCO_3^- efflux is coupled to Cl^- uptake, which is recycled by KCC4. Disruption of KCC4 in mice decreases apical H^+ secretion by increasing the electrochemical gradient against which pumping has to occur. Thus, KCC4 joins the H^+-ATPase and the AE1 anion exchanger as the third transport protein of α-intercalated cells whose mutation entails renal tubular acidosis.

resulting in a reduction of cell volume. Thus swelling-activated K^+-Cl^- cotransport is suggested to play a major role in cellular volume regulation. As the maturation of reticulocytes and young erythrocytes is accompanied by a progressive reduction in cell volume and a decrease in K^+-Cl^- cotransport activity, K^+-Cl^- cotransporters are possibly implicated in RBC maturation (Hall and Ellory, 1986b; Brugnara and Tosteson, 1987). In mature RBCs, K^+-Cl^- cotransport is substantially reduced but can be activated experimentally by various stimuli like NEM or high hydrostatic pressure (Lauf et al., 1985; Hall and Ellory, 1986a). Evidence for the importance of K^+-Cl^- cotransport in erythroid RVD is illustrated by the BDX-31 mouse strain. RBCs of these mice exhibit an increased K^+-Cl^- cotransport activity and display an increased osmotic resistence (Armsby et al., 1995).

KCC1 and KCC3 are expressed in RBCs (Pellegrino et al., 1998; Lauf et al., 2001), whereas KCC4 transcripts have only been detected in erythroid precursor cells and reticulocytes (Crable et al., 2005). Although erythroid K^+-Cl^- cotransport activity has generally been attributed to KCC1 with some contribution of KCC3 (Lauf et al., 2001), various biophysical characteristics better fit to KCC3 than KCC1. For instance, the anion series, that is, $Br^- > Cl^- > I^- > SCN^- = NO_3^-$, coincidences with the order of K^+ transport through KCC3, as demonstrated in heterologous expression systems (Delpire and Mount, 2002).

K^+-Cl^- cotransport activity is increased in RBCs of patients with hemoglobinopathies such as hemoglobin C disease, β-thalassemia, or sickle cell disease (SCD). SCD is caused by a single amino acid exchange in the β-globin gene (Ingram, 1956, 1957), which drastically reduces the solubility of HbS in the deoxygenated state (Harris, 1950; Perutz and Mitchison, 1950). HbS polymerization and hence RBC sickling is thought to be favored by RBC dehydration and hence increase of the HbS concentration due to K^+ efflux via K^+-Cl^- cotransport (Lew and Bookchin, 2005) and the K^+ channel SK4 (IK1/Gardos channel) (Brugnara et al., 1986; Canessa et al., 1986; Ohnishi et al., 1986). RBC sickling causes anemia and vaso-occlusive events in the microcirculation, resulting in acute and chronic organ failure and episodic pain crises. It has been speculated that the binding of positively charged hemoglobin to the inner site of the membrane, either directly to the K^+-Cl^- cotransporter or to one of its regulators, triggers its activation (Brugnara et al., 1985, 1986). Several mouse models for SCD exist like the well-characterized SAD mouse (Trudel et al., 1991; Nagel and Fabry, 2001), which may shed some light onto the role K^+-Cl^- cotransport in this disease in the future. Meanwhile, different experimental approaches confirmed that both K^+-efflux mechanisms are involved in RBC dehydration and contribute to the pathomechanism of the disease. RBC K^+-Cl^- cotransport is modulated by intracellular Mg^{2+} content ($[Mg^{2+}]_i$), which is markedly depleted in RBCs of SCD patients (Olukoga et al., 1989) or SAD mice (De Franceschi et al., 1996). Oral administration of Mg^{2+} was shown to elevate erythrocyte $[Mg^{2+}]_i$, decrease RBC K^+-Cl^- cotransport activity, and prevent RBC dehydration in both, SAD mice and SCD patients (De Franceschi et al., 1996, 1997). Likewise, administration of clotrimazol or its derivate ICA-17043, which blocks SK4 (Ishii et al., 1997; Vandorpe et al., 1998), improves SCD pathology in SAD mice (De Franceschi et al., 1994; Stocker et al., 2003) and SCD patients

(Brugnara et al., 1996). However, the relative contribution of K^+-Cl^- cotransport and SK4 in human SCD is a matter of controversy. Thus K^+-Cl^- cotransport may be responsible for the formation of intermediate-density sickle cells, whereas the SK4 functions primarily in the generation of "hyperdense" cells (Schwartz et al., 1998). The characterization of SCD mouse models with additional targeted deletions of the genes encoding the K^+-Cl^- cotransporters KCC1 or/and KCC3, as well as the Ca^{2+}-activated K^+-channel SK4 promises to yield important tools for the study of K^+ transport processes in SCD.

Much less is known about the physiological relevance of Na^+-dependent cation-chloride cotransport in erythrocytes. Only NKCC1 is expressed in RBCs (Delpire et al., 1999; Matskevich et al., 2005) and is thought to counteract K^+-Cl^- cotransport. Threonine phosphorylation of NKCC1 in ferret erythrocytes correlated with cotransport rate (Matskevich et al., 2005), phosphorylation and transport activity increased 2.5- to 3-fold when cells were treated with the protein phosphatase inhibitor calyculin A. Both low $[Mg^{2+}]_i$ or treatment with kinase inhibitors like staurosporine drastically reduced transport activity. $Nkcc1^{-/-}$ mice did not show any obvious defect in RBC volume regulation. Osmotic resistance and the RBC density distribution profile were unaltered and the hematological parameters of $Nkcc1$-deficient mice were normal (Rust M. B. and Hübner C. A., unpublished data).

VII. CATION-CHLORIDE COTRANSPORTER AND CANCER

Cancer cells undergo rapid proliferation with an increased rate of mitosis and metabolism and therefore need elaborate regulators of cell volume regulation. Indeed, the magnitude of K^+-Cl^- cotransporter activation was significantly increased in several human cervical cancer cell lines and KCC3 mRNA was up-regulated (Shen et al., 2000). KCC3 transfection of NIH-3T3 cell line resulted in enhanced cell growth, which was prevented by DIOA, an inhibitor of K^+-Cl^- cotransport (Shen et al., 2001, 2003). Fluorescence-activated cell sorting measurements and Western blot analysis demonstrated that DIOA caused a significant reduction of the cell fraction in proliferative phase and a change in phosphorylation of retinoblastoma protein (Rb) and cdc2, a cyclin-dependent kinase important for the regulation of the cell cycle, suggesting that KCC3 activity may be important for cell cycle progression. Insulin-like growth factor (IGF)-1 up-regulated KCC3 expression and stimulated cell growth. Tumor necrotic factor-alpha down-regulated KCC3 expression and caused growth arrest. IGF-1 treatment of cervical and ovarian cancer cells triggered phosphatidylinositol 3-kinase and mitogen-activated protein kinase cascades leading to the activation of Akt and extracellular signal-regulated kinase1/2 (Erk1/2), respectively (Shen et al., 2004). Specific reduction of Erk1/2 protein levels with small interference RNA abolished IGF-1-stimulated K^+-Cl^- cotransport activity. Pharmacological inhibition and genetic modification of KCC activity demonstrated that K^+-Cl^- cotransport is necessary for IGF-1-induced cancer cell invasiveness and proliferation. IGF-1 and K^+-Cl^- cotransporters colocalized in the surgical specimens of cervical and ovarian cancer, suggesting autocrine or paracrine IGF-1 stimulation of K^+-Cl^- cotransport. Taken together, these results indicate that K^+-Cl^- cotransport activation by IGF-1 promotes growth and spread of gynecological cancers.

VIII. CATION-CHLORIDE COTRANSPORT AND HYPERTENSION

Blood pressure is determined by cardiac output, which is linked to extracellular volume, and peripheral resistance. Several members of the cation-chloride cotransporter family are either expressed in the kidney or in vascular smooth muscle cells and hence may play a role in the regulation of arterial blood pressure. As already mentioned, the increase of NaCl uptake via NCC in PHAII/Gordons disease results in an expansion of the extracellular volume and elevated blood pressure. On the contrary, salt loss due to loss of function of NCC as observed in Gitelman syndrome or NKCC2 in Bartter syndrome results in hypotension. Furthermore, NCC is the main target for thiazide-type diuretics, which are one component of the pharmacological regimen to treat hypertension. For NKCC2, again its loss of function results in salt loss and hypotension due to volume depletion.

NKCC1 shows a broader expression compared to NCC and NKCC2, including vascular smooth muscle cells. Aldosterone was shown to increase NKCC1 activity in vascular smooth muscle (Jiang *et al.*, 2003). The coupling of Cl^- to Na^+ movement predicts an accumulation of Cl^- in vascular smooth muscle cells. $[Cl^-]_i$ may affect the contractility of vascular smooth muscle cells of small arteries and arterioles. The tone of these "resistance vessels" determines the peripheral resistance to blood flow and thereby arterial blood pressure. Like all muscle cells, vascular smooth muscle cells use $[Ca^{2+}]_i$ as a trigger for contraction. $[Ca^{2+}]_i$ rises due to Ca^{2+} influx via voltage-dependent Ca^{2+} channels in the plasma membrane, or by Ca^{2+} release from intracellular stores. In addition to directly stimulating muscle contraction, $[Ca^{2+}]_i$ may also activate Ca^{2+}-activated Cl^- channels in the plasma membrane. Such channels have been identified functionally in vascular smooth muscle cells and their role in enhancing vasoconstriction was shown in different species (Large and Wang, 1996). The electrochemical Cl^- gradient in vascular smooth muscle cells is such that the opening of plasma membrane Cl^- channels will result in Cl^- efflux and thereby depolarize the cell. If $[Cl^-]_i$ is reduced, as predicted for vascular smooth muscle cells in $Nkcc1^{-/-}$ mice, opening of Ca^{2+}-activated Cl^- channels is expected to lead to a reduced depolarization and hence to a less pronounced Ca^{2+} entry via voltage-gated Ca^{2+} channels. As a consequence, in $Nkcc1^{-/-}$ mice vascular tone may be diminished and thus might cause arterial hypotension. Indeed, one study reported hypotension and a reduction of umbilical vein contractility in $Nkcc1^{-/-}$ mice (Meyer *et al.*, 2002). However, these findings could not be reproduced in another $Nkcc1^{-/-}$ line (Pace *et al.*, 2000). Other factors may modulate hypotension in $Nkcc1^{-/-}$ mice. As shown very recently, the action of vasopressin and the release of aldosterone were attenuated in kidneys from $Nkcc1^{-/-}$ mice (Wall *et al.*, 2005).

KCC1 and KCC3 are expressed in vascular smooth muscle cells as well (Di Fulvio et al., 2003; Rust et al., 2006). $Kcc3^{-/-}$ mice are hypertensive (Boettger et al., 2003) and the intracellular Cl^- concentration in vascular smooth muscle cells is increased (Rust et al., 2006). Isolated saphenous arteries and their third order branches, however, reacted indistinguishably to changes in intravascular pressure, stimulation of α_1-adrenoreceptors, exogenous nitric oxide, or blockade of Ca^{2+}-activated Cl^- channels. Likewise, the responses to α_1-adrenergic stimulation or exogenous nitric oxide in vivo were identical in both genotypes. These results argue against a major vascular-intrinsic component of arterial hypertension in $Kcc3^{-/-}$ mice. Pharmacological intervention with blockers of sympathetic innervation revealed a neurogenic contribution of arterial hypertension in $Kcc3^{-/-}$ mice, which was further supported by an increase of urinary catecholamine excretion. Thus the local control of myogenic tone does not require KCC3 and hypertension in $Kcc3^{-/-}$ mice is most likely a consequence of an elevated sympathetic tone (Rust et al., 2006).

IX. CONCLUDING REMARKS

Since the initial description of electroneutral cation-chloride cotransport and the identification of the first family members at the molecular level, nine members have been identified, two of them being orphans with yet unknown function. For the others, important functions emerged from mutations in humans (NCC, NKCC2, KCC3) or mice (NCC, NKCC2, NKCC1, KCC2, KCC3, KCC4). Whether these transporters are implicated in complex polygenic diseases as for example discussed for arterial hypertension remains to be addressed in the future. Another open question is the elucidation of the structure–function relation, which is still awaiting crystallization of the first member of this important family. Increasingly sophisticated mouse models, as well as a broad spectrum of physiological, biophysical, and cell biological techniques and genetic approaches will be needed to address further aspects.

REFERENCES

Adragna, N. C., Fulvio, M. D., and Lauf, P. K. (2004). Regulation of K-Cl cotransport: From function to genes. *J. Membr. Biol.* **201,** 109–137.

Aguado, F., Carmona, M. A., Pozas, E., Aguilo, A., Martinez-Guijarro, F. J., Alcantara, S., Borrell, V., Yuste, R., Ibanez, C. F., and Soriano, E. (2003). BDNF regulates spontaneous correlated activity at early developmental stages by increasing synaptogenesis and expression of the K^+/Cl^- co-transporter KCC2. *Development* **130,** 1267–1280.

Andermann, F., Andermann, E., Joubert, M., Karpati, G., Carpenter, S., and Melancon, D. (1972). Familial agenesis of the corpus callosum with anterior horn cell disease: A syndrome of mental retardation, areflexia and paraparesis. *Trans. Am. Neurol. Assoc.* **97,** 242–244.

Ando, M. and Takeuchi, S. (1999). Immunological identification of an inward rectifier K^+ channel (Kir4.1) in the intermediate cell (melanocyte) of the cochlear stria vascularis of gerbils and rats. *Cell Tissue Res.* **298,** 179–183.

Armsby, C. C., Brugnara, C., and Alper, S. L. (1995). Cation transport in mouse erythrocytes: Role of K^+-Cl^- cotransport in regulatory volume decrease. *Am. J. Physiol.* **268**, C894–C902.

Arnold, W., Nadol, J. B., Jr., and Weidauer, H. (1981). Ultrastructural histopathology in a case of human ototoxicity due to loop diuretics. *Acta Otolaryngol.* **91**, 399–414.

Attmane-Elakeb, A., Mount, D. B., Sibella, V., Vernimmen, C., Hebert, S. C., and Bichara, M. (1998). Stimulation by *in vivo* and *in vitro* metabolic acidosis of expression of rBSC-1, the Na^+-K^+ (NH_4^+)-$2Cl^-$ cotransporter of the rat medullary thick ascending limb. *J. Biol. Chem.* **273**, 33681–33691.

Attmane-Elakeb, A., Sibella, V., Vernimmen, C., Belenfant, X., Hebert, S. C., and Bichara, M. (2000). Regulation by glucocorticoids of expression and activity of rBSC1, the Na^+-K^+ (NH_4^+)-$2Cl^-$ cotransporter of medullary thick ascending limb. *J. Biol. Chem.* **275**, 33548–33553.

Auger, A. P., Perrot-Sinal, T. S., and McCarthy, M. M. (2001). Excitatory versus inhibitory GABA as a divergence point in steroid-mediated sexual differentiation of the brain. *Proc. Natl. Acad. Sci. USA* **98**, 8059–8064.

Avison, M. J., Gullans, S. R., Ogino, T., and Giebisch, G. (1988). Na^+ and K^+ fluxes stimulated by Na^+-coupled glucose transport: Evidence for a Ba^{2+}-insensitive K^+ efflux pathway in rabbit proximal tubules. *J. Membr. Biol.* **105**, 197–205.

Barry, E. L., Gesek, F. A., Kaplan, M. R., Hebert, S. C., and Friedman, P. A. (1997). Expression of the sodium-chloride cotransporter in osteoblast-like cells: Effect of thiazide diuretics. *Am. J. Physiol.* **272**, C109–C116.

Ben-Ari, Y. (2002). Excitatory actions of GABA during development: The nature of the nurture. *Nat. Rev. Neurosci.* **3**, 728–739.

Ben-Ari, Y., Cherubini, E., Corradetti, R., and Gaiarsa, J. L. (1989). Giant synaptic potentials in immature rat CA3 hippocampal neurones. *J. Physiol.* **416**, 303–325.

Ben-Ari, Y., Khazipov, R., Leinekugel, X., Caillard, O., and Gaiarsa, J. L. (1997). $GABA_A$, NMDA and AMPA receptors: A developmentally regulated 'menage a trois.'. *Trends Neurosci.* **20**, 523–529.

Bergeron, M. J., Gagnon, E., Wallendorff, B., Lapointe, J. Y., and Isenring, P. (2003). Ammonium transport and pH regulation by K^+-Cl^- cotransporters. *Am. J. Physiol. Renal Physiol.* **285**, F68–F78.

Berninger, B., Marty, S., Zafra, F., da Penha Berzaghi, M., Thoenen, H., and Lindholm, D. (1995). GABAergic stimulation switches from enhancing to repressing BDNF expression in rat hippocampal neurons during maturation *in vitro*. *Development* **121**, 2327–2335.

Birkenhäger, R., Otto, E., Schurmann, M. J., Vollmer, M., Ruf, E. M., Maier-Lutz, I., Beekmann, F., Fekete, A., Omran, H., Feldmann, D., Milford, D. V., Jeck, N., *et al.* (2001). Mutation of BSND causes Bartter syndrome with sensorineural deafness and kidney failure. *Nat. Genet.* **29**, 310–314.

Boettger, T., Hübner, C. A., Maier, H., Rust, M. B., Beck, F. X., and Jentsch, T. J. (2002). Deafness and renal tubular acidosis in mice lacking the K-Cl co-transporter Kcc4. *Nature* **416**, 874–878.

Boettger, T., Rust, M. B., Maier, H., Seidenbecher, T., Schweizer, M., Keating, D. J., Faulhaber, J., Ehmke, H., Pfeffer, C., Scheel, O., Lemcke, B., Horst, J., *et al.* (2003). Loss of K-Cl co-transporter KCC3 causes deafness, neurodegeneration and reduced seizure threshold. *EMBO J.* **22**, 5422–5434.

Bradley, J., Reisert, J., and Frings, S. (2005). Regulation of cyclic nucleotide-gated channels. *Curr. Opin. Neurobiol.* **15**, 343–349.

Brugnara, C. and Tosteson, D. C. (1987). Cell volume, K transport, and cell density in human erythrocytes. *Am. J. Physiol.* **252**, C269–C276.

Brugnara, C., Kopin, A. S., Bunn, H. F., and Tosteson, D. C. (1985). Regulation of cation content and cell volume in hemoglobin erythrocytes from patients with homozygous hemoglobin C disease. *J. Clin. Invest.* **75**, 1608–1617.

Brugnara, C., Bunn, H. F., and Tosteson, D. C. (1986). Regulation of erythrocyte cation and water content in sickle cell anemia. *Science* **232**, 388–390.

Brugnara, C., Gee, B., Armsby, C. C., Kurth, S., Sakamoto, M., Rifai, N., Alper, S. L., and Platt, O. S. (1996). Therapy with oral clotrimazole induces inhibition of the Gardos channel and reduction of erythrocyte dehydration in patients with sickle cell disease. *J. Clin. Invest.* **97**, 1227–1234.

Canessa, M., Spalvins, A., and Nagel, R. L. (1986). Volume-dependent and NEM-stimulated K^+, Cl^- transport is elevated in oxygenated SS, SC and CC human red cells. *FEBS Lett.* **200**, 197–202.

Caron, L., Rousseau, F., Gagnon, E., and Isenring, P. (2000). Cloning and functional characterization of a cation-Cl⁻ cotransporter-interacting protein. *J. Biol. Chem.* **275**, 32027–32036.

Casula, S., Shmukler, B. E., Wilhelm, S., Stuart-Tilley, A. K., Su, W., Chernova, M. N., Brugnara, C., and Alper, S. L. (2001). A dominant negative mutant of the KCC1 K-Cl cotransporter: Both N- and C-terminal cytoplasmic domains are required for K-Cl cotransport activity. *J. Biol. Chem.* **276**, 41870–41878.

Chudotvorova, I., Ivanov, A., Rama, S., Hübner, C. A., Pellegrino, C., Ben-Ari, Y., and Medina, I. (2005). Early expression of KCC2 in rat hippocampal cultures augments expression of functional GABA synapses. *J. Physiol.* **566**, 671–679.

Clayton, G. H., Owens, G. C., Wolff, J. S., and Smith, R. L. (1998). Ontogeny of cation-Cl⁻ cotransporter expression in rat neocortex. *Brain Res. Dev. Brain Res.* **109**, 281–292.

Cohen-Salmon, M., Ott, T., Michel, V., Hardelin, J. P., Perfettini, I., Eybalin, M., Wu, T., Marcus, D. C., Wangemann, P., Willecke, K., and Petit, C. (2002). Targeted ablation of connexin26 in the inner ear epithelial gap junction network causes hearing impairment and cell death. *Curr. Biol.* **12**, 1106–1111.

Coull, J. A., Boudreau, D., Bachand, K., Prescott, S. A., Nault, F., Sik, A., De Koninck, P., and De Koninck, Y. (2003). Trans-synaptic shift in anion gradient in spinal lamina I neurons as a mechanism of neuropathic pain. *Nature* **424**, 938–942.

Crable, S. C., Hammond, S. M., Papes, R., Rettig, R. K., Zhou, G. P., Gallagher, P. G., Joiner, C. H., and Anderson, K. P. (2005). Multiple isoforms of the KCl cotransporter are expressed in sickle and normal erythroid cells. *Exp. Hematol.* **33**, 624–631.

Darman, R. B., Flemmer, A., and Forbush, B. (2001). Modulation of ion transport by direct targeting of protein phosphatase type 1 to the Na-K-Cl cotransporter. *J. Biol. Chem.* **276**, 34359–34362.

De Franceschi, L., Saadane, N., Trudel, M., Alper, S. L., Brugnara, C., and Beuzard, Y. (1994). Treatment with oral clotrimazole blocks Ca^{2+}-activated K^+ transport and reverses erythrocyte dehydration in transgenic SAD mice. A model for therapy of sickle cell disease. *J. Clin. Invest.* **93**, 1670–1676.

De Franceschi, L., Beuzard, Y., Jouault, H., and Brugnara, C. (1996). Modulation of erythrocyte potassium chloride cotransport, potassium content, and density by dietary magnesium intake in transgenic SAD mouse. *Blood* **88**, 2738–2744.

De Franceschi, L., Bachir, D., Galacteros, F., Tchernia, G., Cynober, T., Alper, S., Platt, O., Beuzard, Y., and Brugnara, C. (1997). Oral magnesium supplements reduce erythrocyte dehydration in patients with sickle cell disease. *J. Clin. Invest.* **100**, 1847–1852.

de Jong, J. C., Willems, P. H., Mooren, F. J., van den Heuvel, L. P., Knoers, N. V., and Bindels, R. J. (2003). The structural unit of the thiazide-sensitive NaCl cotransporter is a homodimer. *J. Biol. Chem.* **278**, 24302–24307.

DeFazio, R. A., Heger, S., Ojeda, S. R., and Moenter, S. M. (2002). Activation of A-type gamma-aminobutyric acid receptors excites gonadotropin-releasing hormone neurons. *Mol. Endocrinol.* **16**, 2872–2891.

Delpire, E. and Mount, D. B. (2002). Human and murine phenotypes associated with defects in cation-chloride cotransport. *Annu. Rev. Physiol.* **64**, 803–843.

Delpire, E., Lu, J., England, R., Dull, C., and Thorne, T. (1999). Deafness and imbalance associated with inactivation of the secretory Na-K-2Cl co-transporter. *Nat. Genet.* **22**, 192–195.

Di Fulvio, M., Lauf, P. K., Shah, S., and Adragna, N. C. (2003). NONOates regulate KCl cotransporter-1 and -3 mRNA expression in vascular smooth muscle cells. *Am. J. Physiol. Heart Circ. Physiol.* **284**, H1686–H1692.

Dixon, M. J., Gazzard, J., Chaudhry, S. S., Sampson, N., Schulte, B. A., and Steel, K. P. (1999). Mutation of the Na-K-Cl co-transporter gene Slc12a2 results in deafness in mice. *Hum. Mol. Genet.* **8**, 1579–1584.

Dunham, P. B., Stewart, G. W., and Ellory, J. C. (1980). Chloride-activated passive potassium transport in human erythrocytes. *Proc. Natl. Acad. Sci. USA* **77**, 1711–1715.

Dzhala, V. I., Talos, D. M., Sdrulla, D. A., Brumback, A. C., Mathews, G. C., Benke, T. A., Delpire, E., Jensen, F. E., and Staley, K. J. (2005). NKCC1 transporter facilitates seizures in the developing brain. *Nat. Med.* **11**, 1205–1213.

Ellison, D. H., Velazquez, H., and Wright, F. S. (1986). Unidirectional potassium fluxes in renal distal tubule: Effects of chloride and barium. *Am. J. Physiol.* **250**, F885–F894.

Ellison, D. H., Velazquez, H., and Wright, F. S. (1987). Thiazide-sensitive sodium chloride cotransport in early distal tubule. *Am. J. Physiol.* **253**, F546–F554.

Estévez, R., Boettger, T., Stein, V., Birkenhäger, R., Otto, E., Hildebrandt, F., and Jentsch, T. J. (2001). Barttin is a Cl^- channel beta-subunit crucial for renal Cl^- reabsorption and inner ear K^+ secretion. *Nature* **414**, 558–561.

Evans, R. L., Park, K., Turner, R. J., Watson, G. E., Nguyen, H. V., Dennett, M. R., Hand, A. R., Flagella, M., Shull, G. E., and Melvin, J. E. (2000). Severe impairment of salivation in $Na^+/K^+/2Cl^-$ cotransporter (NKCC1)-deficient mice. *J. Biol. Chem.* **275**, 26720–26726.

Eveloff, J. and Warnock, D. G. (1987). K-Cl transport systems in rabbit renal basolateral membrane vesicles. *Am. J. Physiol.* **252**, F883–F889.

Flagella, M., Clarke, L. L., Miller, M. L., Erway, L. C., Giannella, R. A., Andringa, A., Gawenis, L. R., Kramer, J., Duffy, J. J., Doetschman, T., Lorenz, J. N., Yamoah, E. N., et al. (1999). Mice lacking the basolateral Na-K-2Cl cotransporter have impaired epithelial chloride secretion and are profoundly deaf. *J. Biol. Chem.* **274**, 26946–26955.

Gagnon, K. B., England, R., and Delpire, E. (2006a). Characterization of SPAK and OSR1, regulatory kinases of the Na-K-2Cl cotransporter. *Mol. Cell. Biol.* **26**, 689–698.

Gagnon, K. B., England, R., and Delpire, E. (2006b). Volume sensitivity of cation-Cl^- cotransporters is modulated by the interaction of two kinases: Ste20-related proline-alanine-rich kinase and WNK4. *Am. J. Physiol. Cell Physiol.* **290**, C134–C142.

Gamba, G. (2005a). Molecular physiology and pathophysiology of electroneutral cation-chloride cotransporters. *Physiol. Rev.* **85**, 423–493.

Gamba, G. (2005b). Role of WNK kinases in regulating tubular salt and potassium transport and in the development of hypertension. *Am. J. Physiol. Renal Physiol.* **288**, F245–F252.

Gamba, G., Saltzberg, S. N., Lombardi, M., Miyanoshita, A., Lytton, J., Hediger, M. A., Brenner, B. M., and Hebert, S. C. (1993). Primary structure and functional expression of a cDNA encoding the thiazide-sensitive, electroneutral sodium-chloride cotransporter. *Proc. Natl. Acad. Sci. USA* **90**, 2749–2753.

Gamba, G., Miyanoshita, A., Lombardi, M., Lytton, J., Lee, W. S., Hediger, M. A., and Hebert, S. C. (1994). Molecular cloning, primary structure, and characterization of two members of the mammalian electroneutral sodium-(potassium)-chloride cotransporter family expressed in kidney. *J. Biol. Chem.* **269**, 17713–17722.

Ganguly, K., Schinder, A. F., Wong, S. T., and Poo, M. (2001). GABA itself promotes the developmental switch of neuronal GABAergic responses from excitation to inhibition. *Cell* **105**, 521–532.

Ge, S., Goh, E. L. K., Sailor, K. A., Kitabatake, Y., Ming, G., and Song, H. (2006). GABA regulates synaptic integration of newly generated neurons in the adult brain. *Nature* **439**, 589–593.

Gerelsaikhan, T. and Turner, R. J. (2000). Transmembrane topology of the secretory Na^+-K^+-$2Cl^-$ cotransporter NKCC1 studied by *in vitro* translation. *J. Biol. Chem.* **275**, 40471–40477.

Gillen, C. M. and Forbush, B., III. (1999). Functional interaction of the K-Cl cotransporter (KCC1) with the Na-K-Cl cotransporter in HEK-293 cells. *Am. J. Physiol.* **276**, C328–C336.

Gillen, C. M., Brill, S., Payne, J. A., and Forbush, B., III. (1996). Molecular cloning and functional expression of the K-Cl cotransporter from rabbit, rat, and human. A new member of the cation-chloride cotransporter family. *J. Biol. Chem.* **271**, 16237–16244.

Gimenez, I., Isenring, P., and Forbush, B. (2002). Spatially distributed alternative splice variants of the renal Na-K-Cl cotransporter exhibit dramatically different affinities for the transported ions. *J. Biol. Chem.* **277**, 8767–8770.

Gulyas, A. I., Sik, A., Payne, J. A., Kaila, K., and Freund, T. F. (2001). The KCl cotransporter, KCC2, is highly expressed in the vicinity of excitatory synapses in the rat hippocampus. *Eur. J. Neurosci.* **13**, 2205–2217.

Hall, A. C. and Ellory, J. C. (1986a). Effects of high hydrostatic pressure on 'passive' monovalent cation transport in human red cells. *J. Membr. Biol.* **94**, 1–17.

Hall, A. C. and Ellory, J. C. (1986b). Evidence for the presence of volume-sensitive KCl transport in 'young' human red cells. *Biochim. Biophys. Acta* **858**, 317–320.

Han, S. K., Abraham, I. M., and Herbison, A. E. (2002). Effect of GABA on GnRH neurons switches from depolarization to hyperpolarization at puberty in the female mouse. *Endocrinology* **143**, 1459–1466.

Harris, J. W. (1950). Studies on the destruction of red blood cells. VIII. Molecular orientation in sickle cell hemoglobin solutions. *Proc. Soc. Exp. Biol. Med.* **75**, 197–201.

Hebert, S. C., Mount, D. B., and Gamba, G. (2004). Molecular physiology of cation-coupled Cl^- cotransport: The SLC12 family. *Pflugers Arch.* **447**, 580–593.

Hiki, K., D'Andrea, R. J., Furze, J., Crawford, J., Woollatt, E., Sutherland, G. R., Vadas, M. A., and Gamble, J. R. (1999). Cloning, characterization, and chromosomal location of a novel human K^+-Cl^- cotransporter. *J. Biol. Chem.* **274**, 10661–10667.

Howard, H. C., Mount, D. B., Rochefort, D., Byun, N., Dupre, N., Lu, J., Fan, X., Song, L., Riviere, J. B., Prevost, C., Horst, J., Simonati, A., *et al.* (2002). The K-Cl cotransporter KCC3 is mutant in a severe peripheral neuropathy associated with agenesis of the corpus callosum. *Nat. Genet.* **32**, 384–392.

Hübner, C. A. and Jentsch, T. J. (2002). Ion channel diseases. *Hum. Mol. Genet.* **11**, 2435–2445.

Hübner, C. A., Lorke, D. E., and Hermans-Borgmeyer, I. (2001a). Expression of the Na-K-2Cl-cotransporter NKCC1 during mouse development. *Mech. Dev.* **102**, 267–269.

Hübner, C. A., Stein, V., Hermans-Borgmeyer, I., Meyer, T., Ballanyi, K., and Jentsch, T. J. (2001b). Disruption of KCC2 reveals an essential role of K-Cl cotransport already in early synaptic inhibition. *Neuron* **30**, 515–524.

Ikeda, K., Oshima, T., Hidaka, H., and Takasaka, T. (1997). Molecular and clinical implications of loop diuretic ototoxicity. *Hear. Res.* **107**, 1–8.

Inglefield, J. R. and Schwartz-Bloom, R. D. (1998). Optical imaging of hippocampal neurons with a chloride-sensitive dye: Early effects of *in vitro* ischemia. *J. Neurochem.* **70**, 2500–2509.

Ingram, V. M. (1956). A specific chemical difference between the globins of normal human and sickle-cell anaemia haemoglobin. *Nature* **178**, 792–794.

Ingram, V. M. (1957). Gene mutations in human haemoglobin: The chemical difference between normal and sickle cell haemoglobin. *Nature* **180**, 326–328.

Isenring, P. and Forbush, B., III. (1997). Ion and bumetanide binding by the Na-K-Cl cotransporter. Importance of transmembrane domains. *J. Biol. Chem.* **272**, 24556–24562.

Isenring, P. and Forbush, B. (2001). Ion transport and ligand binding by the Na-K-Cl cotransporter, structure-function studies. *Comp. Biochem. Physiol. A Mol. Integr. Physiol.* **130**, 487–497.

Isenring, P., Jacoby, S. C., and Forbush, B., III. (1998). The role of transmembrane domain 2 in cation transport by the Na-K-Cl cotransporter. *Proc. Natl. Acad. Sci. USA* **95**, 7179–7184.

Ishii, T. M., Silvia, C., Hirschberg, B., Bond, C. T., Adelman, J. P., and Maylie, J. (1997). A human intermediate conductance calcium-activated potassium channel. *Proc. Natl. Acad. Sci. USA* **94**, 11651–11656.

Jarolimek, W., Lewen, A., and Misgeld, U. (1999). A furosemide-sensitive K^+-Cl^- cotransporter counteracts intracellular Cl^- accumulation and depletion in cultured rat midbrain neurons. *J. Neurosci.* **19**, 4695–4704.

Jiang, G., Cobbs, S., Klein, J. D., and O'Neill, W. C. (2003). Aldosterone regulates the Na-K-2Cl cotransporter in vascular smooth muscle. *Hypertension* **41**, 1131–1135.

Kahle, K. T., Wilson, F. H., Leng, Q., Lalioti, M. D., O'Connell, A. D., Dong, K., Rapson, A. K., MacGregor, G. G., Giebisch, G., Hebert, S. C., and Lifton, R. P. (2003). WNK4 regulates the balance between renal NaCl reabsorption and K^+ secretion. *Nat. Genet.* **35**, 372–376.

Kahle, K. T., Gimenez, I., Hassan, H., Wilson, F. H., Wong, R. D., Forbush, B., Aronson, P. S., and Lifton, R. P. (2004). WNK4 regulates apical and basolateral Cl^- flux in extrarenal epithelia. *Proc. Natl. Acad. Sci. USA* **101**, 2064–2069.

Kahle, K. T., Rinehart, J., de Los Heros, P., Louvi, A., Meade, P., Vazquez, N., Hebert, S. C., Gamba, G., Gimenez, I., and Lifton, R. P. (2005). WNK3 modulates transport of Cl^- in and out of cells: Implications for control of cell volume and neuronal excitability. *Proc. Natl. Acad. Sci. USA* **102**, 16783–16788.

Kanaka, C., Ohno, K., Okabe, A., Kuriyama, K., Itoh, T., Fukuda, A., and Sato, K. (2001). The differential expression patterns of messenger RNAs encoding K-Cl cotransporters (KCC1,2) and Na-K-2Cl cotransporter (NKCC1) in the rat nervous system. *Neuroscience* **104**, 933–946.

Kaneko, H., Putzier, I., Frings, S., Kaupp, U. B., and Gensch, T. (2004). Chloride accumulation in mammalian olfactory sensory neurons. *J. Neurosci.* **24**, 7931–7938.

Karadsheh, M. F. and Delpire, E. (2001). Neuronal restrictive silencing element is found in the KCC2 gene: Molecular basis for KCC2-specific expression in neurons. *J. Neurophysiol.* **85**, 995–997.

Karadsheh, M. F., Byun, N., Mount, D. B., and Delpire, E. (2004). Localization of the KCC4 potassium-chloride cotransporter in the nervous system. *Neuroscience* **123**, 381–391.

Karet, F. E., Gainza, F. J., Gyory, A. Z., Unwin, R. J., Wrong, O., Tanner, M. J., Nayir, A., Alpay, H., Santos, F., Hulton, S. A., Bakkaloglu, A., Ozen, S., *et al.* (1998). Mutations in the chloride-bicarbonate exchanger gene AE1 cause autosomal dominant but not autosomal recessive distal renal tubular acidosis. *Proc. Natl. Acad. Sci. USA* **95**, 6337–6342.

Katchman, A. N., Vicini, S., and Hershkowitz, N. (1994). Mechanism of early anoxia-induced suppression of the GABAA-mediated inhibitory postsynaptic current. *J. Neurophysiol.* **71**, 1128–1138.

Kharkovets, T., Dedek, K., Maier, H., Schweizer, M., Khimich, D., Nouvian, R., Vardanyan, V., Leuwer, R., Moser, T., and Jentsch, T. J. (2006). Mice with altered KCNQ4 K^+ channels implicate sensory outer hair cells in human progressive deafness. *EMBO J.* **25**, 642–652.

Kikuchi, T., Adams, J. C., Miyabe, Y., So, E., and Kobayashi, T. (2000). Potassium ion recycling pathway via gap junction systems in the mammalian cochlea and its interruption in hereditary nonsyndromic deafness. *Med. Electron Microsc.* **33**, 51–56.

Kim, G. H., Ecelbarger, C. A., Mitchell, C., Packer, R. K., Wade, J. B., and Knepper, M. A. (1999). Vasopressin increases Na-K-2Cl cotransporter expression in thick ascending limb of Henle's loop. *Am. J. Physiol.* **276**, F96–F103.

Lang, F., Busch, G. L., Ritter, M., Volkl, H., Waldegger, S., Gulbins, E., and Haussinger, D. (1998). Functional significance of cell volume regulatory mechanisms. *Physiol. Rev.* **78**, 247–306.

Large, W. A. and Wang, Q. (1996). Characteristics and physiological role of the Ca^{2+}-activated Cl^- conductance in smooth muscle. *Am. J. Physiol.* **271**, C435–C454.

Lauf, P. K. and Adragna, N. C. (2000). K-Cl cotransport: Properties and molecular mechanism. *Cell. Physiol. Biochem.* **10**, 341–354.

Lauf, P. K. and Theg, B. E. (1980). A chloride dependent K^+ flux induced by N-ethylmaleimide in genetically low K^+ sheep and goat erythrocytes. *Biochem. Biophys. Res. Commun.* **92**, 1422–1428.

Lauf, P. K. and Valet, G. (1980). Cation transport in different volume populations of genetically low K^+ lamb red cells. *J. Cell. Physiol.* **104**, 283–293.

Lauf, P. K., Perkins, C. M., and Adragna, N. C. (1985). Cell volume and metabolic dependence of NEM-activated K^+-Cl^- flux in human red blood cells. *Am. J. Physiol.* **249**, C124–C128.

Lauf, P. K., Bauer, J., Adragna, N. C., Fujise, H., Zade-Oppen, A. M., Ryu, K. H., and Delpire, E. (1992). Erythrocyte K-Cl cotransport: Properties and regulation. *Am. J. Physiol.* **263**, C917–C932.

Lauf, P. K., Zhang, J., Delpire, E., Fyffe, R. E., Mount, D. B., and Adragna, N. C. (2001). K-Cl co-transport: Immunocytochemical and functional evidence for more than one KCC isoform in high K and low K sheep erythrocytes. *Comp. Biochem. Physiol. A Mol. Integr. Physiol.* **130**, 499–509.

Lee, M. P., Ravenel, J. D., Hu, R. J., Lustig, L. R., Tomaselli, G., Berger, R. D., Brandenburg, S. A., Litzi, T. J., Bunton, T. E., Limb, C., Francis, H., Gorelikow, M., *et al.* (2000). Targeted disruption of the Kvlqt1 gene causes deafness and gastric hyperplasia in mice. *J. Clin. Invest.* **106**, 1447–1455.

Lew, V. L. and Bookchin, R. M. (2005). Ion transport pathology in the mechanism of sickle cell dehydration. *Physiol. Rev.* **85**, 179–200.

Li, H., Tornberg, J., Kaila, K., Airaksinen, M. S., and Rivera, C. (2002). Patterns of cation-chloride cotransporter expression during embryonic rodent CNS development. *Eur. J. Neurosci.* **16**, 2358–2370.

Liapis, H., Nag, M., and Kaji, D. M. (1998). K-Cl cotransporter expression in the human kidney. *Am. J. Physiol.* **275**, C1432–C1437.

Lifton, R. P., Gharavi, A. G., and Geller, D. S. (2001). Molecular mechanisms of human hypertension. *Cell* **104**, 545–556.

Liu, X., Titz, S., Lewen, A., and Misgeld, U. (2003). KCC2 mediates NH_4^+ uptake in cultured rat brain neurons. *J. Neurophysiol.* **90**, 2785–2790.

Lu, J., Karadsheh, M., and Delpire, E. (1999). Developmental regulation of the neuronal-specific isoform of K-Cl cotransporter KCC2 in postnatal rat brains. *J. Neurobiol.* **39**, 558–568.

Ludwig, A., Li, H., Saarma, M., Kaila, K., and Rivera, C. (2003). Developmental up-regulation of KCC2 in the absence of GABAergic and glutamatergic transmission. *Eur. J. Neurosci.* **18**, 3199–3206.

Lytle, C. (1997). Activation of the avian erythrocyte Na-K-Cl cotransport protein by cell shrinkage, cAMP, fluoride, and calyculin-A involves phosphorylation at common sites. *J. Biol. Chem.* **272**, 15069–15077.

Lytle, C. and Forbush, B., III. (1992). The Na-K-Cl cotransport protein of shark rectal gland. II. Regulation by direct phosphorylation. *J. Biol. Chem.* **267**, 25438–25443.

Lytle, C. and Forbush, B., III. (1996). Regulatory phosphorylation of the secretory Na-K-Cl cotransporter: Modulation by cytoplasmic Cl. *Am. J. Physiol.* **270**, C437–C448.

Lytle, C., McManus, T. J., and Haas, M. (1998). A model of Na-K-2Cl cotransport based on ordered ion binding and glide symmetry. *Am. J. Physiol.* **274**, C299–C309.

Matskevich, I., Hegney, K. L., and Flatman, P. W. (2005). Regulation of erythrocyte Na-K-2Cl cotransport by threonine phosphorylation. *Biochim. Biophys. Acta* **1714**, 25–34.

Mercado, A., Song, L., Vazquez, N., Mount, D. B., and Gamba, G. (2000). Functional comparison of the K^+-Cl^- cotransporters KCC1 and KCC4. *J. Biol. Chem.* **275**, 30326–30334.

Mercado, A., Broumand, V., Zandi-Nejad, K., Enck, A. H., and Mount, D. B. (2006). A carboxy-terminal domain in KCC2 confers constitutive K-Cl cotransport. *J. Biol. Chem.* **281**, 1016–1026.

Meyer, J. W., Flagella, M., Sutliff, R. L., Lorenz, J. N., Nieman, M. L., Weber, C. S., Paul, R. J., and Shull, G. E. (2002). Decreased blood pressure and vascular smooth muscle tone in mice lacking basolateral Na^+-K^+-$2Cl^-$ cotransporter. *Am. J. Physiol. Heart Circ. Physiol.* **283**, H1846–H1855.

Misgeld, U., Deisz, R. A., Dodt, H. U., and Lux, H. D. (1986). The role of chloride transport in postsynaptic inhibition of hippocampal neurons. *Science* **232**, 1413–1415.

Moore-Hoon, M. L. and Turner, R. J. (2000). The structural unit of the secretory Na^+-K^+-$2Cl^-$ cotransporter (NKCC1) is a homodimer. *Biochemistry* **39**, 3718–3724.

Moriguchi, T., Urushiyama, S., Hisamoto, N., Iemura, S., Uchida, S., Natsume, T., Matsumoto, K., and Shibuya, H. (2005). WNK1 regulates phosphorylation of cation-chloride-coupled cotransporters via the STE20-related kinases, SPAK and OSR1. *J. Biol. Chem.* **280**, 42685–42693.

Mount, D. B. and Gamba, G. (2001). Renal potassium-chloride cotransporters. *Curr. Opin. Nephrol. Hypertens.* **10**, 685–691.

Mount, D. B., Baekgaard, A., Hall, A. E., Plata, C., Xu, J., Beier, D. R., Gamba, G., and Hebert, S. C. (1999a). Isoforms of the Na-K-2Cl cotransporter in murine TAL I. Molecular characterization and intrarenal localization. *Am. J. Physiol.* **276**, F347–F358.

Mount, D. B., Mercado, A., Song, L., Xu, J., George, A. L., Jr., Delpire, E., and Gamba, G. (1999b). Cloning and characterization of KCC3 and KCC4, new members of the cation-chloride cotransporter gene family. *J. Biol. Chem.* **274**, 16355–16362.

Nabekura, J., Ueno, T., Okabe, A., Furuta, A., Iwaki, T., Shimizu-Okabe, C., Fukuda, A., and Akaike, N. (2002). Reduction of KCC2 expression and $GABA_A$ receptor-mediated excitation after *in vivo* axonal injury. *J. Neurosci.* **22**, 4412–4417.

Nagel, R. L. and Fabry, M. E. (2001). The panoply of animal models for sickle cell anaemia. *Br. J. Haematol.* **112**, 19–25.

Offner, F. F., Dallos, P., and Cheatham, M. A. (1987). Positive endocochlear potential: Mechanism of production by marginal cells of stria vascularis. *Hear. Res.* **29**, 117–124.

Ohnishi, S. T., Horiuchi, K. Y., and Horiuchi, K. (1986). The mechanism of *in vitro* formation of irreversibly sickled cells and modes of action of its inhibitors. *Biochim. Biophys. Acta* **886**, 119–129.

Olukoga, A. O., Erasmus, R. T., and Adewoye, H. O. (1989). Erythrocyte and plasma magnesium status in Nigerians with diabetes mellitus. *Ann. Clin. Biochem.* **26**(Pt. 1), 74–77.

Pace, A. J., Lee, E., Athirakul, K., Coffman, T. M., O'Brien, D. A., and Koller, B. H. (2000). Failure of spermatogenesis in mouse lines deficient in the Na^+-K^+-$2Cl^-$ cotransporter. *J. Clin. Invest.* **105**, 441–450.

Pace, A. J., Madden, V. J., Henson, O. W., Jr., Koller, B. H., and Henson, M. M. (2001). Ultrastructure of the inner ear of NKCC1-deficient mice. *Hear. Res.* **156**, 17–30.

Payne, J. A. (1997). Functional characterization of the neuronal-specific K-Cl cotransporter: Implications for $[K^+]_o$ regulation. *Am. J. Physiol.* **273**, C1516–C1525.

Payne, J. A. and Forbush, B., III. (1994). Alternatively spliced isoforms of the putative renal Na-K-Cl cotransporter are differentially distributed within the rabbit kidney. *Proc. Natl. Acad. Sci. USA* **91**, 4544–4548.

Payne, J. A., Stevenson, T. J., and Donaldson, L. F. (1996). Molecular characterization of a putative K-Cl cotransporter in rat brain. A neuronal-specific isoform. *J. Biol. Chem.* **271**, 16245–16252.

Pearson, M. M., Lu, J., Mount, D. B., and Delpire, E. (2001). Localization of the K^+-Cl^- cotransporter, KCC3, in the central and peripheral nervous systems: Expression in the choroid plexus, large neurons and white matter tracts. *Neuroscience* **103**, 481–491.

Pellegrino, C. M., Rybicki, A. C., Musto, S., Nagel, R. L., and Schwartz, R. S. (1998). Molecular identification and expression of erythroid K:Cl cotransporter in human and mouse erythroleukemic cells. *Blood Cells Mol. Dis.* **24**, 31–40.

Perutz, M. F. and Mitchison, J. M. (1950). State of haemoglobin in sickle-cell anaemia. *Nature* **166**, 677–679.

Piechotta, K., Lu, J., and Delpire, E. (2002). Cation chloride cotransporters interact with the stress-related kinases Ste20-related proline-alanine-rich kinase (SPAK) and oxidative stress response 1 (OSR1). *J. Biol. Chem.* **277**, 50812–50819.

Plotkin, M. D., Kaplan, M. R., Peterson, L. N., Gullans, S. R., Hebert, S. C., and Delpire, E. (1997). Expression of the Na^+-K^+-$2Cl^-$ cotransporter BSC2 in the nervous system. *Am. J. Physiol.* **272**, C173–C183.

Race, J. E., Makhlouf, F. N., Logue, P. J., Wilson, F. H., Dunham, P. B., and Holtzman, E. J. (1999). Molecular cloning and functional characterization of KCC3, a new K-Cl cotransporter. *Am. J. Physiol.* **277**, C1210–C1219.

Renfro, J. L. (1975). Water and ion transport by the urinary bladder of the teleost Pseudopleuronectes americanus. *Am. J. Physiol.* **228**, 52–61.

Renfro, J. L. (1977). Interdependence of active Na^+ and Cl^- transport by the isolated urinary bladder of the teleost, Pseudopleuronectes americanus. *J. Exp. Zool.* **199**, 383–390.

Reuss, L. (1983). Basolateral KCl co-transport in a NaCl-absorbing epithelium. *Nature* **305**, 723–726.

Rinehart, J., Kahle, K. T., de Los Heros, P., Vazquez, N., Meade, P., Wilson, F. H., Hebert, S. C., Gimenez, I., Gamba, G., and Lifton, R. P. (2005). WNK3 kinase is a positive regulator of NKCC2 and NCC, renal cation-Cl^- cotransporters required for normal blood pressure homeostasis. *Proc. Natl. Acad. Sci. USA* **102**, 16777–16782.

Rivera, C., Voipio, J., Payne, J. A., Ruusuvuori, E., Lahtinen, H., Lamsa, K., Pirvola, U., Saarma, M., and Kaila, K. (1999). The K^+/Cl^- co-transporter KCC2 renders GABA hyperpolarizing during neuronal maturation. *Nature* **397**, 251–255.

Rivera, C., Li, H., Thomas-Crusells, J., Lahtinen, H., Viitanen, T., Nanobashvili, A., Kokaia, Z., Airaksinen, M. S., Voipio, J., Kaila, K., and Saarma, M. (2002). BDNF-induced TrkB activation down-regulates the K^+-Cl^- cotransporter KCC2 and impairs neuronal Cl^- extrusion. *J. Cell Biol.* **159**, 747–752.

Rivera, C., Voipio, J., Thomas-Crusells, J., Li, H., Emri, Z., Sipila, S., Payne, J. A., Minichiello, L., Saarma, M., and Kaila, K. (2004). Mechanism of activity-dependent downregulation of the neuron-specific K-Cl cotransporter KCC2. *J. Neurosci.* **24**, 4683–4691.

Rocha, A. S. and Kokko, J. P. (1973). Sodium chloride and water transport in the medullary thick ascending limb of Henle. Evidence for active chloride transport. *J. Clin. Invest.* **52**, 612–623.

Rust, M. B., Faulhaber, J., Budack, M. K., Pfeffer, C., Maritzen, T., Didie, M., Beck, F. X., Boettger, T., Schubert, R., Ehmke, H., Jentsch, T. J., and Hübner, C. A. (2006). Neurogenic mechanisms contribute to hypertension in mice with disruption of the K-Cl cotransporter KCC3. *Circ. Res.* **98**, 549–556.

Rybak, L. P. (1993). Ototoxicity of loop diuretics. *Otolaryngol. Clin. North Am.* **26**, 829–844.

Schultheis, P. J., Lorenz, J. N., Meneton, P., Nieman, M. L., Riddle, T. M., Flagella, M., Duffy, J. J., Doetschman, T., Miller, M. L., and Shull, G. E. (1998). Phenotype resembling Gitelman's syndrome in mice lacking the apical Na^+-Cl^- cotransporter of the distal convoluted tubule. *J. Biol. Chem.* **273**, 29150–29155.

Schwartz, R. S., Musto, S., Fabry, M. E., and Nagel, R. L. (1998). Two distinct pathways mediate the formation of intermediate density cells and hyperdense cells from normal density sickle red blood cells. *Blood* **92**, 4844–4855.

Seki, G., Taniguchi, S., Uwatoko, S., Suzuki, K., and Kurokawa, K. (1993). Evidence for conductive Cl^- pathway in the basolateral membrane of rabbit renal proximal tubule S3 segment. *J. Clin. Invest.* **92**, 1229–1235.

Shen, M. R., Chou, C. Y., and Ellory, J. C. (2000). Volume-sensitive KCl cotransport associated with human cervical carcinogenesis. *Pflugers Arch.* **440**, 751–760.

Shen, M. R., Chou, C. Y., Hsu, K. F., Liu, H. S., Dunham, P. B., Holtzman, E. J., and Ellory, J. C. (2001). The KCl cotransporter isoform KCC3 can play an important role in cell growth regulation. *Proc. Natl. Acad. Sci. USA* **98**, 14714–14719.

Shen, M. R., Chou, C. Y., Hsu, K. F., Hsu, Y. M., Chiu, W. T., Tang, M. J., Alper, S. L., and Ellory, J. C. (2003). KCl cotransport is an important modulator of human cervical cancer growth and invasion. *J. Biol. Chem.* **278**, 39941–39950.

Shen, M. R., Lin, A. C., Hsu, Y. M., Chang, T. J., Tang, M. J., Alper, S. L., Ellory, J. C., and Chou, C. Y. (2004). Insulin-like growth factor 1 stimulates KCl cotransport, which is necessary for invasion and proliferation of cervical cancer and ovarian cancer cells. *J. Biol. Chem.* **279**, 40017–40025.

Simon, D. B., Karet, F. E., Hamdan, J. M., DiPietro, A., Sanjad, S. A., and Lifton, R. P. (1996a). Bartter's syndrome, hypokalaemic alkalosis with hypercalciuria, is caused by mutations in the Na-K-2Cl cotransporter NKCC2. *Nat. Genet.* **13**, 183–188.

Simon, D. B., Karet, F. E., Rodriguez-Soriano, J., Hamdan, J. H., DiPietro, A., Trachtman, H., Sanjad, S. A., and Lifton, R. P. (1996b). Genetic heterogeneity of Bartter's syndrome revealed by mutations in the K^+ channel, ROMK. *Nat. Genet.* **14**, 152–156.

Simon, D. B., Nelson-Williams, C., Bia, M. J., Ellison, D., Karet, F. E., Molina, A. M., Vaara, I., Iwata, F., Cushner, H. M., Koolen, M., Gainza, F. J., Gitleman, H. J., et al. (1996c). Gitelman's variant of Bartter's syndrome, inherited hypokalaemic alkalosis, is caused by mutations in the thiazide-sensitive Na-Cl cotransporter. *Nat. Genet.* **12**, 24–30.

Simon, D. B., Bindra, R. S., Mansfield, T. A., Nelson-Williams, C., Mendonca, E., Stone, R., Schurman, S., Nayir, A., Alpay, H., Bakkaloglu, A., Rodriguez-Soriano, J., Morales, J. M., et al. (1997). Mutations in the chloride channel gene, CLCNKB, cause Bartter's syndrome type III. *Nat. Genet.* **17**, 171–178.

Smith, R. L., Clayton, G. H., Wilcox, C. L., Escudero, K. W., and Staley, K. J. (1995). Differential expression of an inwardly rectifying chloride conductance in rat brain neurons: A potential mechanism for cell-specific modulation of postsynaptic inhibition. *J. Neurosci.* **15**, 4057–4067.

Song, L., Mercado, A., Vazquez, N., Xie, Q., Desai, R., George, A. L., Jr., Gamba, G., and Mount, D. B. (2002). Molecular, functional, and genomic characterization of human KCC2, the neuronal K-Cl cotransporter. *Brain Res. Mol. Brain Res.* **103**, 91–105.

Stein, V., Hermans-Borgmeyer, I., Jentsch, T. J., and Hübner, C. A. (2004). Expression of the KCl cotransporter KCC2 parallels neuronal maturation and the emergence of low intracellular chloride. *J. Comp. Neurol.* **468**, 57–64.

Steinlein, O. K., Neubauer, B. A., Sander, T., Song, L., Stoodt, J., and Mount, D. B. (2001). Mutation analysis of the potassium chloride cotransporter KCC3 (SLC12A6) in rolandic and idiopathic generalized epilepsy. *Epilepsy Res.* **44**, 191–195.

Stocker, J. W., De Franceschi, L., McNaughton-Smith, G. A., Corrocher, R., Beuzard, Y., and Brugnara, C. (2003). ICA-17043, a novel Gardos channel blocker, prevents sickled red blood cell dehydration *in vitro* and *in vivo* in SAD mice. *Blood* **101**, 2412–2418.

Stokes, J. B. (1984). Sodium chloride absorption by the urinary bladder of the winter flounder. A thiazide-sensitive, electrically neutral transport system. *J. Clin. Invest.* **74**, 7–16.

Strange, K., Singer, T. D., Morrison, R., and Delpire, E. (2000). Dependence of KCC2 K-Cl cotransporter activity on a conserved carboxy terminus tyrosine residue. *Am. J. Physiol. Cell Physiol.* **279**, C860–C867.

Sung, K. W., Kirby, M., McDonald, M. P., Lovinger, D. M., and Delpire, E. (2000). Abnormal GABAA receptor-mediated currents in dorsal root ganglion neurons isolated from Na-K-2Cl cotransporter null mice. *J. Neurosci.* **20,** 7531–7538.

Szabadics, J., Varga, C., Molnar, G., Olah, S., Barzo, P., and Tamas, G. (2006). Excitatory effect of GABAergic axo-axonic cells in cortical microcircuits. *Science* **311,** 233–235.

Takahashi, N., Chernavvsky, D. R., Gomez, R. A., Igarashi, P., Gitelman, H. J., and Smithies, O. (2000). Uncompensated polyuria in a mouse model of Bartter's syndrome. *Proc. Natl. Acad. Sci. USA* **97,** 5434–5439.

Tornberg, J., Voikar, V., Savilahti, H., Rauvala, H., and Airaksinen, M. S. (2005). Behavioural phenotypes of hypomorphic KCC2-deficient mice. *Eur. J. Neurosci.* **21,** 1327–1337.

Trudel, M., Saadane, N., Garel, M. C., Bardakdjian-Michau, J., Blouquit, Y., Guerquin-Kern, J. L., Rouyer-Fessard, P., Vidaud, D., Pachnis, A., Romeo, P. H., Beuzard, Y., and Costantini, F. (1991). Towards a transgenic mouse model of sickle cell disease: Hemoglobin SAD. *EMBO J.* **10,** 3157–3165.

Uvarov, P., Pruunsild, P., Timmusk, T., and Airaksinen, M. S. (2005). Neuronal K/Cl co-transporter (KCC2) transgenes lacking neurone restrictive silencer element recapitulate CNS neurone-specific expression and developmental up-regulation of endogenous KCC2 gene. *J. Neurochem.* **95,** 1144–1155.

van den Pol, A. N., Obrietan, K., and Chen, G. (1996). Excitatory actions of GABA after neuronal trauma. *J. Neurosci.* **16,** 4283–4292.

Vandorpe, D. H., Shmukler, B. E., Jiang, L., Lim, B., Maylie, J., Adelman, J. P., de Franceschi, L., Cappellini, M. D., Brugnara, C., and Alper, S. L. (1998). cDNA cloning and functional characterization of the mouse Ca^{2+}-gated K^+ channel, mIK1. Roles in regulatory volume decrease and erythroid differentiation. *J. Biol. Chem.* **273,** 21542–21553.

Velazquez, H. and Silva, T. (2003). Cloning and localization of KCC4 in rabbit kidney: Expression in distal convoluted tubule. *Am. J. Physiol. Renal Physiol.* **285,** F49–F58.

Velazquez, H., Wright, F. S., and Good, D. W. (1982). Luminal influences on potassium secretion: Chloride replacement with sulfate. *Am. J. Physiol.* **242,** F46–F55.

Velazquez, H., Ellison, D. H., and Wright, F. S. (1992). Luminal influences on potassium secretion: Chloride, sodium, and thiazide diuretics. *Am. J. Physiol.* **262,** F1076–F1082.

Vitari, A. C., Deak, M., Morrice, N. A., and Alessi, D. R. (2005). The WNK1 and WNK4 protein kinases that are mutated in Gordon's hypertension syndrome phosphorylate and activate SPAK and OSR1 protein kinases. *Biochem. J.* **391,** 17–24.

Vu, T. Q., Payne, J. A., and Copenhagen, D. R. (2000). Localization and developmental expression patterns of the neuronal K-Cl cotransporter (KCC2) in the rat retina. *J. Neurosci.* **20,** 1414–1423.

Wagner, S., Castel, M., Gainer, H., and Yarom, Y. (1997). GABA in the mammalian suprachiasmatic nucleus and its role in diurnal rhythmicity. *Nature* **387,** 598–603.

Wall, S. M., Knepper, M. A., Hassell, K. A., Fischer, M. P., Shodeinde, A., Shin, W., Pham, T. D., Meyer, J. W., Lorenz, J. N., Beierwaltes, W. H., Dietz, J. R., Shull, G. E., et al. (2005). Hypotension in NKCC1 null mice: Role of the kidneys. *Am. J. Physiol. Renal Physiol.* **290,** F409–F416.

Welling, P. A. and Linshaw, M. A. (1988). Importance of anion in hypotonic volume regulation of rabbit proximal straight tubule. *Am. J. Physiol.* **255,** F853–F860.

Williams, J. R. and Payne, J. A. (2004). Cation transport by the neuronal K^+-Cl^- cotransporter KCC2: Thermodynamics and kinetics of alternate transport modes. *Am. J. Physiol. Cell Physiol.* **287,** C919–C931.

Williams, J. R., Sharp, J. W., Kumari, V. G., Wilson, M., and Payne, J. A. (1999). The neuron-specific K-Cl cotransporter, KCC2. Antibody development and initial characterization of the protein. *J. Biol. Chem.* **274,** 12656–12664.

Wilson, F. H., Disse-Nicodeme, S., Choate, K. A., Ishikawa, K., Nelson-Williams, C., Desitter, I., Gunel, M., Milford, D. V., Lipkin, G. W., Achard, J. M., Feely, M. P., Dussol, B., et al. (2001). Human hypertension caused by mutations in WNK kinases. *Science* **293,** 1107–1112.

Wilson, F. H., Kahle, K. T., Sabath, E., Lalioti, M. D., Rapson, A. K., Hoover, R. S., Hebert, S. C., Gamba, G., and Lifton, R. P. (2003). Molecular pathogenesis of inherited hypertension with hyperkalemia: The Na-Cl cotransporter is inhibited by wild-type but not mutant WNK4. *Proc. Natl. Acad. Sci. USA* **100,** 680–684.

Wingo, C. S. (1989). Reversible chloride-dependent potassium flux across the rabbit cortical collecting tubule. *Am. J. Physiol.* **256,** F697–F704.

Woo, N. S., Lu, J., England, R., McClellan, R., Dufour, S., Mount, D. B., Deutch, A. Y., Lovinger, D. M., and Delpire, E. (2002). Hyperexcitability and epilepsy associated with disruption of the mouse neuronal-specific K-Cl cotransporter gene. *Hippocampus* **12,** 258–268.

Xu, J. C., Lytle, C., Zhu, T. T., Payne, J. A., Benz, E., Jr., and Forbush, B., III. (1994). Molecular cloning and functional expression of the bumetanide-sensitive Na-K-Cl cotransporter. *Proc. Natl. Acad. Sci. USA* **91,** 2201–2205.

Yamada, J., Okabe, A., Toyoda, H., Kilb, W., Luhmann, H. J., and Fukuda, A. (2004). Cl^- uptake promoting depolarizing GABA actions in immature rat neocortical neurones is mediated by NKCC1. *J. Physiol.* **557,** 829–841.

Yan, Y., Dempsey, R. J., and Sun, D. (2001). Na^+-K^+-Cl^- cotransporter in rat focal cerebral ischemia. *J. Cereb. Blood Flow Metab.* **21,** 711–721.

Yan, Y., Dempsey, R. J., Flemmer, A., Forbush, B., and Sun, D. (2003). Inhibition of Na^+-K^+-Cl^- cotransporter during focal cerebral ischemia decreases edema and neuronal damage. *Brain Res.* **961,** 22–31.

Yang, C. L., Angell, J., Mitchell, R., and Ellison, D. H. (2003). WNK kinases regulate thiazide-sensitive Na-Cl cotransport. *J. Clin. Invest.* **111,** 1039–1045.

Zhu, L., Lovinger, D., and Delpire, E. (2005). Cortical neurons lacking KCC2 expression show impaired regulation of intracellular chloride. *J. Neurophysiol.* **93,** 1557–1568.

Chapter 11
Plasma Membrane Cl^-/HCO_3^- Exchange Proteins

Haley J. Shandro[1] and Joseph R. Casey[1,2]

[1]*Membrane Protein Research Group, Department of Biochemistry,*
University of Alberta, Edmonton, Alberta, Canada T6G 2H7
[2]*Membrane Protein Research Group, Department of Physiology,*
University of Alberta, Edmonton, Alberta, Canada T6G 2H7

I. Introduction
II. The Cl^-/HCO_3^- Exchangers
    A. AE1
    B. AE2
    C. AE3
    D. NDCBE1
    E. SLC26A3
    F. SLC26A4
    G. SLC26A6
    H. SLC26A7
    I. SLC26A9
III. Disputed Cl^-/HCO_3^- Exchangers
IV. Structure
    A. Structure of SLC4 Cl^-/HCO_3^- Exchangers
    B. Structure of SLC26 Cl^-/HCO_3^- Exchangers
V. Regulation and Inhibition of Cl^-/HCO_3^- Exchange
    A. HCO_3^- Transport Metabolon
    B. Kinases
    C. PDZ Domain-Mediated Interactions
    D. STAS Domain
    E. Inhibitors of Cl^-/HCO_3^- Exchange
VI. Pathophysiology
    A. AE1
    B. AE2
    C. AE3
    D. SLC26A3
    E. SLC26A4
    F. SLC26A6
    References

I. INTRODUCTION

Metabolic activity continuously produces acid, a waste product that must be removed to allow respiratory oxidation to proceed. The body employs several mechanisms to remove acid equivalents: CO_2 is exhaled at the lungs, H^+ is excreted in the urine accompanied by reabsorption of HCO_3^-, and buffers throughout the body neutralize acid buildup. HCO_3^- transporters in the body are essential to maintain pH homeostasis as they move the base HCO_3^- across biological membranes. One major class of HCO_3^- transporters are Cl^-/HCO_3^- exchangers. Cl^-/HCO_3^- exchange in mammalian cells is mediated by members of two families of polytopic membrane proteins: the SLC4 family, and the SLC26 family (Table I). The Human Genome Organization (HUGO) has given solute carrier transport proteins the nomenclature "SLC" (http://www.hugo-international.org, 2004).

The SLC4 family includes both Na^+-dependent [Na^+/bicarbonate cotransporters (NBCs), and Na^+-dependent Cl^-/HCO_3^- exchanger (NDCBE1)] and Na^+-independent (AEs) members. Cl^-/HCO_3^- exchangers in the SLC4 family are called "AE" for *A*nion *E*xchanger. AE1–AE3 are electroneutral Na^+-independent Cl^-/base exchangers (Casey and Reithmeier, 1998; Alper et al., 2002). AE1, AE3, and NDCBE1 have restricted expression patterns in the body, while AE2 is more widely expressed. In epithelial tissues, AE proteins are found on the basolateral surface of cells.

Formerly known as sulfate transporters, there are 10 mammalian SLC26 transporters, with wide substrate selectivity and tissue expression. In this chapter, human orthologues will be denoted SLC26, while murine orthologues will be denoted Slc26.

This chapter will focus on the SLC4 and SLC26 family members that together comprise the set of Cl^-/HCO_3^- exchangers found in mammals (Fig. 1). Saier's group has analyzed the SLC4 and SLC26 transport families, and found no significant evolutionary relationship between them. This indicates that, although members of both families function as Cl^-/HCO_3^- exchangers, the families are distinct (Milton Saier, personal communication). The distinct relationship is underscored by the separate phylogenetic clustering of SLC4 and SLC26 transporters (Fig. 1).

Cl^-/HCO_3^- exchangers have a range of physiological roles. In general terms, they regulate cell pH by transmembrane movement of the base, HCO_3^- (Section II.B), regulate cell volume (Section II.B), and function to remove the metabolic waste product, CO_2 (Section II.A). Many other examples of specific physiological roles will be discussed for particular transporters. Several diseases are associated with mutations or defects in Cl^-/HCO_3^- exchange proteins. This chapter will provide an overview of molecular, structural, and physiological characteristics of mammalian Cl^-/HCO_3^- exchangers.

II. THE Cl^-/HCO_3^- EXCHANGERS

Table I provides an overview of the tissue localization, substrate specificity, and net charge movement for the Cl^-/HCO_3^- exchangers discussed in this chapter.

Table 1
Properties of Cl^-/HCO_3^- Exchangers

Transport protein	Other names	Tissue distribution	Mechanism	Net charge movement	References
AE1	SLC4A1, band 3	Erythrocyte, kidney, heart	Cl^-/HCO_3^- exchange	0	Kopito and Lodish (1985)
AE2	SLC4A2	Widespread	Cl^-/HCO_3^- exchange	0	Alper et al. (1988)
AE3	SLC4A3	Brain, heart, retina, gastrointestinal tract, kidney, smooth muscle	Cl^-/HCO_3^- exchange	0	Kudrycki et al. (1990)
NDCBE1	SLC4A8, NDAE1, NCBE, SLC4A10, (NBC3)[a]	Neurons, kidney, fibroblasts	Na^+-dependent Cl^-/HCO_3^- exchange	0	Romero et al. (2000), Wang et al. (2000), Grichtchenko et al. (2001)
SLC26A3	DRA, CLD	Colon, ileum, eccrine sweat gland, seminal vesicles, pancreas	Cl^-/HCO_3^- exchange, sulfate, OH^-	−1 or −2	Schweinfest et al. (1993), Melvin et al. (1999)
SLC26A4	Pendrin, PDS	Inner ear, thyroid, kidney	Cl^-/HCO_3^- exchange, also I^-, formate	0	Scott et al. (1999), Everett et al. (2001), Royaux et al. (2001), Petrovic et al. (2003c), Wall et al. (2003)
SLC26A6	PAT-1, CFEX	Kidney, heart, pancreas, liver, skeletal muscle, intestine, placenta	Cl^-/HCO_3^- exchange, also oxalate, formate, sulfate, OH^-	−1 (2 HCO_3^-: 1 Cl^-) or 0	Knauf et al. (2001), Waldegger et al. (2001), Wang et al. (2002), Xie et al. (2002)
SLC26A7	SLC26A7	Kidney, stomach	Cl^-/HCO_3^- exchange, also SO_4^{2-}, oxalate	Electrogenic Cl^-/HCO_3^- or Cl^- exchange	Lohi et al. (2002a), Vincourt et al. (2002), Petrovic et al. (2003a, 2004), Kim et al. (2005a); Morgan, Pastoreková, and Casey (unpublished observations)
SLC26A9	SLC26A9	Lung, stomach, trachea	Cl^-/HCO_3^- exchange, also SO_4^{2-}, oxalate	Unknown	Lohi et al. (2002a), Xu et al. (2005)

[a]Bracketed names are not generally accepted.

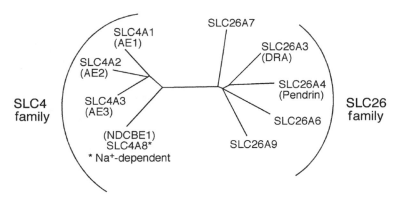

Figure 1. Phylogenetic tree of the SLC4 and SLC26 gene families. The tree was generated by analysis of the amino acid sequences of the human orthologues that exchange Cl^- for HCO_3^-, using the program Phylip from the Clustal W website. The degree of sequence similarity is proportional to the length of the line between proteins. The bracketed names are commonly used names. The solid curves enclosing the transporters indicate different groups: the SLC4 family, which includes the Na^+-dependent SLC4A8, and the SLC26 family.

A. AE1

1. Cloning

AE1 has been cloned from several species including human (Tanner et al., 1988; Lux et al., 1989), mouse (Kopito and Lodish, 1985), rat (Kudrycki and Shull, 1989), chicken (Cox and Lazarides, 1988), and trout (Hubner et al., 1992). Human erythrocyte AE1 (eAE1) has been purified and extensively characterized (Casey et al., 1989; Casey and Reithmeier, 1991a). eAE1 is a 911 amino acid polytopic glycoprotein. kAE1, transcribed from an alternate promoter, is the kidney isoform. kAE1 is a truncated form of eAE1, lacking the N-terminal 65 amino acids of eAE1 (Brosius et al., 1989).

2. Expression

AE1 makes up 50% of the integral membrane protein of the red blood cell (Steck et al., 1971). The original name for AE1 was band 3, reflecting the electrophoretic position of the protein in SDS-polyacrylamide gel electrophoresis of erythrocyte membranes. kAE1 is localized on the basolateral membrane of α-intercalated cells in the renal distal tubule (Verlander et al., 1988, 1991, 1994; Madsen et al., 1991; Tisher et al., 1991; Schuster, 1993). AE1 is targeted specifically to the basolateral surface of renal α-intercalated cells by information in the C-terminal tail. AE1n, a truncated version of eAE1, is observed in rat neonatal cardiomyocytes (Richards et al., 1999) and transcripts are seen in rat heart tissue (Alvarez et al., 2001b).

3. Transport Function

AE1 mediates the electroneutral exchange of Cl^- for HCO_3^- across the plasma membrane, with a turnover rate of about $5 \times 10^4 \, s^{-1}$ (Jennings, 1989). AE1 functions by a "ping-pong" mechanism where a single transport site alternately faces inward and outward, but can only reorient when carrying substrate (Gunn and Frohlich, 1979; Furuya et al., 1984). Electroneutral Cl^-/Cl^- exchange is blocked on mutation of E699 to Q in mouse AE1, although the mutant exchanger mediates Cl^-/SO_4^{2-} exchange (Chernova et al., 1997). The corresponding residue in human AE1, E681, is important for anion exchange activity (Jennings and Anderson, 1987; Jennings and Smith, 1992). AE1 is also capable of Cl^-/OH^- exchange when expressed in HEK 293 cells (Ko et al., 2002b). Residues involved in anion exchange activity, the translocation pore, and the anion selectivity filter have been identified (Wieth et al., 1982; Matsuyama et al., 1986; Passow, 1986; Hamasaki et al., 1990; Okubo et al., 1994; Müller-Berger et al., 1995; Tang et al., 1999; Zhu and Casey, 2004).

4. Physiological Role

eAE1 has a vital role in CO_2 metabolism and the HCO_3^- transport cycle in erythrocytes is well characterized (Fig. 2). Metabolically active tissues produce CO_2 (HCO_3^-) as the endpoint product of respiratory oxidation. Failure to remove CO_2 from the body results in metabolic acidosis. In peripheral capillaries, CO_2 diffuses into the erythrocyte where the cytosolic enzyme, carbonic anhydrase II (CAII), rapidly hydrates CO_2 to HCO_3^- plus H^+. Driven by the outward HCO_3^- gradient, the Cl^-/HCO_3^- exchanger, AE1, exports membrane-impermeant HCO_3^- into the plasma in exchange for Cl^-. The low CO_2 environment at the lungs drives the process in reverse. HCO_3^- moves into the erythrocyte in exchange for Cl^-, CAII catalyzes CO_2 formation, and CO_2 is exhaled. The erythrocyte HCO_3^- transport cycle is essential to whole body pH

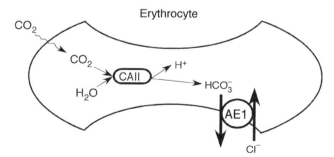

Figure 2. Role of AE1 in the erythrocyte. In peripheral capillaries, CO_2 diffuses across the erythrocyte plasma membrane. Cytosolic CAII catalyzes the hydration of CO_2 to $HCO_3^- + H^+$. HCO_3^- is moved out of the erythrocyte by AE1 in exchange for Cl^-. At the lungs, the process is reversed and CO_2 is exhaled. HCO_3^- is a more soluble form of CO_2 and this process increases the blood's carrying capacity for CO_2.

homeostasis since CO_2, if accumulated, will cause acidosis and limit sustained respiratory oxidation. HCO_3^- is a more soluble form of CO_2, and conversion of CO_2 to HCO_3^- increases the blood's carrying capacity for CO_2 (Sterling and Casey, 2002). The physiological significance of erythrocyte Cl^-/HCO_3^- exchange activity is that AE1 may be rate limiting to cardiovascular performance (Brahm, 1988).

The primary functions of the kidney are to maintain acid–base homeostasis and fluid and electrolyte balance. This is accomplished by NaCl absorption accompanied by secretion of acid or base equivalents throughout the nephron (Alpern et al., 1991). kAE1 has a role in urinary acidification and HCO_3^- reabsorption in the kidney. Most HCO_3^- is reabsorbed in the proximal tubule, in a mechanism requiring basolateral Na^+-coupled HCO_3^- transporters. In the final segments of the renal tubule, the distal tubule and collecting duct, Cl^-/HCO_3^- exchangers are involved in final control of HCO_3^- levels, involving about 5% of total filtered HCO_3^- (O'Callaghan and Brenner, 2000). Two types of intercalated cells have opposing function in either reabsorbing or secreting HCO_3^-, depending on whole body acid status. Type alpha (α-, A-type, acid secreting) intercalated cells reabsorb HCO_3^-; type beta (β-, B-type, base secreting) intercalated cells secrete HCO_3^-, although only under the rare circumstance of metabolic alkalosis (Fig. 3). HCO_3^- absorption occurs via basolateral Cl^-/HCO_3^- exchangers (kAE1) in the collecting duct.

A novel splicing variant of AE1, AE1n, is a truncated isoform of eAE1 observed in rat adult heart and neonatal cardiomyocytes (Richards et al., 1999; Alvarez et al., 2001b), but no unique role has been assigned to AE1n.

5. Regulation

Neither NH_4^+-induced acidification nor hypertonicity affects the transport activity of AE1 (Humphreys et al., 1994, 1997; Chernova et al., 2003). AE1 transport activity is relatively insensitive to changes in pH (Passow, 1986; Zhang et al., 1996; Sterling and Casey, 1999; Stewart et al., 2001, 2002).

B. AE2

1. Cloning

AE2 was cloned in 1988 from a mouse kidney library (Alper et al., 1988). Human AE2 was cloned nearly a decade later (Medina et al., 1997). Both mouse and human AE2 have three alternative promoters (a, b, and c) which result in multiple splicing variants, five in mouse (a, b1, b2, c1, and c2) and four in human (a, b1, b2, and c) (Wang et al., 1996).

2. Expression

AE2 is the most widely expressed SLC4 Cl^-/HCO_3^- exchanger. AE2 is expressed basolaterally in a variety of tissues; the highest expression of AE2 is in the stomach in

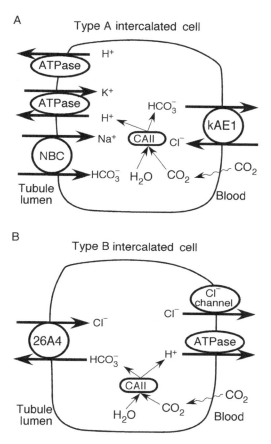

Figure 3. Role of Cl^-/HCO_3^- exchanger in the intercalated cells of the kidney. (A) Type A intercalated cells (α-intercalated cells) are acid secreting and express basolateral kAE1, which extrudes HCO_3^- into the blood in exchange for Cl^-. At the apical surface, a sodium bicarbonate cotransporter imports HCO_3^- and Na^+ from the distal tubular lumen. Cytosolic CAII converts CO_2 to HCO_3^- and H^+; apical ATPases secrete H^+ into the tubular lumen. (B) Type B intercalated cells (β-intercalated cells) are base secreting. CAII catalyzes the generation of HCO_3^- and H^+. Apical SLC26A4 (Pendrin) secretes HCO_3^- in exchange for Cl^-, while a basolateral H^+-ATPase reabsorbs H^+ to prevent cellular acidification. Cl^- channels remove Cl^- imported by SLC26A4 into the blood.

gastric parietal cells (Stuart-Tilley et al., 1994), in choroid plexus epithelial cells (Alper et al., 1994), in surface enterocytes in the colon (Alper et al., 1999), and in the kidney in the renal collecting duct (Alper et al., 1997; Stuart-Tilley et al., 1998) and the thick ascending limb of the loop of Henle (Alper et al., 1997; Frische et al., 2004). AE2 is expressed in osteoclasts, where it plays a role in bone remodeling (Alper, 1991). AE2 splice variants exhibit tissue specificity. AE2a is widely expressed, while AE2b1 and AE2b2 are expressed in epithelial cells (Wang et al., 1996; Medina et al., 1997). AE2a, AE2b1, and AE2b2 localize apically and subapically in hepatocytes (Tietz et al., 2003;

Aranda et al., 2004), although the mechanism for this abnormal localization is not known. AE2c is expressed primarily in stomach (Wang et al., 1996). AE2 mRNA is present in the heart (Kudrycki et al., 1990), and protein is expressed in rat neonatal cardiomyocytes (Richards et al., 1999). The epididymis and developing spermatozoa in the male reproductive system also exhibit AE2 protein expression (Holappa et al., 1999; Jensen et al., 1999).

3. Transport Function

AE2 is a Na^+-independent Cl^-/HCO_3^- exchanger (Humphreys et al., 1994; Rossmann et al., 2001). AE2 functions by an electroneutral anion exchange mechanism, with 1 Cl^-:1 HCO_3^- stoichiometry.

4. Physiological Role

AE2, the most broadly expressed Cl^-/HCO_3^- exchanger, has a wide range of roles. AE2 is the "housekeeping" Cl^-/HCO_3^- exchanger, responsible for regulation of steady state pH. By transporting HCO_3^- across the plasma membrane, Cl^-/HCO_3^- exchangers modify intracellular, and consequently, whole organism pH. HCO_3^- efflux causes a drop in intracellular pH (pHi), while HCO_3^- influx induces cellular alkalinization. In acid-secreting cells such as gastric parietal cells of the stomach, dangerous intracellular alkalinization would rapidly occur if not for Cl^-/HCO_3^- exchangers exporting the base HCO_3^- into the blood. AE2, in the gastric epithelium, plays a major role in gastric HCl secretion (Muallem et al., 1988; Machen et al., 1989; Paradiso et al., 1989; Thomas and Machen, 1991; Gawenis et al., 2004). In gastric parietal cells, AE2 exports HCO_3^- into the blood in exchange for Cl^-. The apical H^+/K^+-ATPase pumps H^+ into the stomach lumen and Cl^- channels export Cl^-, resulting in a net secretion of HCl (Fig. 4). AE2 not only facilitates gastric acid secretion, but also plays a role in regulation of pHi and cell volume in parietal cells. When acid is secreted across the apical membrane into the lumen, intracellular hydroxyl ions are generated. By extruding HCO_3^-, AE2 prevents intracellular alkalinization that occurs due to the export of H^+ by the H^+/K^+-ATPase. Import of Cl^- is followed by an influx of H_2O, and in this way AE2 contributes to control of cell volume. Although SLC26A7 is also expressed in gastric parietal cells (Petrovic et al., 2003a), major HCl secretion defects in AE2-null mice indicate that AE2, specifically isoform AE2c, is the primary Cl^-/HCO_3^- exchanger involved in gastric HCl secretion (Medina et al., 2003; Gawenis et al., 2004).

In the process of bone remodeling, osteoclasts digest old bone to allow for the synthesis of new bone. At the apical surface (toward the bone), osteoclasts express the vacuolar H^+-ATPase, which acidifies the periosteal lacunar space, digesting bone matrix (Blair et al., 1989). AE2 expressed in the contralacunar membrane (i.e., the basolateral surface, facing the blood) removes HCO_3^- in exchange for Cl^-. The role of AE2 in osteoclasts is thus essentially the same as its role in the gastric parietal cell (Alper, 1991).

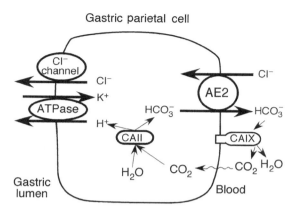

Figure 4. Hydrochloric acid secretion in the stomach. CO_2 that diffuses into the gastric parietal cell is hydrated to $H^+ + HCO_3^-$ by CAII. HCO_3^- is removed in exchange for Cl^- by AE2. H^+ is pumped into the stomach lumen by the H^+-ATPase, along with Cl^- via Cl^- channels, resulting in a net secretion of HCl into the stomach lumen. At the basolateral surface, CAIX converts HCO_3^-, exported by AE2, back to CO_2 to feed the cycle.

In the brain's choroid plexus, AE2 works in concert with Na^+/H^+ exchanger (NHE) to facilitate NaCl uptake and to regulate cerebrospinal fluid production (Mason et al., 1989). Regulatory volume increase (RVI) can be accomplished by concomitant activation of a Cl^-/HCO_3^- exchanger with an NHE, resulting in net NaCl loading, with associated osmotic water loading resulting in RVI. The process has been well defined in lymphocytes (Mason et al., 1989), where AE2 contributes Cl^-/HCO_3^- exchange function.

In the renal tubule, basolateral AE2 likely has only a housekeeping role in pH regulation and is not involved in renal tubular HCO_3^- reabsorption (Alper et al., 2002).

5. Regulation

pH regulates AE2 ion transport activity in both mammalian cells (Lee et al., 1991; Jiang et al., 1994) and *Xenopus* oocytes (Humphreys et al., 1995; Zhang et al., 1996; Stewart et al., 2001). AE2 is inhibited by acidic pH (Humphreys et al., 1994; Jiang et al., 1994; Zhang et al., 1996), and stimulated by an increase in pHi or extracellular pH (pHo) (Stewart et al., 2002). When studied in HEK 293 cells, AE2 was 65% active at pH 7.3, and activity dropped to 37% at pH 6.0, when pHi and pHo were changed together (Sterling and Casey, 1999). AE2 is also activated by hypertonicity (Humphreys et al., 1995; Chernova et al., 2003) and NH_4^+ (Humphreys et al., 1997). Activation by hypertonicity and NH_4^+-induced alkalinization is inhibited by the imidazole calmodulin antagonist calmidazolium (Chernova et al., 2003). Mutations in a highly conserved region (WRETARWIKFEE) in the cytoplasmic N-terminal domain of AE2 can modify the response of AE2 to pH, hypertonicity, and NH_4^+ (Romero et al., 1997; Stewart et al., 2002; Chernova et al., 2003). Regulation by pH and NH_4^+ is mediated through the cytoplasmic, or "modifier" domain (Humphreys et al., 1997) including the conserved cytoplasmic domain residues 336–347, 357–362, and 391–510 (Kopito et al., 1989; Zhang et al., 1996; Alper et al., 2001; Stewart et al., 2001, 2002, 2004).

C. AE3

1. Cloning

AE3 was first cloned in 1989 (Kopito et al., 1989). There are two AE3 isoforms, which result from alternative promoter usage: AE3 full-length (AE3fl) and AE3 cardiac (AE3c) (Kopito et al., 1989; Kudrycki et al., 1990; Linn et al., 1992). The N-terminal 270 residues of AE3fl are replaced with a unique region of 73 amino acids in AE3c (Yannoukakos et al., 1994).

2. Expression

AE3 is expressed mainly in excitable tissues, in particular heart (Linn et al., 1992; Yannoukakos et al., 1994), brain (Kopito et al., 1989; Kudrycki et al., 1990), retina (Kobayashi et al., 1994), and smooth muscle (Brosius et al., 1997). AE3c and AE3fl are expressed in the retina at similar levels but are localized to different retinal cell layers (Kobayashi et al., 1994). Both are expressed in the heart (Richards et al., 1999; Alvarez et al., 2004), in ventricles but not atria (Alvarez et al., 2004). AE3c is more highly expressed than AE3fl in the heart (Linn et al., 1995). The physiological reason for AE3 expression in excitable cells is not known.

3. Transport Function

AE3 is a Na^+-independent Cl^-/HCO_3^- exchanger (Lee et al., 1991) that does not mediate Cl^-/OH^- exchange (Alvarez et al., 2004). AE3 has one-third the Cl^-/HCO_3^- exchanger activity of AE1 and AE2 when expressed in HEK 293 cells due to a decrease in targeting to the plasma membrane (Lee et al., 1991; Fujinaga et al., 2003).

4. Physiological Role

AE3 is responsible for 50% of HCO_3^- extrusion in myocardial tissue (Cingolani et al., 2003) and therefore plays an important role in pHi regulation in the myocardium.

AE3 is expressed in retina (Kobayashi et al., 1994; Brosius et al., 1997; Richards et al., 1999). Interestingly, two splice forms (AE3c and AE3fl) are differentially expressed in retinal layers, and developmental stages (Kobayashi et al., 1994). This suggests that the alternate AE3 transcripts differ in their function. AE3 may regulate pH and HCO_3^- levels since retina is the most energy-utilizing tissue of the body.

5. Regulation

AE3 hyperactivity occurs in spontaneously hypertensive rats (SHRs), which has been linked to development of myocardial hypertrophy (Perez et al., 1995). AE3 hyperactivity can be normalized in SHRs by treatment with the angiotensin-converting enzyme inhibitor, enalapril (Ennis et al., 1998). These findings indicate a role for AE3 in the development of hypertrophy through a PKC-dependent pathway since angiotensin II (Ang II) works through PKC activation (Camilion de Hurtado et al., 2000). In HEK 293 cells, Cl^-/HCO_3^- exchange activity by AE3fl is activated by PKC and also by the PKC activators, Ang II and endothelin I (Alvarez et al., 2001a). In cultured rat neonatal cardiomyocytes, treatment with Ang II stimulates anion exchange via PKC phosphorylation of AE3fl at serine 67 (Alvarez et al., 2001a). These data suggest that Ang II signaling causes PKC to phosphorylate AE3, which increases Cl^-/HCO_3^- exchange and leads to hypertrophy. In contrast, AE3c anion exchange activity, which is about 50% of AE3fl (Sterling and Casey, 1999), is decreased in response to PKC stimulation (Alvarez et al., 2001a). Differences in the N-termini of AE3fl and AE3c therefore lead to differential regulation.

The pH regulation of AE3 remains controversial. Sterling and Casey (1999) determined that AE3 activity was not sensitive to changes in pH over the range of pH 6.0–9.0, although Alper's group found that AE3c is stimulated by an increase in pH (Stewart et al., 2001). Importantly, the two experiments differed in that the first paper concurrently altered pHi and pHo while the latter changed only cytosolic pH. If AE3 is active independently of pH, it may be able to function in response to an acid load to reestablish cellular pH. AE3 also contains two potential SH3-binding sites in the N-terminal domain, which may mediate regulatory protein–protein interactions.

D. NDCBE1

1. Cloning

A human Na^+-driven anion exchanger (NDCBE1) was cloned from brain (Romero et al., 2000). Na^+-driven Cl^-/HCO_3^- exchangers had previously been described in *Drosophila* (NDAE1) (Romero et al., 2000) and mouse (NCBE) (Wang et al., 2000), and also more recently in squid (sqNDCBE) (Virkki et al., 2003). Identification of this exchanger provided an identity to the protein responsible for Na^+-dependent Cl^-/HCO_3^- anion exchange (NDAE) activity that had been seen previously in neurons (Russell and Boron, 1976; Thomas, 1977; Schwiening and Boron, 1994) and kidney (Guggino et al., 1983; Ganz et al., 1989). NDCBE1 (SLC4A8) shares roughly 30% sequence similarity with the classical Cl^-/HCO_3^- exchangers, 47% to *Drosophila* NDAE1, and 71% to mouse NCBE. The mouse orthologue has a unique C-terminal region with 66 amino acids replacing the 17 C-terminal amino acids of NDCBE1 (Wang et al., 2001). This variation arises as a result of alternative splicing in the 3' end of NCBE (Wang et al., 2001).

2. Expression

NDAE1 is found primarily in the nervous system of *Drosophila*, and NCBE localizes to mouse brain with lower levels in ileum and testis (Romero et al., 2000; Wang et al., 2000). mRNA transcripts of NDCBE1 (~12 kb) are found mainly in brain and testis, with low levels also in kidney and ovary (Grichtchenko et al., 2001).

3. Transport Function

Functional studies performed on NDCBE1 expressed in *Xenopus* oocytes indicate that NDCBE1 reversibly exchanges two extracellular HCO_3^- and one extracellular Na^+ for one intracellular Cl^-, and is therefore an electroneutral exchanger (Grichtchenko et al., 2001). NDAE1 mediates exchange of one extracellular Na^+ and two HCO_3^- for one intracellular Cl^- and H^+ (Romero et al., 2000). NDAE1 can also transport OH^- in place of HCO_3^- (Romero et al., 2000). Therefore, NDCBE1 anion exchange activity is dependent on HCO_3^-, but NDAE1 activity is not (Romero et al., 2000; Grichtchenko et al., 2001). In addition to Cl^-/HCO_3^- exchange, NDCBE1 also mediates Cl^-/Cl^- exchange; $^{36}Cl^-$ efflux increases threefold in oocytes expressing NDCBE1 as compared to water-injected controls (Grichtchenko et al., 2001).

4. Physiological Role

Localization of NDCBE1 to brain tissues indicates a potentially vital role for the exchanger in the regulation of pHi in neurons. Regulation of pHi in neurons is an important process, since control of neuronal firing has been linked to changes in pHi (Bonnet and Wiemann, 1999; Bonnet et al., 2000; Meyer et al., 2000). Also, formation of neurites is a process that requires HCO_3^- (Kostenko et al., 1983). As NDCBE1 is further studied, a role in neuronal regulation may be revealed.

E. SLC26A3

1. Cloning

SLC26A3 was cloned from a colon subtraction library (Schweinfest et al., 1993) in a screen for genes down-regulated in adenoma, hence its original name, down-regulated in adenoma (DRA). It has also been termed CLD as the SLC26A3 gene is mutated in the autosomal recessive disorder congenital chloride diarrhea (CLD) (OMIM: #214700 and #126650) (Silberg et al., 1995; Byeon et al., 1996; Moseley et al., 1999). SLC26A3 shares 44% identity and 60% similarity to SLC26A4. It is related to the sulfate transporters DTSDT (60% similar, 30% identical) (Hastbacka et al., 1994), and SAT-1 (59% similar, 32% identical) (Bissig et al., 1994). SLC26A3 has 10–14 TM (Byeon et al., 1996) and contains a *s*ulfate *t*ransporters and bacterial *a*nti-sigma factor antagonists (STAS) domain in the C-terminal cytoplasmic region that is an absolute requirement for transport activity (Chernova *et al.*, 2003). There is a Type 1 PDZ domain and a putative site for tyrosine phosphorylation in the C-terminus (Chernova *et al.*, 2003). Y756F exhibits lowered activity, suggesting SLC26A3 regulation through tyrosine phosphorylation (Chernova *et al.*, 2003).

2. Expression

SLC26A3 is expressed along the gastrointestinal tract, primarily in the duodenum and colon, and also in the ileum and cecum (Silberg *et al.*, 1995; Byeon *et al.*, 1996; Melvin *et al.*, 1999; Lohi *et al.*, 2000; Jacob *et al.*, 2002). In these tissues, SLC26A3 is localized to the mucosal epithelium, specifically columnar epithelial cells in the brush border (Byeon *et al.*, 1996; Jacob *et al.*, 2002). In the colon, SLC26A3 is found at the apical membrane of surface and crypt cells (Byeon *et al.*, 1996; Mahajan *et al.*, 1996; Moseley *et al.*, 1999; Haila *et al.*, 2000; Rajendran *et al.*, 2000). SLC26A3 is also found in seminal vesicles, sweat glands (Haila *et al.*, 2000), and pancreas (Greeley *et al.*, 2001). Apical localization of SLC26A3 is seen in the duodenum and colon of rat, rabbit, and human (Silberg *et al.*, 1995; Jacob *et al.*, 2002).

3. Transport Function

SLC26A3 is a Cl^-/base exchanger (Melvin *et al.*, 1999; Alper *et al.*, 2001; Ko *et al.*, 2002b). In the ileum and colon, SLC26A3 functions as an apical Cl^-/HCO_3^- exchanger (Melvin *et al.*, 1999; Rajendran *et al.*, 2000). SLC26A3 mediates Cl^-/Cl^- and Cl^-/HCO_3^- exchange when heterologously expressed in *Xenopus* oocytes (Moseley *et al.*, 1999; Chernova *et al.*, 2003), and in cultured mammalian cells exhibits Cl^-/HCO_3^- exchange (Melvin *et al.*, 1999; Greeley *et al.*, 2001; Sterling *et al.*, 2002b). Cl^-/HCO_3^- exchange by SLC26A3 is electroneutral (Lamprecht *et al.*, 2005) and Na independent (Silberg *et al.*, 1995). SLC26A3 is reported to transport sulfate and oxalate, although at low levels (Silberg *et al.*, 1995; Scott *et al.*, 1999; Karniski, 2001; Xie *et al.*, 2002; Chernova *et al.*, 2003), and sulfate transport capacity is controversial (Chernova *et al.*, 2003). SLC26A3 may also transport bromide, nitrate, and acetate (Mahajan *et al.*, 1996).

4. Physiological Role

The pancreatic duct secretes up to 140-mM HCO_3^- (Sohma *et al.*, 1996) via the mechanism proposed in Fig. 5. At the pancreatic luminal surface, HCO_3^- secretion is dependent on Cl^-. CO_2 that diffuses into pancreatic duct cells basolaterally is hydrated to HCO_3^- and H^+ by cytosolic CAII. H^+ move out of the cell into the blood by a basolateral NHE, and HCO_3^- is secreted into the pancreatic lumen by a Cl^-/HCO_3^- exchanger. SLC26A3, an apical Cl^-/HCO_3^- exchanger, is a candidate for secreting HCO_3^- into the pancreatic duct lumen (Greeley *et al.*, 2001). Interactions with other

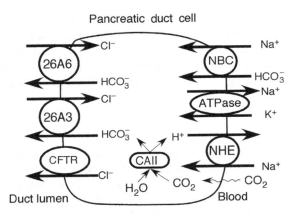

Figure 5. HCO_3^- secretion in the pancreas. Two apical SLC26 Cl^-/HCO_3^- exchangers, SLC26A3 and SLC26A6, are expressed in pancreatic duct cells. HCO_3^- is secreted in exchange for Cl^-. To prevent intracellular Cl^- accumulation, the cystic fibrosis transmembrane conductance regulator (CFTR) leaks Cl^- back into the duct lumen. HCO_3^- is provided both by CAII and also by a Na^+/HCO_3^- cotransporter (NBC) at the basolateral surface. H^+ generated by CAII catalysis is extruded by the Na^+/H^+ exchanger (NHE) and imported Na^+ is removed by the Na^+/K^+-ATPase. The pancreas secretes a fluid containing up to 140-mM HCO_3^-.

proteins such as the *c*ystic *f*ibrosis *t*ransmembrane *c*onductance *r*egulator (CFTR) stimulate Cl^-/HCO_3^- exchange activity mediated by the SLC26 transporters (Ko et al., 2002b; Chapter 12, this volume). The exact mechanism by which the pancreas can secrete such high levels of HCO_3^- remains unclear.

SLC26A3 mediates Cl^-/HCO_3^- exchange in the intestinal lumen, which is a major site of Cl^- reabsorption (Melvin et al., 1999; Rajendran et al., 2000), by exchanging intracellular HCO_3^- for luminal Cl^-. In the duodenum, Cl^-/HCO_3^- exchange results in HCO_3^- secretion and Cl^- reabsorption (Brown et al., 1989; Isenberg et al., 1993; Safsten, 1993; Ainsworth et al., 1998). Working in concert with apically colocalized NHE3, SLC26A3 works in the process of luminal NaCl reabsorption.

SLC26A3 may play a role in regulation of cell growth, since colon tumor progression is inversely correlated with SLC26A3 mRNA levels (Antalis et al., 1998), and inducible overexpression of SLC26A3 in cultured cell lines suppresses cell growth (Chapman et al., 2002).

5. Regulation

SLC26A3 is insensitive to changes in pHo (Chernova et al., 2003), but inhibited at acidic pHi and activated at alkaline pHi, a regulatory mechanism that is mediated by the C-terminal cytoplasmic domain (Chernova et al., 2003). Modulation of SLC26A3 activity by pHi is important both for Cl^- uptake from the intestinal lumen, and also for pHi regulation. A pH sensor in SLC26A3 may sense adequate substrate availability for Cl^-/HCO_3^- exchange and help to avoid intracellular acidification (Chernova et al., 2003). NaCl absorption is inhibited by luminal acidification, and this effect could be mediated through SLC26A3. NHE3 in the apical membrane secretes H^+ during NaCl reabsorption, and SLC26A3 and NHE3 could be coupled (Lamprecht et al., 2002).

SLC26A3 may also be regulated by the cAMP-sensitive Cl^- channel, the CFTR. Cystic fibrosis (CF) is an autosomal recessive disease caused by an inactivation or misprocessing of CFTR (Quinton, 1999). CF patients suffer from defective fluid secretion, and reduced HCO_3^- secretion (Welsh and Fick, 1987; Quinton, 1990), which lead to respiratory, pancreatic, and hepatobiliary problems (Quinton, 1999). SLC26A3 contributes to HCO_3^- secretion in the duodenum (Jacob et al., 2002) and is associated physically and functionally with CFTR (Lee et al., 1999; Lohi et al., 2000, 2003; Greeley et al., 2001; Jacob et al., 2002; Ko et al., 2004). SLC26A3 expression in trachea epithelial cells and cultured pancreatic duct cells increases in the presence of active CFTR, and an increase in Cl^-/HCO_3^- exchange occurs (Wheat et al., 2000; Greeley et al., 2001). The defect in HCO_3^- secretion in CF patients has been proposed to be due in part to the down-regulation of Cl^-/HCO_3^- exchange by SLC26A3 (Wheat et al., 2000; Greeley et al., 2001). In CFTR-null and CF mice, however, there is no decrease in SLC26A3 expression in the ileum and colon (Chernova et al., 2003; Simpson et al., 2005), so it is unclear what effect CFTR has on SLC26A3 expression in human CF patients.

SLC26A3 transport is stimulated by NH_4^+ in the intestinal lumen; this regulation is mediated by the C-terminal cytoplasmic domain (Chernova et al., 2003).

F. SLC26A4

1. Cloning

The gene linked with Pendred syndrome (SLC26A4, PDS) was identified by positional cloning (Everett et al., 1999). The gene, located on chromosome 7, encodes SLC26A4 (Pendrin), which is mutated in PDS (OMIM: #274600) (Everett et al., 1997; Royaux et al., 2001).

2. Expression

SLC26A4 is found in apical membranes (Royaux et al., 2001; Mount and Romero, 2004) of the kidney collecting ducts (Royaux et al., 2000; Soleimani et al., 2001) and thyroid follicular cells (Royaux et al., 2000). It is also expressed in mammary glands (Rillema and Hill, 2003) and in endometrium and placenta (Bidart et al., 2000), as well as the developing ear (Everett et al., 1999). In the inner ear, it is found in regions where endolymphatic fluid reabsorption occurs suggesting a role of SLC26A4 in this process (Everett et al., 1999). SLC26A4 is localized to the type B and non-A non-B intercalated cells in the kidney cortical collecting ducts (Royaux et al., 2001; Kim et al., 2002; Wall et al., 2003). Apical staining of SLC26A4 increases during alkalosis (Wagner et al., 2002) indicating a role for SLC26A4 in maintaining pH homeostasis.

3. Transport Function

van Adelsberg et al. (1993), proposed that the Cl^-/HCO_3^- exchanger responsible for HCO_3^- efflux at the apical membrane of renal β-intercalated cells was the kAE1. It has since been determined that SLC26A4 is the responsible Cl^-/HCO_3^- exchanger. In β-intercalated cells, SLC26A4 carries out Cl^-/HCO_3^- exchange (Royaux et al., 2001; Soleimani et al., 2001) at the apical membrane (Petrovic et al., 2003b). SLC26A4 transports monovalent anions including HCO_3^-, I^-, Cl^-, and formate, but not sulfate or oxalate (Scott et al., 1999; Scott and Karniski, 2000). SLC26A4 also mediates Cl^-/OH^- exchange (Xie et al., 2002).

4. Physiological Role

HCO_3^- secretion from the β-intercalated cells occurs via the apical Cl^-/HCO_3^- exchanger, Pendrin (SLC26A4) (Fig. 3).

5. Regulation

SLC26A4 is up-regulated on administration of aldosterone analogues and restriction of dietary NaCl (Verlander et al., 2003; Wall et al., 2004). SLC26A4 expression is regulated by changes in Cl^- balance in the rat kidney (Quentin et al., 2004) and is important not only in pH homeostasis, but also for conservation of Cl^- and water during NaCl restriction (Wall et al., 2004). mRNA expression is decreased in the cortical collecting duct of rats with metabolic acidosis, thereby down-regulating the activity of apical Cl^-/HCO_3^- activity in β-intercalated cells of rat cortical collecting ducts, demonstrating the role of SLC26A4 in pH homeostasis (Petrovic et al., 2003c).

G. SLC26A6

1. Cloning

The full-length human cDNA of SLC26A6 (PAT-1) and the mouse orthologue Slc26a6 (CFEX) have both been cloned (Lohi et al., 2000; Knauf et al., 2001). SLC26A6 is located on chromosome 3, and encodes an 85-kDa protein (Lohi et al., 2000; Alvarez et al., 2004). There is 78% amino acid identity between the human and mouse orthologues. Alternative splicing of both the mouse and human genes gives rise to 2 major transcripts, 1 with a truncation of the first 23 N-terminal amino acids

(Lohi et al., 2000; Waldegger et al., 2001; Xie et al., 2002). Three other human splicing variants have been described (Lohi et al., 2002b).

2. Expression

Slc26a6 is the predominant anion exchanger in the adult mouse heart at both the mRNA and protein level, expressed primarily in ventricles (Alvarez et al., 2004). Slc26A3 is the only other SLC26 family Cl^-/HCO_3^- exchanger expressed in heart (Alvarez et al., 2004). Cardiac localization implicates Slc26a6 as a regulator of myocardial pHi. Slc26a6 is also expressed in the brush border membrane of duodenal villus cells (Wang et al., 2002), in pancreatic duct cells (Lohi et al., 2000), and in the stomach (Petrovic et al., 2002). Slc26a6 mRNA has been found in the brain and liver (Petrovic et al., 2003b). While Slc26a6 is found in the kidney only in the brush border membrane of proximal tubule cells (Knauf et al., 2001), in humans SLC26A6 has a wider renal distribution, with expression in distal parts of the proximal tubule, the ascending loop of Henle, macula densa cells, and intercalated cells in the collecting duct (Kujala et al., 2005).

3. Transport Function

Both SLC26A6 and Slc26a6 transport Cl^-, SO_4^-, HCO_3^-, OH^-, and oxalate (Jiang et al., 2002; Ko et al., 2002b; Wang et al., 2002; Xie et al., 2002; Lohi et al., 2003; Chernova et al., 2005). Slc26a6 mediates Cl^-/formate exchange (Knauf et al., 2001; Xie et al., 2002), and SLC26A6 in HEK 293 cells exhibits Cl^-/OH^- exchange and Cl^-/HCO_3^- exchange (Alvarez et al., 2004). SLC26A6 mediates Cl^-/HCO_3^- exchange in the duodenum (Wang et al., 2002). Cl^-/HCO_3^- exchange by SLC26A6 is reported to be electrogenic (Ko et al., 2002b; Mount and Romero, 2004), with a stoichiometry of $2\,HCO_3^-:1\,Cl^-$ (Ko et al., 2002b). There remains some disagreement on the stoichiometry of SLC26A6.

4. Physiological Role

SLC26A6 is the predominant Cl^-/HCO_3^- exchanger in the heart and may contribute either to regulation of myocyte pH or to handling of metabolic HCO_3^- load (Alvarez et al., 2004). Slc26a6 is more highly expressed in the ventricles than the atria (Alvarez et al., 2004). Ventricular muscle works against a greater load than atrial muscle, and the ventricle is thus more metabolically active, suggesting that expression of anion exchangers correlates with the higher level of metabolic load. Changes in pHi dramatically affect the contractile ability of the heart (Fabiato and Fabiato, 1978). Plasma membrane Cl^-/HCO_3^- exchange in cardiomyocytes (Xu and Spitzer, 1994; Leem et al., 1999) is important for pH homeostasis (Leem et al., 1999) as well as recovery from ischemia (Vandenberg et al., 1993), and during disease states such as cardiac hypertrophy (Perez *et al.*, 1995; Ennis *et al.*, 1998). It has also been proposed that cardiac Cl^-/HCO_3^- exchangers help to maintain cardiac Cl^- levels above the resting equilibrium potential (Vaughan-Jones, 1982). SLC26A6 and SLC26A3 are both also capable of mediating Cl^-/OH^- exchange, which has been observed in ventricular myocytes (Sun *et al.*, 1996; Alvarez *et al.*, 2004), although the relevance of this capability is not clear in the context of the heart (Alvarez *et al.*, 2004).

The proximal tubule is the renal site where the majority of filtered Cl^- is reabsorbed. Since SLC26A6 is expressed in the proximal tubule, and it is capable of transporting a variety of substrates that are involved in the process of NaCl reabsorption in the proximal tubule (Jiang *et al.*, 2002; Xie *et al.*, 2002), it may have an important role in Cl^- reabsorption (Knauf *et al.*, 2001). It may also play a role in NaCl transport in the distal collecting duct (Wang *et al.*, 1993; Reilly and Ellison, 2000; Kujala *et al.*, 2005). Slc26a6-null mice, despite normal kidney function, growth, blood pressure, and serum electrolyte levels, exhibit renal and intestinal transport defects (Wang *et al.*, 2005). They are deficient in apical Cl^-/base exchange in the proximal tubule, as well as decreased HCO_3^- secretion, and oxalate-stimulated absorption of NaCl (Wang *et al.*, 2005). Slc26a6 is clearly the major apical Cl^-/HCO_3^- exchanger in the proximal tubule as evidenced by the transport defects in Slc26a6-null mice. There remains, however, residual Cl^-/HCO_3^- activity in Slc26a6-null mice, indicating that Slc26a6 is not the only Cl^-/HCO_3^- exchanger involved in HCO_3^- secretion at the apical surface in kidney epithelial cells (Wang *et al.*, 2005). Slc26a6 may also function as an apical Cl^-/formate exchanger in the murine proximal tubule (Knauf *et al.*, 2001).

A similar phenomenon is observed in the duodenum, where residual HCO_3^- secretion is found in Slc26a6-null mice, indicating the presence of other apical anion exchangers (Wang *et al.*, 2005). In fact, there was no difference in forskolin-stimulated HCO_3^- secretion in the duodenum between Slc26a6-null mice and wild-type mice (Wang *et al.*, 2005). Slc26a6 may play a major role in intestinal Cl^-/oxalate exchange, since Slc26a6-null mice exhibit oxalate transport defects (Freel *et al.*, 2006).

SLC26A6, an apical Cl^-/HCO_3^- exchanger, is a major contributor to pancreatic HCO_3^- secretion (Lohi *et al.*, 2000; Ko *et al.*, 2002b; Steward *et al.*, 2005).

5. Regulation

Transport activity of SLC26A6 is decreased by PKC phosphorylation at S553 (Alvarez *et al.*, 2005). Phosphorylation of SLC26A6 will be discussed in more detail in the following section concerning regulation and inhibition of Cl^-/HCO_3^- exchangers.

The CFTR may stimulate SLC26A6 to secrete HCO_3^- in the pancreas (Ko *et al.*, 2002b; Simpson *et al.*, 2005). CF, caused by mutations in CFTR, is characterized by a defect in HCO_3^- secretion in the pancreas (Kulczycki *et al.*, 2003; Steward *et al.*, 2005). PKC activation in pancreatic duct cells decreases anion secretion (Cheng *et al.*, 1999). This effect could be explained by an inhibition of SLC26A6-mediated HCO_3^- secretion due to regulation by PKC. Alvarez *et al.* (2005), have proposed that if this is the case, insufficient HCO_3^- secretion could be treated using therapies designed to increase

SLC26A6–CAII interaction in the pancreas. SLC26A6 is associated physically (Lohi et al., 2003) and functionally with CFTR (Lee et al., 1999; Lohi et al., 2000, 2003; Greeley et al., 2001; Ko et al., 2004). Activation of Slc26a6 is accompanied by an activation of CFTR when coexpressed in cultured cells (Ko et al., 2004). CFTR activates Cl$^-$/OH$^-$ exchange by SLC26A6 (Ko et al., 2002b), by a mechanism that is not clear.

H. SLC26A7

1. Cloning

SLC26A7 was cloned from human high endothelial venules and kidney (Vincourt et al., 2002). Sequence analysis reveals that SLC26A7 shares 30% similarity with SLC26A3 (DRA). SLC26A7 has 656 amino acids, which comprise 12 putative transmembrane segments (Vincourt et al., 2002). Two protein isoforms result from alternative splicing; SLC26A7-1 and SLC26A7-2 differ in the 11 C-terminal residues (Vincourt et al., 2002). There are splicing variants found in mouse that are roughly 80% identical to human SLC26A7 (Vincourt et al., 2002). The peptide sequence indicates two consensus sites of N-glycosylation in the second putative extracellular loop (Vincourt et al., 2002). One of the sites is conserved in two other SLC26 transporters, A3 and A4 (Vincourt et al., 2002). Two functional domains characteristic of SLC26 family transporters are found in SLC26A7: the sulfate transporter domain (Aravind and Koonin, 2000) in the N-terminal region (amino acids 164–471) (Vincourt et al., 2002), and the STAS domain (Aravind and Koonin, 2000) in the C-terminal region (amino acids 571–637) (Vincourt et al., 2002). The STAS domain of SLC26A7 contains a conserved serine (serine 585), which is found in SLC26A3 and the *Bacillus subtilis* anti-antisigma factor SPOIIAA at equivalent positions (Vincourt et al., 2002). Phosphorylation of this conserved serine is involved in protein–protein interaction regulation (Vincourt et al., 2002). There is a PDZ motif in the C-terminal region (Lohi et al., 2002a).

2. Expression

SLC26A7 is expressed in the human kidney, and mouse kidney and stomach (Vincourt et al., 2002; Petrovic et al., 2003a), with mRNA transcripts also found in mouse testes (Lohi et al., 2002a). Expression of SLC26A7 in the kidney has been localized to the basolateral membrane of α-intercalated cells of the human collecting duct (Kujala et al., 2005), and rat collecting duct where it colocalizes with kAE1 (Petrovic et al., 2004, 2005). Lower levels of SLC26A7 are found in the rat inner medulla (Petrovic et al., 2004), and human cortex (Kujala et al., 2005) although it is absent in the cortex in rat and mouse (Petrovic et al., 2004). SLC26A7 expression is also observed in renal extraglomerular mesangial cells (Kujala et al., 2005), suggesting a broader role for the protein in kidney. Recently, SLC26A7 was found in the proximal tubule and basolaterally in the thick ascending limb in mouse and rat kidney (Dudas et al., 2005).

3. Transport Function

The transport function of SLC26A7 is currently unresolved. Initial functional studies showed that SLC26A7 facilitates Na^+-independent Cl^-/HCO_3^- exchange when expressed in *Xenopus* oocytes (Petrovic et al., 2003a; Petrovic et al., 2004). Similarly, SLC26A7 has Cl^-/HCO_3^- exchange activity when expressed in HEK 293 (Morgan, Pastorékova, and Casey, unpublished observations). In contrast, SLC26A7 expressed in *Xenopus* oocytes has also been reported to act as a Cl^- channel with minimal HCO_3^- transport (Kim et al., 2005a). Adding to the complexity, SLC26A7 also transports oxalate and sulfate (Lohi et al., 2002a).

4. Physiological Role

SLC26A7 expression at the basolateral membrane of gastric parietal cells (Petrovic et al., 2003a) suggests a role in acid secretion. Although the Cl^-/HCO_3^- exchanger, AE2, is also expressed at the basolateral membrane of gastric parietal cells (Cox et al., 1996; Jöns and Drenckhahn, 1998; Rossmann et al., 2001), there is gastric Cl^-/HCO_3^- exchange activity at the basolateral membrane not attributed to AE2 (Paradiso et al., 1989; Seidler et al., 1992, 1994). The identification of SLC26A7 expression in the stomach provided an explanation for this unidentified Cl^-/HCO_3^- exchange activity and suggested a role for SLC26A7 in gastric acid secretion (Lohi et al., 2002a; Petrovic et al., 2003a).

SLC26A7 may also play an important role in renal Cl^-/base exchange in the distal segments of the nephron. In SLC26A6-null mice, there are unidentified Cl^-/base exchangers in the proximal tubule contributing to the considerable Cl^- reabsorption via apical Cl^-/base exchange (Dudas et al., 2005; Wang et al., 2005). SLC26A7 may function in the mouse and rat kidney to help reabsorb Cl^-.

5. Regulation

SLC26A7 activity in α-intercalated cells is up-regulated by hypertonic conditions (Barone et al., 2004; Petrovic et al., 2004). The regulation of SLC26A7 by changes in osmolarity provides a mechanism by which α-intercalated cells can regulate their volume (Barone et al., 2004). The basolateral NHE1, found in α-intercalated cells (Biemesderfer et al., 1992), is also up-regulated by hypertonicity (Demaurex and Grinstein, 1994; Grinstein and Wieczorek, 1994; Grinstein et al., 1994). If these two transporters work in concert, Na^+ and Cl^- influx into α-intercalated cells will increase, thereby restoring the water balance (Barone et al., 2004). SLC26A7 expression in the collecting duct cells has been described as endosomal, with increased targeting to the basolateral membrane with hypertonicity and hypokalemia, an effect mediated by the C-terminus of the exchanger (Xu *et al.*, 2006).

Since the activity of AE1 and AE2 are limited in acidic pH (Alper, 1991), but SLC26A7 activity is pH insensitive (Petrovic *et al.*, 2003a, 2004), the colocalization of SLC26A7 with AE1 in the kidney (Petrovic *et al.*, 2004), and AE2 in the stomach, may ensure that Cl^-/HCO_3^- exchange occur regardless of the pHi.

I. SLC26A9

1. Cloning

Recently cloned, SLC26A9 has high sequence similarity to the other SLC26 Cl^-/HCO_3^- exchangers, particularly SLC26A6 (Lohi *et al.*, 2002a) (Fig. 1).

2. Expression

SLC26A9 is expressed in the lung in alveolar, bronchial, and tracheal epithelial cells (Lohi *et al.*, 2002a; Xu *et al.*, 2005), and in the stomach in gastric pits and at the apical surface of gastric epithelial cells (Xu *et al.*, 2005). Some expression is also seen in the pancreas and prostate (Lohi *et al.*, 2002a).

3. Transport Function

SLC26A9 has been subjected to little study since its recent identification. When expressed in *Xenopus* oocytes, SLC26A9 mediates Cl^-, sulfate, and oxalate transport that is inhibited by thiosulfate (Lohi *et al.*, 2002a). SLC26A9 mediates Cl^-/HCO_3^- exchange and Cl^--independent HCO_3^- transport by an unknown catalytic mechanism when expressed in HEK 293 cells (Xu *et al.*, 2005).

4. Physiological Role

Maintenance of the correct composition of lung surface fluid is critical for lung defense mechanisms (Noone *et al.*, 1994) and may be regulated by ion transport systems (Tarran *et al.*, 2001). Expression of SLC26A9 in the lung could indicate a role for the exchanger in airway surface liquid composition regulation (Lohi *et al.*, 2002a).

SLC26A9 at the apical surface of stomach epithelial cells may play a protective role. Since the stomach lumen can reach a pH below 2 (Garner *et al.*, 1984; Allen *et al.*, 1993; Feldman, 1998), an HCO_3^--rich fluid is secreted at the epithelial surface

(Engel et al., 1995; Synnerstad et al., 2001; Phillipson et al., 2002; Phillipson, 2004) to neutralize stomach acid and protect against acid digestion of the gastric mucosa. SLC26A9 may contribute to the secretion of this protective fluid (Xu et al., 2005).

5. Regulation

SLC26A9-mediated Cl^-/HCO_3^- exchange is inhibited by NH_4^+. SLC26A9 may be inhibited by stomach pathogens such as the ulcer-causing *Helicobacter pylori*, which produces NH_3/NH_4^+ (Sachs et al., 2003; Hoang et al., 2004; Said et al., 2004; van Vliet et al., 2004). This susceptibility may decrease the protective HCO_3^- secretion mediated by SLC26A9, thereby increasing acidic injury to the stomach lining associated with ulcers (Xu et al., 2005).

III. DISPUTED Cl^-/HCO_3^- EXCHANGERS

On the basis of sequence homology to SLC4 or SLC26 Cl^-/HCO_3^- exchangers, other proteins, including AE4 (SLC4A9), Btr-1 (SLC4A11), and Tat-1 (SLC26A8), have been postulated to be Cl^-/HCO_3^- exchangers (Parker et al., 2001; Toure et al., 2001; Tsuganezawa et al., 2001; Lohi et al., 2002a), but the substrates transported by these proteins have not been conclusively determined.

Rabbit Slc4a9 (AE4) was first identified in rabbit kidney cells (Parker et al., 2001), and was reported to exhibit Na^+-independent, inhibitor-insensitive Cl^-/HCO_3^- exchange activity (Tsuganezawa et al., 2001). On this basis, Slc4a9 was called AE4 as a potential fourth Cl^-/HCO_3^- exchanger of the SLC4 family. Subsequent contrary studies have indicated that while rat Slc4a9 is capable of inhibitor-sensitive Cl^-/HCO_3^- exchange activity in HEK 293 cells (Ko et al., 2002a), no measurable Cl^-/HCO_3^- exchange activity is found in *Xenopus* oocytes expressing the protein. Rather, human SLC4A9, which phylogenetically clusters strongly with Na^+/HCO_3^- cotransporters (NBCs), not Cl^-/HCO_3^- exchangers, exhibits electroneutral Na^+/HCO_3^- cotransport in *Xenopus* oocytes (Romero et al., 2004). The human orthologue, at least, is thus likely an NBC, not a Cl^-/HCO_3^- exchanger.

Both the human SLC4A11 (Btr-1) and the plant orthologue have been described as boron transporters (Takano et al., 2002; Park et al., 2004). Although SLC4A11 has sequence similarity to Cl^-/HCO_3^- exchangers (Zhao and Reithmeier, 2001), no Cl^-/HCO_3^- exchange activity mediated by SLC4A11 has been described.

SLC26A8 (Tat-1) is a newly identified SLC26 transporter, whose function has not been extensively investigated. SLC26A8 has been reported to transport Cl^-, SO_4^{2-}, and oxalate (Toure et al., 2001; Lohi et al., 2002a), and is closely related to SLC26A3, SLC26A4, and SLC26A6 (Lohi et al., 2002a). Cl^-/HCO_3^- exchange activity mediated by SLC26A8 has not yet been reported.

IV. STRUCTURE

A. Structure of SLC4 Cl^-/HCO_3^- Exchangers

Cl^-/HCO_3^- exchangers of the SLC4 family share a common architecture (Fig. 6). The general structure consists of three domains. The N-terminal cytoplasmic domain varies in length from 400–700 amino acids, depending on the isoform. AE2 and AE3 share 53% identity in the cytoplasmic domain; the AE1 cytoplasmic domain is 300 residues shorter than that of AE2 or AE3. The transmembrane domain of SLC4 transporters, which carries out Cl^-/HCO_3^- exchange, consists of 10–14 transmembrane segments (~500 amino acids). AE2 and AE3 share 90% identity in the membrane domain. Finally, the short C-terminal domain (~40 residues) contains the acidic motif required for the binding of CAII. AE1 shares 60% sequence similarity with AE2 in the C-terminal domain. The interaction between Cl^-/HCO_3^- exchangers and CAII will be discussed later in this chapter.

The SLC4 members are posttranslationally modified. AE1 receives a well-characterized polylactosaminoglycan carbohydrate structure at Asn642 in the fourth extracellular loop (Fukuda et al., 1984; Kopito and Lodish, 1985). Glycosylation is not required for protein structure or function but may stabilize the protein (Casey and Reithmeier, 1991b). The largest extracellular loop in AE2 and AE3 is loop 3, which is the site of glycosylation (Zolotarev et al., 1996; Alper et al., 2002). AE1 is palmitoylated at Cys843 (Okubo et al., 1991; Cheung and Reithmeier, 2004), but the functional significance of this modification is unclear.

AE1 is the most intensively studied Cl^-/HCO_3^- exchanger, whose structure is consequently best understood. The N-terminal domain of AE1 binds to glycolytic enzymes and components of the erythroid cytoskeleton. Reentrant loops have been described in the C-terminal and N-terminal region of AE1 (Zhu et al., 2003; Cheung et al., 2005) (Fig. 6). Although no high-resolution structure exists for the SLC4 exchangers, information about the transport mechanism of AE1 has been determined. Residues involved in the anion selectivity filter and the permeability barrier have been identified (Zhu et al., 2003; Zhu and Casey, 2004). E681 in transmembrane segment 8 is a key residue in the permeability barrier and anion translocation pathway (Chernova et al., 1997; Jennings, 2005). AE1 forms a tight dimer through interactions within the membrane domain (Casey and Reithmeier, 1991a), which can only be disrupted to monomers with protein denaturation (Boodhoo and Reithmeier, 1984). Nonetheless, the basic functional unit is the monomer, since inhibition of one monomer within a dimeric unit leaves the uninhibited monomer with normal transport activity (Jennings, 1982). The extracellular loops of AE1 form blood group antigens (Blumenfeld, 1999).

The crystal structure of the N-terminal cytoplasmic domain of AE1 (residues 1–379) has been solved to 2.6-Å resolution (Zhang et al., 2000). The structure consists of a dimer, with dimensions of $75 \times 55 \times 45$ Å. The membrane domain of AE1 has been examined at a low resolution (20-Å resolution), using electron microscopy (Wang et al., 1994). The membrane domain, like the cytoplasmic domain, exists as a dimer with a thickness of 80 Å.

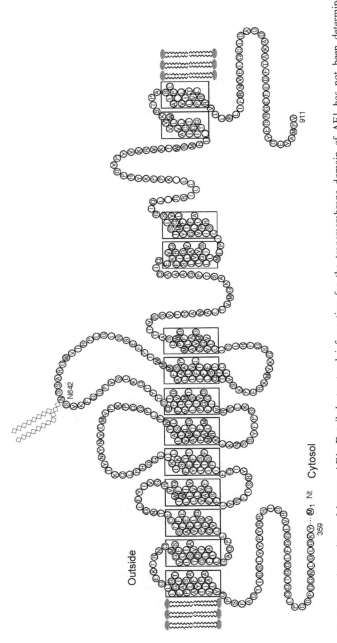

Figure 6. Proposed topology of human AE1. Detailed structural information for the transmembrane domain of AE1 has not been determined. The topology of the loop between transmembrane segment 11 and 12 (residues 803–836) is unclear. This region may be a short transmembrane segment, and could form part of the translocation pore. The branched structure is the glycosylation site (Asn642). Model is based on a previous determination of AE1 topology (Zhu and Casey, 2004).

B. Structure of SLC26 Cl$^-$/HCO$_3^-$ Exchangers

SLC26 family members are membrane proteins that either transport sulfate or share homology with sulfate transporters (Markovich, 2001). Members of the SLC26 family share several common structural features, although the residue conservation between isoforms and species orthologues is varied (Makalowski and Boguski, 1998). The median amino acid identity between mouse and human SLC26 transporters is 86%. *Drosophila* and *Caenorhabditis elegans* SLC26 transporters are 25–40% identical to mammalian orthologues.

SLC26 family members have been subjected to little experimental study and most structural information has been inferred from sequence analysis. The SLC26 transporters have 10–14 membrane-spanning segments (Moseley *et al.*, 1999; Saier *et al.*, 1999), and the N- and C-termini are predicted to be intracellular (Lohi *et al.*, 2003; Zheng *et al.*, 2003). One region of homology between the mammalian family members is the 22-residue "sulfate transport" motif, which is localized near the N-terminal end of the transmembrane segments and hypothesized to be required for anion transport function. A motif near the C-terminal end of the transmembrane domain contains the triplet NQE and other conserved residues (Saier *et al.*, 1999), although the significance of this region to general SLC26 transporter structure or function is not well understood. Several of the SLC26 members contain a class 1 PDZ motif at the extreme C-terminus which is associated with mediating interactions with proteins containing a conserved PDZ motif (Songyang *et al.*, 1997; Hung and Sheng, 2002). The only SLC26 Cl$^-$/HCO$_3^-$ transporter without this motif is SLC26A4. While not required for transport activity (Alper *et al.*, 2001; Lohi *et al.*, 2003), the PDZ motif may be critical in mediation of protein–protein interactions. All SLC26 proteins contain a C-terminal cytoplasmic domain that includes the STAS domain (Aravind and Koonin, 2000). Although the exact function of this domain is not yet known, mutations in this domain in SLC26A2, SLC26A3, and SLC26A4 cause disease (Everett *et al.*, 1997; Aravind and Koonin, 2000; Rossi and Superti-Furga, 2001; Makela *et al.*, 2002; Lohi *et al.*, 2003), indicating an important role in correct processing or function of SLC26 proteins. The STAS domains of SLC26A6 and SLC26A3 interact with the regulatory domain of CFTR, with R domain phosphorylation enhancing this interaction (Ko *et al.*, 2004; Chapter 12, this volume). The SLC26A6 STAS domain also forms the binding site for CAII (Alvarez *et al.*, 2005).

V. REGULATION AND INHIBITION OF Cl$^-$/HCO$_3^-$ EXCHANGE

A. HCO$_3^-$ Transport Metabolon

Weakly associated complexes of metabolic enzymes, or metabolons (Srere, 1985, 1987), exist for the glycolytic chain, the citric acid cycle, and the urea cycle (Reithmeier, 2001). Metabolons facilitate substrate channeling (Miles *et al.*, 1999; Reithmeier, 2001) from the active site of one enzyme to the next by limiting diffusional distances between active sites, thereby minimizing loss of intermediate metabolites and maximizing flux through a chain of enzymes (Srere, 1985, 1987; Miles et al., 1999; Sterling et al., 2001).

The HCO_3^- transport metabolon is a complex between a HCO_3^- transporter and carbonic anhydrase (CA) enzymes (Sterling et al., 2001; McMurtrie et al., 2004). Carbonic anhydrases (CAs) are Zn metalloenzymes that carry out the reversible hydration of CO_2 to HCO_3^- plus H^+ (Brown et al., 1990). In the metabolon, the physical interaction of an HCO_3^- transport protein with CAs results in a high local concentration of HCO_3^- at the active site of the transporter. The product of the CA is fed into the active site of the transporter due to close proximity. The interaction increases the total flux through the transporter. Indeed, interaction between CAII and the Cl^-/HCO_3^- exchangers activate the rate of Cl^-/HCO_3^- exchange by about 40% (Sterling et al., 2001).

At the extracellular side of the plasma membrane, CAs that are linked to the membrane, such as the glycosyl phosphatidylinositol (GPI)-linked CAIV, interact with HCO_3^- transporters to convert HCO_3^- that has been transported across the membrane to CO_2 and H_2O. This enrichment of HCO_3^- at the intracellular face of the transporter and the depletion of HCO_3^- at the extracellular face provides a "push-pull" mechanism to increase the overall transmembrane HCO_3^- movement by maximizing the magnitude of the transmembrane HCO_3^- gradient. Localization of HCO_3^--metabolizing CAs to the surface of an HCO_3^- transporter maximizes the transmembrane HCO_3^- concentration gradient local to the transporter, which maximizes the transport rate. The recent finding that the interaction of CAII with SLC26A6 could be inhibited by PKC-mediated phosphorylation suggests that the HCO_3^- transport metabolon is a key regulatory mechanism for HCO_3^- transport function.

There are 15 CA isoforms in mammals, with different tissue and subcellular distribution (Sly and Hu, 1995; Schwartz, 2002; Supuran et al., 2003; Hilvo et al., 2005). The ubiquitous CAII is a highly active cytosolic CA, with a turnover rate of 10^6 s^{-1} (Maren, 1967). CAII forms the intracellular component of the HCO_3^- transport metabolon. The CAII-binding motif is a hydrophobic residue followed by four residues, at least two of which are acidic (Sterling et al., 2001). All Cl^-/HCO_3^- transporters, with the exception of SLC26A3, possess this motif (Sterling et al., 2001). CAII, which is present in stoichiometric amounts to AE1 ($\sim 10^6$ copies per cell) (Steck et al., 1971; Tashian and Carter, 1976; Ship et al., 1977), interacts with AE1 to enhance Cl^-/HCO_3^- exchange activity (Keilen and Mann, 1941; Jacobs, 1942; Maren and Wiley, 1970; Cousin et al., 1975). Vince and Reithmeier characterized the binding between AE1 and CAII (Vince and Reithmeier, 1998, 2000; Vince et al., 2000). A basic patch in the N-terminal region of CAII binds to the short cytoplasmic C-terminal domain of AE1 at L886DADD (Vince and Reithmeier, 2000; Vince et al., 2000). When the sequence is mutated to the homologous sequence of AE2 (DANE), binding to CAII still occurs (Vince and Reithmeier, 2000). Since CAII binds AE1 at a site close to the membrane domain, which is responsible for transport activity, CAII is in close proximity to the active site of AE1. Since AE1 can transport a variety of small anion substrates, localizing HCO_3^- to the active site will favor HCO_3^- transport over the transport of other substrates. The binding between AE1 and CAII is dependent on ionic strength and pH (Vince and Reithmeier, 1998). Transport is enhanced by acidic pH and low ionic strength. The binding motif is conserved among AE1 orthologues: mouse, rat, chicken, and bovine AE1 have DXDD, and trout AE1 has DASD. The AE2 binding sequence is also conserved between mouse, rat, guinea pig, and human. AE2 and AE3 form an intracellular HCO_3^- transport metabolon with CAII (Sterling et al., 2001).

The sulfonamide, acetazolamide, inhibits CAII enzymatic activity but does not directly affect anion exchange (Cousin et al., 1975). HEK 293 cells transiently transfected with cDNA for CAII and AE1 and treated with acetazolamide exhibit a decreased anion exchange activity compared with cells expressing CAII and AE1 but not treated with acetazolamide (Sterling et al., 2001). A dominant negative form of CAII, V143Y, is able to bind AE1 but is enzymatically inactive (Fierke et al., 1991; Sterling et al., 2001). V143Y is expressed at levels 20-fold higher than endogenous wild-type CAII, and displaces the endogenously expressed active CAII from binding to AE1 (Sterling et al., 2001). Expression of V143Y in HEK 293 cells expressing AE1 also decreased AE1 anion exchange activity (Sterling et al., 2001). Expression of V143Y decreased transport activity by up to 60%, and treatment with acetazolamide decreased activity by a similar amount (Sterling et al., 2001). V143Y inhibits AE3c and AE2 by 40% (Sterling et al., 2001). CAII is more catalytically active than AE1 and the rate-limiting step in HCO_3^- transport is the transport event. This means that the metabolon increases the rate-limiting step by decreasing diffusion rates. These findings support the presence of an HCO_3^- transport metabolon, where Cl^-/HCO_3^- exchangers are maximally active when complexed to CA. Further support for the importance of the CAII/AE1 interaction was found in co-expression experiments with wild-type AE1 (with a CAII binding mutation) and transport defective AE1 (with a functional CAII binding site) (Dahl et al., 2003). Binding of CAII at the inactive monomer within the dimeric AE1 unit was sufficient to activate transport activity in the functionally active (but CAII binding incompetent) AE1 monomer.

The extracellular component of the metabolon incorporates membrane-associated CAs. CAIV is anchored to the plasma membrane via a GPI anchor (Waheed et al., 1992). CAIX, CAXII, and CAXIV have a transmembrane region. CAIV binds to the fourth extracellular loop of AE1 to increase the Cl^-/HCO_3^- activity of AE1 (Sterling et al., 2002a). AE2 and AE3 also form a metabolon with CAIV (Sterling et al., 2002a). CAIV is the extracellular CA in the heart (Vandenberg et al., 1996), and has a role in pH recovery following ischemic reperfusion (Vandenberg et al., 1996), possibly due to an interaction with AE3. CAIX interacts with AE2 and AE3 at the extracellular face of the plasma membrane (Morgan, Pastorékova and Casey, unpublished observations).

SLC26A6 was the first SLC26 transporter shown to interact with a CA, and to form a transport metabolon (Alvarez et al., 2005). SLC26A6 binds CAII through the SLC26A6 STAS domain (Alvarez et al., 2005). Transport is decreased with acetazolamide treatment, which inhibits endogenous CAII (Alvarez et al., 2005).

SLC26A3 is unable to bind CAII, but still requires CAII for optimal activity, suggesting that perhaps CFTR is required to mediate the interaction between CAII and SLC26A3 (Sterling et al., 2002b). In CF, CFTR is mistargeted or nonfunctional, and is perhaps unable to mediate the interaction between CAII and SLC26A3, and therefore SLC26A3 HCO_3^- secretion is impaired (Sterling et al., 2002b). CFTR regulates

SLC26A3 by conferring sensitivity to cAMP to the normally cAMP-insensitive SLC26A3-mediated Cl^-/HCO_3^- exchange (Chernova *et al.*, 2003). The regulation may be conferred through scaffolding proteins that can bind to PDZ domains (Lamprecht *et al.*, 2002; Ko *et al.*, 2004) and through binding of SLC26A3's STAS domain to the PKA-phosphorylated R domain of CFTR (Ko *et al.*, 2004).

B. Kinases

There is considerable evidence for phosphorylation of Cl^-/HCO_3^- exchangers, but less data on how phosphorylation can alter transport activity. AE1 is phosphorylated at tyrosine 904 (Yannoukakos *et al.*, 1991); this phosphorylation site is not involved in CAII binding (Vince and Reithmeier, 2000), and its function is unknown. Skate AE1, however, may be regulated through phosphorylation (Musch and Goldstein, 2005). Phosphorylation of SLC26A6 by PKC at S553 results in a decrease in the transport activity of SLC26A6 (Alvarez *et al.*, 2005). The cause of this is a displacement of CAII from SLC26A6 (Alvarez *et al.*, 2005). The STAS domain of SLC26A6 spans the region from E530 to A471 (Ko *et al.*, 2004). Contained in this domain are the CAB site (D546–F549) which binds CAII (Ko *et al.*, 2004) and S553, the site of PKC phosphorylation (Ko *et al.*, 2004). On the basis of sequence alignments, the CAB domain and PKC phosphorylation site of SLC26A6 are not conserved regions between other STAS domain-containing transporters, indicating that this variable region may confer variable regulation on transport activity (Alvarez *et al.*, 2005). PKC has the opposite effect on AE3, which is also expressed in the heart. AE2 has a conserved PKC consensus site in its N-terminus, and also a PKA consensus site (Wang *et al.*, 1996; Rossmann *et al.*, 2001). The PKC activator, TPA, increases acid/base flux in gastric mucous cells, where AE2a is expressed, indicating that PKC activates AE2a, potentially through phosphorylation (Rossmann *et al.*, 2001).

C. PDZ Domain-Mediated Interactions

SLC26 transporters have a Type 1 PDZ domain at their extreme cytoplasmic C-terminus. The interacting motif for PDZ proteins is C-terminal T/SXϕ, where ϕ is a hydrophobic amino acid (Caplan, 1997; Songyang *et al.*, 1997). PDZ-containing proteins help to maintain cell polarity and function (Aroeti *et al.*, 1998; Fanning and Anderson, 1999). The function of PDZ motifs in SLC26 proteins is not fully understood. In the kidney, SLC26A6 may interact with other transporters, such as NHE3, through PDZ scaffolding proteins (Ko *et al.*, 2002b; Gisler *et al.*, 2003; Lohi *et al.*, 2003). The two proteins may function in concert in the kidney to maintain NaCl homeostasis and HCO_3^- transport (Gisler *et al.*, 2003). NHE3 imports Na^+ in exchange for protons. Once in the lumen, the protons react with HCO_3^-, leading to the influx of Cl^- into kidney cells by SLC26A6 (Knauf *et al.*, 2001; Schwartz, 2002; Wang *et al.*, 2002). PDZ scaffolding proteins may allow these transporters to be physically complexed. The SLC26 family transporters may also interact with CFTR through PDZ motifs (Ko *et al.*, 2004). The interaction between SLC26 transporters and CFTR will be discussed in Section V.D.

D. STAS Domain

There is a similarity between the C-terminal cytoplasmic domains of the SLC26 transporters and the bacterial anti-sigma factor antagonists (ASA) such as SPOIIAA from *B. subtilis* (Aravind and Koonin, 2000). This region of similarity is known as the STAS domain (Aravind and Koonin, 2000). Bacteria such as gram-positive bacteria, Actinomycetes, Cyanobacteria, chlamydiae, *Treponema*, and *Thermotoga* have ASA-like proteins, which interact with a protein kinase (the anti-sigma factor) to "positively regulate" sigma factors (Duncan *et al.*, 1996). The kinase phosphorylates the ASA-like protein and inactivates it (Duncan *et al.*, 1996). Activation of the ASA-like protein occurs when the phosphorylation is removed by a phosphatase (Duncan *et al.*, 1996). The STAS domain may bind NTP, based on sequence conservation and the ability of SPOIIAA to bind GTP and ATP (Najafi *et al.*, 1996). The conserved loop where phosphorylation occurs contains a serine that is phosphorylated. When this serine is mutated or dephosphorylated, SPOIIAA can no longer bind NTP (Najafi *et al.*, 1996). The interaction has been mapped to the predicted phosphate-binding loop and probably results in its disruption. If the STAS domain binds NTP, anion transport could be regulated by intracellular nucleotides (Aravind and Koonin, 2000). The conserved serine (serine 58 in SPOIIAA) is also conserved in SLC26A7 (Ser585) and SLC26A3 (Vincourt *et al.*, 2002).

An interaction between SLC26A3 and SLC26A6 and CFTR is mediated by both PDZ domains and the SLC26 STAS domain, which binds the R domain of CFTR (Ko *et al.*, 2004). Protein kinase A-mediated phosphorylation of the R domain augments the interaction between the STAS domain of SLC26 and the R domain of CFTR (Ko *et al.*, 2004). CFTR is a cAMP-regulated Cl^- channel (Ratjen and Doring, 2003) with some HCO_3^- conductance (Hug *et al.*, 2003). SLC26 transporters and CFTR colocalize in the epithelial luminal membrane (Ko *et al.*, 2004) and CFTR activates Cl^-/OH^- exchange mediated by SLC26A3, SLC26A4, and SLC26A6 (Ko *et al.*, 2002b). Since cAMP activation of Cl^-/base exchange appears to be mediated by CFTR (Ko *et al.*, 2002b), a role for SLC26 transporters is indicated in CF (Choi *et al.*, 2001). SLC26A3 expression is up-regulated in tracheal epithelial cells and cultured pancreatic duct cells also expressing CFTR (Wheat *et al.*, 2000; Greeley *et al.*, 2001), and SLC26A3 can effect an increase the open probability of CFTR (Ko *et al.*, 2004). Recombinant STAS was able to activate CFTR in duct cells (Ko *et al.*, 2004).

CFTR has been suggested to enhance anion exchange in villous epithelia by acting as a Cl^- leak channel, enabling sustained $Cl^-_{in}/HCO_{3\ out}^-$ exchange activity (Simpson *et al.*, 2005). Cl^-/HCO_3^- exchange activity was reduced in CF duodenal villous epithelial cells (Simpson *et al.*, 2005), and $Cl^-_{in}/HCO_{3\ out}^-$ exchange was decreased in wild-type duodenal epithelium on treatment with glibenclamide, a CFTR-selective channel blocker (Simpson *et al.*, 2005). CFTR/Cl^-/HCO_3^- exchanger interaction may be pH-dependent since NH_4^+ disrupts the interaction between CFTR and Cl^-/HCO_3^- exchangers (Worrell *et al.*, 2005).

E. Inhibitors of Cl^-/HCO_3^- Exchange

One of the challenges to the study of Cl^-/HCO_3^- exchangers is the absence of highly selective inhibitors. The most important class of inhibitors of Cl^-/HCO_3^- exchangers is the stilbene disulfonates (Cabantchik and Greger, 1992). These compounds inhibit by virtue of their dual character. The hydrophobic contribution of the stilbene favors localization in the hydrophobic membrane environment, while the negatively charged sulfonate groups have affinity for anion-binding sites in proteins. Combined, this results in high affinity (binding constants <1 µM) interaction between stilbene disulfonates and Cl^-/HCO_3^- exchangers. Many derivatives of stilbene disulfonates are available, differing in whether they are covalently reactive, and the size and charge of the substituent groups (Cabantchik and Greger, 1992). The best known of the anion exchange inhibitors is DIDS (4,4′-diisothiocyanatostilbene-2,2′-disulfonic acid), which will covalently react with free lysine residues in Cl^-/HCO_3^- exchangers (Okubo et al., 1994). Stilbene disulfonates potently and effectively inhibit Cl^-/HCO_3^- exchange, but it needs to be borne in mind that Cl^- channels and other anion transporters (e.g., lactate transporters) share a sensitivity to these compounds.

AE1 is inhibited by the stilbene inhibitor DIDS, which binds to the outward-facing conformation of the protein (Kaplan et al., 1976; Cabantchik and Greger, 1992; Jennings et al., 1998). AE2, AE3, and NDAE1 are also inhibited by DIDS (Humphreys et al., 1994; Romero et al., 2000; Grichtchenko et al., 2001; Rossmann et al., 2001; Pushkin and Kurtz, 2006). Sensitivity of SLC26 Cl^-/HCO_3^- exchangers to inhibitors is not well established, but it appears that they share the property of sensitivity to inhibition by stilbene disulfonates. SLC26A6, SLC26A7, and SLC26A9 are DIDS sensitive (Ko et al., 2002b; Lohi et al., 2002a; Wang et al., 2002; Xie et al., 2002; Petrovic et al., 2004). SLC26A7 has also been reported as sensitive to inhibition by diphenylamine-2-carboxylic acid and glybenclamide (Kim et al., 2005a). The sensitivity of SLC26A3 to stilbene inhibitors remains controversial (Silberg et al., 1995; Byeon et al., 1998; Melvin et al., 1999; Moseley et al., 1999; Chernova et al., 2003); sulfate transport by SLC26A3 is DIDS sensitive (Silberg et al., 1995).

Six other classes of compounds have been used as Cl^-/HCO_3^- exchange inhibitors. Oxonol dyes inhibit their transport activity, with inhibitory constants around 1 µM (Knauf et al., 1995). Oxonol dyes also inhibit Cl^- channels, so they cannot be used as selective inhibitors of AE transport. The second class of Cl^-/HCO_3^- exchange inhibitors are squalamine-related polyaminosterols, which inhibit transport with low micromolar affinity (Alper et al., 1998). Differential sensitivity of AE isoforms has been reported for some of these compounds (Alper et al., 1998). The compound, S20787, has been reported as a selective inhibitor of Cl^-/HCO_3^- exchange in cardiomyocytes (Lagadic-Gossmann et al., 1996; Loh et al., 2001), but absolute determination of the specificity of the compound awaits further study. Niflumate (also called flufenamic acid) has been widely used as a Cl^-/HCO_3^- exchange inhibitor, which inhibits SLC4 transporters only from the intracellular surface (Knauf et al., 1996). SLC26A3 is inhibited by both niflumate (IC_{50} 7 µM) and tenidap (IC_{50} 10 µM) (Chernova et al., 2003). Finally antibodies directed at the extracellular region of Cl^-/HCO_3^- exchangers have been used successfully as transport inhibitors (Cingolani et al., 2003; Chiappe de Cingolani, 2006).

The role of CA inhibitors in the inhibition of Cl^-/HCO_3^- exchanger needs also be noted. CAs provide the substrate for transport by Cl^-/HCO_3^- exchangers, so inhibition of these enzymes will inhibit the rate of transport activity substantially (Sterling et al., 2001). Interestingly, since both CAs and Cl^-/HCO_3^- exchangers share a binding site for HCO_3^-, they also share sensitivity to some inhibitors that bind to the active site, including some sulfonamide compounds (Cousin and Motais, 1976; Morgan et al., 2004).

VI. PATHOPHYSIOLOGY

Human diseases associated with the Cl^-/HCO_3^- exchangers are listed in Table II, and further information can be accessed at: http://www.ncbi.nlm.nih.gov/entrez/query.fcgi?db = OMIM

A. AE1

AE1-null mice exhibit severe hemolytic anemia secondary to fragile membranes (Peters et al., 1996; Southgate et al., 1996). The fragility arises from the loss of plasma membrane cytoskeletal interactions in the $AE1^{-/-}$ mice, which normally greatly stabilize the erythrocyte. The AE1 cytoplasmic domain binds ankyrin to link the plasma membrane to the cytoskeleton. Mutations in AE1 cause hereditary spherocytosis (HS) (OMIM: +109270), hereditary stomatocytosis (Hst) (OMIM: +109270), Southeast Asian Ovalocytosis (SAO) (OMIM: +109270), and distal renal tubular acidosis (dRTA) (OMIM: #179800 and #602722).

Table II
Human Diseases Associated with Cl^-/HCO_3^- Exchangers

Transport protein	Other names	Disease association	OMIM[a]
AE1	SLC4A1, band 3	Distal renal tubular acidosis, hereditary spherocytosis (HS), hereditary stomatocytosis (Hst) hemolytic anemia, Southeast Asian Ovalocytosis (SAO)	#179800, #602722, and +109270
AE2	SLC4A2	None identified	
AE3	SLC4A3	None identified	
NDCBE1	SLC4A8, NDAE1, NCBE, SLC4A10	None identified	
SLC26A3	DRA, CLD	Congenital chloride diarrhea	#214700 and #126650
SLC26A4	Pendrin, PDS	Pendred syndrome, deafness (DFNB4)	#274600
SLC26A6	PAT-1, CFEX	None identified	
SLC26A7	SLC26A7	None identified	
SLC26A9	SLC26A9	None identified	

[a]OMIM citations can be accessed at http://www.ncbi.nlm.nih.gov/entrez/query.fcgi?db = OMIM.

Missense, nonsense, and frameshift mutations throughout AE1 cause 20% of HS cases. Autosomal dominant HS results from mistargeting of mutant AE1 to the erythrocyte membrane as a result of misfolding, accompanied by weakened contacts between the plasma membrane and underlying cytoskeleton (Peters et al., 1996). HS is characterized by reduced erythrocyte stability and surface area (Lux and Palek, 1995; Tse and Lux, 1999). A dominant negative effect is seen with R290C (band 3Bicetrel), which is retained in the ER (Dhermy et al., 1999). Mutations in the membrane domain causing HS include L707P, R760Q, R808C, H834P, T837B, and R870W. Proteins carrying these mutations are retained intracellularly (Quilty and Reithmeier, 2000). E40K, G130R, and P327R are mutations in the N-terminal cytoplasmic domain that may cause AE1 to fail to interact with erythrocyte band 4.2, thereby weakening contacts between the cytoskeleton and plasma membrane (Bustos and Reithmeier, 2006). Hst results in cellular leakage of Na and K (Bruce, 2006).

Deletion of amino acids 400–408 of human AE1, spanning the end of the cytoplasmic domain and beginning of the first transmembrane segment, causes SAO (Liu et al., 1990). SAO results in a nonfunctional AE1 protein (Dahl et al., 2003), which binds the cytoskeleton more strongly (Sarabia et al., 1993). The enhanced cytoskeletal interaction rigidifies the erythrocyte membrane, limiting malarial parasite invasion. People are not born homozygous for the SAO mutation as it is lethal *in utero*.

AE1 mutations cause dRTA by loss of Cl^-/HCO_3^- exchange activity in α-intercalated cells of the distal tubule (Fig. 3). dRTA-causing mutations are distinct from those causing erythroid phenotypes. dRTA is characterized by impaired urinary acid secretion, and patients may suffer from metabolic acidosis, growth retardation, hypercalciuria, and hypokalemia (Morris and Ives, 1996; Han et al., 2002). dRTA can be either dominant or recessive. In the case of dominant dRTA, the mutants, such as R589H, are not processed to the plasma membrane, and are retained in the ER (Jarolim et al., 1998; Kittanakom et al., 2004). While the mutants have normal transport activity, there is an overall deficiency in Cl^-/HCO_3^- exchange activity resulting from loss of protein at the basolateral membrane. Other mutants, such as 901X which lacks the 11 C-terminal amino acids, reach the plasma membrane, but are mistargeted, being expressed both apically and basolaterally (Devonald et al., 2003; Toye et al., 2004). Mutants that cause recessive dRTA, such as G701D, are retained intracellularly. Patients with autosomal recessive dRTA are found in Thailand (Tanphaichitr et al., 1998), New Guinea, and Malaysia (Bruce et al., 2000). The high incidence in these countries results from a population predisposed by a high underlying incidence of SAO, which causes loss of the transport function of one AE1 allele.

B. AE2

No hereditary human diseases are associated with mutations in the AE2 gene (OMIM website). AE2-null mice exhibit major defects in gastric HCl secretion and parietal cell development (Gawenis et al., 2004). Male mice lacking all but the AE2c isoform (but not female littermates) are infertile (Medina et al., 2003), and AE2 is essential for spermiogenesis in mice (Medina *et al.*, 2003). Loss of AE2 in some mice is embryonic lethal (Gawenis *et al.*, 2004), and loss of AE2 activity in humans may lead to embryonic lethality (Pushkin and Kurtz, 2006).

C. AE3

Although some susceptibility to idiopathic generalized epilepsy is seen with the substitution polymorphism Ala867Asp (Sander *et al.*, 2002), no human diseases have been linked to AE3 mutations (OMIM website). AE3$^{-/-}$ mice exhibit a lowered seizure threshold when exposed to seizure-inducing drugs such as bicuculline, pilocarpine, or pentylenetetrazole (Hentschke *et al.*, 2006), and seizure-induced mortality is increased in AE3$^{-/-}$ mice compared to wild-type mice (Hentschke *et al.*, 2006). AE3 is strongly expressed in the hippocampal CA3 region, and Cl^-/HCO_3^- exchange in this region is abrogated in AE3-null mice (Hentschke *et al.*, 2006). Studying the role of AE3 in seizure susceptibility may reveal a role for pHi regulation in epilepsy (Hentschke *et al.*, 2006).

D. SLC26A3

Mutations in SLC26A3 cause CLD (OMIM: #214700 and #126650) (Silberg *et al.*, 1995; Byeon *et al.*, 1996; Moseley *et al.*, 1999). The disorder is characterized by watery diarrhea accompanied by metabolic alkalosis, a high fecal Cl^- content, and acidic stool (Darrow, 1945; Gamble *et al.*, 1945; Kere *et al.*, 1999). The majority of cases are found in Finland (Holmberg *et al.*, 1977), Poland (Tomaszewski *et al.*, 1987), Kuwait (Lubani *et al.*, 1989), and Saudi Arabia (Kagalwalla, 1994). If untreated, CLD results in severe dehydration, renal failure, or death due to water and electrolyte loss (Holmberg *et al.*, 1977). Treatment for CLD involves lifetime fluid and electrolyte replacement. Early studies of CLD discovered a defect in Cl^-/HCO_3^- exchange in the ileum and colon (Holmberg *et al.*, 1975). In this disease, a functionally defective SLC26A3 protein inserts into the membrane, but can still move Cl^- down the electrochemical gradient (Moseley *et al.*, 1999). The severity of the Cl^- reabsorption defect in this disease indicates that SLC26A3 is the primary Cl^-/HCO_3^- transport protein in the ileum and colon (Hoglund *et al.*, 1996, 1998). Apical NHE3 and SLC26A3 function in parallel to accomplish net NaCl reabsorption (Melvin *et al.*, 1999) and failed NaCl reabsorption results in osmotic water loss and the symptoms of CLD. Down-regulation of SLC26A3 may also cause subfertility in males with CLD and CF (Hihnala *et al.*, 2006) as sperm motility is dependent on correct HCO_3^- secretion, which regulates pH of the seminal fluid (Okamura *et al.*, 1985).

E. SLC26A4

PDS (OMIM: #274600), an autosomal recessive disorder characterized by congenital sensorineural deafness and goiter, is caused by mutations in SLC26A4

(Everett et al., 2001). It is the most common hereditary cause of syndromic deafness. Four mutations in European families make up 74% of the PDS cases (Coyle et al., 1998). Mice with a targeted disruption of SLC26A4 do not exhibit thyroid abnormalities or renal pathology (Everett et al., 2001), however, they exhibit a defect in HCO_3^- secretion when alkali loaded (Royaux et al., 2001). The mice still exhibit I^- transport, which must be sufficient to prevent thyroid problems (Scott et al., 2000). Disruption in SLC26A4 also results in deafness and vestibular defects (Everett et al., 2001). The SLC26A4 gene is located close to the DFNB4 gene, which is responsible for an autosomal recessive form of nonsyndromic deafness (OMIM: #600791) (Coyle et al., 1996), and mutations in SLC26A4 are associated with DFNB4 (Li et al., 1998). SLC26A4 mutants associated with PDS lack transport activity, while those associated with only DFNB4 have residual activity (Scott et al., 2000). $SLC26A4^{-/-}$ mice and PDS patients exhibit normal arterial pH, renal function, and fluid balance, but suffer from goiter and deafness (Royaux et al., 2001). $Slc26a4^{-/-}$ mice have lowered urinary pH and P_{CO_2} due to an HCO_3^- secretion deficiency (Kim et al., 2005b).

F. SLC26A6

Although no human diseases are associated with mutations in SLC26A6, Slc26a6-null mice develop calcium oxalate urolithiasis (Jiang et al., 2006). This condition develops from an elevated plasma oxalate concentration, resulting from loss of the Cl^-/oxalate exchange activity of SLC26A6, which is needed for oxalate secretion in the intestine (Jiang et al., 2006).

ACKNOWLEDGMENTS

J.R.C. is a scientist of the Alberta Heritage Foundation for Medical Research. H.J.S. is supported by the Canadian Institute of Health Research Strategic Training Initiative in Cardiovascular Membrane Proteins. We thank members of the Casey laboratory, Bernardo Alvarez, Danielle Johnson, and Patricio Morgan, for helpful comments on the chapter.

REFERENCES

Ainsworth, M. A., Hogan, D. L., Rapier, R. C., Amelsberg, M., Dreilinger, A. D., and Isenberg, J. I. (1998). Acid/base transporters in human duodenal enterocytes. *Scand. J. Gastroenterol.* **33,** 1039–1046.
Allen, A., Flemstrom, G., Garner, A., and Kivilaakso, E. (1993). Gastroduodenal mucosal protection. *Physiol. Rev.* **73,** 823–857.
Alper, S. L. (1991). The band 3-related anion exchanger family. *Annu. Rev. Physiol.* **53,** 549–564.
Alper, S. L., Kopito, R. R., Libresco, S. M., and Lodish, H. F. (1988). Cloning and characterization of a murine band 3-related cDNA from kidney and a lymphoid cell line. *J. Biol. Chem.* **263,** 17092–17099.
Alper, S. L., Stuart-Tilley, A., Simmons, C. F., Brown, D., and Drenckhahn, D. (1994). The fodrin-ankyrin cytoskeleton of choroid plexus preferentially colocalizes with apical Na^+/K^+-ATPase rather than with basolateral anion exchanger AE2. *J. Clin. Invest.* **93,** 1430–1438.

Alper, S. L., Stuart-Tilley, A. K., Biemesderfer, D., Shmukler, B. E., and Brown, D. (1997). Immunolocalization of AE2 anion exchanger in rat kidney. *Am. J. Physiol.* **273**, F601–F614.
Alper, S. L., Chernova, M. N., Williams, J., Zasloff, M., Law, F. Y., and Knauf, P. A. (1998). Differential inhibition of AE1 and AE2 anion exchangers by oxonol dyes and by novel polyaminosterol analogs of the shark antibiotic squalamine. *Biochem. Cell Biol.* **76**, 799–806.
Alper, S. L., Rossmann, H., Wilhelm, S., Stuart-Tilley, A. K., Shmukler, B. E., and Seidler, U. (1999). Expression of AE2 anion exchanger in mouse intestine. *Am. J. Physiol.* **277**, G321–G332.
Alper, S. L., Chernova, M. N., and Stewart, A. K. (2001). Regulation of Na^+-independent Cl^-/HCO_3^- exchangers by pH. *JOP* **2**, 171–175.
Alper, S. L., Darman, R. B., Chernova, M. N., and Dahl, N. K. (2002). The AE gene family of Cl^-/HCO_3^- exchangers. *J. Nephrol.* **15**, S41–S53.
Alpern, R. J., Stone, D. K., and Rector, F. C., Jr. (1991). Renal acidification mechanisms. In: *The Kidney* (B. M. Brenner and F. C. Rector, Jr., Eds.), pp. 318–379. Saunders, Philadelphia.
Alvarez, B. V., Fujinaga, J., and Casey, J. R. (2001a). Angiotensin II stimulates cardiac chloride/bicarbonate exchange activity through phosphorylation of Serine 67 (ser67) of AE3. *J. Mol. Cell. Cardiol.* **33**, A3.
Alvarez, B. V., Fujinaga, J., and Casey, J. R. (2001b). Molecular basis for angiotensin II-induced increase of chloride/bicarbonate exchange in the myocardium. *Circ. Res.* **89**, 1246–1253.
Alvarez, B. V., Kieller, D. M., Quon, A. L., Markovich, D., and Casey, J. R. (2004). Slc26a6: A cardiac chloride/hydroxyl exchanger and predominant chloride/bicarbonate exchanger of the heart. *J. Physiol.* **561**, 721–734.
Alvarez, B. V., Vilas, G. L., and Casey, J. R. (2005). Metabolon disruption: A mechanism that regulates bicarbonate transport. *EMBO J.* **24**, 2499–2511.
Antalis, T. M., Reeder, J. A., Gotley, D. C., Byeon, M. K., Walsh, M. D., Henderson, K. W., Papas, T. S., and Schweinfest, C. W. (1998). Down-regulation of the down-regulated in adenoma (DRA) gene correlates with colon tumor progression. *Clin. Cancer Res.* **4**, 1857–1863.
Aranda, V., Martinez, I., Melero, S., Lecanda, J., Banales, J. M., Prieto, J., and Medina, J. F. (2004). Shared apical sorting of anion exchanger isoforms AE2a, AE2b1, and AE2b2 in primary hepatocytes. *Biochem. Biophys. Res. Commun.* **319**, 1040–1046.
Aravind, L. and Koonin, E. V. (2000). The STAS domain—a link between anion transporters and antisigma-factor antagonists. *Curr. Biol.* **10**, R53–R55.
Aroeti, B., Okhrimenko, H., Reich, V., and Orzech, E. (1998). Polarized trafficking of plasma membrane proteins: Emerging roles for coats, SNAREs, GTPases and their link to the cytoskeleton. *Biochim. Biophys. Acta* **1376**, 57–90.
Barone, S., Amlal, H., Xu, J., Kujala, M., Kere, J., Petrovic, S., and Soleimani, M. (2004). Differential regulation of basolateral Cl^-/HCO_3^- exchangers SLC26A7 and AE1 in kidney outer medullary collecting duct. *J. Am. Soc. Nephrol.* **15**, 2002–2011.
Bidart, J. M., Mian, C., Lazar, V., Russo, D., Filetti, S., Caillou, B., and Schlumberger, M. (2000). Expression of pendrin and the Pendred syndrome (PDS) gene in human thyroid tissues. *J. Clin. Endocrinol. Metab.* **85**, 2028–2033.
Biemesderfer, D., Reilly, R. F., Exner, M., Igarashi, P., and Aronson, P. S. (1992). Immunocytochemical characterization of Na^+-H^+ exchanger isoform NHE-1 in rabbit kidney. *Am. J. Physiol.* **263**, F833–F840.
Bissig, M., Hagenbuch, B., Stieger, B., Koller, T., and Meier, P. J. (1994). Functional expression cloning of the canalicular sulfate transport system of rat hepatocytes. *J. Biol. Chem.* **269**, 3017–3021.
Blair, H. C., Teitelbaum, S. L., Ghiselli, R., and Gluck, S. (1989). Osteoclastic bone resorption by a polarized vacuolar proton pump. *Science* **245**, 855–857.
Blumenfeld, O. (1999). http://www.bioc.aecom.yu.edu/bgmut/index.php, *Blood Group Antigen Gene Mutation Database. 2006.*
Bonnet, U. and Wiemann, M. (1999). Ammonium prepulse: Effects on intracellular pH and bioelectric activity of CA3-neurones in guinea pig hippocampal slices. *Brain Res.* **840**, 16–22.
Bonnet, U., Leniger, T., and Wiemann, M. (2000). Alteration of intracellular pH and activity of CA3-pyramidal cells in guinea pig hippocampal slices by inhibition of transmembrane acid extrusion. *Brain Res.* **872**, 116–124.

Boodhoo, A. and Reithmeier, R. A. F. (1984). Characterization of matrix-bound band 3, the anion transport protein from human erythrocyte membranes. *J. Biol. Chem.* **259**, 785–790.

Brahm, J. (1988). The red cell anion-transport system: Kinetics and physiological implications. *Soc. Gen. Physiol. Ser.* **43**, 141–150.

Brosius, F. C., III, Alper, S. L., Garcia, A. M., and Lodish, H. F. (1989). The major kidney band 3 gene transcript predicts an amino-terminal truncated band 3 polypeptide. *J. Biol. Chem.* **264**, 7784–7787.

Brosius, F. C., III, Pisoni, R. L., Cao, X., Deshmukh, G., Yannoukakos, D., Stuart-Tilley, A. K., Haller, C., and Alper, S. L. (1997). AE anion exchanger mRNA and protein expression in vascular smooth muscle cells, aorta, and renal microvessels. *Am. J. Physiol.* **273**, F1039–F1047.

Brown, C. D., Dunk, C. R., and Turnberg, L. A. (1989). Cl^-/HCO_3^- exchange and anion conductance in rat duodenal apical membrane vesicles. *Am. J. Physiol.* **257**, G661–G667.

Brown, D., Zhu, X. L., and Sly, W. S. (1990). Localization of membrane-associated carbonic anhydrase type IV in kidney epithelial cells. *Proc. Natl. Acad. Sci. USA* **87**, 7457–7461.

Bruce, L. (2006). Mutations in band 3 and cation leaky red cells. *Blood Cells Mol. Dis.* **36**, 331–336.

Bruce, L. J., Wrong, O., Toye, A. M., Young, M. T., Ogle, G., Ismail, Z., Sinha, A. K., McMaster, P., Hwaihwanje, I., Nash, G. B., Hart, S., Lavu, E., et al. (2000). Band 3 mutations, renal tubular acidosis and South-East Asian ovalocytosis in Malaysia and Papua New Guinea: Loss of up to 95% band 3 transport in red cells. *Biochem. J.* **350**, 41–51.

Bustos, S. P. and Reithmeier, R. A. (2006). Structure and stability of hereditary spherocytosis mutants of the cytosolic domain of the erythrocyte anion exchanger 1 protein. *Biochemistry* **45**, 1026–1034.

Byeon, M. K., Westerman, M. A., Maroulakou, I. G., Henderson, K. W., Suster, S., Zhang, X. K., Papas, T. S., Vesely, J., Willingham, M. C., Green, J. E., and Schweinfest, C. W. (1996). The down-regulated in adenoma (DRA) gene encodes an intestine-specific membrane glycoprotein. *Oncogene* **12**, 387–396.

Byeon, M. K., Frankel, A., Papas, T. S., Henderson, K. W., and Schweinfest, C. W. (1998). Human DRA functions as a sulfate transporter in Sf9 insect cells. *Protein Expr. Purif.* **12**, 67–74.

Cabantchik, Z. I. and Greger, R. (1992). Chemical probes for anion transporters of mammalian membranes. *Am. J. Physiol.* **262**, C803–C827.

Camilion de Hurtado, M. C., Alvarez, B. V., Ennis, I. L., and Cingolani, H. E. (2000). Stimulation of myocardial Na^+-independent Cl^--HCO_3^- exchanger by angiotensin II is mediated by endogenous endothelin. *Circ. Res.* **86**, 622–627.

Caplan, M. J. (1997). Membrane polarity in epithelial cells: Protein sorting and establishment of polarized domains. *Am. J. Physiol.* **272**, F425–F429.

Casey, J. R. and Reithmeier, R. A. F. (1991a). Analysis of the oligomeric state of band 3, the anion transport protein of the human erythrocyte membrane, by size exclusion high performance liquid chromatography: Oligomeric stability and origin of heterogeneity. *J. Biol. Chem.* **266**, 15726–15737.

Casey, J. R. and Reithmeier, R. A. F. (1991b). Transport activity of deglycosylated band 3, the anion exchange protein of the erythrocyte membrane. *Glycoconj. J.* **8**, 138.

Casey, J. R. and Reithmeier, R. A. F. (1998). Anion exchangers in the red cell and beyond. *Biochem. Cell Biol.* **76**, 709–713.

Casey, J. R., Lieberman, D. M., and Reithmeier, R. A. F. (1989). Purification and characterization of band 3 protein. *Meth. Enzymol.* **173**, 494–512.

Chapman, J. M., Knoepp, S. M., Byeon, M. K., Henderson, K. W., and Schweinfest, C. W. (2002). The colon anion transporter, down-regulated in adenoma, induces growth suppression that is abrogated by E1A. *Cancer Res.* **62**, 5083–5088.

Cheng, H. S., Wong, W. S., Chan, K. T., Wang, X. F., Wang, Z. D., and Chan, H. C. (1999). Modulation of Ca^{2+}-dependent anion secretion by protein kinase C in normal and cystic fibrosis pancreatic duct cells. *Biochim. Biophys. Acta* **1418**, 31–38.

Chernova, M. N., Jiang, L., Crest, M., Hand, M., Vandorpe, D. H., Strange, K., and Alper, S. L. (1997). Electrogenic sulfate/chloride exchange in *Xenopus* oocytes mediated by murine AE1 E699Q. *J. Gen. Physiol.* **109**, 345–360.

Chernova, M. N., Stewart, A. K., Jiang, L., Friedman, D. J., Kunes, Y. Z., and Alper, S. L. (2003). Structure-function relationships of AE2 regulation by the Ca^{2+}-sensitive stimulators, NH_4^+ and hypertonicity. *Am. J. Physiol. Cell Physiol.* **284**, C1235–C1246.

Chernova, M. N., Jiang, L., Friedman, D. J., Darman, R. B., Lohi, H., Kere, J., Vandorpe, D. H., and Alper, S. L. (2005). Functional comparison of mouse slc26a6 anion exchanger with human SLC26A6 polypeptide variants: Differences in anion selectivity, regulation, and electrogenicity. *J. Biol. Chem.* **280**, 8564–8580.

Cheung, J. C. and Reithmeier, R. A. (2004). Palmitoylation is not required for trafficking of human anion exchanger 1 to the cell surface. *Biochem. J.* **378**, 1015–1021.

Cheung, J. C., Li, J., and Reithmeier, R. A. (2005). Topology of transmembrane segments 1–4 in the human chloride/bicarbonate anion exchanger 1 (AE1) by scanning N-glycosylation mutagenesis. *Biochem. J.* **390**, 137–144.

Chiappe de Cingolani, G. E., Ennis, I. L., Morgan, P. E., Alvarez, B. V., Casey, J. R., and Camilion de Hurtado, M. C. (2006). Involvement of AE3 isoform of Na^+-independent Cl^-/HCO_3^- exchanger in myocardial pHi recovery from intracellular alkalization. *Life Sci.* **78**, 3018–3026.

Choi, J. Y., Muallem, D., Kiselyov, K., Lee, M. G., Thomas, P. J., and Muallem, S. (2001). Aberrant CFTR-dependent HCO_3^- transport in mutations associated with cystic fibrosis. *Nature* **410**, 94–97.

Cingolani, H. E., Chiappe, G. E., Ennis, I. L., Morgan, P. G., Alvarez, B. V., Casey, J. R., Dulce, R. A., Perez, N. G., and Camilion de Hurtado, M. C. (2003). Influence of Na^+-independent Cl^--HCO_3^- exchange on the slow force response to myocardial stretch. *Circ. Res.* **93**, 1082–1088.

Cousin, J. L. and Motais, R. (1976). The role of carbonic anhydrase inhibitors on anion permeability into ox red blood cells. *J. Physiol.* **256**, 61–80.

Cousin, J. L., Motais, R., and Sola, F. (1975). Transmembrane exchange of chloride with bicarbonate ion in mammalian red blood cells: Evidence for a sulphonamide-sensitive "carrier". *J. Physiol.* **253**, 385–399.

Cox, J. V. and Lazarides, E. (1988). Alternative primary structures in the transmembrane domain of the chicken erythroid anion transporter. *Mol. Cell. Biol.* **8**, 1327–1335.

Cox, K. H., Adair-Kirk, T. L., and Cox, J. V. (1996). Variant AE2 anion exchanger transcripts accumulate in multiple cell types in the chicken gastric epithelium. *J. Biol. Chem.* **271**, 8895–8902.

Coyle, B., Coffey, R., Armour, J. A., Gausden, E., Hochberg, Z., Grossman, A., Britton, K., Pembrey, M., Reardon, W., and Trembath, R. (1996). Pendred syndrome (goitre and sensorineural hearing loss) maps to chromosome 7 in the region containing the nonsyndromic deafness gene DFNB4. *Nat. Genet.* **12**, 421–423.

Coyle, B., Reardon, W., Herbrick, J. A., Tsui, L. C., Gausden, E., Lee, J., Coffey, R., Grueters, A., Grossman, A., Phelps, P. D., Luxon, L., Kendall-Taylor, P., *et al.* (1998). Molecular analysis of the PDS gene in Pendred syndrome. *Hum. Mol. Genet.* **7**, 1105–1112.

Dahl, N. K., Jiang, L., Chernova, M. N., Stuart-Tilley, A. K., Shmukler, B. E., and Alper, S. L. (2003). Deficient HCO_3^- transport in an AE1 mutant with normal Cl^- transport can be rescued by carbonic anhydrase II presented on an adjacent AE1 protomer. *J. Biol. Chem.* **278**, 44949–44958.

Darrow, D. C. (1945). Congenital alkalosis with diarrhea. *J. Pediatr.* **26**, 519–532.

Demaurex, N. and Grinstein, S. (1994). Na^+/H^+ antiport: Modulation by ATP and role in cell volume regulation. *J. Exp. Biol.* **196**, 389–404.

Devonald, M. A., Smith, A. N., Poon, J. P., Ihrke, G., and Karet, F. E. (2003). Non-polarized targeting of AE1 causes autosomal dominant distal renal tubular acidosis. *Nat. Genet.* **33**, 125–127.

Dhermy, D., Burnier, O., Bourgeois, M., and Grandchamp, B. (1999). The red blood cell band 3 variant (band 3Biceetrel:R490C) associated with dominant hereditary spherocytosis causes defective membrane targeting of the molecule and a dominant negative effect. *Mol. Membr. Biol.* **16**, 305–312.

Dudas, P. L., Mentone, S., Greineder, C. F., Biemesderfer, D., and Aronson, P. S. (2005). Immunolocalization of anion transporter Slc26a7 in mouse kidney. *Am. J. Physiol. Renal Physiol.* **290**, F937–F945.

Duncan, L., Alper, S., and Losick, R. (1996). SpoIIAA governs the release of the cell-type specific transcription factor sigma F from its anti-sigma factor SpoIIAB. *J. Mol. Biol.* **260**, 147–164.

Engel, E., Guth, P. H., Nishizaki, Y., and Kaunitz, J. D. (1995). Barrier function of the gastric mucus gel. *Am. J. Physiol.* **269**, G994–G999.

Ennis, I. L., Alvarez, B. V., Camilion de Hurtado, M. C., and Cingolani, H. E. (1998). Enalapril induces regression of cardiac hypertrophy and normalization of pHi regulatory mechanisms. *Hypertension* **31**, 961–967.

Everett, L. A., Glaser, B., Beck, J. C., Idol, J. R., Buchs, A., Heyman, M., Adawi, F., Hazani, E., Nassir, E., Baxevanis, A. D., Sheffield, V. C., and Green, E. D. (1997). Pendred syndrome is caused by mutations in a putative sulphate transporter gene (PDS). *Nat. Genet.* **17**, 411–442.

Everett, L. A., Morsli, H., Wu, D. K., and Green, E. D. (1999). Expression pattern of the mouse ortholog of the Pendred's syndrome gene (Pds) suggests a key role for pendrin in the inner ear. *Proc. Natl. Acad. Sci. USA* **96**, 9727–9732.

Everett, L. A., Belyantseva, I. A., Noben-Trauth, K., Cantos, R., Chen, A., Thakkar, S. I., Hoogstraten-Miller, S. L., Kachar, B., Wu, D. K., and Green, E. D. (2001). Targeted disruption of mouse Pds provides insight about the inner-ear defects encountered in Pendred syndrome. *Hum. Mol. Genet.* **10**, 153–161.

Fabiato, A. and Fabiato, F. (1978). Effects of pH on the myofilaments and the sarcoplasmic reticulum of skinned cells from cardiac and skeletal muscles. *J. Physiol.* **276**, 233–255.

Fanning, A. S. and Anderson, J. M. (1999). PDZ domains: Fundamental building blocks in the organization of protein complexes at the plasma membrane. *J. Clin. Invest.* **103**, 767–772.

Feldman, M. D. (1998). Normal and abnormal. In: *Sleisenger and Fordtran's Gastrointestinal and Liver Disease: Pathophysiology/Diagnosis/Management* (M. D. Feldman, M. H. Sleisenger, B. F. Scharschmidt, and S. Klein, Eds.), Vol. 1, pp. 587–603. Saunders, Philadelphia.

Fierke, C. A., Calderone, T. L., and Krebs, J. F. (1991). Functional consequences of engineering the hydrophobic pocket of carbonic anhydrase II. *Biochemistry* **30**, 11054–11063.

Freel, R. W., Hatch, M., Green, M., and Soleimani, M. (2006). Ileal oxalate absorption and urinary oxalate excretion are enhanced in Slc26a6 null mice. *Am. J. Physiol. Gastrointest. Liver Physiol.* **290**, G719–G728.

Frische, S., Zolotarev, A. S., Kim, Y. H., Praetorius, J., Alper, S., Nielsen, S., and Wall, S. M. (2004). AE2 isoforms in rat kidney: Immunohistochemical localization and regulation in response to chronic NH_4Cl loading. *Am. J. Physiol. Renal Physiol.* **286**, F1163–F1170.

Fujinaga, J., Loiselle, F. B., and Casey, J. R. (2003). Transport activity of chimaeric AE2-AE3 chloride/bicarbonate anion exchange proteins. *Biochem. J.* **371**, 687–696.

Fukuda, M., Dell, A., Oates, J. E., and Fukuda, M. N. (1984). Structure of the branched lactosaminoglycan, the carbohydrate moiety of band 3 isolated from adult human erythrocytes. *J. Biol. Chem.* **259**, 8260–8273.

Furuya, W., Tarshis, T., Law, F. Y., and Knauf, P. A. (1984). Transmembrane effects of intracellular chloride on the inhibitory potency of extracellular H_2DIDS. Evidence for two conformations of the transport site of the human erythrocyte anion exchange protein. *J. Gen. Physiol.* **83**, 657–681.

Gamble, J. L., Fahey, K. R., Appleton, J., and MacLachlan, E. A. (1945). Congenital alkalosis with diarrhea. *J. Pediatr.* **26**, 509–518.

Ganz, M. B., Boyarsky, G., Sterzel, R. B., and Boron, W. F. (1989). Arginine vasopressin enhances pHi regulation in the presence of HCO_3^- by stimulating three acid-base transport systems. *Nature* **337**, 648–651.

Garner, A., Flemstrom, G., Allen, A., Heylings, J. R., and McQueen, S. (1984). Gastric mucosal protective mechanisms: Roles of epithelial bicarbonate and mucus secretions. *Scand. J. Gastroenterol. Suppl.* **101**, 79–86.

Gawenis, L. R., Ledoussal, C., Judd, L. M., Prasad, V., Alper, S. L., Stuart-Tilley, A. K., Woo, A. L., Grisham, C., Sanford, L. P., Doetschman, T., Miller, M. L., and Shull, G. E. (2004). Mice with a targeted disruption of the AE2 Cl^-/HCO_3^- exchanger are achlorhydric. *J. Biol. Chem.* **279**, 30531–30539.

Gisler, S. M., Pribanic, S., Bacic, D., Forrer, P., Gantenbein, A., Sabourin, L. A., Tsuji, A., Zhao, Z. S., Manser, E., Biber, J., and Murer, H. (2003). PDZK1: I. A major scaffolder in brush borders of proximal tubular cells. *Kidney Int.* **64**, 1733–1745.

Greeley, T., Shumaker, H., Wang, Z., Schweinfest, C. W., and Soleimani, M. (2001). Downregulated in adenoma and putative anion transporter are regulated by CFTR in cultured pancreatic duct cells. *Am. J. Physiol. Gastrointest. Liver Physiol.* **281**, G1301–G1308.

Grichtchenko, I. I., Choi, I., Zhong, X., Bray-Ward, P., Russell, J. M., and Boron, W. F. (2001). Cloning, characterization, and chromosomal mapping of a human electroneutral Na^+-driven Cl^-/HCO_3^- exchanger. *J. Biol. Chem.* **276,** 8358–8363.

Grinstein, S. and Wieczorek, H. (1994). Cation antiports of animal plasma membranes. *J. Exp. Biol.* **196,** 307–318.

Grinstein, S., Woodside, M., Goss, G. G., and Kapus, A. (1994). Osmotic activation of the Na^+/H^+ antiporter during volume regulation. *Biochem. Soc. Trans.* **22,** 512–516.

Guggino, W. B., London, R., Boulpaep, E. L., and Giebisch, G. (1983). Chloride transport across the basolateral cell membrane of the Necturus proximal tubule: Dependence on bicarbonate and sodium. *J. Membr. Biol.* **71,** 227–240.

Gunn, R. B. and Frohlich, O. (1979). Asymmetry in the mechanism for anion exchange in human red blood cell membranes. Evidence for reciprocating sites that react with one transported anion at a time. *J. Gen. Physiol.* **74,** 351–374.

Haila, S., Saarialho-Kere, U., Karjalainen-Lindsberg, M. L., Lohi, H., Airola, K., Holmberg, C., Hastbacka, J., Kere, J., and Hoglund, P. (2000). The congenital chloride diarrhea gene is expressed in seminal vesicle, sweat gland, inflammatory colon epithelium, and in some dysplastic colon cells. *Histochem. Cell Biol.* **113,** 279–286.

Hamasaki, N., Izuhara, K., Okubo, K., Kanazawa, Y., Omachi, A., and Kleps, R. A. (1990). Inhibition of chloride binding to the anion transport site by diethylpyrocarbonate modification of band 3. *J. Membr. Biol.* **116,** 87–91.

Han, J. S., Kim, G. H., Kim, J., Jeon, U. S., Joo, K. W., Na, K. Y., Ahn, C., Kim, S., Lee, S. E., and Lee, J. S. (2002). Secretory-defect distal renal tubular acidosis is associated with transporter defect in H^+-ATPase and anion exchanger-1. *J. Am. Soc. Nephrol.* **13,** 1425–1432.

Hastbacka, J., de la Chapelle, A., Mahtani, M. M., Clines, G., Reeve-Daly, M. P., Daly, M., Hamilton, B. A., Kusumi, K., Trivedi, B., and Weaver, A. (1994). The diastrophic dysplasia gene encodes a novel sulfate transporter: Positional cloning by fine-structure linkage disequilibrium mapping. *Cell* **78,** 1073–1087.

Hentschke, M., Wiemann, M., Hentschke, S., Kurth, I., Hermans-Borgmeyer, I., Seidenbecher, T., Jentsch, T. J., Gal, A., and Hubner, C. A. (2006). Mice with a targeted disruption of the Cl^-/HCO_3^- exchanger AE3 display a reduced seizure threshold. *Mol. Cell. Biol.* **26,** 182–191.

Hihnala, S., Kujala, M., Toppari, J., Kere, J., Holmberg, C., and Hoglund, P. (2006). Expression of SLC26A3, CFTR and NHE3 in the human male reproductive tract: Role in male subfertility caused by congenital chloride diarrhoea. *Mol. Hum. Reprod.* **12,** 107–111.

Hilvo, M., Tolvanen, M., Clark, A., Shen, B., Shah, G. N., Waheed, A., Halmi, P., Hanninen, M., Hamalainen, J. M., Vihinen, M., Sly, W. S., and Parkkila, S. (2005). Characterization of CAXV, a new GPI-anchored form of carbonic anhydrase. *Biochem. J.* **392,** 83–92.

Hoang, T. T., Wheeldon, T. U., Bengtsson, C., Phung, D. C., Sorberg, M., and Granstrom, M. (2004). Enzyme-linked immunosorbent assay for *Helicobacter pylori* needs adjustment for the population investigated. *J. Clin. Microbiol.* **42,** 627–630.

Hoglund, P., Haila, S., Socha, J., Tomaszewski, L., Saarialho-Kere, U., Karjalainen-Lindsberg, M. L., Airola, K., Holmberg, C., de la Chapelle, A., and Kere, J. (1996). Mutations of the down-regulated in adenoma (DRA) gene cause congenital chloride diarrhoea. *Nat. Genet.* **14,** 316–319.

Hoglund, P., Haila, S., Gustavson, K. H., Taipale, M., Hannula, K., Popinska, K., Holmberg, C., Socha, J., de la Chapelle, A., and Kere, J. (1998). Clustering of private mutations in the congenital chloride diarrhea/down-regulated in adenoma gene. *Hum. Mutat.* **11,** 321–327.

Holappa, K., Mustonen, M., Parvinen, M., Vihko, P., Rajaniemi, H., and Kellokumpu, S. (1999). Primary structure of a sperm cell anion exchanger and its messenger ribonucleic acid expression during spermatogenesis. *Biol. Reprod.* **61,** 981–986.

Holmberg, C., Perheentupa, J., and Launiala, K. (1975). Colonic electrolyte transport in health and in congenital chloride diarrhea. *J. Clin. Invest.* **56,** 302–310.

Holmberg, C., Perheentupa, J., Launiala, K., and Hallman, N. (1977). Congenital chloride diarrhoea. Clinical analysis of 21 Finnish patients. *Arch. Dis. Child.* **52,** 255–267.

http://www.hugo-international.org (2004). The Human Genome Organization. 2006.

Hubner, S., Michel, F., Rudloff, V., and Appelhans, H. (1992). Amino acid sequence of band 3 protein from rainbow trout erythrocytes derived from cDNA. *Biochem. J.* **285**, 17–23.

Hug, M. J., Tamada, T., and Bridges, R. J. (2003). CFTR and bicarbonate secretion by epithelial cells. *News Physiol. Sci.* **18**, 38–42.

Humphreys, B. D., Jiang, L., Chernova, M. N., and Alper, S. L. (1994). Functional characterization and regulation by pH of murine AE2 anion exchanger expressed in *Xenopus* oocytes. *Am. J. Physiol. Cell Physiol.* **267**, C1295–C1307.

Humphreys, B. D., Jiang, L., Chernova, M. N., and Alper, S. L. (1995). Hypertonic activation of AE2 anion exchanger in *Xenopus* oocytes via NHE-mediated intracellular alkalinization. *Am. J. Physiol.* **268**, C201–C209.

Humphreys, B. D., Chernova, M. N., Jiang, L., Zhang, Y., and Alper, S. L. (1997). NH_4Cl activates AE2 anion exchanger in *Xenopus* oocytes at acidic pHi. *Am. J. Physiol.* **272**, C1232–C1240.

Hung, A. Y. and Sheng, M. (2002). PDZ domains: Structural modules for protein complex assembly. *J. Biol. Chem.* **277**, 5699–5702.

Isenberg, J. I., Ljungstrom, M., Safsten, B., and Flemstrom, G. (1993). Proximal duodenal enterocyte transport: Evidence for Na^+-H^+ and Cl^--HCO_3^- exchange and $NaHCO_3$ cotransport. *Am. J. Physiol.* **265**, G677–G685.

Jacob, P., Rossmann, H., Lamprecht, G., Kretz, A., Neff, C., Lin-Wu, E., Gregor, M., Groneberg, D. A., Kere, J., and Seidler, U. (2002). Down-regulated in adenoma mediates apical Cl^-/HCO_3^- exchange in rabbit, rat, and human duodenum. *Gastroenterology* **122**, 709–724.

Jacobs, M. H. and Stewart, D. R. (1942). The role of carbonic anhydrase in certain ionic exchanges involving the erythrocyte. *J. Gen. Physiol.* **25**, 539–552.

Jarolim, P., Shayakul, C., Prabakaran, D., Jiang, L., Stuart-Tilley, A., Rubin, H. L., Simova, S., Zavadil, J., Herrin, J. T., Brouillette, J., Somers, M. J., Seemanova, E., *et al.* (1998). Autosomal dominant distal renal tubular acidosis is associated in three families with heterozygosity for the R589H mutation in the AE1 (band 3) Cl^-/HCO_3^- exchanger. *J. Biol. Chem.* **273**, 6380–6388.

Jennings, M. L. (1982). Stoichiometry of a half-turnover of band 3, the chloride transport protein of human erythrocytes. *J. Gen. Physiol.* **79**, 169–185.

Jennings, M. L. (1989). Structure and function of the red blood cell anion transport protein. *Annu. Rev. Biophys. Biophys. Chem.* **18**, 397–430.

Jennings, M. L. (2005). Evidence for a second binding/transport site for chloride in erythrocyte anion transporter AE1 modified at glutamate 681. *Biophys. J.* **88**, 2681–2691.

Jennings, M. L. and Anderson, M. P. (1987). Chemical modification and labelling of glutamate residues at the stilbenedisulonate site of human red cell band 3 protein. *J. Biol. Chem.* **262**, 1691–1697.

Jennings, M. L. and Smith, J. S. (1992). Anion-proton cotransport through the human red blood cell band 3 protein. Role of glutamate 681. *J. Biol. Chem.* **267**, 13964–13971.

Jennings, M. L., Whitlock, J., and Shinde, A. (1998). Pre-steady state transport by erythrocyte band 3 protein: Uphill countertransport induced by the impermeant inhibitor H_2DIDS. *Biochem. Cell Biol.* **76**, 807–813.

Jensen, L. J., Stuart-Tilley, A. K., Peters, L. L., Lux, S. E., Alper, S. L., and Breton, S. (1999). Immunolocalization of AE2 anion exchanger in rat and mouse epididymis. *Biol. Reprod.* **61**, 973–980.

Jiang, L., Stuart-Tilley, A., Parkash, J., and Alper, S. (1994). pH_i and serum regulate AE2-mediated Cl^-/HCO_3^- exchange in CHOP cells of defined transient transfection status. *Am. J. Physiol. Cell Physiol.* **36**, C845–C856.

Jiang, Z., Grichtchenko, I. I., Boron, W. F., and Aronson, P. S. (2002). Specificity of anion exchange mediated by mouse Slc26a6. *J. Biol. Chem.* **277**, 33963–33967.

Jiang, Z., Asplin, J. R., Evan, A. P., Rajendran, V. M., Velazquez, H., Nottoli, T. P., Binder, H. J., and Aronson, P. S. (2006). Calcium oxalate urolithiasis in mice lacking anion transporter Slc26a6. *Nat. Genet.* **38**, 474–478.

Jöns, T. and Drenckhahn, D. (1998). Anion exchanger 2 (AE2) binds to erythrocyte ankyrin and is colocalized with ankyrin along the basolateral plasma membrane of human gastric parietal cells. *Eur. J. Cell Biol.* **75**, 232–236.

Kagalwalla, A. F. (1994). Congenital chloride diarrhea. A study in Arab children. *J. Clin. Gastroenterol.* **19,** 36–40.

Kaplan, J. H., Scorah, K., Fasold, H., and Passow, H. (1976). Sidedness of the inhibitory action of disulfonic acids on chloride equilibrium exchange and net transport across the human erythrocyte membrane. *FEBS Lett.* **62,** 182–185.

Karniski, L. P. (2001). Mutations in the diastrophic dysplasia sulfate transporter (DTDST) gene: Correlation between sulfate transport activity and chondrodysplasia phenotype. *Hum. Mol. Genet.* **10,** 1485–1490.

Keilen, D. and Mann, T. (1941). Activity of carbonic anhydrase within red blood corpuscles. *Nature* **148,** 493–496.

Kere, J., Lohi, H., and Hoglund, P. (1999). Genetic disorders of membrane transport III. Congenital chloride diarrhea. *Am. J. Physiol.* **276,** G7–G13.

Kim, Y. H., Kwon, T. H., Frische, S., Kim, J., Tisher, C. C., Madsen, K. M., and Nielsen, S. (2002). Immunocytochemical localization of pendrin in intercalated cell subtypes in rat and mouse kidney. *Am. J. Physiol. Renal Physiol.* **283,** F744–F754.

Kim, K. H., Shcheynikov, N., Wang, Y., and Muallem, S. (2005a). SLC26A7 is a Cl^- channel regulated by intracellular pH. *J. Biol. Chem.* **280,** 6463–6470.

Kim, Y. H., Verlander, J. W., Matthews, S. W., Kurtz, I., Shin, W., Weiner, I. D., Everett, L. A., Green, E. D., Nielsen, S., and Wall, S. M. (2005b). Intercalated cell H^+/OH^- transporter expression is reduced in Slc26a4 null mice. *Am. J. Physiol. Renal Physiol.* **289,** F1262–F1272.

Kittanakom, S., Cordat, E., Akkarapatumwong, V., Yenchitsomanus, P. T., and Reithmeier, R. A. (2004). Trafficking defects of a novel autosomal recessive distal renal tubular acidosis mutant (S773P) of the human kidney anion exchanger (kAE1). *J. Biol. Chem.* **279,** 40960–40971.

Knauf, F., Yang, C. L., Thomson, R. B., Mentone, S. A., Giebisch, G., and Aronson, P. S. (2001). Identification of a chloride-formate exchanger expressed on the brush border membrane of renal proximal tubule cells. *Proc. Natl. Acad. Sci. USA* **98,** 9425–9430.

Knauf, P. A., Law, F.-Y., and Hahn, K. (1995). An oxonol dye is the most potent known inhibitor of band 3-mediated anion exchange. *Am. J. Physiol.* **269,** C1073–C1077.

Knauf, P. A., Gasbjerg, P. K., and Brahm, J. (1996). The asymmetry of chloride transport at 38 degrees C in human red blood cell membranes. *J. Gen. Physiol.* **108,** 577–589.

Ko, S. B., Luo, X., Hager, H., Rojek, A., Choi, J. Y., Licht, C., Suzuki, M., Muallem, S., Nielsen, S., and Ishibashi, K. (2002a). AE4 is a DIDS-sensitive Cl^-/HCO_3^- exchanger in the basolateral membrane of the renal CCD and the SMG duct. *Am. J. Physiol. Cell Physiol.* **283,** C1206–C1218.

Ko, S. B., Shcheynikov, N., Choi, J. Y., Luo, X., Ishibashi, K., Thomas, P. J., Kim, J. Y., Kim, K. H., Lee, M. G., Naruse, S., and Muallem, S. (2002b). A molecular mechanism for aberrant CFTR-dependent HCO_3^- transport in cystic fibrosis. *EMBO J.* **21,** 5662–5672.

Ko, S. B., Zeng, W., Dorwart, M. R., Luo, X., Kim, K. H., Millen, L., Goto, H., Naruse, S., Soyombo, A., Thomas, P. J., and Muallem, S. (2004). Gating of CFTR by the STAS domain of SLC26 transporters. *Nat. Cell Biol.* **6,** 343–350.

Kobayashi, S., Morgans, C. W., Casey, J. R., and Kopito, R. R. (1994). AE3 Anion exchanger isoforms in the vertebrate retina: Developmental regulation and differential expression in neurons and glia. *J. Neurosci.* **14,** 6266–6279.

Kopito, R. R. and Lodish, H. F. (1985). Primary structure and transmembrane orientation of the murine anion exchange protein. *Nature* **316,** 234–238.

Kopito, R. R., Lee, B. S., Simmons, D. M., Lindsey, A. E., Morgans, C. W., and Schneider, K. (1989). Regulation of intracellular pH by a neuronal homolog of the erythrocyte anion exchanger. *Cell* **59,** 927–937.

Kostenko, M. A., Musienko, V. S., and Smolikhina, T. I. (1983). Ca^{2+} and pH affect the neurite formation in cultured mollusc isolated neurones. *Brain Res.* **276,** 43–50.

Kudrycki, K. E. and Shull, G. E. (1989). Primary structure of the rat kidney band 3 anion exchange protein deduced from cDNA. *J. Biol. Chem.* **264,** 8185–8192.

Kudrycki, K. E., Newman, P. R., and Shull, G. E. (1990). cDNA cloning and tissue distribution of mRNAs for two proteins that are related to the band 3 Cl^-/HCO_3^- exchanger. *J. Biol. Chem.* **265,** 462–471.

Kujala, M., Tienari, J., Lohi, H., Elomaa, O., Sariola, H., Lehtonen, E., and Kere, J. (2005). SLC26A6 and SLC26A7 anion exchangers have a distinct distribution in human kidney. *Nephron. Exp. Nephrol.* **101,** e50–e58.

Kulczycki, L. L., Kostuch, M., and Bellanti, J. A. (2003). A clinical perspective of cystic fibrosis and new genetic findings: Relationship of CFTR mutations to genotype-phenotype manifestations. *Am. J. Med. Genet.* **116A,** 262–267.

Lagadic-Gossmann, D., Le Prigent, K., Baut, G. L., Caignard, D. H., Renard, P., Scalbert, E., and Feuvray, D. (1996). Effects of S20787 on pHi-regulating mechanisms in isolated rat ventricular myocytes. *J. Cardiovasc. Pharmacol.* **28,** 547–552.

Lamprecht, G., Heil, A., Baisch, S., Lin-Wu, E., Yun, C. C., Kalbacher, H., Gregor, M., and Seidler, U. (2002). The down regulated in adenoma (dra) gene product binds to the second PDZ Domain of the NHE3 kinase a regulatory protein (E3KARP), potentially linking intestinal Cl^-/HCO_3^- exchange to Na^+/H^+ exchange. *Biochemistry* **41,** 12336–12342.

Lamprecht, G., Baisch, S., Schoenleber, E., and Gregor, M. (2005). Transport properties of the human intestinal anion exchanger DRA (down-regulated in adenoma) in transfected HEK293 cells. *Pflügers Arch.* **449,** 479–490.

Lee, B. S., Gunn, R. B., and Kopito, R. R. (1991). Functional differences among nonerythroid anion exchangers expressed in a transfected human cell line. *J. Biol. Chem.* **266,** 11448–11454.

Lee, M. G., Choi, J. Y., Luo, X., Strickland, E., Thomas, P. J., and Muallem, S. (1999). Cystic fibrosis transmembrane conductance regulator regulates luminal Cl^-/HCO_3^- exchange in mouse submandibular and pancreatic ducts. *J. Biol. Chem.* **274,** 14670–14677.

Leem, C. H., Lagadic-Gossmann, D., and Vaughan-Jones, R. D. (1999). Characterization of intracellular pH regulation in the guinea-pig ventricular myocyte. *J. Physiol. (Lond.)* **517,** 159–180.

Li, X. C., Everett, L. A., Lalwani, A. K., Desmukh, D., Friedman, T. B., Green, E. D., and Wilcox, E. R. (1998). A mutation in PDS causes non-syndromic recessive deafness. *Nat. Genet.* **18,** 215–217.

Linn, S. C., Kudrycki, K. E., and Shull, G. E. (1992). The predicted translation product of a cardiac AE3 mRNA contains an N-terminus distinct from that of the brain AE3 Cl^-/HCO_3^- exchanger. *J. Biol. Chem.* **267,** 7927–7935.

Linn, S. C., Askew, G. R., Menon, A. G., and Shull, G. E. (1995). Conservation of an AE3 Cl^-/HCO_3^- exchanger cardiac-specific exon and promotor region and AE3 mRNA expression patterns in murine and human hearts. *Circ. Res.* **76,** 584–591.

Liu, S.-C., Zhai, S., Palek, J., Golan, D. E., Amato, D., Hassan, K., Nurse, G. T., Babano, D., Coetzer, T., Jarolim, P., Zaik, M., and Borwein, S. (1990). Molecular defect of the band 3 protein in Southeast Asian ovalocytosis. *N. Engl. J. Med.* **323,** 1530–1538.

Loh, S. H., Tsai, C. S., Lin, C. I., Jin, J. S., and Vaughan-Jones, R. D. (2001). Effect of S20787, a Novel Cl^--HCO_3^- exchange inhibitor, on intracellular ph regulation in guinea pig ventricular myocytes. *J. Biomed. Sci.* **8,** 395–405.

Lohi, H., Kujala, M., Kerkela, E., Saarialho-Kere, U., Kestila, M., and Kere, J. (2000). Mapping of five new putative anion transporter genes in human and characterization of SLC26A6, a candidate gene for pancreatic anion exchanger. *Genomics* **70,** 102–112.

Lohi, H., Kujala, M., Makela, S., Lehtonen, E., Kestila, M., Saarialho-Kere, U., Markovich, D., and Kere, J. (2002a). Functional characterization of three novel tissue-specific anion exchangers: SLC26A7, A8 and A9. *J. Biol. Chem.* **277,** 14246–14254.

Lohi, H., Makela, S., Pulkkinen, K., Hoglund, P., Karjalainen-Lindsberg, M. L., Puolakkainen, P., and Kere, J. (2002b). Upregulation of CFTR expression but not SLC26A3 and SLC9A3 in ulcerative colitis. *Am. J. Physiol. Gastrointest. Liver Physiol.* **283,** G567–G575.

Lohi, H., Lamprecht, G., Markovich, D., Heil, A., Kujala, M., Seidler, U., and Kere, J. (2003). Isoforms of SLC26A6 mediate anion transport and have functional PDZ interaction domains. *Am. J. Physiol. Cell Physiol.* **284,** C769–C779.

Lubani, M. M., Doudin, K. I., Sharda, D. C., Shaltout, A. A., Al-Shab, T. S., Abdul Al, Y. K., Said, M. A., Salhi, M. M., and Ahmed, S. A. (1989). Congenital chloride diarrhoea in Kuwaiti children. *Eur. J. Pediatr.* **148,** 333–336.

Lux, S. E. and Palek, J. (1995). Disorders of the red cell membrane. In: *Blood: Principles and Practice of Hematology* (R. I. Handin, S. E. Lux, and T. P. Stossel, Eds.), Lippincott Co., Philadelphia.

Lux, S. E., John, K. M., Kopito, R. R., and Lodish, H. F. (1989). Cloning and characterization of band 3, the human erythrocyte anion-exchange protein. *Proc. Natl. Acad. Sci. USA* **86**, 9089–9093.

Machen, T. E., Townsley, M. C., Paradiso, A. M., Wenzl, E., and Negulescu, P. A. (1989). H^+ and HCO_3^- transport across the basolateral membrane of the parietal cell. *Ann. NY. Acad. Sci.* **574**, 447–462.

Madsen, K. M., Verlander, J. W., Kim, J., and Tisher, C. C. (1991). Morphological adaptation of the collecting duct to acid-base disturbances. *Kidney Int. Suppl* **33**, S57–S63.

Mahajan, R. J., Baldwin, M. L., Harig, J. M., Ramaswamy, K., and Dudeja, P. K. (1996). Chloride transport in human proximal colonic apical membrane vesicles. *Biochim. Biophys. Acta* **1280**, 12–18.

Makalowski, W. and Boguski, M. S. (1998). Evolutionary parameters of the transcribed mammalian genome: An analysis of 2,820 orthologous rodent and human sequences. *Proc. Natl. Acad. Sci. USA* **95**, 9407–9412.

Makela, S., Kere, J., Holmberg, C., and Hoglund, P. (2002). SLC26A3 mutations in congenital chloride diarrhea. *Hum. Mutat.* **20**, 425–438.

Maren, T. H. (1967). Carbonic anhydrase: Chemistry, physiology, and inhibition. *Physiol. Rev.* **47**, 595–781.

Maren, T. H. and Wiley, C. W. (1970). Kinetics of carbonic anhydrase in whole red cells as measured by transfer of carbon dioxide and ammonia. *Mol. Pharmacol.* **6**, 430–440.

Markovich, D. (2001). Physiological roles and regulation of mammalian sulfate transporters. *Physiol. Rev.* **81**, 1499–1533.

Mason, M. J., Smith, J. D., Garcia-Soto, J. J., and Grinstein, S. (1989). Internal pH-sensitive site couples Cl^--HCO_3^- exchange to Na^+-H^+ antiport in lymphocytes. *Am. J. Physiol.* **256**, C428–C433.

Matsuyama, H., Kawano, Y., and Hamasaki, N. (1986). Involvement of a histidine residue in inorganic phosphate and phosphoenolpyruvate transport across the human erythrocyte membrane. *J. Biochem. (Tokyo)* **99**, 495–501.

McMurtrie, H. L., Cleary, H. J., Alvarez, B. V., Loiselle, F. B., Sterling, D., Morgan, P. E., Johnson, D. E., and Casey, J. R. (2004). The bicarbonate transport metabolon. *J. Enzyme Inhib. Med. Chem.* **19**, 231–236.

Medina, J. F., Acin, A., and Prieto, J. (1997). Molecular cloning and characterization of the human AE2 anion exchanger (SLC4A2) gene. *Genomics* **39**, 74–85.

Medina, J. F., Recalde, S., Prieto, J., Lecanda, J., Saez, E., Funk, C. D., Vecino, P., van Roon, M. A., Ottenhoff, R., Bosma, P. J., Bakker, C. T., and Elferink, R. P. (2003). Anion exchanger 2 is essential for spermiogenesis in mice. *Proc. Natl. Acad. Sci. USA* **100**, 15847–15852.

Melvin, J. E., Park, K., Richardson, L., Schultheis, P. J., and Shull, G. E. (1999). Mouse down-regulated in adenoma (DRA) is an intestinal Cl^-/HCO_3^- exchanger and is up-regulated in colon of mice lacking the NHE3 Na^+/H^+ exchanger. *J. Biol. Chem.* **274**, 22855–22861.

Meyer, T. M., Munsch, T., and Pape, H. C. (2000). Activity-related changes in intracellular pH in rat thalamic relay neurons. *Neuroreport* **11**, 33–37.

Miles, E. W., Rhee, S., and Davies, D. R. (1999). The molecular basis of substrate channeling. *J. Biol. Chem.* **274**, 12193–12196.

Morgan, P. E., Supuran, C. T., and Casey, J. R. (2004). Direct inhibition of the human AE1 Cl^-/HCO_3^- exchanger by novel sulfonamide compounds. *Mol. Mem. Biol.* **21**, 423–433.

Morris, R. C. and Ives, H. E. (1996). Inherited disorders of the renal tubule. In: *The Kidney* (B. M. Brenner, Ed.), pp. 1764–1827. W. B. Saunders, Philadelphia, PA.

Moseley, R. H., Hoglund, P., Wu, G. D., Silberg, D. G., Haila, S., de la Chapelle, A., Holmberg, C., and Kere, J. (1999). Downregulated in adenoma gene encodes a chloride transporter defective in congenital chloride diarrhea. *Am. J. Physiol.* **276**, G185–G192.

Mount, D. B. and Romero, M. F. (2004). The SLC26 gene family of multifunctional anion exchangers. *Pflügers Arch.* **447**, 710–721.

Muallem, S., Blissard, D., Cragoe, E. J., Jr., and Sachs, G. (1988). Activation of the Na^+/H^+ and Cl^-/HCO_3^- exchange by stimulation of acid secretion in the parietal cell. *J. Biol. Chem.* **263**, 14703–14711.

Müller-Berger, S., Karbach, D., Konig, J., Lepke, S., Wood, P. G., Appelhans, H., and Passow, H. (1995). Inhibition of mouse erythroid band 3-mediated chloride transport by site-directed mutagenesis of histidine residues and its reversal by second site mutation of Lys 558, the locus of covalent H_2DIDS binding. *Biochemistry* **34**, 9315–9324.

Musch, M. W. and Goldstein, L. (2005). Tyrosine kinase inhibition affects skate anion exchanger isoform I alterations after volume expansion. *Am. J. Physiol. Regul. Integr. Comp. Physiol.* **288**, R885–R890.

Najafi, S. M., Harris, D. A., and Yudkin, M. D. (1996). The SpoIIAA protein of Bacillus subtilis has GTP-binding properties. *J. Bacteriol.* **178**, 6632–6634.

Noone, P. G., Olivier, K. N., and Knowles, M. R. (1994). Modulation of the ionic milieu of the airway in health and disease. *Annu. Rev. Med.* **45**, 421–434.

O'Callaghan, C. and Brenner, B. M. (2000). *The Kidney at a Glance.* Blackwell Science, Malden.

Okamura, N., Tajima, Y., Soejima, A., Masuda, H., and Sugita, Y. (1985). Sodium bicarbonate in seminal plasma stimulates the motility of mammalian spermatozoa through direct activation of adenylate cyclase. *J. Biol. Chem.* **260**, 9699–9705.

Okubo, K., Hamasaki, N., Hara, K., and Kageura, M. (1991). Palmitoylation of cysteine 69 from the C-terminal of band 3 protein in the human erythrocyte membrane. *J. Biol. Chem.* **266**, 16420–16424.

Okubo, K., Kang, D., Hamasaki, N., and Jennings, M. L. (1994). Red blood cell band 3. Lysine 539 and lysine 851 react with the same H_2DIDS (4,4'-diisothiocyanodihydrostilbene-2,2'-disulfonic acid) molecule. *J. Biol. Chem.* **269**, 1918–1926.

Paradiso, A. M., Townsley, M. C., Wenzl, E., and Machen, T. E. (1989). Regulation of intracellular pH in resting and in stimulated parietal cells. *Am. J. Physiol.* **257**, C554–C561.

Park, M., Li, Q., Shcheynikov, N., Zeng, W., and Muallem, S. (2004). NaBC1 is a ubiquitous electrogenic Na^+-coupled borate transporter essential for cellular boron homeostasis and cell growth and proliferation. *Mol. Cell* **16**, 331–341.

Parker, M. D., Ourmozdi, E. P., and Tanner, M. J. (2001). Human BTR1, a new bicarbonate transporter superfamily member and human AE4 from kidney. *Biochem. Biophys. Res. Commun.* **282**, 1103–1109.

Passow, H. (1986). Molecular aspects of band 3-mediated anion transport across the red blood cell membrane. *Rev. Physiol. Biochem. Pharmacol.* **103**, 61–223.

Perez, N. G., Alvarez, B. V., Camilion de Hurtado, M. C., and Cingolani, H. E. (1995). pHi regulation in myocardium of the spontaneously hypertensive rat. Compensated enhanced activity of the Na^+-H^+ exchanger. *Circ. Res.* **77**, 1192–1200.

Peters, L. L., Shivdasani, R. A., Liu, S. C., Hanspal, M., John, K. M., Gonzalez, J. M., Brugnara, C., Gwynn, B., Mohandas, N., Alper, S. L., Orkin, S. H., and Lux, S. E. (1996). Anion exchanger 1 (band 3) is required to prevent erythrocyte membrane surface loss but not to form the membrane skeleton. *Cell* **86**, 917–927.

Petrovic, S., Wang, Z., Ma, L., Seidler, U., Forte, J. G., Shull, G. E., and Soleimani, M. (2002). Colocalization of the apical Cl^-/HCO_3^- exchanger PAT1 and gastric H^+K^+-ATPase in stomach parietal cells. *Am. J. Physiol. Gastrointest. Liver Physiol.* **283**, G1207–G1216.

Petrovic, S., Ju, X., Barone, S., Seidler, U., Alper, S. L., Lohi, H., Kere, J., and Soleimani, M. (2003a). Identification of a basolateral Cl^-/HCO_3^- exchanger specific to gastric parietal cells. *Am. J. Physiol. Gastrointest. Liver Physiol.* **284**, G1093–G1103.

Petrovic, S., Ma, L., Wang, Z., and Soleimani, M. (2003b). Identification of an apical Cl^-/HCO_3^- exchanger in rat kidney proximal tubule. *Am. J. Physiol. Cell Physiol.* **285**, C608–C617.

Petrovic, S., Wang, Z., Ma, L., and Soleimani, M. (2003c). Regulation of the apical Cl^-/HCO_3^- exchanger pendrin in rat cortical collecting duct in metabolic acidosis. *Am. J. Physiol. Renal Physiol.* **284**, F103–F112.

Petrovic, S., Barone, S., Xu, J., Conforti, L., Ma, L., Kujala, M., Kere, J., and Soleimani, M. (2004). SLC26A7: A basolateral Cl^-/HCO_3^- exchanger specific to intercalated cells of the outer medullary collecting duct. *Am. J. Physiol. Renal Physiol.* **286**, F161–F169.

Petrovic, S., Amlal, H., Sun, X., Karet, F., Barone, S., and Soleimani, M. (2005). Vasopressin induces the expression of the Cl^-/HCO_3^- exchanger SLC26A7 in the kidney medullary collecting duct of Brattleboro rats. *Am. J. Physiol. Renal Physiol.* **290**, F1194–F1201.

Phillipson, M. (2004). Acid transport through gastric mucus. *Ups. J. Med. Sci.* **109**, 1–24.

Phillipson, M., Atuma, C., Henriksnas, J., and Holm, L. (2002). The importance of mucus layers and bicarbonate transport in preservation of gastric juxtamucosal pH. *Am. J. Physiol. Gastrointest. Liver Physiol.* **282**, G211–G219.

Pushkin, A. and Kurtz, I. (2006). SLC4 base (HCO_3^-, CO_3^{2-}) transporters: Classification, function, structure, genetic diseases, and knockout models. *Am. J. Physiol. Renal Physiol.* **290**, F580–F599.

Quentin, F., Eladari, D., Frische, S., Cambillau, M., Nielsen, S., Alper, S. L., Paillard, M., and Chambrey, R. (2004). Regulation of the Cl^-/HCO_3^- exchanger AE2 in rat thick ascending limb of Henle's loop in response to changes in acid-base and sodium balance. *J. Am. Soc. Nephrol.* **15**, 2988–2997.

Quilty, J. A. and Reithmeier, R. A. (2000). Trafficking and folding defects in hereditary spherocytosis mutants of the human red cell anion exchanger. *Traffic* **1**, 987–998.

Quinton, P. M. (1990). Cystic fibrosis: A disease in electrolyte transport. *FASEB J.* **4**, 2709–2717.

Quinton, P. M. (1999). Physiological basis of cystic fibrosis: A historical perspective. *Physiol. Rev.* **79**(Suppl. 1), S3–S22.

Rajendran, V. M., Black, J., Ardito, T. A., Sangan, P., Alper, S. L., Schweinfest, C., Kashgarian, M., and Binder, H. J. (2000). Regulation of DRA and AE1 in rat colon by dietary Na depletion. *Am. J. Physiol. Gastrointest. Liver Physiol.* **279**, G931–G942.

Ratjen, F. and Doring, G. (2003). Cystic fibrosis. *Lancet* **361**, 681–689.

Reilly, R. F. and Ellison, D. H. (2000). Mammalian distal tubule: Physiology, pathophysiology, and molecular anatomy. *Physiol. Rev.* **80**, 277–313.

Reithmeier, R. A. F. (2001). A membrane metabolon linking carbonic anhydrase with chloride/bicarbonate anion exchangers. *Blood cells Mol. Dis.* **27**, 85–89.

Richards, S. M., Jaconi, M. E., Vassort, G., and Puceat, M. (1999). A spliced variant of AE1 gene encodes a truncated form of band 3 in heart: The predominant anion exchanger in ventricular myocytes. *J. Cell Sci.* **112**, 1519–1528.

Rillema, J. A. and Hill, M. A. (2003). Pendrin transporter carries out iodide uptake into MCF-7 human mammary cancer cells. *Exp. Biol. Med. (Maywood)* **228**, 1078–1082.

Romero, M. F., Hediger, M. A., Boulpaep, E. L., and Boron, W. F. (1997). Expression cloning and characterization of a renal electrogenic Na^+/HCO_3^- cotransporter. *Nature* **387**, 409–413.

Romero, M. F., Henry, D., Nelson, S., Harte, P. J., Dillon, A. K., and Sciortino, C. M. (2000). Cloning and characterization of a Na^+ driven anion exchanger (NDAE1): A new bicarbonate transporter. *J. Biol. Chem.* **275**, 24552–24559.

Romero, M. F., Fulton, C. M., and Boron, W. F. (2004). The SLC4 family of HCO_3^- transporters. *Pflügers Arch.* **447**, 495–509.

Rossi, A. and Superti-Furga, A. (2001). Mutations in the diastrophic dysplasia sulfate transporter (DTDST) gene (SLC26A2): 22 novel mutations, mutation review, associated skeletal phenotypes, and diagnostic relevance. *Hum. Mutat.* **17**, 159–171.

Rossmann, H., Bachmann, O., Wang, Z., Shull, G. E., Obermaier, B., Stuart-Tilley, A., Alper, S. L., and Seidler, U. (2001). Differential expression and regulation of AE2 anion exchanger subtypes in rabbit parietal and mucous cells. *J. Physiol.* **534**, 837–848.

Royaux, I. E., Suzuki, K., Mori, A., Katoh, R., Everett, L. A., Kohn, L. D., and Green, E. D. (2000). Pendrin, the protein encoded by the Pendred syndrome gene (PDS), is an apical porter of iodide in the thyroid and is regulated by thyroglobulin in FRTL-5 cells. *Endocrinology* **141**, 839–845.

Royaux, I. E., Wall, S. M., Karniski, L. P., Everett, L. A., Suzuki, K., Knepper, M. A., and Green, E. D. (2001). Pendrin, encoded by the Pendred syndrome gene, resides in the apical region of renal intercalated cells and mediates bicarbonate secretion. *Proc. Natl. Acad. Sci. USA* **98**, 4221–4226.

Russell, J. M. and Boron, W. F. (1976). Role of choloride transport in regulation of intracellular pH. *Nature* **264**, 73–74.

Sachs, G., Weeks, D. L., Melchers, K., and Scott, D. R. (2003). The gastric biology of *Helicobacter pylori*. *Annu. Rev. Physiol.* **65**, 349–369.

Safsten, B. (1993). Duodenal bicarbonate secretion and mucosal protection. Neurohumoral influence and transport mechanisms. *Acta Physiol. Scand. Suppl.* **613**, 1–43.

Said, R. M., Cheah, P. L., Chin, S. C., and Goh, K. L. (2004). Evaluation of a new biopsy urease test: Pronto Dry, for the diagnosis of *Helicobacter pylori* infection. *Eur. J. Gastroenterol. Hepatol.* **16**, 195–199.

Saier, M. H., Jr., Eng, B. H., Fard, S., Garg, J., Haggerty, D. A., Hutchinson, W. J., Jack, D. L., Lai, E. C., Liu, H. J., Nusinew, D. P., Omar, A. M., Pao, S. S., *et al.* (1999). Phylogenetic characterization of novel transport protein families revealed by genome analyses. *Biochim. Biophys. Acta* **1422**, 1–56.

Sander, T., Toliat, M. R., Heils, A., Leschik, G., Becker, C., Ruschendorf, F., Rohde, K., Mundlos, S., and Nurnberg, P. (2002). Association of the 867Asp variant of the human anion exchanger 3 gene with common subtypes of idiopathic generalized epilepsy. *Epilepsy Res.* **51**, 249–255.

Sarabia, V. E., Casey, J. R., and Reithmeier, R. A. (1993). Molecular characterization of the band 3 protein from Southeast Asian ovalocytes. *J. Biol. Chem.* **268**, 10676–10680.

Schuster, V. L. (1993). Function and regulation of collecting duct intercalated cells. *Annu. Rev. Physiol.* **55**, 267–288.

Schwartz, G. J. (2002). Physiology and molecular biology of renal carbonic anhydrase. *J. Nephrol.* **15**, S61–S74.

Schweinfest, C. W., Henderson, K. W., Suster, S., Kondoh, N., and Papas, T. S. (1993). Identification of a colon mucosa gene that is down-regulated in colon adenomas and adenocarcinomas. *Proc. Natl. Acad. Sci. USA* **90**, 4166–4170.

Schwiening, C. J. and Boron, W. F. (1994). Regulation of intracellular pH in pyramidal neurones from the rat hippocampus by Na^+-dependent Cl^--HCO_3^- exchange. *J. Physiol.* **475**, 59–67.

Scott, D. A. and Karniski, L. P. (2000). Human pendrin expressed in *Xenopus laevis* oocytes mediates chloride/formate exchange. *Am. J. Physiol. Cell Physiol.* **278**, C207–C211.

Scott, D. A., Wang, R., Kreman, T. M., Sheffield, V. C., and Karnishki, L. P. (1999). The Pendred syndrome gene encodes a chloride-iodide transport protein. *Nat. Genet.* **21**, 440–443.

Scott, D. A., Wang, R., Kreman, T. M., Andrews, M., McDonald, J. M., Bishop, J. R., Smith, R. J., Karniski, L. P., and Sheffield, V. C. (2000). Functional differences of the PDS gene product are associated with phenotypic variation in patients with Pendred syndrome and non-syndromic hearing loss (DFNB4). *Hum. Mol. Genet.* **9**, 1709–1715.

Seidler, U., Roithmaier, S., Classen, M., and Silen, W. (1992). Influence of acid secretory state on Cl^--base and Na^+-H^+ exchange and pHi in isolated rabbit parietal cells. *Am. J. Physiol.* **262**, G81–G91.

Seidler, U., Hubner, M., Roithmaier, S., and Classen, M. (1994). pHi and HCO_3^- dependence of proton extrusion and Cl^--base exchange rates in isolated rabbit parietal cells. *Am. J. Physiol.* **266**, G759–G766.

Ship, S., Shami, Y., Breuer, W., and Rothstein, A. (1977). Synthesis of tritiated 4,4′-diisothiocyano-2,2′-stilbene disulfonic acid ([^{3}H]DIDS) and its covalent reaction with sites related to anion transport in human red blood cells. *J. Membr. Biol.* **33**, 311–323.

Silberg, D. G., Wang, W., Moseley, R. H., and Traber, P. G. (1995). The down regulated in adenoma (dra) gene encodes an intestine-specific membrane sulfate transport protein. *J. Biol. Chem.* **270**, 11897–11902.

Simpson, J. E., Gawenis, L. R., Walker, N. M., Boyle, K. T., and Clarke, L. L. (2005). Chloride conductance of CFTR facilitates basal Cl^-/HCO_3^- exchange in the villous epithelium of intact murine duodenum. *Am. J. Physiol. Gastrointest. Liver Physiol.* **288**, G1241–G1251.

Sly, W. S. and Hu, P. Y. (1995). Human carbonic anhydrases and carbonic anhydrase deficiencies. *Annu. Rev. Biochem.* **64**, 375–401.

Sohma, Y., Imai, Y., Gray, M. A., and Argent, B. E. (1996). A mathematical model of the pancreatic ductal epithelium. *J. Membr. Biol.* **154**, 53–67.

Soleimani, M., Greeley, T., Petrovic, S., Wang, Z., Amlal, H., Kopp, P., and Burnham, C. E. (2001). Pendrin: An apical Cl^-/OH^-/HCO_3^- exchanger in the kidney cortex. *Am. J. Physiol. Renal Physiol.* **280**, F356–F364.

Songyang, Z., Fanning, A. S., Fu, C., Xu, J., Marfatia, S. M., Chishti, A. H., Crompton, A., Chan, A. C., Anderson, J. M., and Cantley, L. C. (1997). Recognition of unique carboxyl-terminal motifs by distinct PDZ domains. *Science* **275**, 73–77.

Southgate, C. D., Chishti, A. H., Mitchell, B., Yi, S. J., and Palek, J. (1996). Targeted disruption of the murine erythroid band 3 gene results in spherocytosis and severe haemolytic anaemia despite a normal membrane skeleton. *Nat. Genet.* **14**, 227–230.

Srere, P. A. (1985). The metabolon. *Trends Biochem. Sci.* **10**, 109–110.
Srere, P. A. (1987). Complexes of sequential metabolic enzymes. *Annu. Rev. Biochem.* **56**, 89–124.
Steck, T. L., Fairbanks, G., and Wallach, D. F. (1971). Disposition of the major proteins in the isolated erythrocyte membrane. Proteolytic dissection. *Biochemistry* **10**, 2617–2624.
Sterling, D. and Casey, J. R. (1999). Transport activity of AE3 chloride/bicarbonate anion-exchange proteins and their regulation by intracellular pH. *Biochem. J.* **344**, 221–229.
Sterling, D. and Casey, J. R. (2002). Bicarbonate transport proteins. *Biochem. Cell Biol.* **80**, 483–497.
Sterling, D., Reithmeier, R. A., and Casey, J. R. (2001). A transport metabolon. Functional interaction of carbonic anhydrase II and chloride/bicarbonate exchangers. *J. Biol. Chem.* **276**, 47886–47894.
Sterling, D., Alvarez, B. V., and Casey, J. R. (2002a). The extracellular component of a transport metabolon: Extracellular loop 4 of the human AE1 Cl^-/HCO_3^- exchanger binds carbonic anhydrase IV. *J. Biol. Chem.* **277**, 25239–25246.
Sterling, D., Brown, N., Supuran, C. T., and Casey, J. R. (2002b). The functional and physical relationship between the downregulated in adenoma bicarbonate transporter and carbonic anhydrase II. *Am. J. Physiol.* **283**, C1522–C1529.
Steward, M. C., Ishiguro, H., and Case, R. M. (2005). Mechanisms of bicarbonate secretion in the pancreatic duct. *Annu. Rev. Physiol.* **67**, 377–409.
Stewart, A. K., Chernova, M. N., Kunes, Y. Z., and Alper, S. L. (2001). Regulation of AE2 anion exchanger by intracellular pH: Critical regions of the NH_2-terminal cytoplasmic domain. *Am. J. Physiol. Cell Physiol.* **281**, C1344–C1354.
Stewart, A. K., Chernova, M. N., Shmukler, B. E., Wilhelm, S., and Alper, S. L. (2002). Regulation of AE2-mediated Cl^- transport by intracellular or by extracellular pH requires highly conserved amino acid residues of the AE2 NH_2-terminal cytoplasmic domain. *J. Gen. Physiol.* **120**, 707–722.
Stewart, A. K., Kerr, N., Chernova, M. N., Alper, S. L., and Vaughan-Jones, R. D. (2004). Acute pH-dependent regulation of AE2-mediated anion exchange involves discrete local surfaces of the NH_2-terminal cytoplasmic domain. *J. Biol. Chem.* **279**, 52664–52676.
Stuart-Tilley, A., Sardet, C., Pouysségur, J., Schwartz, M. A., Brown, D., and Alper, S. L. (1994). Immunolocalization of anion exchanger AE2 and cation exchanger NHE-1 in distinct adjacent cells of gastric mucosa. *Am. J. Physiol.* **35**, C559–C568.
Stuart-Tilley, A. K., Shmukler, B. E., Brown, D., and Alper, S. L. (1998). Immunolocalization and tissue-specific splicing of AE2 anion exchanger in mouse kidney. *J. Am. Soc. Nephrol.* **9**, 946–959.
Sun, B., Leem, C. H., and Vaughan-Jones, R. D. (1996). Novel chloride-dependent acid loader in the guinea-pig ventricular myocyte: Part of a dual acid-loading mechanism. *J. Physiol.* **495**, 65–82.
Supuran, C. T., Scozzafava, A., and Casini, A. (2003). Carbonic anhydrase inhibitors. *Med. Res. Rev.* **23**, 146–189.
Synnerstad, I., Johansson, M., Nylander, O., and Holm, L. (2001). Intraluminal acid and gastric mucosal integrity: The importance of blood-borne bicarbonate. *Am. J. Physiol. Gastrointest. Liver Physiol.* **280**, G121–G129.
Takano, J., Noguchi, K., Yasumori, M., Kobayashi, M., Gajdos, Z., Miwa, K., Hayashi, H., Yoneyama, T., and Fujiwara, T. (2002). Arabidopsis boron transporter for xylem loading. *Nature* **420**, 337–340.
Tang, X.-B., Kovacs, M., Sterling, D., and Casey, J. R. (1999). Identification of residues lining the translocation pore of human AE1, plasma membrane anion exchange protein. *J. Biol. Chem.* **274**, 3557–3564.
Tanner, M. J., Martin, P. G., and High, S. (1988). The complete amino acid sequence of the human erythrocyte membrane anion-transport protein deduced from the cDNA sequence. *Biochem. J.* **256**, 703–712.
Tanphaichitr, V. S., Sumboonnanonda, A., Ideguchi, H., Shayakul, C., Brugnara, C., Takao, M., Veerakul, G., and Alper, S. L. (1998). Novel AE1 mutations in recessive distal renal tubular acidosis. Loss-of-function is rescued by glycophorin A. *J. Clin. Invest.* **102**, 2173–2179.
Tarran, R., Grubb, B. R., Gatzy, J. T., Davis, C. W., and Boucher, R. C. (2001). The relative roles of passive surface forces and active ion transport in the modulation of airway surface liquid volume and composition. *J. Gen. Physiol.* **118**, 223–236.

Tashian, R. E. and Carter, N. D. (1976). Biochemical genetics of carbonic anhydrase. *Adv. Hum. Genet.* **7**, 1–56.

Thomas, H. A. and Machen, T. E. (1991). Regulation of Cl^-/HCO_3^- exchange in gastric parietal cells. *Cell Regul.* **2**, 727–737.

Thomas, R. C. (1977). The role of bicarbonate, chloride and sodium ions in the regulation of intracellular pH in snail neurones. *J. Physiol.* **273**, 317–338.

Tietz, P. S., Marinelli, R. A., Chen, X. M., Huang, B., Cohn, J., Kole, J., McNiven, M. A., Alper, S., and LaRusso, N. F. (2003). Agonist-induced coordinated trafficking of functionally related transport proteins for water and ions in cholangiocytes. *J. Biol. Chem.* **278**, 20413–20419.

Tisher, C. C., Madsen, K. M., and Verlander, J. W. (1991). Structural adaptation of the collecting duct to acid-base disturbances. *Contrib. Nephrol.* **95**, 168–177.

Tomaszewski, L., Kulesza, E., and Socha, J. (1987). Congenital chloride diarrhoea in Poland. *Mater. Med. Pol.* **19**, 271–277.

Toure, A., Morin, L., Pineau, C., Becq, F., Dorseuil, O., and Gacon, G. (2001). Tat1, a novel sulfate transporter specifically expressed in human male germ cells and potentially linked to rhoGTPase signaling. *J. Biol. Chem.* **276**, 20309–20315.

Toye, A. M., Banting, G., and Tanner, M. J. (2004). Regions of human kidney anion exchanger 1 (kAE1) required for basolateral targeting of kAE1 in polarised kidney cells: Mis-targeting explains dominant renal tubular acidosis (dRTA). *J. Cell Sci.* **117**, 1399–1410.

Tse, W. T. and Lux, S. E. (1999). Red blood cell membrane disorders. *Br. J. Haematol.* **104**, 2–13.

Tsuganezawa, H., Kobayashi, K., Iyori, M., Araki, T., Koizumi, A., Watanabe, S. I., Kaneko, A., Fukao, T., Monkawa, T., Yoshida, T., Kim, D. K., Kanai, Y., et al. (2001). A new member of the HCO_3^- transporter superfamily is an apical anion exchanger of beta-intercalated cells in the kidney. *J. Biol. Chem.* **276**, 8180–8189.

van Adelsberg, J., Edwards, J. C., and Al-Awqati, Q. (1993). The apical Cl^-/HCO_3^- exchanger of b intercalated cells. *J. Biol. Chem.* **268**, 11283–11289.

van Vliet, A. H., Kuipers, E. J., Stoof, J., Poppelaars, S. W., and Kusters, J. G. (2004). Acid-responsive gene induction of ammonia-producing enzymes in *Helicobacter pylori* is mediated via a metal-responsive repressor cascade. *Infect. Immun.* **72**, 766–773.

Vandenberg, J. I., Metcalfe, J. C., and Grace, A. A. (1993). Mechanisms of pHi recovery after global ischemia in the perfused heart. *Circ. Res.* **72**, 993–1003.

Vandenberg, J. I., Carter, N. D., Bethell, H. W., Nogradi, A., Ridderstrale, Y., Metcalfe, J. C., and Grace, A. A. (1996). Carbonic anhydrase and cardiac pH regulation. *Am. J. Physiol.* **271**, C1838–C1846.

Vaughan-Jones, R. D. (1982). Chloride activity and its control in skeletal and cardiac muscle. *Philos. Trans. R. Soc. Lond.* **299**, 537–548.

Verlander, J. W., Madsen, K. M., Low, P. S., Allen, D. P., and Tisher, C. C. (1988). Immunocytochemical localization of band 3 protein in the rat collecting duct. *Am. J. Physiol.* **255**, F115–F125.

Verlander, J. W., Madsen, K. M., and Tisher, C. C. (1991). Structural and functional features of proton and bicarbonate transport in the rat collecting duct. *Semin. Nephrol.* **11**, 465–477.

Verlander, J. W., Madsen, K. M., Cannon, J. K., and Tisher, C. C. (1994). Activation of acid-secreting intercalated cells in rabbit collecting duct with ammonium chloride loading. *Am. J. Physiol.* **266**, F633–F645.

Verlander, J. W., Hassell, K. A., Royaux, I. E., Glapion, D. M., Wang, M. E., Everett, L. A., Green, E. D., and Wall, S. M. (2003). Deoxycorticosterone upregulates PDS (Slc26a4) in mouse kidney: Role of pendrin in mineralocorticoid-induced hypertension. *Hypertension* **42**, 356–362.

Vince, J. W. and Reithmeier, R. A. (2000). Identification of the carbonic anhydrase II binding site in the Cl^-/HCO_3^- anion exchanger AE1. *Biochemistry* **39**, 5527–5533.

Vince, J. W. and Reithmeier, R. A. F. (1998). Carbonic anhydrase II binds to the carboxyl-terminus of human band 3, the erythrocyte Cl^-/HCO_3^- exchanger. *J. Biol. Chem.* **273**, 28430–28437.

Vince, J. W., Carlsson, U., and Reithmeier, R. A. (2000). Localization of the Cl^-/HCO_3^- anion exchanger binding site to the amino-terminal region of carbonic anhydrase II. *Biochemistry* **39**, 13344–13349.

Vincourt, J. B., Jullien, D., Kossida, S., Amalric, F., and Girard, J. P. (2002). Molecular cloning of SLC26A7, a novel member of the SLC26 sulfate/anion transporter family, from high endothelial venules and kidney. *Genomics* **79**, 249–256.

Virkki, L. V., Choi, I., Davis, B. A., and Boron, W. F. (2003). Cloning of a Na^+-driven Cl^-/HCO_3^- exchanger from squid giant fiber lobe. *Am. J. Physiol. Cell Physiol.* **285**, C771–C780.

Wagner, C. A., Finberg, K. E., Stehberger, P. A., Lifton, R. P., Giebisch, G. H., Aronson, P. S., and Geibel, J. P. (2002). Regulation of the expression of the Cl^-/anion exchanger pendrin in mouse kidney by acid-base status. *Kidney Int.* **62**, 2109–2117.

Waheed, A., Zhu, X. L., Sly, W. S., Wetzel, P., and Gros, G. (1992). Rat skeletal muscle membrane associated carbonic anhydrase is 39-kDa, glycosylated, GPI-anchored CA IV. *Arch. Biochem. Biophys.* **294**, 550–556.

Waldegger, S., Moschen, I., Ramirez, A., Smith, R. J., Ayadi, H., Lang, F., and Kubisch, C. (2001). Cloning and characterization of SLC26A6, a novel member of the solute carrier 26 gene family. *Genomics* **72**, 43–50.

Wall, S. M., Hassell, K. A., Royaux, I. E., Green, E. D., Chang, J. Y., Shipley, G. L., and Verlander, J. W. (2003). Localization of pendrin in mouse kidney. *Am. J. Physiol. Renal Physiol.* **284**, F229–F241.

Wall, S. M., Kim, Y. H., Stanley, L., Glapion, D. M., Everett, L. A., Green, E. D., and Verlander, J. W. (2004). NaCl restriction upregulates renal Slc26a4 through subcellular redistribution: Role in Cl^- conservation. *Hypertension* **44**, 982–987.

Wang, C. Z., Yano, H., Nagashima, K., and Seino, S. (2000). The Na^+-driven Cl^-/HCO_3^- exchanger: Cloning, tissue distribution, and functional characterization. *J. Biol. Chem.* **275**, 35486–35490.

Wang, D. N., Sarabia, V. E., Reithmeier, R. A., and Kuhlbrandt, W. (1994). Three-dimensional map of the dimeric membrane domain of the human erythrocyte anion exchanger, band 3. *EMBO J.* **13**, 3230–3235.

Wang, T., Agulian, S. K., Giebisch, G., and Aronson, P. S. (1993). Effects of formate and oxalate on chloride absorption in rat distal tubule. *Am. J. Physiol.* **264**, F730–F736.

Wang, Z., Schultheis, P. J., and Shull, G. E. (1996). Three N-terminal variants of the AE2 Cl^-/HCO_3^- exchanger are encoded by mRNAs transcribed from alternative promoters. *J. Biol. Chem.* **271**, 7835–7843.

Wang, Z., Conforti, L., Petrovic, S., Amlal, H., Burnham, C. E., and Soleimani, M. (2001). Mouse Na^+: HCO_3^- cotransporter isoform NBC-3 (kNBC-3): Cloning, expression, and renal distribution. *Kidney Int.* **59**, 1405–1414.

Wang, Z., Petrovic, S., Mann, E., and Soleimani, M. (2002). Identification of an apical Cl^-/HCO_3^- exchanger in the small intestine. *Am. J. Physiol. Gastrointest. Liver Physiol.* **282**, G573–G579.

Wang, Z., Wang, T., Petrovic, S., Tuo, B., Riederer, B., Barone, S., Lorenz, J. N., Seidler, U., Aronson, P. S., and Soleimani, M. (2005). Renal and intestinal transport defects in Slc26a6-null mice. *Am. J. Physiol. Cell Physiol.* **288**, C957–C965.

Welsh, M. J. and Fick, R. B. (1987). Cystic fibrosis. *J. Clin. Invest.* **80**, 1523–1526.

Wheat, V. J., Shumaker, H., Burnham, C., Shull, G. E., Yankaskas, J. R., and Soleimani, M. (2000). CFTR induces the expression of DRA along with Cl^-/HCO_3^- exchange activity in tracheal epithelial cells. *Am. J. Physiol. Cell Physiol.* **279**, C62–C71.

Wieth, J. O., Andersen, O. S., Brahm, J., Bjerrum, P. J., and Borders, C. L., Jr. (1982). Chloride–bicarbonate exchange in red blood cells: Physiology of transport and chemical modification of binding sites. *Philos. Trans. R. Soc. Lond. B Biol. Sci.* **299**, 383–399.

Worrell, R. T., Best, A., Crawford, O. R., Xu, J., Soleimani, M., and Matthews, J. B. (2005). Apical ammonium inhibition of cAMP-stimulated secretion in T84 cells is bicarbonate dependent. *Am. J. Physiol. Gastrointest. Liver Physiol.* **289**, G768–G778.

Xie, Q., Welch, R., Mercado, A., Romero, M. F., and Mount, D. B. (2002). Molecular characterization of the murine Slc26a6 anion exchanger: Functional comparison with Slc26a1. *Am. J. Physiol. Renal Physiol.* **283**, F826–F838.

Xu, J., Henriksnas, J., Barone, S., Witte, D., Shull, G. E., Forte, J. G., Holm, L., and Soleimani, M. (2005). SLC26A9 is expressed in gastric surface epithelial cells, mediates Cl^-/HCO_3^- exchange, and is inhibited by NH_4^+. *Am. J. Physiol. Cell Physiol.* **289**, C493–C505.

Xu, J., Worrell, R. T., Li, H. C., Barone, S. L., Petrovic, S., Amlal, H., and Soleimani, M. (2006). Chloride/bicarbonate exchanger SLC26A7 is localized in endosomes in medullary collecting duct cells and is targeted to the basolateral membrane in hypertonicity and potassium depletion. *J. Am. Soc. Nephrol.* **17**, 956–967.

Xu, P. and Spitzer, K. W. (1994). Na-independent Cl^--HCO_3^- exchange mediates recovery from alkalosis in guinea pig ventricular myocytes. *Am. J. Physiol.* **267**, H85–H91.

Yannoukakos, D., Vasseur, C., Piau, J.-P., Wajcman, H., and Bursaux, E. (1991). Phosphorylation sites in human erythrocyte band 3 protein. *Biochim. Biophys. Acta* **1061**, 252–266.

Yannoukakos, D., Stuart-Tilley, A., Fernandez, H. A., Fey, P., Duyk, G., and Alper, S. L. (1994). Molecular cloning, expression and chromosomal localization of two isoforms of AE3 anion exchanger from human heart. *Circ. Res.* **75**, 603–614.

Zhang, D., Kiyatkin, A., Bolin, J. T., and Low, P. S. (2000). Crystallographic structure and functional interpretation of the cytoplasmic domain of erythrocyte membrane band 3. *Blood* **96**, 2925–2933.

Zhang, Y., Chernova, M. N., Stuart-Tilley, A. K., Jiang, L., and Alper, S. L. (1996). The cytoplasmic and transmembrane domains of AE2 both contribute to regulation of anion exchange by pH. *J. Biol. Chem.* **271**, 5741–5749.

Zhao, R. and Reithmeier, R. A. (2001). Expression and characterization of the anion transporter homologue YNL275w in saccharomyces cerevisiae. *Am. J. Physiol. Cell Physiol.* **281**, C33–C45.

Zheng, J., Long, K. B., Matsuda, K. B., Madison, L. D., Ryan, A. D., and Dallos, P. D. (2003). Genomic characterization and expression of mouse prestin, the motor protein of outer hair cells. *Mamm. Genome* **14**, 87–96.

Zhu, Q. and Casey, J. R. (2004). The substrate anion selectivity filter in the human erythrocyte Cl^-/HCO_3^- exchange protein, AE1. *J. Biol. Chem.* **279**, 23565–23573.

Zhu, Q., Lee, D. W. K., and Casey, J. R. (2003). Novel topology in C-terminal region of the human plasma membrane anion exchanger, AE1. *J. Biol. Chem.* **278**, 3112–3120.

Zolotarev, A. S., Townsend, R. R., Stuart-Tilley, A., and Alper, S. L. (1996). HCO_3^--dependent conformational change in gastric parietal cell AE2, a glycoprotein naturally lacking sialic acid. *Am. J. Physiol.* **271**, G311–G321.

Chapter 12
Orchestration of Vectorial Chloride Transport by Epithelia

Peying Fong[1] and Michael A. Gray[2]

[1]*The Department of Physiology, The Johns Hopkins University School of Medicine, Baltimore, Maryland 21205*
[2]*Institute for Cell and Molecular Biosciences and School of Biomedical Sciences, University Medical School, Newcastle-Upon-Tyne, United Kingdom*

I. Introduction
II. Epithelial Cell Structure and Organization
    A. Morphological Specializations of Epithelial Membrane Domains
    B. Review of Vectorial Chloride Transport
III. Junctional Proteins: An Overview
    A. Zona Occludens
    B. Zona Adherens
IV. Chloride Transporting Proteins in Epithelia: Channels, Transporters, and All Things in Between
    A. Uptake Mechanisms
    B. Exit Mechanisms
V. PDZ Proteins: What Are They? Where Are They? What Do They Do?
    A. Introduction to PDZ Proteins
    B. PDZ Proteins and the Organization of Transducisomes
VI. Two Diseases Involving Epithelial Chloride Transport
    A. Cystic Fibrosis
    B. Polycystic Kidney Disease
VII. Summary
    References

I. INTRODUCTION

Chloride transport plays a fundamental role in all cells. However, it is in epithelial cells that chloride movement becomes critical to the primary specialized task of the entire tissue: the vectorial transport of salt and water. Effective directional movement of chloride in turn fulfills the diverse fundamental physiological roles of epithelia: acting as barriers against pathogens, regulating overall fluid and electrolyte balance, determining the milieu for efficient sensory transduction, providing the optimal environment for digestive processes, and driving development. The specialized structure and organization of epithelial cells work together with the expression and activity of specific, polarized chloride transporting proteins to orchestrate the coordinated movement of chloride. In this chapter we consider the intrinsic properties of epithelia that enable directional chloride transport, thus permitting epithelia to accomplish their diverse functions. We discuss the established mechanisms by which chloride crosses epithelial tissues and offer an introduction to the molecular species that participate in these mechanisms. Finally, we discuss recent developments in our understanding of how defects in chloride transport likely underlie two epithelial diseases.

II. EPITHELIAL CELL STRUCTURE AND ORGANIZATION

Polarity constitutes the hallmark of epithelial cells. That stated, epithelial cells must produce, traffic, and retain suitable levels of transporters in an asymmetric fashion. Only when these requirements are satisfied can there be directional movements of solutes and water. Thus, in order to transport chloride ions in a vectorial fashion, chloride channels and transporters must move to the appropriate membrane domains. Moreover, many of their associated regulatory factors must be similarly delivered in order to ensure efficient action.

Figure 1 summarizes the organization of epithelial cells within an epithelium. The two distinct membrane regions of epithelial cells are the *apical* (mucosal) and *basolateral* (serosal) membranes. With a few notable exceptions, the former comes into contact with the external environment whereas the latter faces the interstitium, deriving sustenance in the form of nutrients and oxygen from constant vascular perfusion while also ridding the tissue of waste products. Epithelial *junctional complexes* determine the structural integrity of an epithelium by snugly bundling individual cells together, which in most cases preserves the pristine internal milieu. In addition, the junctional complexes delineate the apical and basolateral membranes, as well as contribute to the paracellular shunt pathway.

Epithelial sheets assume diverse configurations. They may be organized into sac-like structures of varying sizes ranging from very small (e.g., glandular acini) to large (e.g., urinary bladder). Similarly, epithelial sheets may arrange as small (e.g., glandular ducts, nephron segments) to large (e.g., colon, trachea) tubes. In select cases, epithelia assume parenchymal (e.g., liver) and follicular (e.g., thyroid) structures.

A. Morphological Specializations of Epithelial Membrane Domains

The respective functions of a given epithelium can be predicted by examining the morphological specializations of the apical membrane. In cases such as the epithelia

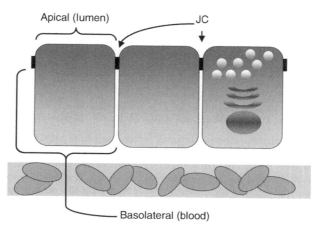

Figure 1. Schematic diagram illustrating the three critical domains of epithelial cells: the apical and basolateral membrane domains and the junctional complex (JC). The apical and basolateral membranes face the lumen and blood, respectively. The JC consists of four distinct structures: the zona occludens, zona adherens, macula adherens, and gap junctions. The apical and basolateral membrane, together with the JC, are responsible for functional polarity with respect to transepithelial ion transport. The rightmost cell depicts the polarized arrangement of some intracellular organelles (*top to bottom*: secretory granules, Golgi apparatus, nucleus), important with respect to cellular and tissue biology.

lining the airways, the apical membrane interfaces with the external world and contributes to defense from pathogens. This function is supported by the motile cilia that elaborate these membranes, which synchronously sweep away invasive and irritating agents. Epithelia such as the renal proximal tubule have brush border apical membranes, composed of specializations (microvilli) that facilitate bulk absorption by massively increasing the effective apical contact area. The bases of the microvilli also act as sites for endocytosis. The long, apical membrane elaborations of the retinal pigment epithelium maintain an uncluttered subretinal space by phagocytosing shed photoreceptor disc segments.

While motile cilia are found in a restricted number of epithelial tissues, nonmotile, primary cilia (PC) are much more common and have been observed in many different types of epithelial cells, particularly those in the kidney (Davenport and Yoder, 2005). These cilia lack the central pair of microtubules that are required to generate motile force but they can be shown to bend in response to shear forces generated by fluid flow for example (Fig. 2). Most epithelial cells possess only a single apically located PC, and these extensions of the plasma membrane protrude into the extracellular environment by up to 30 μm (Fig. 2). Despite being identified more than 100 years ago, the precise function of the PC remains unclear. Their location hints that they likely act as a "sensor" that relays information about the extracellular environment to the cell, and thereby signal changes in processes as diverse as differentiation, proliferation, and membrane ion transport. A well-studied example can be found, not altogether

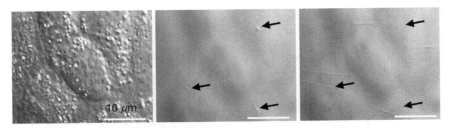

Figure 2. Differential interference contrast images of primary cilia from MDCK II cells. *Left panel* shows the image at the apical surface of MDCK cells (×100 magnification). Cells were grown for 5 days on a glass coverslip and were fully confluent. The arrows indicate the top of three unbent primary cilia (*middle panel*) visualized 14–16 μm above the cell surface under low-flow conditions (1 ml/min). The *right panel* shows the bending of the cilia on application of higher flow to the bath chamber (6.0 ml/min, right to left). Note that after bending, not all of the primary cilia remain in the focal plane and therefore the entire length is not always observed. Scale bars as indicated. (Unpublished observations, Verdon, Gray, Argent, and Simmons.)

surprisingly, in a sensory system. The retina, a tissue derived from neural epithelium, elaborates the photoreceptor outer segments from a nonmotile cilium. Indeed, vestiges of those origins can be found in the connecting ciliary bridge that joins the inner and outer segments. Photoreceptor outer segments detect light, and subsequent to the isomerization of rhodopsin generate the ionic/electrical changes that underlie vision. In many other epithelia, however, exactly what the PC sense (chemicals, osmolarity, flow) and how they transduce these sensations into a cellular response remain major, unanswered questions.

Basolateral specializations also exist. Extensive invaginations are characteristic of robustly transporting epithelial cells; predictably, this optimizes uptake of transported substrate and provides more real estate to accommodate greater numbers of the Na^+/K^+-ATPase. Mitochondria populate the cytosolic regions adjacent to these basolateral infoldings, ostensibly to provide energy necessary to power the sodium pump.

In pig pancreatic ducts, stimulation of fluid secretion has been associated with the exocytosis and insertion of tubulovesicles containing bafilomycin-sensitive H^+-ATPase pumps into the basolateral membrane. This insertion, which was associated with a marked expansion of the basolateral domain of the duct cell, could be blocked by microtubule inhibitors (Buanes *et al.*, 1987; Veel *et al.*, 1990). Thus, the dynamic regulation of the basolateral membrane area and resident protein population contributes to the prosecretory response of this epithelium. It is likely that other epithelial tissues exploit analogous mechanisms.

Other organelles also distribute in a polarized fashion (Fig. 1). Epithelial cells have Golgi stacks that organize in a region apical to the nuclei. Those that secrete a presynthesized protein product, such as the acinar cells of the exocrine pancreas, accumulate secretory granules in the subapical region, where they are well positioned for luminal release on reception of the appropriate signal.

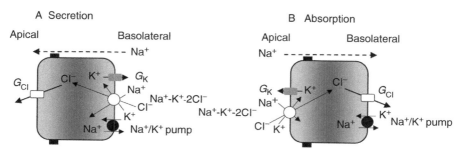

Figure 3. Summary of key transport processes mediating transepithelial secretion (A; left) and absorption (B; right). Both processes rely on a cation-coupled chloride transporter (here, shown as the Na^+-K^+-$2Cl^-$ cotransporter) for the accumulation of intracellular chloride to levels permitting its passive exit through an appropriately positioned conductive route (G_{Cl}). The cotransport mechanism depends absolutely on the activity of the Na^+/K^+-ATPase, which maintains low intracellular sodium levels. Note that the net transport of NaCl in both secreting and absorbing epithelia results from the paracellular movement of sodium.

B. Review of Vectorial Chloride Transport

1. Secretion

The shark rectal gland represents the classical model for chloride secretion (Epstein and Silva, 1985). This tissue secretes salt efficiently by a mechanism that is echoed in chloride-secreting epithelia, such as airway submucosal glands, and that is mirrored in absorptive epithelia like the thick ascending limb of the nephron (Fig. 3). Briefly, an inwardly directed sodium gradient permits the electroneutral, sodium-coupled uptake of chloride across the basolateral membrane. In the rectal gland, this uptake is facilitated by the $Na^+/K^+/2Cl^-$ cotransporter, NKCC1. As pointed out by Epstein and Silva (1985), a potassium-dependent chloride uptake mechanism, such as NKCC1, allows the epithelium to double the efficiency of the sodium pump. As a result, chloride levels rise above electrochemical equilibrium. On opening of apically situated chloride channels, chloride exits into the luminal or mucosal compartment. Potassium cycles across the basolateral membrane via potassium channel(s), whereas sodium is removed from the cell by the action of the Na^+/K^+-ATPase. Driven by the lumen-negative potential difference created by chloride exit, sodium enters the lumen by moving through cation-selective (and presumably anion-impermeant) paracellular routes, ultimately resulting in the net secretion of NaCl.

2. Absorption

The mammalian thick ascending limb of Henle's loop (TAL) absorbs NaCl avidly. The $Na^+/K^+/2Cl^-$ cotransporter, NKCC2, and the potassium channel that permits recycling (KCNJ1; the renal outer medullary potassium channel, ROMK) are localized to the apical membrane of TAL cells. CLC-Kb/barttin, the chloride conductance mediating exit, resides in the basolateral membrane. The exit of chloride generates a lumen-positive transepithelial potential difference that also drives the movement of sodium from the lumen, across the junctions and ultimately to the peritubular capillaries. Thus, with the exception of the basolateral Na^+/K^+ pump, absorption by the TAL can be described as the mirror image of secretion by the rectal gland. The distal tubule utilizes a similar mechanism, with the primary difference being that rather than a potassium-coupled mechanism, the thiazide-sensitive Na^+/Cl^--cotransporter (NCC) mediates luminal salt uptake.

III. JUNCTIONAL PROTEINS: AN OVERVIEW

The apical and basolateral regions are delimited by discrete points of contact between adjacent epithelial cells. Histologists have long appreciated their existence and designated them as four distinct structures: the zona occludens, zona adherens, macula adherens, and gap junctions. Together, these form a diffusional obstacle between the interstitial fluid and the apical aspect of the epithelium. Molecular components have been identified for all of these structures. We highlight in the following two sections several critical components in the function of the zona occludens and the zona adherens. Together with their associated signaling molecules, these structures regulate paracellular transport and epithelial polarization, functions highly relevant to our topic.

A. Zona Occludens

Diverse proteins contribute to this structure, which is also known as the tight junction (TJ). These components include: (1) the related, transmembrane spanning claudins and occludins; (2) membrane-associated, cytoplasmic PDZ proteins such as ZO-1, ZO-2, and ZO-3; (3) membrane-spanning PDZ proteins (CRB1); (4) junctional adhesion molecules (JAMs); (5) signaling molecules such as protein kinase A (PKA) and PKC, Rho-family GTPases and heterotrimeric G-proteins; and (6) putative nuclear transcription factors. Here, we focus primarily on the claudins, a large family of proteins that recently have been shown to mediate paracellular permeation pathways. The relationships between various PDZ proteins with junctional components and signaling molecules are discussed further in Section V.

Furuse et al. (1998) first identified claudins as integral membrane proteins distinct from occludin, which up to then was the only such tight junctional component. To date, at least 20 claudin genes have been identified. Their gene products have a molecular mass of ~22 kDa and share a common structural motif consisting of four membrane-spanning domains, intracellular N- and C-termini, and two extracellular loops.

With respect to the level of ionic flux (leakiness) and selectivity, the paracellular permeability differs markedly between different epithelia. Claudins are expressed in an epithelial tissue-specific manner. Recent evidence suggests their role in determining the characteristic paracellular selectivity of each tissue, behaving as *bona fide* paracellular ionic channels.

Furuse *et al.* (1999) raised the possibility that heterogeneous claudins may interact and determine the overall permeability of the junctions. Subsequently, they demonstrated that expression of claudin-2 converted a model tight epithelium, MDCK I, to one with leaky junctions (Furuse *et al.*, 2001). Structure-function analysis points to a role for the extracellular loops in conferring this attribute; notably, the charge at position 65 within the first extracellular loop dictated the relative preference for cations or anions (Colegio *et al.*, 2002). Moreover, the overexpression of different claudins not only alters the charge selectivity of the paracellular pathway in MDCK II and LLC-PK1 renal epithelial cell lines (Van Itallie *et al.*, 2003) but also affects the transepithelial resistance. These lines of evidence raise the possibility that hormonally responsive epithelia such as the kidney collecting duct may regulate junctional permeability in the transition from basal to stimulated states. For example, the massive sodium absorption that results from aldosterone stimulation also generates a steep transtubular sodium gradient. This would impede further absorption were it not for stringent regulation of the paracellular pathway. Collecting ducts express claudin-4 and claudin-8, species known to select against cations. Thus, in addition to the well-studied increases involving the epithelial sodium channel (ENaC), aldosterone status may also signal elevated expression of claudin-4 and claudin-8.

Importantly, the mutation of claudin-16, paracellin, is associated with the human disease, familial hypercalciuric hypomagnesemia (Simon *et al.*, 1999). Salt absorption by the TAL normally results in a lumen-positive potential difference (Section B.2) that drives divalent cations across the junctions. Disease-causing mutations result in the inability of this segment to reabsorb calcium and magnesium and hence renal calcium and magnesium wasting.

B. Zona Adherens

The adherens junction serves as a focal point for cell–cell adhesion that is mediated primarily by E-cadherin. E-cadherin is a critical adhesion molecule that contains five extracellular repeats, a single membrane-spanning domain, and an intracellular C-terminus. Extracellular calcium promotes homophilic interactions between juxtaposing E-cadherin molecules. Associated with E-cadherin on the cytosolic aspect are the catenins (α and β), α-actinin, and vinculin. Catenins bind to the C-terminus of E-cadherin and link it to the actin cytoskeleton. β-Catenin also localizes in the cytosol, but in the absence of regulatory factors these levels are generally kept low by continuous degradation.

A homologue of the *Drosophila armadillo* gene product, β-catenin plays critical roles in adhesion, signaling, and transcription (Nelson and Nusse, 2004; Harris and Peifer, 2005). One well-studied pathway signaling and transcriptional regulation involves the binding of Wnt ligands to their receptors, the seven transmembrane domain (TMD) receptor, Frizzled, and the lipoprotein-related receptors 5 and 6 (LPR5/LPR6).

This ligand-receptor binding activates Disheveled (Dsh). Activated Dsh rescues cytosolic β-catenin from degradation, making it available for complex formation with the transcription factor, TEF/LEF, and thereby inducing target gene expression. Disruption of this, the *canonical* Wnt signaling pathway, by Wnt mutations results in gross developmental abnormalities (Cadigan and Nusse, 1997).

The zinc-finger transcription factor, Snail, reportedly represses the transcription of critical proteins in both the adherens and occludens junctions (Cano *et al.*, 2000; Ikenouchi *et al.*, 2003). Affected proteins include occludin, claudin-3, claudin-4, and claudin-7, and E-cadherin. However, Snail-mediated interference of not only transcriptional but also posttranscriptional events paves the transition of epithelia to mesenchyme (Ohkubo and Ozawa, 2004).

In summary, both the zona occludens and the zona adherens host signaling proteins that are important to epithelial polarity and the dynamic regulation of morphology.

IV. CHLORIDE TRANSPORTING PROTEINS IN EPITHELIA: CHANNELS, TRANSPORTERS, AND ALL THINGS IN BETWEEN

A. Uptake Mechanisms

Together with the potassium-coupled chloride transporters (KCCs) (Section IV.B.1) that mediate cellular chloride efflux, NCCs and NKCCs comprise a distinct molecular superfamily of 12 TMD proteins, the cation-chloride cotransporters (CCC). Detailed information on these molecules, so critical to diverse cellular homeostatic mechanisms, can be found in another chapter of this volume (Chapter 10). Here, we provide a brief overview, followed by special emphasis on the special role of these molecules in transepithelial transport.

As we discussed in Sections II.B.1 and II.B.2, sodium-coupled transporters, such as the NKCCs and the NCC, mediate the uptake of chloride in both secretory and absorptive epithelia. These mechanisms operate analogously to those mediating the sodium-coupled uptake of substrates such as glucose and amino acids: they are all *secondary active transporters*. Secondary active transport hinges upon the activity of the Na^+/K^+-ATPase, which generates and sustains an inwardly directed sodium gradient. In the case of NCC and NKCCs, this gradient enables the accumulation of chloride to intracellular levels exceeding electrochemical equilibrium.

The thiazide-sensitive NCC was first isolated from flounder urinary bladder and rat kidney (Gamba *et al.*, 1993, 1994). The identification of the bumetanide-sensitive $Na^+/K^+/Cl^-$ cotransporter from a dogfish shark rectal gland cDNA library (NKCC1) (Xu *et al.*, 1994) was followed shortly thereafter by the isolation of the kidney-specific NKCC2 from rabbit and rat kidney (Gamba *et al.*, 1994; Payne and Forbush, 1994).

Overall, NCC and NKCCs share approximately 45–48% amino acid identity, with the greatest divergence within the large N- and C-termini. These regions harbor potential serine and threonine regulatory sites. In the following sections, we discuss the physiological regulation of CCCs.

1. NKCC

The elucidation of the primary structures of NKCC1 and NKCC2 immediately verified the close molecular relationship between these two species, a relationship that originally was predicted based on functional data. NKCC1 and NKCC2 bear roughly 60% identity on the amino acid level, with deglycosylated molecular masses of approximately 120–130 kDa. NKCC1 shows a broad distribution not only in diverse transporting epithelia but also in excitable tissues such as neurons, as well as cardiac and skeletal muscle. It likely plays a critical role in regulatory volume increase, the recovery of cellular volume in response to shrinkage. Thus, it alternatively is referred to as either the "secretory" or "housekeeping" isoform of the NKCCs. In chloride-secreting epithelia such as the shark rectal gland and the serous cells of mammalian airway submucosal glands, NKCC1 localizes to the basolateral membrane and functions as the principal chloride uptake pathway. In contrast, NKCC2 is expressed exclusively in the kidney, albeit as three splice variants that distribute in a distinctive pattern (Payne and Forbush, 1994). Whereas one is found in both cortex and medulla, the second localizes to the cortex (cortical TAL) and the third to the medulla (mTAL). This overall distribution pattern suggests that each splice variant is optimally suited to regulation in the context of their respective environments. Moreover, these observations afford a starting point for structure–function studies.

Decreases in intracellular chloride at constant cell volume increase phosphorylation of NKCC and activates NKCC-mediated transport (Lytle and Forbush, 1996). Recent evidence points to a role for a novel serine-threonine kinase, Ste20-related $p$roline-$a$lanine-rich $k$inase (SPAK), alternatively referred to as $p$roline $a$lanine-rich STE20-related $k$inase (PASK), in the regulation of both NKCC1 and NKCC2 (Piechotta *et al.*, 2002; Dowd and Forbush, 2003).

2. NCC

Previous *in vivo* functional studies of rat distal tubules identified a cation-coupled chloride cotransporter distinct from the NKCC in its lack of requirement for external potassium, as well as in its inhibition by thiazide diuretics (Velazquez *et al.*, 1984; Ellison *et al.*, 1987). Stokes (1984) also discovered such a mechanism in teleost bladder epithelium. Using a *Xenopus* oocyte expression cloning strategy, Gamba *et al.* (1993) isolated the cDNA encoding a transporter with precisely these properties from flounder urinary bladder, the first member of the CCC family. A clone from rat kidney was subsequently isolated (Gamba *et al.*, 1994), and high stringency riboprobe analysis indicated primarily a renal distribution.

Mutations in the WNK [$w$ith no lysine(K) kinases, WNK1 and WNK4, regulate NCC activity and result in the autosomal dominant disease, pseudohypoaldosteronism type II (PHA II; Yang *et al.*, 2003), that is characterized by hypertension and hyperkalemia. PHA II responds to thiazide diuretic treatment, consistent with the notion that disease-associated WNK mutations cause inappropriate NCC activation. Insights into the underlying regulatory mechanism recently were provided by studies showing that WNK1 acts via SPAK (or PASK) and OSR1 (for oxidative stress response 1). Moriguchi et al. (2005) suggested that WNK1-mediated phosphorylation of SPAK and OSR1 in turn activates these kinases to phosphorylate directly the regulatory regions at the N-termini of NCC, as well as NKCC1 and NKCC2.

Although NCC is expressed primarily in the distal tubule of the nephron, its functional importance in extra renal epithelia is poorly understood. A recent study demonstrates its presence in the ileum and jejunum (Bazzini et al., 2005), where it may interact functionally with an epithelial calcium channel by a mechanism similar to that described for the mammalian distal tubule (Reilly and Ellison, 2000).

3. Anion Exchangers

Many epithelial cells also possess an electroneutral, DIDS-sensitive Cl^--HCO_3^- anion exchanger (AE) activity which can be functionally detected as a rise in pH_i when chloride is removed from the basolateral side, and which is independent of Na^+. Under normal physiological conditions, this transporter would act to load the cells with chloride while removing HCO_3^-, and thus help in generating a larger driving force for chloride secretion. These electroneutral AEs play important roles in normal acid/base balance and cell volume regulation (see Alper, 2006 for a review). Electroneutral AEs belong to the SLC4 family of AE transporters that, in addition to four electroneutral Cl^--HCO_3^- exchangers (AE1–4), also includes a number of Na^+-dependent HCO_3^- transporters that are critical to the uptake of HCO_3^- into epithelial cells (NBCs) (Section IV.B.5). AE2 is the most widely expressed AE, and is found in the basolateral membrane of most epithelial cells.

B. Exit Mechanisms

In the case of chloride-secreting epithelia, the gating of apically localized chloride channels permits the passive movement of chloride into the luminal compartment. Salt-absorbing epithelia, on the other hand, must have mechanisms that permit the exit of chloride across the basolateral membrane. In the following sections we review a selection of these pathways.

1. KCC

Electroneutral KCl transport has a well-recognized role in regulatory volume decrease in response to cellular swelling. Also established is its involvement in transepithelial transport (Greger and Schlatter, 1983; Reuss, 1983; Guggino, 1986).

Gillen et al. (1996) first identified the sequences for human, rabbit, and rat KCC1s. These ubiquitously expressed proteins, having deglycosylated molecular masses of ~120 kDa, share significant amino acid identity (~67%) with the neuronal form, KCC2, cloned from rat brain (Payne et al., 1996). To date, two more isoforms, KCC3 and KCC4, have been identified and studied (Mount et al., 1999). KCC3 is more similar to KCC1 (~77% homology), whereas KCC4 bears greatest similarity to KCC2 (~69%). KCCs have ~25% identity with the NKCCs and NCCs. Although KCCs share the same general predicted structure as the other CCCs, with 12 transmembrane spanning domains and intracellular N- and C-termini, the large extracellular loop of this protein rests between TMD 5 and TMD 6 rather than TMD 7 and TMD 8. The subtle, observed differences in activation by cell swelling and affinities for extracellular substrates likely are due to the substitutions of amino acids at critical TMDs, as well as at phosphorylation sites at the intracellular termini.

Predictably, ^{86}Rb efflux carried by KCC requires chloride, but is unaffected by sodium replacement. KCC has a higher affinity for the loop diuretic furosemide than bumetanide. It also can be distinguished from NKCC on the basis of its markedly poorer bumetanide sensitivity ($K>100$ μM vs ~1 μM for NKCC), as well as its activation (rather than inhibition) by N-ethylmaleimide.

Analyses of KCC2, KCC3, and KCC4 knockout mice have elucidated the diverse roles of these transporters, and speak to their importance in regulating intracellular chloride and in epithelial ion transport. The knockout of the genes encoding the neuronally expressed forms, KCC2 and KCC3 resulted in, respectively, (1) neonatal death due to elevated intraneuronal chloride levels and the associated loss of GABAergic inhibition and (2) deafness and neurodegeneration (Hübner et al., 2001; Boettger et al., 2003). Knockout of the gene encoding KCC4 resulted in epithelial transport disturbances, such as deafness due to interference of potassium recycling in the inner ear and renal tubular acidosis due to an indirect impairment of chloride recycling at the basolateral membrane of α-intercalated cells (Boettger et al., 2002).

2. CLC Chloride Channels and Transporters

The CLC family of transmembrane proteins represents a diverse group of ion channels and transporters involved in chloride transport across the plasma membrane and intracellular organelles of a variety of epithelial cells (discussed in depth in Chapters 2–4) (Jentsch et al., 2002). Of the nine mammalian family members, eight are expressed in the kidney and two of these are linked to human renal disease: ClC-Kb to Bartter syndrome and ClC-5 to Dent's disease (Jentsch et al., 2005). The four family members expressed in the plasma membrane (ClC-1, ClC-2, and ClC-Ka/Kb) share the most sequence homology with ClC-0, the prototypic CLC protein (Jentsch et al., 1990, 2002). ClC-1 is only found in skeletal muscle, and ClC-Ka and ClC-Kb are restricted to the kidney and inner ear. On the other hand, ClC-2 is expressed in most cells of the body (Thiemann et al., 1992; Cid et al., 1995). Analysis of ClC-2-expressing oocytes suggests that this channel likely plays a role in volume regulation, as well as in neuronal excitability (Clark et al., 1998). However, its specific role in epithelial function is obscure, as the $ClCn2^{-/-}$ mouse only shows severe retinal and testicular degeneration.

ClC-3, ClC-4, and ClC-5 are broadly expressed and localize to endosomal compartments. Here they help in endosomal acidification through a mechanism that is not fully established but is now likely to be through their ability to act as Cl^-/H^+ exchangers (see following paragraph). The remaining family members, ClC-6 and ClC-7, are also found in intracellular organelles where they are also likely to regulate the pH of intracellular compartments.

To date, attempts at heterologous expression have not produced functional currents. ClC-6 functionally complements the *gef1* yeast phenotype (Kida *et al.*, 2001). When overexpressed in COS cells, the c-myc-tagged splice variants ClC-6a and ClC-6c colocalize with the sarco/endoplasmic reticulum calcium pump, SERCA2b (Buyse *et al.*, 1998). The ClC-7 knockout mouse ($Clcn7^{-/-}$) has osteopetrosis, thus revealing the importance of this protein in the acidification of the resorptive lacunae (Kornak *et al.*, 2001). More recent work shows that the loss of ClC-7 function results in lysosomal storage abnormalities, even in the absence of gross lysosomal pH changes. $Clcn7^{-/-}$ mice also show severe degeneration of the central nervous system (Kasper *et al.*, 2005). Taken together, these studies suggest a role for ClC-6 and ClC-7 in modulating the composition of intracellular compartments.

Perhaps the most recent significant development in this field was the demonstration in 2004 from Chris Miller's group (Accardi and Miller, 2004) that the bacterial ClC homologue, ClC-ec1, functions as a Cl^-/H^+ exchanger and not an ion channel as originally assumed (Accardi *et al.*, 2004). Elegant electrophysiological experiments showed that the transporter exchanges two chloride ions for every proton, thus making the transporter electrogenic, and that preestablished gradients of either chloride or protons were able to induce uphill transport of the other exchange ion, identifying it as a *bona fide* secondary active transporter. Since then two mammalian members ClC-4 and ClC-5, but not ClC-0, ClC-2, or ClC-Ka, have also been shown to act as coupled chloride-proton exchangers, after expression in heterologous expression systems (Picollo and Pusch, 2005; Scheel *et al.*, 2005). In addition, the neutralization of a key glutamate residue, which was known to regulate ion passage through the "pore" of the channel (Dutzler *et al.*, 2003), eliminated coupling of anion transport to proton movement (Picollo and Pusch, 2005; Scheel *et al.*, 2005), providing strong evidence that ion channel and Cl^-/H^+ exchange activity resided in the same protein. However, how this new function of ClC-4 or ClC-5 relates to the regulation of endosomal pH is still unclear.

Because CLCs often are expressed in the same cells, the potential exists for the formation of heterodimeric channels. Coexpression studies of CLC channels using the *Xenopus* oocyte system support this concept (Lorenz *et al.*, 1996; Weinreich and Jentsch, 2001). Coimmunoprecipitation experiments using rat kidney suggest that ClC-4 and ClC-5 form heterodimers (Mohammad-Panah *et al.*, 2003). In principle, the presence of heteromeric assemblies would lead to a variety of dimeric configurations, thus accounting for the diverse spectrum of characteristics exhibited by CLC-like channels in native cell systems. This could explain the range of acidification capacities characteristic of different intracellular, organellar compartments. Because ClC-3, ClC-4, and ClC-5 all localize intracellularly in vesicular and tubulovesicular structures, their heterodimerization conceivably could generate novel assemblies that promote acidification to different degrees. Thus, whether heterodimers indeed assemble and function in native tissues poses a provocative question for future studies.

Within their C-termini, all mammalian CLCs contain two copies of a highly conserved structural motif, called the CBS domain (for cystathionine β-synthase, a protein in which the motif was originally described) (Bateman, 1997; Ponting, 1997). Another protein in which CBS domain duos exist is inosine monophosphate dehydrogenase (IMPDH), and the structure of this protein is known (Sintchak et al., 1996). Scott et al. (2004) recently demonstrated that tandem pairs of CBS domains bind AMP, ATP, and S-adenosyl methionine, thus suggesting a role in sensing the cellular metabolic status. These findings suggest that the CBS domains may provide a link between the activity of mammalian CLCs and cellular metabolism.

Assessment of ionic channel and transporter physiology historically has been facilitated by the availability of pharmacological agents that dramatically alter a particular aspect of function. However, until very recently, the lack of specific blockers for chloride channels in general, and specifically CLC transport proteins, has hampered progress in understanding their physiology, as well as in developing useful therapeutic interventions of their associated disease states. Important developments in this field are discussed in more detail in Chapter 4 in this volume.

3. CFTR

The cystic fibrosis transmembrane conductance regulator (CFTR) is a member of the ATP-Binding Cassette (ABC) transporter family that functions as an anion channel in a wide variety of epithelial cells (Riordan et al., 1989; Sheppard and Welsh, 1999). Mutations in the gene encoding CFTR (*CFTR*) causes cystic fibrosis (CF), a life-limiting, autosomal recessive human disease that is characterized by impaired ion transport in the airways, pancreas, and other tissues (Doull, 2001). CFTR dysfunction also underlies many forms of secretory diarrhea, where its overactivity produces a profound intestinal loss of salt and fluid. *CFTR* encodes a 160-kDa protein, comprising two 6-transmembrane spanning domain regions (TMD1 and TMD2), two cytoplasmic nucleotide-binding domains (NBD1 and NBD2), and a regulatory (R) domain. From extensive molecular, biochemical, and electrophysiological analyses, we know that CFTR activation requires not only the phosphorylation of the R domain by PKA and PKC, but also the binding of ATP to NBD1 and NBD2. However, one major question that is still not fully answered regarding this unusual ion channel is exactly how phosphorylation and nucleotide binding promote channel opening and closing. The motivated reader is encouraged to refer to Chapters 5 and 6 for detailed treatments of this topic.

In addition to the above-mentioned functional domains, the CFTR C-terminus ends with four amino acids, DTRL, which interact with PDZ domains. Although PDZ interactions have been proposed to facilitate interactions critical to gating (Wang et al., 2000), their functionality primarily may rest in their spatial sequestration and stabilization of macromolecular signaling components. PDZ domain interactions are discussed extensively in Section V of this chapter.

In addition to acting as a chloride channel, CFTR plays a role in regulating other ionic channels. These include an outwardly rectifying chloride channel (ORCC) that remains to be identified on the molecular level (Guggino, 1993). Expression studies have shown its influence on cation channels such as the renal outer medullary potassium channel, ROMK2 (Kir1.1a; McNicholas et al., 1996, 1997; Lu et al., 2006), as well as on the ENaC. Because it forms an important conceptual basis for understanding the pathology of CF disease, we devote the next paragraphs to a short description of the relationship between CFTR and ENaC.

In the airways, CFTR is thought to act as an inhibitor of ENaC and thus limit sodium reabsorption. Hence, the absence or dysfunction of CFTR in CF airways results in the observed sodium hyperabsorption (Knowles et al., 1981; Mall et al., 2004). Using a heterologous oocyte expression system, Kunzelmann showed that this inhibitory effect of CFTR on ENaC required functionally active NBD1 and R domains (Kunzelmann et al., 1997) and moreover depended on the ability of CFTR to conduct chloride, since the effects were not observed when ΔF508-CFTR was coexpressed with ENaC (Mall et al., 1996; Kunzelmann et al., 1997). Similar conclusions were also reached in studies conducted on M-1 mouse cortical collecting duct (CCD) cells, suggesting that in the kidney, as in the airways, CFTR normally inhibits sodium reabsorption (Stutts et al., 1995; Kunzelmann et al., 1997; Letz and Korbmacher, 1997; Briel et al., 1998; Bachhuber et al., 2005).

However, the jury is still out. Recent work on the sweat duct (Reddy and Quinton, 2006) concludes that the interaction of CFTR and ENaC depends critically on the net direction of chloride permeation through CFTR. Physiologically, this is an important finding as it suggests that ENaC is not inhibited by CFTR activity in epithelial tissues which actively absorb chloride transcellularly (e.g., sweat duct). This contrasts with the situation found in tissues which secrete chloride via CFTR, which show inhibition of ENaC.

4. CaCCs

In contrast to both CLCs and CFTR, the molecular identity of the channels that underlie calcium-activated chloride currents (CaCCs) (see also Chapter 7), present in the majority of chloride-secreting epithelial tissues, is still not fully resolved. Part of this problem may relate to the heterogeneity in the functional properties of CaCCs in different epithelial cells (Kidd and Thorn, 2000; Hartzell et al., 2005). CaCCs can be categorized on the basis of their requirements for activation (Pusch, 2004). Thus, they can be either (1) strictly and exclusively dependent on intracellular calcium levels, (2) regulated by the calcium/calmodulin-dependent protein kinase II (CaMKII), or (3) regulated by cGMP. In the context of epithelial transport, activation of CaCCs usually leads to relatively large, but transient, increases in transepithelial chloride secretion. This contrasts with the more sustained secretion observed with activation of

CFTR, and suggests that the role of CaCCs may be to fine-tune chloride secretion dominated by CFTR-dependent mechanisms (Tarran et al., 2002, 2006). However, in many types of acinar cells, CFTR is absent and CaCCs dominate the chloride exit pathway across the apical membrane. In these cells, CaCC activity is exquisitely controlled by local (apical) changes in cytosolic calcium generated through IP_3^--dependent mechanisms (Petersen, 2005).

To date, three families of putative CaCCs have been described. These include the CLCAs, the bestrophins, and, most recently, human homologues of the maxi-Cl^- channels originally described in *Drosophila* (Suzuki, 2006).

a. CLCAs

In 1995, the cloning of the first putative CaCC was reported from bovine trachea (Cunningham et al., 1995). Now known as bCLCA1, the cDNA encoded a 100- to140-kDa membrane protein of 903 amino acids. The expression of the bCLCA1 cRNA in *Xenopus* oocytes yielded channels with similar functional properties to those originally reported for a 38-kDa protein purified from bovine tracheal epithelium, and subsequently studied in planar lipid bilayers. Posttranslational cleavage may underlie the discrepancy between the size of this product and the purified 38-kDa protein, but this has not yet been addressed rigorously. Importantly, like the reconstituted protein (Fuller et al., 1994), the expressed bCLCA1 currents showed increased sensitivity to intracellular Ca^{2+} levels after phosphorylation by CaMKII (Cunningham et al., 1995), and were inhibited by DIDS and DTT. Interestingly, bCLCA1 currents were insensitive to niflumic acid, a compound known to block the endogenous oocyte CaCC, as well as many endogenous CaCCs previously characterized in epithelial cells (Cunningham et al., 1995).

It soon became clear that bCLCA1 founded a diverse molecular family (Elble et al., 1997; Gandhi et al., 1998; Gruber et al., 1998; Romio et al., 1999; Evans et al., 2004). Many roles have been attributed to bovine, human, porcine and mouse members of the CLCA family. These include: (1) regulation of ionic channels (Loewen and Forsyth, 2005); (2) action as extracellular matrix proteins (Abdel-Ghany et al., 2001, 2002, 2003), as well as tumor suppressors (Gruber et al., 1999c; Bustin et al., 2001; Elble and Pauli, 2001); (3) function as targets for cytokine signaling (Zhou et al., 2001; Hoshino et al., 2002; Toda et al., 2002; Atherton et al., 2003); and (4) up-regulation of mucin secretion by induction of *MUC2* (McNamara et al., 2004). Overall agreement on CLCA topology is lacking, with predictions ranging between 1–5 membrane-spanning domains (Gruber et al., 1999b; Romio et al., 1999). More perplexing is the paucity of structure–function information pinpointing which of several proteolytic fragments forms the anion-conducting domain. Moreover, hCLCA3 encodes a truncated protein that lacks TMDs and is secreted, further complicating the picture (Gruber et al., 1999a). Those CLCA isoforms that express functional chloride currents show considerable heterogeneity in biophysical and pharmacological properties. Particularly important is that several of these characteristics are at odds with those documented for the endogenous CaCCs in most epithelial cells, and may even depend on the choice of expression system. These include calcium (and CaMKII) dependence and rectification properties, as well as sensitivity to inhibitors such as niflumic acid and DTT (discussed in Eggermont, 2004). However, at least in the mouse renal IMCD cell line (mIMCD-3), the endogenous CaCC has most of the characteristics expected of CLCA channels, as well as expressing a number of CLCA isoforms by RT-PCR (Shindo et al., 1996; Stewart et al., 2001). But in many cases, CLCA expression is undetectable in some cell types that clearly show CaCC activity (Papassotiriou et al., 2001; Fong et al., 2003). Therefore, at this time, whether CLCAs encode the channels comprising CaCCs or regulators of yet-to-be-identified native proteins remains controversial and will require continued research.

b. Bestrophins

The first member of the bestrophin family was identified by positional cloning as the product of *VMD2*, the gene that is disrupted in the autosomal dominant retinopathy, Best disease, or vitelloform macular dystrophy (Marquardt et al., 1998; Petrukhin et al., 1998; Stohr et al., 2005). Individuals with Best disease were predicted to lack a calcium-regulated chloride conductance from physiological measurements made on retinal pigment epithelial cells, leading Sun et al. (2002) to hypothesize that *VMD2* encodes a chloride channel. Functional analysis of the two human homologues, hBest1 and hBest2, in HEK 293 cells (Sun et al., 2002) showed whole cell currents with an anion permeability sequence of $NO_3^- > I^- > Br^- > Cl^-$, a linear *I-V* relationship, and block by 500-µM DIDS and DTT as well as the sulfhydryl-reactive agents, MTSEA and MTSET. The expression of cysteine-less hBest1 mutants resulted in currents that resisted block by MTSEA and MTSET. Photorelease of calcium augmented the current, thus further characterizing bestrophin as a calcium-activated channel. Moreover, Best disease-associated mutants yielded diminished whole cell currents. Work from Tsunenari et al. (2003) gave a detailed topographical mapping of hBest1 that predicted three extracellular loops, four TMDs and cytosolic N- and C-termini.

To date, four human and four mouse bestrophin family members have been identified (Stohr et al., 2002; Tsunenari et al., 2003). A recent in depth functional analysis further identifies mouse bestrophin-2 (mBest2) as a time-independent, calcium-activated chloride channel, with an EC_{50} for calcium of 230 nM (Qu et al., 2003). Structure–function analysis of residues in the second putative TMD, together with sulfhydryl modification studies, identified S79 as a pore-lining residue that acts as an anion binding site and determines anion selectivity by a mechanism independent from electrostatic interaction. Thus, these findings appear to confirm that bestrophin family members do indeed function as chloride channels (Qu et al., 2004).

One key question that remains is whether the bestrophins underlie the endogenous CaCCs that have been functionally characterized in many different, nonretinal, epithelial cells (Hartzell et al., 2005). This remains an unanswered question as molecular data is still limited. Nonetheless, RT-PCR evidence speaks to this possibility at least in the kidney (Petrukhin et al., 1998), as well as in the mouse IMCD-K2 cell line (Qu et al., 2003), the latter cells having well-characterized CaCCs (Boese et al., 2000;

Qu et al., 2003; Boese et al., 2004). However, the biophysical properties of CaCCs measured in IMCD-K2 cells differ substantially from those of bestrophins. Thus, an involvement of bestrophins in mediating the CaCCs in IMCD-K2 cells is highly unlikely. In addition, recent work by Rosenthal et al. (2006) found that expression of hBest1 in retinal pigment epithelial cells led to a change in L-type Ca^{2+} channel kinetics and voltage dependence, but did not induce a Ca^{2+}-activated Cl^- conductance per se, suggesting that Best proteins may have multiple functional roles in epithelial cells. Certainly, the advent of siRNA technology will enable attempts to knock down CaCC. This will permit identification of the responsible molecular species as well as exclusion of those that are irrelevant.

c. *Large-Conductance Ca^{2+}-Activated Cl^- Channels*

In 2004, a novel chloride channel family was discovered that was related to the *Drosophila* gene known as *tweety* (Suzuki and Mizuno, 2004; Suzuki, 2006). These proteins are predicted to contain five or six transmembrane segments and are widely expressed in excitable tissues. One family member, hTTYH3, when expressed in CHO cells, generated a linear, time-independent chloride current only after exposure of the cells to ionomycin. The activated current had an anion permeability sequence of $I > Br > Cl = SCN > NO_3 >$ gluconate, and was blocked by DIDS, but was insensitive to niflumic acid, SITS, and DTT. Single channel studies revealed a unitary conductance of 260 pS, with channels displaying complex gating kinetics. Whether similar channels are expressed in epithelial cells awaits elucidation.

5. Other Pathways

The efflux of chloride across the apical membrane can also be utilized to mediate net vectorial transport (secretion) of other anions such as HCO_3^-, SO_4^-, formate, and oxalate (Fig. 4). In this case, the parallel operation of an apical chloride channel with a Cl^-/anion exchanger results in chloride recycling across the apical membrane, with the other transported anion being secreted luminally. This mode of transport is well documented in the case of HCO_3^- secretion from the ductal epithelium of the pancreas and liver (Steward et al., 2005; Argent et al., 2006) as well as in the small intestine (Simpson, et al., 2005). In the case of the ductal cells of the pancreas, the net secretion of HCO_3^- is a two-stage process. The initial step of HCO_3^- secretion is uptake of HCO_3^- into the duct cells from the extracellular space. HCO_3^- can enter the epithelium either by the forward transport of HCO_3^- via a basolateral Na^+/HCO_3^--cotransporter (NBC) (Fig. 4), or by the diffusion of CO_2 into the cells, subsequent hydration to H_2CO_3 by carbonic anhydrase (CA), and backward transport of protons via Na^+/H^+ exchangers (NHEs) and/or H^+ pumps. HCO_3^- secretion across the apical membrane is thought to occur via CFTR operating in conjunction with anion exchangers (see below). However, the relative importance of each of these apical transporters in HCO_3^- secretion is a controversial issue (Steward et al., 2005; Argent et al., 2006;

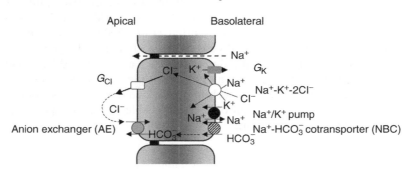

Figure 4. Model of Cl^--dependent HCO_3^- secretion. Under stimulated conditions, an anion exchanger (AE) drives the recycling of secreted chloride at the apical membrane and also mediates HCO_3^- secretion. A basolateral, sodium gradient-driven cotransporter (Na^+-*b*icarbonate *c*otransporter, NBC) facilitates the uptake of HCO_3^- from the blood. Note that a negative transepithelial potential difference also results under these conditions, thereby driving paracellular sodium movement as in Fig. 3A. For simplicity, basolateral AE pathways (e.g., AE2) do not appear in this illustration; note, however, that these must be inhibited during secretion (Section IV.B.5).

Section V.B.2). Note that ductal cells are known to contain a basolateral AE that needs to be switched off under stimulated conditions, in order to facilitate net Cl^--dependent HCO_3^- secretion.

Recently our understanding of the transporters that mediate apical AE has greatly improved with the identification of a new AE gene family (SLC26) (Mount and Romero, 2004), which are structurally and functionally distinct from the classical AE (SLC4) family (Alper, 2006). Mutations in at least three of the family members are associated with human disease; SLC26A2 (diastrophic dysplasia), SLC26A3 (congenital chloride diarrhea), and SLC26A4 (Pendred syndrome) (Mount and Romero, 2004), and recent knockout studies of slc26a6 in mice indicate that this transporter is important for renal and intestinal function (Wang *et al.*, 2005). To date, 11 family members have been identified (SLC26A1–SLC26A11), although SLC26A10 is a pseudogene. All function as anion exchangers, except SLC26A5 and SLC26A7. The exchangers transport both monovalent and divalent anions, but with a diversity of anion specificity and stoichiometry (Mount and Romero, 2004). Interestingly, recent work showed that SLC26A7 functions as an ion channel and not an exchanger (Kim *et al.*, 2005), which suggests that, like CLCs, these proteins may have both ion channel and transporter capabilities.

A number of laboratories have shown that four members of the family function as dedicated Cl^-/HCO_3^- exchangers, but may transport other anions as well (e.g., SO_4^{2-} and oxalate). These are SLC26A3 (DRA or CLD), SLC26A4 (pendrin), SLC26A6 (PAT-1 or CFEX), and SLC26A9 (Melvin *et al.*, 1999; Royaux *et al.*, 2001; Lohi *et al.*, 2003; Xu *et al.*, 2005). In addition, transport by both SLC26A3 and SLC26A6 is now thought to be electrogenic (Ko *et al.*, 2002; Xie *et al.*, 2002; Ko *et al.*, 2004), but with opposite stoichiometries (A3, $1HCO_3^-:2Cl^-$; A6, $2HCO_3^-:1Cl^-$). However, this finding is still controversial (Chernova *et al.*, 2003, 2005; Lamprecht *et al.*, 2005) and may be dependent on expression system or indeed the particular splice variant studied (Chernova *et al.*, 2005).

The apical plasma membrane location of these transporters enables their important role in regulating luminal chloride absorption and HCO_3^- secretion in a number of tissues, including the kidney, gut, and pancreas.

V. PDZ PROTEINS: WHAT ARE THEY? WHERE ARE THEY? WHAT DO THEY DO?

A. Introduction to PDZ Proteins

In the last decade, PDZ domain proteins have emerged as important adaptor proteins that organize and stabilize supramolecular complexes. Of particular relevance to our topic is that some of these complexes contain ion channels, transporters, and associated regulatory molecules important for vectorial transport. We discuss key aspects of PDZ proteins in the following section and highlight their particularly critical role in the cellular status of CFTR, an important epithelial chloride channel.

1. Nomenclature

PDZ proteins are characterized by the presence of at least one distinct structural motif, the PDZ domain. PDZ domain regions consist of approximately 80–90 amino acids. Structural studies indicate that these amino acids are organized as six antiparallel β-strands and two α-helices that form a ligand binding groove. The currently favored nomenclature, PDZ, arises from the fact that these motifs were originally recognized in proteins from rat brain postsynaptic densities (*P*SD-95; Cho *et al.*, 1992), the septate junctions linking *Drosophila* epithelia (*D*iscs Large protein, the product of the *dlg* tumor suppressor gene) (Woods and Bryant, 1991), and the TJs of mammalian epithelia (*Z*O-1) (Itoh *et al.*, 1993; Willott *et al.*, 1993). However, PDZ domains have also have been referred to as DHR (for *D*iscs Large *H*omologous *R*egion) and GLGF (after a conserved amino acid sequence motif) domains in the early literature.

Cho *et al.* (1992) first recognized the structural homology existing between PSD-95 and the previously identified Discs Large protein. Interestingly, not only do both possess three PDZ domains at their respective N-termini, but they also share a similar overall domain structure: the PDZ domains are followed by an *src* homology 3 (SH3) domain as well as a guanylate kinase (GK) domain. Soon thereafter, elucidation of the primary structure of the epithelial TJ protein, ZO-1, likewise revealed the same general body plan. It is important to keep in mind that although these three defining members of PDZ domain proteins also constitute a class of *m*embrane-*a*ssociated *gu*anylate *k*inases (MAGUKs), not all PDZ proteins are MAGUKs. As will be discussed, diverse proteins contain PDZ domains. Through their binding to specific C-terminal peptide sequences, internally localized sequences and even to PDZ regions within ligands, PDZ domains ultimately enable a diverse range of highly specialized functions.

2. Prototypical PDZ Proteins Are Localized to Areas of Cell–Cell Contact

Another attribute that PSD-95, Discs Large, and ZO-1 share is their association with membranes that come into close contact with those of neighboring cells. For example, synapses are the highly specialized points of contact between neurons, the site at which action potentials are transmitted between the pre- and postsynaptic cells. The sequence of PSD-95 was originally obtained from the postsynaptic density fraction of rat brain synaptosomes (Cho et al., 1992). Independently, Kistner et al. (1993) cloned and sequenced a 90-kDa protein that they christened SAP90 (for *s*ynapse-*a*ssociated *p*rotein). Like Cho et al. (1992), Kistner and coworkers noted the striking similarity in the domain structure of SAP90 to that previously described for Discs Large. It is now clear that PSD-95 and SAP90 are the same protein. In fact, the immunoelectron microscopic localization studies of Kistner et al. (1993) demonstrated the presence of SAP90 in the *pre*synaptic termini of basket cells that form GABAergic synapses on cerebellar Purkinje cells. Thus, despite the origins of its name, the distribution of PSD-95 is not restricted to the postsynaptic densities.

Like PSD-95, the product of the *dlg* gene, Discs Large, appears at intercellular interfaces, the septate junctions (Woods and Bryant, 1991). These regions represent the invertebrate counterpart to the vertebrate TJs. Indeed, the molecular identification of ZO-1 in both mouse and human revealed significant homology to Discs Large and strongly support the conjectured relationship between these structures (Itoh et al., 1993; Willott et al., 1993). The *dlg* gene possesses tumor-suppressor activity, and mutations of this gene disrupt the polarization of epithelial cells, thus promoting neoplastic growth (Woods and Bryant, 1991). Whether mutations in ZO-1 and other PDZ proteins that mediate epithelial polarization similarly underlie carcinogenesis remains a topic of scrutiny (De Lorenzo et al., 1999; Reichert et al., 2000; Nakagawa et al., 2004). However, morphological disruptions such as those caused by mutation of the *dlg* gene clearly would impair epithelial barrier function.

To summarize, the concentration of the above-mentioned prototypic PDZ proteins at synapses and junctional complexes suggests one of their most important functions: the positioning and organization of proteins that are important to tissue-specific intercellular communication.

B. PDZ Proteins and the Organization of Transducisomes

1. Diversity of PDZ Proteins: Molecular LEGO

The prototypic PDZ proteins, PSD-95, Discs Large, and ZO-1, as well as the related MAGUK, p55 (Lue et al., 1994; Marfatia et al., 1996), interact with transmembrane proteins as well as the microtubule cytoskeleton (Yap et al., 1995; Woods et al., 1996; Niethammer et al., 1998; Passafaro et al., 1999). Thus, they act as linkers and have, moreover, the potential to convey information between the cell membrane and the cytoskeleton. Their SH3 and GK moieties also functionally link them with signaling pathways involving small-GTPase activity. Apart from these MAGUKs, in which PDZ domains are found in combination with SH3 and GK motifs, combinations of PDZ domains with other interaction modules exist. Another MAGUK, Lin-2/CASK, contains an N-terminus, calmodulin-dependent protein kinase-like domain, followed by one PDZ domain, an SH3 region, and a C-terminus GK (Simske et al., 1996; Kim, 1997).

Some proteins lack obvious signaling modules but contain other protein–protein interaction modules besides PDZ domains. Scribble, a protein important for the establishment and maintenance of epithelial polarity, contains an N-terminus, leucine-rich repeat (LRR) that is followed by four PDZ domains (Bilder and Perrimon, 2000; Bilder et al., 2003). In some instances, the presence of other protein interaction modules can also link PDZ proteins to signaling pathways. The CFTR-associated ligand (CAL) contains two coiled-coil domains in addition to a PDZ domain (Cheng et al., 2002). The PDZ domain interacts with the C-terminus of CFTR, whereas the second coiled-coil domain can interact specifically with TC10, a Rho-family GTPase, as well as mediate the formation of homodimers (Neudauer et al., 2001).

Other PDZ proteins contain multiple PDZ repeats but no other protein–protein interaction motifs or signaling domains. MUPP1 is one example, having no apparent catalytic regions, but no fewer than 13 PDZ domains. Thus, MUPP1 offers multiple sites for potential protein–protein interactions. Indeed, binding of MUPP1 PDZ10 to the C-termini of claudins, as well as its PDZ9 to that of JAM, suggests a role in the organization of TJs (Hamazaki et al., 2002). The identification of ligands for the 11 other PDZ domains of MUPP1 no doubt will provide insights into the establishment and regulation of TJs.

Multiple PDZ domain proteins that are particularly germane to the regulation of CFTR function include the sodium–hydrogen exchange regulatory factor (NHE-RF) (Weinman et al., 1995) and the closely related E3KARP (Yun et al., 1997). NHE-RF and E3KARP each contain two PDZ domains. Importantly, not only do NHE-RF and E3KARP bind NHE3, but NHE-RF also can interact with the C-terminus of CFTR through its first PDZ domain (Hall et al., 1998; Wang et al., 1998), whereas E3KARP does so via its second PDZ region (Sun et al., 2000). Similarly, the β2-adrenergic and P2Y1 receptors, as well as the G–protein-coupled receptor kinase 6A (Hall et al., 1999), also interact with NHE-RF. Moreover, both NHE-RF and E3KARP bind the actin-binding protein, ezrin (Lamprecht et al., 1998; Yun et al., 1998), which in turn anchors PKA (Dransfield et al., 1997). It quickly becomes evident that multiple PDZ domain–containing proteins also can nucleate elaborate macromolecular signaling complexes.

Thus, the modular organization of PDZ proteins, either alone or in conjunction with other protein interaction or signal transduction moieties, makes these proteins exceptionally dynamic and plastic scaffolds upon which complex regulatory systems can be built.

2. PDZ Proteins Cluster Channels and Transporters

Transepithelial transport requires precise synchronization of channel and transporter activity. Channels and transporters comprise some of the most intensively studied membrane signal transduction proteins. How these proteins are organized closely therefore presents a logical question for further study, and PDZ proteins therefore are good candidates. The earliest reports of PDZ proteins such as PSD-95 implicate an involvement in the clustering of ionic channels. This first was convincingly shown for PSD-95 and the presynaptic, Shaker-type potassium ($K_{V1.1-1.4}$) channels (Kim et al., 1995), as well as for PSD-95 and the postsynaptic NMDA receptor channels (Kornau et al., 1995).

Recent findings illustrate the importance of PDZ protein clustering of epithelial chloride channels and transporters. The C-terminus of CFTR (D-T-R-L) is a Class I PDZ domain interaction motif (X-S/T-X-L), increasing its scope as a regulatory molecule. Multiple PDZ domain proteins facilitate homomeric CFTR-CFTR interactions as well as those between CFTR and other molecular species. Wang et al. (2000) proposed that the multiple PDZ domain protein, CAP-70, potentiates CFTR activation by promoting PDZ3/PDZ4-mediated CFTR–CFTR intermolecular interactions.

PDZ proteins facilitate CFTR interactions with Cl^-/HCO_3^- exchangers of the SLC26 family (Lamprecht et al., 2002; Ko et al., 2004; for a perspective see Gray, 2004). These findings support a model whereby PDZ interactions with EBP50 (NHE-RF) and E3KARP maintain CFTR and SLC26 transporters in close proximity of each other (Fig. 5). On PKA-mediated phosphorylation, the R domain of CFTR then interacts directly with the STAS domain (for *s*ulfate *t*ransporter and *a*nti-*s*igma-factor antagonist) Aravind and Koonin, 2000), and up-regulates SLC26A3 and SLC26A6 transporter activity. Moreover, binding of the STAS domain stimulates CFTR activity, suggesting reciprocal regulation. Mutations in the STAS domain cause congenital chloride diarrhea, thus suggesting a role for CFTR in genesis of this disease. In addition, it appears that expression of CFTR is also associated with an increase in mRNA levels for both SLC26A3 and SLC26A6 in pancreatic and airway cells (Wheat et al., 2000; Greeley et al., 2001), suggesting that CFTR exerts both short-term control (transport activity) and long-term control (gene expression) of SLC26 function.

Ogura et al. (2002) reported a functionally significant association between CFTR and ClC-3b via the PDZ domains of EBP50. Furthermore, the study implied that ClC-3b functions as the molecular correlate of the elusive ORCC. However, these conclusions were vigorously challenged (Gentzsch et al., 2003) and remain debatable.

Thus, we are now starting to appreciate the potential of cells to weave intricate signaling and regulatory mechanisms by exploiting the diversity of PDZ-domain proteins and the plethora of functionally important binding partners. Moreover, the modular

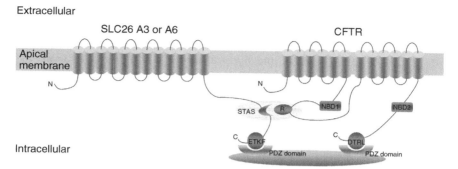

Figure 5. Model depicting the putative mechanism underlying reciprocal regulation of the anion exchangers, SLC26A3 and SLC26A6, by CFTR. The two PDZ domains of either EBP50 or E3KARP corral either SLC26 transporter in a complex with CFTR. After protein kinase A-mediated phosphorylation of CFTR's R domain, the STAS (sulfate transporter and anti-sigma-factor antagonist) domain of the anion exchanger interacts with the phosphorylated R domain (indicated by encircling, lightly shaded oval), resulting in activation of anion exchange. Activated SLC26 in turn can enhance CFTR activity, possibly by regulating the extent of R domain phosphorylation. For additional details, see Ko et al. (2004).

nature of PDZ proteins lends additional flexibility and dimensions to regulatory interactions.

3. Membrane Proteins Are Stabilized by PDZ Proteins

The trafficking and membrane retention of CFTR provides an example of how PDZ proteins regulate the stability of a chloride channel at the plasma membrane. In the schema proposed by Cheng et al. (2004, 2005), the terminal D-T-R-L of CFTR binds the PDZ protein, CAL. CAL retains CFTR within the Golgi and promotes its trafficking to the lysosome. However, binding of the Rho-GTPase, TC10, promotes movement of the CFTR–CAL complex to the cell membrane. There, the PDZ interaction switches to one between CFTR and NHE-RF. Thus, in essence, CAL "hands-off" CFTR to NHE-RF; the interaction between D-T-R-L and PDZ1 of NHE-RF in turn stabilizes CFTR at the membrane.

VI. TWO DISEASES INVOLVING EPITHELIAL CHLORIDE TRANSPORT

Several human diseases result from the disruption of genes encoding proteins that mediate critical epithelial transport processes. These include relatively rare conditions such as Bartter syndrome, Gitelman syndrome, and Dent's disease. Bartter syndrome is a renal salt-wasting disorder that presents in five genetically distinct forms, each resulting from mutations in the genes encoding the NKCC2, the ROMK channel,

ClC-Kb, the accessory protein, barrtin, and the calcium-sensing receptor. Gitelman syndrome results from mutations in NCC that ultimately impairs the absorption of NaCl by the distal convoluted tubule, and is distinguished from Bartter's by the presentation of hypomagnesemia and hypocalciuria. Mutations in *CLCN5*, the gene encoding the intracellular CLC transporter, ClC-5, cause the X-linked nephrolithiasis and low molecular weight proteinuria characteristic of Dent's disease. All three diseases have been reviewed extensively (Simon and Lifton, 1998; Jentsch *et al.*, 2005), and preceding chapters of this volume (especially Chapters 2 and 10) discuss the current state of research into these clinical problems.

In the following section, we highlight recent progress toward understanding two, comparatively more common, diseases in which the disruption of transepithelial chloride transport may play a role: CF and polycystic kidney disease (PKD).

A. Cystic Fibrosis

As discussed in Section IV.B.3, CF is caused by mutations in the *CFTR* gene, which encodes an ATP-gated chloride channel regulated by PKA phosphorylation. The protein is expressed mainly in the apical plasma membrane of epithelial tissues, where it plays a crucial role in regulating fluid secretion in the airways, salivary glands, intestine, and genital tract. When CFTR fails to work properly, it produces an imbalance in salt and fluid transport, resulting in thickened secretions in all these tissues. In the airways, there is accumulation of sticky mucous that leads eventually to respiratory failure. Blockages in the gut, pancreas, and liver lead to poor digestion and malnutrition, and defects in the genital tract lead to infertility (Doull, 2001).

To date, more than 1400 mutations in *CFTR* have been identified (http://www.genet.sickkids.on.ca). CF occurs at a frequency of 1 in 2500 live births. A deletion of phenylalanine at position 508 (ΔF508) is the most common mutation and occurs in over 85% of alleles. Loss of F508 leads to misprocessing of the protein in the synthetic pathway and a failure in normal delivery of the mutant channel to the plasma membrane. CFTR mutations range in their severity and can be categorized roughly into five classes (Choo-Kang and Zeitlin, 2000). On the severe end of the spectrum are the Class I–III mutations which cause a near complete loss of CFTR channel activity, and which are generally associated with severe lung disease and pancreatic insufficiency. On the other end of the spectrum are the Class IV and V mutations which lead to reduced, but measurable CFTR channel function; these patients have better lung function and are pancreatic sufficient. Individuals affected with Class V mutations are often asymptomatic, but are identified when presenting with idiopathic pancreatitis or congenital bilateral absence of the vas deferens (Chillon *et al.*, 1995; Choudari *et al.*, 1999).

Although reduced epithelial chloride transport has been the "hallmark" of CF for many years, work published in 2001 by the Muallem group turned this theory on its head (Choi *et al.*, 2001). Their findings were consistent with the primary defect in CF being, not a failure in chloride transport per se, but a problem with HCO_3^--driven fluid secretion, caused by the inability of mutant forms of CFTR to activate Cl^-/HCO_3^- exchange. Indeed, we have known for many years that most, if not all, CF-affected epithelia show aberrant HCO_3^- and fluid transport, and this work provided an explanation for this (Gray, 2004). Subsequent studies from the same group (Ko *et al.*, 2002, 2004) extended the link between CFTR and HCO_3^- by demonstrating that CFTR specifically increased the activity of three members of the SLC26 gene family, and was in turn itself up-regulated by a physical interaction with the SLC26 transporter (Section V.B.2). Consistent with these findings inhibiting Cl^-/HCO_3^- exchange activity in isolated pancreatic ducts (where CFTR is highly expressed) inhibits cAMP-stimulated HCO_3^- secretion (Hegyi *et al.*, 2003, 2005). Also, basal HCO_3^- secretion in the mouse duodenum, which is predominantly via Cl^-/HCO_3^- exchange, is also dependent on CFTR chloride transport activity (Simpson *et al.*, 2005). However, in the rat and rabbit duodenum this strict coupling does not seem to apply (Spiegel *et al.*, 2003), and most stimulated HCO_3^- secretion appears to be via a conductive pathway only, presumably CFTR. In a further twist to the story, recent *in vitro* Ussing chamber experiments on mouse duodenum from SLC26A6 and CFTR knockout mice (Tuo *et al.*, 2006) have shown that HCO_3^- secretion stimulated by prostaglandins is dependent on SLC26A6 activity but independent of CFTR, whereas forskolin-stimulated HCO_3^- secretion has the opposite profile. These results suggest that the mechanism of HCO_3^- secretion is likely to be agonist/second messenger-dependent, and importantly that HCO_3^- secretion is not totally reliant on CFTR expression. This later conclusion does have important implications for therapeutic treatment of impaired HCO_3^- transport in CF-affected epithelia.

Overall, there is still much debate about the basic defect in CF, and the role that CFTR plays in HCO_3^- secretion. Nonetheless, recent research has led many to conclude that a defect in CFTR-dependent HCO_3^- transport is likely to be of major importance in the majority of CF-affected epithelia. Thus, a better understanding of the role of SLC26 transporters in epithelial anion transport is an important area of future CF research.

B. Polycystic Kidney Disease

PKD is the most common monogenic inherited human disease, and a major cause of end-stage renal failure in adults (Wilson, 2004). Its symptoms include hypertension, cardiovascular abnormalities, and cysts in other organs (liver and pancreas). PKD is characterized by progressively enlarging cysts derived from the massive enlargement of fluid-filled renal tubules and/or collecting ducts. This, together with interstitial fibrosis, destroys normal kidney architecture and eventually leads to renal failure (Gabow, 1990; Ravine *et al.*, 1992). Two forms of PKD have been identified: autosomal dominant PKD (ADPKD) and autosomal recessive PKD (ARPKD). ADPKD is the most common cause of kidney failure and occurs with an incidence of 1:1000 in Caucasians. It is caused by mutations in either the *PKD1* or *PKD2* genes, which encode, respectively, the integral membrane proteins polycystin 1 and polycystin 2.

Polycystin 2, a member of the TrpP cation channel family, also functions as a calcium-permeable channel. Mutations in PKD1 account for over 85% of PKD cases.

Many other forms of cystic disease are also known (Hildebrandt and Otto, 2005). The last decade has presented the exciting realization that the majority of proteins linked to cystic disease (renal and extrarenal) appear to localize to the primary cilium (PC)-basal body complex, making this structure a major focus of cystic disease research (Davenport and Yoder, 2005). Because cyst expansion at least in part involves fluid secretion, a better understanding of how the PC regulates transepithelial chloride transport is clearly of major importance in cystic disease research.

Recent studies have helped to shed new light about the role of the PC in renal tubules and PKD (Nauli and Zhou, 2005; Praetorius and Spring, 2005). Work from these two labs indicates that the PC functions as a mechanoreceptor which responds to alterations in tubular flow by bending, and transduces this response into an increase in $[Ca^{2+}]_i$. Bending of the PC is linked to the opening of a plasma membrane calcium-permeable channel (a complex of polycystin-1 and polycystin-2) (Delmas, 2005), which induces further calcium release from internal stores, through a process of calcium-induced calcium-release. So a current hypothesis posits a disruption of flow-induced changes in $[Ca^{2+}]_i$ in patients with ADPKD, predisposing them to renal cyst development.

Normal renal cells increase $[Ca^{2+}]_i$ in response to flow. This leads to a decrease in cAMP levels through two distinct mechanisms: (1) via activation of a calcium-sensitive cAMP-dependent phosphodiesterase (PDE1) and (2) by suppression of a calcium-inhibitable adenylate cyclase (AC6). Because of defective PC-mediated mechanotransduction in ADPKD cells, the appropriate changes in cytosolic calcium do not occur and inappropriately high intracellular levels of cAMP levels result. Based on current models of transepithelial chloride transport (Section IV), a rise in cAMP, via PKA, would activate CFTR and hence stimulate fluid secretion. Consistent with this hypothesis, transepithelial potential difference measurements of forskolin-stimulated fluid secretion using excised, intact ADPKD cysts, as well as polarized monolayers grown from cysts, indicate that fluid secretion is driven by chloride secretion, and most likely involves CFTR (Grantham, 1994, 1995; Belibi et al., 2004).

Immunolocalization studies have also shown that CFTR is found in the apical membrane of primary cultures of ADPKD cells and the luminal membrane of ADPKD cyst sections (Brill et al., 1996; Hanaoka et al., 1996; Wilson, 1999; Lebeau et al., 2002). Furthermore, patch-clamp studies in primary cultures of ADPKD cells identified whole cells currents that had the expected biophysical and pharmacological profile for CFTR (Hanaoka et al., 1996) and the down-regulation of CFTR in ADPKD cells by antisense oligonucleotides, attenuated fluid secretion (Davidow et al., 1996). Interestingly, patients affected with both ADPKD and CF have less severe renal disease compared to non-CF, ADPKD patients (Cotton and Avner, 1998; O'Sullivan et al., 1998). Collectively, all these data strongly support a role for CFTR in fluid expansion of cysts, and further suggest that modulating its function could be a target for treatment of ADPKD. A recent study by Li et al. (2004) tested the ability of a dozen classical and novel inhibitors of CFTR to block transepithelial secretion, cell proliferation, and cyst growth in MDCK cells grown as electrically tight monolayers and as cysts in collagen gels. The data strongly correlate the inhibition of transepithelial chloride current with suppression of cyst growth, yet at best only a marginal correlation between decreased cell proliferation and diminishment of cyst growth could be drawn.

Although CFTR likely plays an important role in cystic fluid secretion, immunolocalization studies revealed that not all cystic cells express CFTR (Davidow et al., 1996), and patch-clamp measurements revealed that only ~50% cells responded to cAMP-stimulation (Hanaoka et al., 1996). These data suggest that non-CFTR chloride channels may also be involved in cyst expansion, and research has also focused on the role of CaCCs in polycystic kidney disease (Hooper et al., 2003; Wildman et al., 2003; Section IV.B.4). These channels are activated by an increase in $[Ca^{2+}]_i$, and in the context of the kidney, this could occur via autocrine and/or paracrine purinergic receptor-mediated mechanisms (Schwiebert et al., 2002). Wilson et al. (1999) measured ATP levels in cyst fluid as high as 10 μM, sufficient amounts to activate purinoceptors. Overall, their findings support a scenario whereby ATP is released into the cyst lumen, activates either or both ionotropic P2X or metabotropic P2Y receptors, elevates $[Ca^{2+}]_i$, and activates CaCCs, thus driving fluid secretion. Furthermore, stable transfection of the polycystin 1 C-terminus in the M-1 mouse renal collecting duct line produces a dominant negative effect and recapitulates the ATP-stimulated ADPKD chloride and fluid transport phenotype (Hooper et al., 2003). Detailed analysis indicates that the purinergic response in these M-1 transfectants involves a prolongation of both the ATP-stimulated calcium transient and the whole cell chloride current, in an extracellular calcium-dependent manner (Wildman et al., 2003). More recent findings have also shown that heterologous expression of full-length polycystin 1 in MDCK cells accelerated the decay of the ATP-induced calcium response probably through an upregulation of SERCA pump activity in the ER (and the consequent inhibition of calcium-influx through store-operated channels; Hooper et al., 2005). Thus polycystin 1 appears to function as a brake on ATP-stimulated chloride secretion under normal circumstances. In ADPKD the loss of polycystin1 function is therefore predicted to prolong the cytoplasmic calcium signal and so cause an exaggerated response to calcium-mobilizing agonists (such as ATP).

In contrast to ADPKD, it is not clear whether dysregulated chloride transport also occurs in autosomal recessive PKD. ARPKD is caused by mutations in the *PKHD1* (for *p*olycystic *k*idney and *h*epatic *d*isease) gene, which encodes a novel protein dubbed polyductin (Onuchic et al., 2002) or, alternatively, fibrocystin (Ward et al., 2002). The structure of fibrocystin suggests that it functions as a cell surface receptor, and it too localizes to the PC (Wang et al., 2004). In contrast to ADPKD, encapsulated autonomous cysts are not commonly found in ARPKD. Instead, the dilated and enlarged collecting ducts remain contiguous with the nephron (Wilson, 2004). Previous work on the *cpk* transgenic mouse model of ARPKD (Nakanishi et al., 2001) found no evidence for the involvement of CFTR in the pathogenesis of ARPKD. These results agree with recent findings of Olteanu et al. (2006), using another mouse model of ARPKD, the *orpk* mouse (see Guay-Woodford, 2003 for further details of mouse models of cystic disease). Here, a change in the absorptive (not secretory) capacity of cultured renal CCD cells was observed, which appeared to be due to an up-regulation of ENaC-mediated sodium absorption. However, this potentially important finding contrasts with the work of Veizis and Cotton (2005) using cultured CD cells from the *bpk* mouse. Here a pronounced *decrease* in ENaC-mediated absorption was observed upon chronic exposure to apical EGF, a result not observed in normal cells and probably explained by the mislocalization of the EGF receptor to the apical plasma membrane. Thus, a reduction (and not an increase) in sodium absorption may contribute to ARPKD pathophysiology (Veizis and Cotton, 2005).

Although both findings indeed may be correct, further work is required to understand the role of abnormal salt transport in ARPKD. In this context, it is noteworthy that recent patch-clamp studies performed on isolated renal PC suggest that ENaC resides and functions in the ciliary membrane itself (Raychowdhury *et al.*, 2005), thus establishing a direct link between the PC and ENaC. In addition, Liu *et al.* (2005) showed that flow-induced changes in $[Ca^{2+}]_i$ are blunted markedly in the *orpk* mutant cells, a finding similar to that which has been reported in cells lacking polycystin 1 or polycystin 2 (Nauli and Zhou, 2004). Dysregulated calcium signaling thus also may contribute to the change in ENaC-mediated transport. Clearly, key areas of future research will focus on how dysfunction in PC-mediated calcium signaling underlies disordered sodium and chloride transport in renal cells.

VII. SUMMARY

Epithelial chloride transport is of vital physiological importance to many tissues and is a process that is tightly regulated by the cell. In the last decade, a combination of biochemical, molecular, and physiological research has furthered our understanding of the major players in epithelial chloride transport, and we are beginning to understand at the atomic level how some of these transporters actually work. It is becoming increasingly clear that effective transepithelial movement of chloride not only requires the coordinated action of a variety of plasma membrane transporters, but also a myriad of associated binding partners. Many of these partners have only recently been discovered, and many more are likely to be found. The dynamic interaction of these multiprotein complexes is now becoming fully appreciated and in the future likely will drive intense research efforts. Dysfunction in salt and fluid absorption or secretion can have dire consequences for the organism, and is known to be the underlying cause of a number of important human diseases. Our current understanding of the physiological function of epithelial chloride channels and transporters, together with the advent of transgenic animals and other molecular/genetic approaches, form the basis for achieving the future insights into the pathophysiology of a variety of diseases. We anticipate these insights will produce new therapies to treat disorders of chloride transport and enhance the quality of life for affected individuals.

REFERENCES

Abdel-Ghany, M., Cheng, H. C., Elble, R. C., and Pauli, B. U. (2001). The breast cancer beta 4 integrin and endothelial human CLCA2 mediate lung metastasis. *J. Biol. Chem.* **276**, 25438–25446.

Abdel-Ghany, M., Cheng, H. C., Elble, R. C., and Pauli, B. U. (2002). Focal adhesion kinase activated by beta(4) integrin ligation to mCLCA1 mediates early metastatic growth. *J. Biol. Chem.* **277**, 34391–34400.

Abdel-Ghany, M., Cheng, H. C., Elble, R. C., Lin, H., DiBiasio, J., and Pauli, B. U. (2003). The interacting binding domains of the beta(4) integrin and calcium-activated chloride channels (CLCAs) in metastasis. *J. Biol. Chem.* **278**, 49406–49416.

Accardi, A. and Miller, C. (2004). Secondary active transport mediated by a prokaryotic homologue of ClC Cl^- channels. *Nature* **427**, 803–807.

Accardi, A., Kolmakova-Partensky, L., Williams, C., and Miller, C. (2004). Ionic currents mediated by a prokaryotic homologue of CLC Cl^- channels. *J. Gen. Physiol.* **123**, 109–119.

Alper, S. L. (2006). Molecular physiology of SLC4 anion exchangers. *Exp. Physiol.* **91**, 153–161.

Aravind, L. and Koonin, E. V. (2000). The STAS domain—a link between anion transporters and antisigma-factor antagonists. *Curr. Biol.* **10**, R53–R55.

Argent, B. E., Gray, M. A., Steward, M. C., and Case, R. M. (2006). Cell physiology of pancreatic ducts. In: *Physiology of the Gastrointestinal Tract* (Johnson, L. R., Ed.), 4th edn. Elsevier, San Diego.

Atherton, H., Mesher, J., Poll, C. T., and Danahay, H. (2003). Preliminary pharmacological characterisation of an interleukin-13-enhanced calcium-activated chloride conductance in the human airway epithelium. *Naunyn Schmiedebergs Arch. Pharmacol.* **367**, 214–217.

Bachhuber, T., König, J., Voelcker, T., Murle, B., Schreiber, R., and Kunzelmann, K. (2005). Cl interference with the epithelial Na^+ channel ENaC. *J. Biol. Chem.* **280**, 31587–31594.

Bateman, A. (1997). The structure of a domain common to archaebacteria and the homocystinuria disease protein. *Trends Biochem. Sci.* **22**, 12–13.

Bazzini, C., Vezzoli, V., Sironi, C., Dossena, S., Ravasio, A., De, B. S., Garavaglia, M., Rodighiero, S., Meyer, G., Fascio, U., Furst, J., Ritter, M., et al. (2005). Thiazide-sensitive NaCl–cotransporter in the intestine: Possible role of hydrochlorothiazide in the intestinal Ca^{2+} uptake. *J. Biol. Chem.* **280**, 19902–19910.

Belibi, F. A., Reif, G., Wallace, D. P., Yamaguchi, T., Olsen, L., Li, H., Helmkamp, G. M., Jr., and Grantham, J. J. (2004). Cyclic AMP promotes growth and secretion in human polycystic kidney epithelial cells. *Kidney Int.* **66**, 964–973.

Bilder, D. and Perrimon, N. (2000). Localization of apical epithelial determinants by the basolateral PDZ protein scribble. *Nature* **403**, 676–680.

Bilder, D., Schober, M., and Perrimon, N. (2003). Integrated activity of PDZ protein complexes regulates epithelial polarity. *Nat. Cell Biol.* **5**, 53–58.

Boese, S. H., Glanville, M., Aziz, O., Gray, M. A., and Simmons, N. L. (2000). Co-expression of Ca^{2+} and cAMP-activated Cl^- conductances mediating Cl^- secretion in a mouse renal IMCD cell-line. *J. Physiol.* **523**(Pt. 2), 325–338.

Boese, S. H., Aziz, O., Simmons, N. L., and Gray, M. A. (2004). Kinetics and regulation of a Ca^{2+}-activated Cl^- conductance in mouse renal inner medullary collecting duct (IMCD) cells. *Am. J. Physiol. Renal Physiol.* **286**, F682–F692.

Boettger, T., Hübner, C. A., Maier, H., Rust, M. B., Beck, F. X., and Jentsch, T. J. (2002). Deafness and renal tubular acidosis in mice lacking the K-Cl co-transporter KCC4. *Nature* **416**, 874–878.

Boettger, T., Rust, M. B., Maier, H., Seidenbecher, T., Schweizer, M., Keating, D. J., Faulhaber, J., Ehmke, H., Pfeffer, C., Scheel, O., Lemcke, B., Horst, J., et al. (2003). Loss of K-Cl co-transporter KCC3 causes deafness, neurodegeneration and reduced seizure threshold. *EMBO J.* **22**, 5422–5434.

Briel, M., Greger, R., and Kunzelmann, K. (1998). Cl^- transport by cystic fibrosis transmembrane conductance regulator (CFTR) contributes to the inhibition of epithelial Na^+ channels (ENaCs) in *Xenopus* oocytes co-expressing CFTR and ENaC. *J. Physiol.* **508**, 825–836.

Brill, S. R., Ross, K. E., Davidow, C. J., Ye, M., Grantham, J. J., and Caplan, M. J. (1996). Immunolocalization of ion transport proteins in human autosomal dominant polycystic kidney epithelial cells. *Proc. Natl. Acad. Sci. USA* **93**, 10206–10211.

Buanes, T., Grotmol, T., Landsverk, T., and Raeder, M. G. (1987). Ultrastructure of pancreatic duct cells at secretory rest and during secretin-dependent $NaHCO_3$ secretion. *Acta Physiol. Scand.* **131**, 55–62.

Bustin, S. A., Li, S. R., and Dorudi, S. (2001). Expression of the Ca^{2+}-activated chloride channel genes CLCA1 and CLCA2 is downregulated in human colorectal cancer. *DNA Cell Biol.* **20**, 331–338.

Buyse, G., Trouet, D., Voets, T., Missiaen, L., Droogmans, G., Nilius, B., and Eggermont, J. (1998). Evidence for the intracellular location of chloride channel (ClC)-type proteins: Co-localization of ClC-6a and ClC-6c with the sarco/endoplasmic-reticulum Ca^{2+} pump SERCA2b. *Biochem. J.* **330**, 1015–1021.

Cadigan, K. M. and Nusse, R. (1997). Wnt signaling: A common theme in animal development. *Genes Dev.* **11**, 3286–3305.

Cano, A., Perez-Moreno, M. A., Rodrigo, I., Locascio, A., Blanco, M. J., del Barrio, M. G., Portillo, F., and Nieto, M. A. (2000). The transcription factor snail controls epithelial-mesenchymal transitions by repressing E-cadherin expression. *Nat. Cell Biol.* **2**, 76–83.

Cheng, J., Moyer, B. D., Milewski, M., Loffing, J., Ikeda, M., Mickle, J. E., Cutting, G. R., Li, M., Stanton, B. A., and Guggino, W. B. (2002). A Golgi-associated PDZ domain protein modulates cystic fibrosis transmembrane regulator plasma membrane expression. *J. Biol. Chem.* **277**, 3520–3529.

Cheng, J., Wang, H., and Guggino, W. B. (2004). Modulation of mature cystic fibrosis transmembrane regulator protein by the PDZ domain protein CAL. *J. Biol. Chem.* **279**, 1892–1898.

Cheng, J., Wang, H., and Guggino, W. B. (2005). Regulation of cystic fibrosis transmembrane regulator trafficking and protein expression by a Rho family small GTPase TC10. *J. Biol. Chem.* **280**, 3731–3739.

Chernova, M. N., Jiang, L., Shmukler, B. E., Schweinfest, C. W., Blanco, P., Freedman, S. D., Stewart, A. K., and Alper, S. L. (2003). Acute regulation of the SLC26A3 congenital chloride diarrhoea anion exchanger (DRA) expressed in *Xenopus* oocytes. *J. Physiol.* **549**, 3–19.

Chernova, M. N., Jiang, L., Friedman, D. J., Darman, R. B., Lohi, H., Kere, J., Vandorpe, D. H., and Alper, S. L. (2005). Functional comparison of mouse slc26a6 anion exchanger with human SLC26A6 polypeptide variants: Differences in anion selectivity, regulation, and electrogenicity. *J. Biol. Chem.* **280**, 8564–8580.

Chillon, M., Casals, T., Mercier, B., Bassas, L., Lissens, W., Silber, S., Romey, M. C., Ruiz-Romero, J., Verlingue, C., Claustres, M., Nunes, V., Férec, C., *et al.* (1995). Mutations in the cystic fibrosis gene in patients with congenital absence of the vas deferens. *N. Engl. J. Med.* **332**, 1475–1480.

Cho, K. O., Hunt, C. A., and Kennedy, M. B. (1992). The rat brain postsynaptic density fraction contains a homolog of the Drosophila discs-large tumor suppressor protein. *Neuron* **9**, 929–942.

Choi, J. Y., Muallem, D., Kiselyov, K., Lee, M. G., Thomas, P. J., and Muallem, S. (2001). Aberrant CFTR-dependent HCO3-transport in mutations associated with cystic fibrosis. *Nature* **410**, 94–97.

Choo-Kang, L. R. and Zeitlin, P. L. (2000). Type I, II, III, IV, and V cystic fibrosis transmembrane conductance regulator defects and opportunities for therapy. *Curr. Opin. Pulm. Med.* **6**, 521–529.

Choudari, C. P., Lehman, G. A., and Sherman, S. (1999). Pancreatitis and cystic fibrosis gene mutations. *Gastroenterol. Clin. North Am.* **28**, 543–549.

Cid, L. P., Montrose-Rafizadeh, C., Smith, D. I., Guggino, W. B., and Cutting, G. R. (1995). Cloning of a putative human voltage-gated chloride channel (ClC-2) cDNA widely expressed in human tissues. *Hum. Mol. Genet.* **4**, 407–413.

Clark, S., Jordt, S. E., Jentsch, T. J., and Mathie, A. (1998). Characterization of the hyperpolarization-activated chloride current in dissociated rat sympathetic neurons. *J. Physiol.* **506**, 665–678.

Colegio, O. R., Van Itallie, C. M., McCrea, H. J., Rahner, C., and Anderson, J. M. (2002). Claudins create charge-selective channels in the paracellular pathway between epithelial cells. *Am. J. Physiol. Cell Physiol.* **283**, C142–C147.

Cotton, C. U. and Avner, E. D. (1998). PKD and CF: An interesting family provides insight into the molecular pathophysiology of polycystic kidney disease. *Am. J. Kidney Dis.* **32**, 1081–1083.

Cunningham, S. A., Awayda, M. S., Bubien, J. K., Ismailov, I. I., Arrate, M. P., Berdiev, B. K., Benos, D. J., and Fuller, C. M. (1995). Cloning of an epithelial chloride channel from bovine trachea. *J. Biol. Chem.* **270,** 31016–31026.

Davenport, J. R. and Yoder, B. K. (2005). An incredible decade for the primary cilium: A look at a once-forgotten organelle. *Am. J. Physiol. Renal Physiol.* **289,** F1159–F1169.

Davidow, C. J., Maser, R. L., Rome, L. A., Calvet, J. P., and Grantham, J. J. (1996). The cystic fibrosis transmembrane conductance regulator mediates transepithelial fluid secretion by human autosomal dominant polycystic kidney disease epithelium *in vitro*. *Kidney Int.* **50,** 208–218.

De Lorenzo, C., Mechler, B. M., and Bryant, P. J. (1999). What is Drosophila telling us about cancer? *Cancer Metastasis Rev.* **18,** 295–311.

Delmas, P. (2005). Polycystins: Polymodal receptor/ion-channel cellular sensors. *Pflügers Arch.* **451,** 264–276.

Doull, I. J. (2001). Recent advances in cystic fibrosis. *Arch. Dis. Child* **85,** 62–66.

Dowd, B. F. and Forbush, B. (2003). PASK (proline-alanine-rich STE20-related kinase), a regulatory kinase of the Na-K-Cl cotransporter (NKCC1). *J. Biol. Chem.* **278,** 27347–27353.

Dransfield, D. T., Bradford, A. J., Smith, J., Martin, M., Roy, C., Mangeat, P. H., and Goldenring, J. R. (1997). Ezrin is a cyclic AMP-dependent protein kinase anchoring protein. *EMBO J.* **16,** 35–43.

Dutzler, R., Campbell, E. B., and MacKinnon, R. (2003). Gating the selectivity filter in ClC chloride channels. *Science* **300,** 108–112.

Eggermont, J. (2004). Calcium-activated chloride channels (un)known, (un)loved? *Proc. Am. Thorac. Soc.* **1,** 22–27.

Elble, R. C. and Pauli, B. U. (2001). Tumor suppression by a proapoptotic calcium-activated chloride channel in mammary epithelium. *J. Biol. Chem.* **276,** 40510–40517.

Elble, R. C., Widom, J., Gruber, A. D., Abdel-Ghany, M., Levine, R., Goodwin, A., Cheng, H. C., and Pauli, B. U. (1997). Cloning and characterization of lung-endothelial cell adhesion molecule-1 suggest it is an endothelial chloride channel. *J. Biol. Chem.* **272,** 27853–27861.

Ellison, D. H., Velazquez, H., and Wright, F. S. (1987). Thiazide-sensitive sodium chloride cotransport in early distal tubule. *Am. J. Physiol.* **253,** F546–F554.

Epstein, F. H. and Silva, P. (1985). Na-K-Cl cotransport in chloride-transporting epithelia. *Ann. NY Acad. Sci.* **456,** 187–197.

Evans, S. R., Thoreson, W. B., and Beck, C. L. (2004). Molecular and functional analyses of two new calcium-activated chloride channel family members from mouse eye and intestine. *J. Biol. Chem.* **279,** 41792–41800.

Fong, P., Argent, B. E., Guggino, W. B., and Gray, M. A. (2003). Characterization of vectorial chloride transport pathways in the human pancreatic duct adenocarcinoma cell line HPAF. *Am. J. Physiol. Cell Physiol.* **285,** C433–C445.

Fuller, C. M., Ismailov, I. I., Keeton, D. A., and Benos, D. J. (1994). Phosphorylation and activation of a bovine tracheal anion channel by Ca^{2+}/calmodulin-dependent protein kinase II. *J. Biol. Chem.* **269,** 26642–26650.

Furuse, M., Fujita, K., Hiiragi, T., Fujimoto, K., and Tsukita, S. (1998). Claudin-1 and -2: Novel integral membrane proteins localizing at tight junctions with no sequence similarity to occludin. *J. Cell Biol.* **141,** 1539–1550.

Furuse, M., Sasaki, H., and Tsukita, S. (1999). Manner of interaction of heterogeneous claudin species within and between tight junction strands. *J. Cell Biol.* **147,** 891–903.

Furuse, M., Furuse, K., Sasaki, H., and Tsukita, S. (2001). Conversion of zonulae occludentes from tight to leaky strand type by introducing claudin-2 into Madin-Darby canine kidney I cells. *J. Cell Biol.* **153,** 263–272.

Gabow, P. A. (1990). Autosomal dominant polycystic kidney disease—more than a renal disease. *Am. J. Kidney Dis.* **16,** 403–413.

Gamba, G., Saltzberg, S. N., Lombardi, M., Miyanoshita, A., Lytton, J., Hediger, M. A., Brenner, B. M., and Hebert, S. C. (1993). Primary structure and functional expression of a cDNA encoding the thiazide-sensitive, electroneutral sodium-chloride cotransporter. *Proc. Natl. Acad. Sci. USA* **90,** 2749–2753.

Gamba, G., Miyanoshita, A., Lombardi, M., Lytton, J., Lee, W. S., Hediger, M. A., and Hebert, S. C. (1994). Molecular cloning, primary structure, and characterization of two members of the mammalian electroneutral sodium-(potassium)-chloride cotransporter family expressed in kidney. *J. Biol. Chem.* **269,** 17713–17722.

Gandhi, R., Elble, R. C., Gruber, A. D., Schreur, K. D., Ji, H. L., Fuller, C. M., and Pauli, B. U. (1998). Molecular and functional characterization of a calcium-sensitive chloride channel from mouse lung. *J. Biol. Chem.* **273,** 32096–32101.

Gentzsch, M., Cui, L., Mengos, A., Chang, X. B., Chen, J. H., and Riordan, J. R. (2003). The PDZ-binding chloride channel ClC-3B localizes to the Golgi and associates with cystic fibrosis transmembrane conductance regulator-interacting PDZ proteins. *J. Biol. Chem.* **278,** 6440–6449.

Gillen, C. M., Brill, S., Payne, J. A., and Forbush, B., III (1996). Molecular cloning and functional expression of the K-Cl cotransporter from rabbit, rat, and human. A new member of the cation-chloride cotransporter family. *J. Biol. Chem.* **271,** 16237–16244.

Grantham, J. J. (1994). Pathogenesis of renal cyst expansion: Opportunities for therapy. *Am. J. Kidney Dis.* **23,** 210–218.

Grantham, J. J., Ye, M., Gattone, V. H., and Sullivan, L. P. (1995). In vitro fluid secretion by epithelium from polycystic kidneys. *J. Clin. Invest.* **95,** 195–202.

Gray, M. A. (2004). Bicarbonate secretion: It takes two to tango. *Nat. Cell Biol.* **6,** 292–294.

Greger, R. and Schlatter, E. (1983). Properties of the basolateral membrane of the cortical thick ascending limb of Henle's loop of rabbit kidney. A model for secondary active chloride transport. *Pflügers Arch.* **396,** 325–334.

Greeley, T., Shumaker, H., Wang, Z., Schweinfest, C. W., and Soleimani, M. (2001). Downregulated in adenoma and putative anion transporter are regulated by CFTR in cultured pancreatic duct cells. *Am. J. Physiol. Gastrointest. Liver Physiol.* **281,** G1301–G1308.

Gruber, A. D. and Pauli, B. U. (1999a). Molecular cloning and biochemical characterization of a truncated, secreted member of the human family of Ca^{2+}-activated Cl^- channels. *Biochim. Biophys Acta* **1444,** 418–423.

Gruber, A. D. and Pauli, B. U. (1999c). Tumorigenicity of human breast cancer is associated with loss of the Ca^{2+}-activated chloride channel CLCA2. *Cancer Res.* **59,** 5488–5491.

Gruber, A. D., Elble, R. C., Ji, H. L., Schreur, K. D., Fuller, C. M., and Pauli, B. U. (1998). Genomic cloning, molecular characterization, and functional analysis of human CLCA1, the first human member of the family of Ca^{2+}-activated Cl^- channel proteins. *Genomics* **54,** 200–214.

Gruber, A. D., Schreur, K. D., Ji, H. L., Fuller, C. M., and Pauli, B. U. (1999b). Molecular cloning and transmembrane structure of hCLCA2 from human lung, trachea, and mammary gland. *Am. J. Physiol.* **276,** C1261–C1270.

Guay-Woodford, L. M. (2003). Murine models of polycystic kidney disease: Molecular and therapeutic insights. *Am. J. Physiol. Renal Physiol.* **285,** F1034–F1049.

Guggino, W. B. (1986). Functional heterogeneity in the early distal tubule of the Amphiuma kidney: Evidence for two modes of Cl^- and K^+ transport across the basolateral cell membrane. *Am. J. Physiol.* **250,** F430–F440.

Guggino, W. B. (1993). Outwardly rectifying chloride channels and CF: A divorce and remarriage. *J. Bioenerg. Biomembr.* **25,** 27–35.

Hall, R. A., Ostedgaard, L. S., Premont, R. T., Blitzer, J. T., Rahman, N., Welsh, M. J., and Lefkowitz, R. J. (1998). A C-terminal motif found in the beta2-adrenergic receptor, P2Y1 receptor and cystic fibrosis transmembrane conductance regulator determines binding to the Na^+/H^+ exchanger regulatory factor family of PDZ proteins. *Proc. Natl. Acad. Sci. USA* **95,** 8496–8501.

Hall, R. A., Spurney, R. F., Premont, R. T., Rahman, N., Blitzer, J. T., Pitcher, J. A., and Lefkowitz, R. J. (1999). G protein-coupled receptor kinase 6A phosphorylates the Na^+/H^+ exchanger regulatory factor via a PDZ domain-mediated interaction. *J. Biol. Chem.* **274,** 24328–24334.

Hamazaki, Y., Itoh, M., Sasaki, H., Furuse, M., and Tsukita, S. (2002). Multi-PDZ domain protein 1 (MUPP1) is concentrated at tight junctions through its possible interaction with claudin-1 and junctional adhesion molecule. *J. Biol. Chem.* **277**, 455–461.

Hanaoka, K., Devuyst, O., Schwiebert, E. M., Wilson, P. D., and Guggino, W. B. (1996). A role for CFTR in human autosomal dominant polycystic kidney disease. *Am. J. Physiol.* **270**, C389–C399.

Harris, T. J. and Peifer, M. (2005). Decisions, decisions: Beta-catenin chooses between adhesion and transcription. *Trends Cell Biol.* **15**, 234–237.

Hartzell, C., Putzier, I., and Arreola, J. (2005). Calcium-activated chloride channels. *Annu. Rev. Physiol.* **67**, 719–758.

Hegyi, P., Gray, M. A., and Argent, B. E. (2003). Substance P inhibits bicarbonate secretion from guinea pig pancreatic ducts by modulating an anion exchanger. *Am. J. Physiol. Cell Physiol.* **285**, C268–C276.

Hegyi, P., Rakonczay, Z., Jr., Tiszlavicz, L., Varro, A., Toth, A., Racz, G., Varga, G., Gray, M. A., and Argent, B. E. (2005). Protein kinase C mediates the inhibitory effect of substance P on HCO_3^- secretion from guinea pig pancreatic ducts. *Am. J. Physiol. Cell Physiol.* **288**, C1030–C1041.

Hildebrandt, F. and Otto, E. (2005). Cilia and centrosomes: A unifying pathogenic concept for cystic kidney disease? *Nat. Rev. Genet.* **6**, 928–940.

Hooper, K. M., Unwin, R. J., and Sutters, M. (2003). The isolated C-terminus of polycystin-1 promotes increased ATP-stimulated chloride secretion in a collecting duct cell line. *Clin. Sci. (Lond.)* **104**, 217–221.

Hooper, K. M., Boletta, A., Germino, G. G., Hu, Q., Ziegelstein, R. C., and Sutters, M. (2005). Expression of polycystin-1 enhances endoplasmic reticulum calcium uptake and decreases capacitative calcium entry in ATP-stimulated MDCK cells. *Am. J. Physiol. Renal Physiol.* **289**, F521–F530.

Hoshino, M., Morita, S., Iwashita, H., Sagiya, Y., Nagi, T., Nakanishi, A., Ashida, Y., Nishimura, O., Fujisawa, Y., and Fujino, M. (2002). Increased expression of the human Ca^{2+}–activated Cl^- channel 1 (CaCC1) gene in the asthmatic airway. *Am. J. Respir. Crit. Care Med.* **165**, 1132–1136.

Hübner, C. A., Stein, V., Hermans-Borgmeyer, I., Meyer, T., Ballanyi, K., and Jentsch, T. J. (2001). Disruption of KCC2 reveals an essential role of K-Cl cotransport already in early synaptic inhibition. *Neuron* **30**, 515–524.

Ikenouchi, J., Matsuda, M., Furuse, M., and Tsukita, S. (2003). Regulation of tight junctions during the epithelium-mesenchyme transition: Direct repression of the gene expression of claudins/occludin by Snail. *J. Cell Sci.* **116**, 1959–1967.

Itoh, M., Nagafuchi, A., Yonemura, S., Kitani-Yasuda, T., Tsukita, S., and Tsukita, S. (1993). The 220-kD protein colocalizing with cadherins in non-epithelial cells is identical to ZO-1, a tight junction-associated protein in epithelial cells: cDNA cloning and immunoelectron microscopy. *J. Cell Biol.* **121**, 491–502.

Jentsch, T. J., Steinmeyer, K., and Schwarz, G. (1990). Primary structure of torpedo marmorata chloride channel isolated by expression cloning in *Xenopus* oocytes. *Nature* **348**, 510–514.

Jentsch, T. J., Stein, V., Weinreich, F., and Zdebik, A. A. (2002). Molecular structure and physiological function of chloride channels. *Physiol. Rev.* **82**, 503–568.

Jentsch, T. J., Maritzen, T., and Zdebik, A. A. (2005). Chloride channel diseases resulting from impaired transepithelial transport or vesicular function. *J. Clin. Invest.* **115**, 2039–2046.

Kasper, D., Planells-Cases, R., Fuhrmann, J. C., Scheel, O., Zeitz, O., Ruether, K., Schmitt, A., Poët, M., Steinfeld, R., Schweizer, M., Kornak, U., and Jentsch, T. J. (2005). Loss of the chloride channel ClC-7 leads to lysosomal storage disease and neurodegeneration. *EMBO J.* **24**, 1079–1091.

Kida, Y., Uchida, S., Miyazaki, H., Sasaki, S., and Marumo, F. (2001). Localization of mouse CLC-6 and CLC-7 mRNA and their functional complementation of yeast CLC gene mutant. *Histochem. Cell Biol.* **115**, 189–194.

Kidd, J. F. and Thorn, P. (2000). Intracellular Ca^{2+} and Cl^- channel activation in secretory cells. *Annu. Rev. Physiol.* **62**, 493–513.

Kim, E., Niethammer, M., Rothschild, A., Jan, Y. N., and Sheng, M. (1995). Clustering of Shaker-type K^+ channels by interaction with a family of membrane-associated guanylate kinases. *Nature* **378**, 85–88.

Kim, K. H., Shcheynikov, N., Wang, Y., and Muallem, S. (2005). SLC26A7 is a Cl$^-$ channel regulated by intracellular pH. *J. Biol. Chem.* **280**, 6463–6470.

Kim, S. K. (1997). Polarized signaling: Basolateral receptor localization in epithelial cells by PDZ-containing proteins. *Curr. Opin. Cell Biol.* **9**, 853–859.

Kistner, U., Wenzel, B. M., Veh, R. W., Cases-Langhoff, C., Garner, A. M., Appeltauer, U., Voss, B., Gundelfinger, E. D., and Garner, C. C. (1993). SAP90, a rat presynaptic protein related to the product of the Drosophila tumor suppressor gene dlg-A. *J. Biol. Chem.* **268**, 4580–4583.

Knowles, M., Gatzy, J., and Boucher, R. (1981). Increased bioelectric potential difference across respiratory epithelia in cystic fibrosis. *N. Engl. J. Med.* **305**, 1489–1495.

Ko, S. B., Shcheynikov, N., Choi, J. Y., Luo, X., Ishibashi, K., Thomas, P. J., Kim, J. Y., Kim, K. H., Lee, M. G., Naruse, S., and Muallem, S. (2002). A molecular mechanism for aberrant CFTR-dependent HCO(3)(−) transport in cystic fibrosis. *EMBO J.* **21**, 5662–5672.

Ko, S. B., Zeng, W., Dorwart, M. R., Luo, X., Kim, K. H., Millen, L., Goto, H., Naruse, S., Soyombo, A., Thomas, P. J., and Muallem, S. (2004). Gating of CFTR by the STAS domain of SLC26 transporters. *Nat. Cell Biol.* **6**, 343–350.

Kornak, U., Kasper, D., Bosl, M. R., Kaiser, E., Schweizer, M., Schulz, A., Friedrich, W., Delling, G., and Jentsch, T. J. (2001). Loss of the ClC-7 chloride channel leads to osteopetrosis in mice and man. *Cell* **104**, 205–215.

Kornau, H. C., Schenker, L. T., Kennedy, M. B., and Seeburg, P. H. (1995). Domain interaction between NMDA receptor subunits and the postsynaptic density protein PSD-95. *Science* **269**, 1737–1740.

Kunzelmann, K., Kiser, G. L., Schreiber, R., and Riordan, J. R. (1997). Inhibition of epithelial Na$^+$ currents by intracellular domains of the cystic fibrosis transmembrane conductance regulator. *FEBS Lett.* **400**, 341–344.

Lamprecht, G., Weinman, E. J., and Yun, C. H. (1998). The role of NHERF and E3KARP in the cAMP-mediated inhibition of NHE3. *J. Biol. Chem.* **273**, 29972–29978.

Lamprecht, G., Heil, A., Baisch, S., Lin-Wu, E., Yun, C. C., Kalbacher, H., Gregor, M., and Seidler, U. (2002). The down regulated in adenoma (DRA) gene product binds to the second PDZ domain of the NHE3 kinase A regulatory protein (E3KARP), potentially linking intestinal Cl$^-$. *Biochemistry* **41**, 12336–12342.

Lamprecht, G., Baisch, S., Schoenleber, E., and Gregor, M. (2005). Transport properties of the human intestinal anion exchanger DRA (down-regulated in adenoma) in transfected HEK293 cells. *Pflügers Arch.* **449**, 479–490.

Lebeau, C., Hanaoka, K., Moore-Hoon, M. L., Guggino, W. B., Beauwens, R., and Devuyst, O. (2002). Basolateral chloride transporters in autosomal dominant polycystic kidney disease. *Pflügers Arch.* **444**, 722–731.

Letz, B. and Korbmacher, C. (1997). cAMP stimulates CFTR-like Cl$^-$ channels and inhibits amiloride-sensitive Na$^+$ channels in mouse CCD cells. *Am. J. Physiol.* **272**, C657–C666.

Li, H., Findlay, I. A., and Sheppard, D. N. (2004). The relationship between cell proliferation, Cl$^-$ secretion, and renal cyst growth: A study using CFTR inhibitors. *Kidney Int.* **66**, 1926–1938.

Liu, W., Murcia, N. S., Duan, Y., Weinbaum, S., Yoder, B. K., Schwiebert, E., and Satlin, L. M. (2005). Mechanoregulation of intracellular Ca^{2+} concentration is attenuated in collecting duct of monocilium-impaired orpk mice. *Am. J. Physiol. Renal Physiol.* **289**, F978–F988.

Loewen, M. E. and Forsyth, G. W. (2005). Structure and function of CLCA proteins. *Physiol. Rev.* **85**, 1061–1092.

Lohi, H., Lamprecht, G., Markovich, D., Heil, A., Kujala, M., Seidler, U., and Kere, J. (2003). Isoforms of SLC26A6 mediate anion transport and have functional PDZ interaction domains. *Am. J. Physiol. Cell Physiol.* **284**, C769–C779.

Lorenz, C., Pusch, M., and Jentsch, T. J. (1996). Heteromultimeric CLC chloride channels with novel properties. *Proc. Natl. Acad. Sci. USA* **93**, 13362–13366.

Lu, M., Leng, Q., Egan, M. E., Caplan, M. J., Boulpaep, E. L., Giebisch, G. H., and Hebert, S. C. (2006). CFTR is required for PKA-regulated ATP sensitivity of Kir1.1 potassium channels in mouse kidney. *J. Clin. Invest.* **116**, 797–807.

Lue, R. A., Marfatia, S. M., Branton, D., and Chishti, A. H. (1994). Cloning and characterization of hdlg: The human homologue of the Drosophila discs large tumor suppressor binds to protein 4.1. *Proc. Natl. Acad. Sci. USA* **91**, 9818–9822.

Lytle, C. and Forbush, B., III (1996). Regulatory phosphorylation of the secretory Na-K-Cl cotransporter: Modulation by cytoplasmic Cl. *Am. J. Physiol.* **270**, C437–C448.

Mall, M., Hipper, A., Greger, R., and Kunzelmann, K. (1996). Wild type but not deltaF508 CFTR inhibits Na^+ conductance when coexpressed in *Xenopus* oocytes. *FEBS Lett.* **381**, 47–52.

Mall, M., Grubb, B. R., Harkema, J. R., O'Neal, W. K., and Boucher, R. C. (2004). Increased airway epithelial Na^+ absorption produces cystic fibrosis-like lung disease in mice. *Nat. Med.* **10**, 487–493.

Marfatia, S. M., Morais Cabral, J. H., Lin, L., Hough, C., Bryant, P. J., Stolz, L., and Chishti, A. H. (1996). Modular organization of the PDZ domains in the human discs-large protein suggests a mechanism for coupling PDZ domain-binding proteins to ATP and the membrane cytoskeleton. *J. Cell Biol.* **135**, 753–766.

Marquardt, A., Stohr, H., Passmore, L. A., Kramer, F., Rivera, A., and Weber, B. H. (1998). Mutations in a novel gene, VMD2, encoding a protein of unknown properties cause juvenile-onset vitelliform macular dystrophy (Best's disease). *Hum. Mol. Genet.* **7**, 1517–1525.

McNamara, N., Gallup, M., Khong, A., Sucher, A., Maltseva, I., Fahy, J., and Basbaum, C. (2004). Adenosine up-regulation of the mucin gene, MUC2, in asthma. *FASEB J.* **18**, 1770–1772.

McNicholas, C. M., Guggino, W. B., Schwiebert, E. M., Hebert, S. C., Giebisch, G., and Egan, M. E. (1996). Sensitivity of a renal K^+ channel (ROMK2) to the inhibitory sulfonylurea compound glibenclamide is enhanced by coexpression with the ATP-binding cassette transporter cystic fibrosis transmembrane regulator. *Proc. Natl. Acad. Sci. USA* **93**, 8083–8088.

McNicholas, C. M., Nason, M. W., Jr., Guggino, W. B., Schwiebert, E. M., Hebert, S. C., Giebisch, G., and Egan, M. E. (1997). A functional CFTR-NBF1 is required for ROMK2-CFTR interaction. *Am. J. Physiol.* **273**, F843–F848.

Melvin, J. E., Park, K., Richardson, L., Schultheis, P. J., and Shull, G. E. (1999). Mouse down-regulated in adenoma (DRA) is an intestinal Cl^-/HCO_3^- exchanger and is up-regulated in colon of mice lacking the NHE3 Na^+/H^+ exchanger. *J. Biol. Chem.* **274**, 22855–22861.

Mohammad-Panah, R., Harrison, R., Dhani, S., Ackerley, C., Huan, L. J., Wang, Y., and Bear, C. E. (2003). The chloride channel ClC-4 contributes to endosomal acidification and trafficking. *J. Biol. Chem.* **278**, 29267–29277.

Moriguchi, T., Urushiyama, S., Hisamoto, N., Iemura, S., Uchida, S., Natsume, T., Matsumoto, K., and Shibuya, H. (2005). WNK1 regulates phosphorylation of cation-chloride-coupled cotransporters via the STE20-related kinases, SPAK and OSR1. *J. Biol. Chem.* **280**, 42685–42693.

Mount, D. B. and Romero, M. F. (2004). The SLC26 gene family of multifunctional anion exchangers. *Pflügers Arch.* **447**, 710–721.

Mount, D. B., Mercado, A., Song, L., Xu, J., George, A. L., Jr., Delpire, E., and Gamba, G. (1999). Cloning and characterization of KCC3 and KCC4, new members of the cation-chloride cotransporter gene family. *J. Biol. Chem.* **274**, 16355–16362.

Nakagawa, S., Yano, T., Nakagawa, K., Takizawa, S., Suzuki, Y., Yasugi, T., Huibregtse, J. M., and Taketani, Y. (2004). Analysis of the expression and localisation of a LAP protein, human scribble, in the normal and neoplastic epithelium of uterine cervix. *Br. J. Cancer* **90**, 194–199.

Nakanishi, K., Sweeney, W. E., Jr., Macrae, D. K., Cotton, C. U., and Avner, E. D. (2001). Role of CFTR in autosomal recessive polycystic kidney disease. *J. Am. Soc. Nephrol.* **12**, 719–725.

Nauli, S. M. and Zhou, J. (2004). Polycystins and mechanosensation in renal and nodal cilia. *Bioessays* **26**, 844–856.

Nelson, W. J. and Nusse, R. (2004). Convergence of Wnt, beta-catenin, and cadherin pathways. *Science* **303**, 1483–1487.

Neudauer, C. L., Joberty, G., and Macara, I. G. (2001). PIST: A novel PDZ/coiled-coil domain binding partner for the Rho-family GTPase TC10. *Biochem. Biophys. Res. Commun.* **280**, 541–547.

Niethammer, M., Valtschanoff, J. G., Kapoor, T. M., Allison, D. W., Weinberg, T. M., Craig, A. M., and Sheng, M. (1998). CRIPT, a novel postsynaptic protein that binds to the third PDZ domain of PSD-95/SAP90. *Neuron* **20**, 693–707.

O'Sullivan, D. A., Torres, V. E., Gabow, P. A., Thibodeau, S. N., King, B. F., and Bergstralh, E. J. (1998). Cystic fibrosis and the phenotypic expression of autosomal dominant polycystic kidney disease. *Am. J. Kidney Dis.* **32**, 976–983.

Ogura, T., Furukawa, T., Toyozaki, T., Yamada, K., Zheng, Y. J., Katayama, Y., Nakaya, H., and Inagaki, N. (2002). ClC-3B, a novel ClC-3 splicing variant that interacts with EBP50 and facilitates expression of CFTR-regulated ORCC. *FASEB J.* **16**, 863–865.

Ohkubo, T. and Ozawa, M. (2004). The transcription factor Snail downregulates the tight junction components independently of E-cadherin downregulation. *J. Cell Sci.* **117**, 1675–1685.

Olteanu, D., Yoder, B. K., Liu, W., Croyle, M. J., Welty, E. A., Rosborough, K., Wyss, J. M., Bell, P. D., Guay-Woodford, L. M., Bevensee, M. O., Satlin, L. M., Schwiebert, E. M., et al. (2006). Heightened ENaC-mediated sodium absorption in a murine polycystic kidney disease model epithelium lacking apical monocilia. *Am. J. Physiol. Cell Physiol.* **290**, C952–C963.

Onuchic, L. F., Furu, L., Nagasawa, Y., Hou, X., Eggermann, T., Ren, Z., Bergmann, C., Senderek, J., Esquivel, E., Zeltner, R., Rudnik-Schoneborn, S., Mrug, M., et al. (2002). PKHD1, the polycystic kidney and hepatic disease 1 gene, encodes a novel large protein containing multiple immunoglobulin-like plexin-transcription-factor domains and parallel beta-helix 1 repeats. *Am. J. Hum. Genet.* **70**, 1305–1317.

Papassotiriou, J., Eggermont, J., Droogmans, G., and Nilius, B. (2001). Ca^{2+}-activated Cl^- channels in Ehrlich ascites tumor cells are distinct from mCLCA1, 2 and 3. *Pflügers Arch.* **442**, 273–279.

Passafaro, M., Sala, C., Niethammer, M., and Sheng, M. (1999). Microtubule binding by CRIPT and its potential role in the synaptic clustering of PSD-95. *Nat. Neurosci.* **2**, 1063–1069.

Payne, J. A. and Forbush, B., III (1994). Alternatively spliced isoforms of the putative renal Na-K-Cl cotransporter are differentially distributed within the rabbit kidney. *Proc. Natl. Acad. Sci. USA* **91**, 4544–4548.

Payne, J. A., Stevenson, T. J., and Donaldson, L. F. (1996). Molecular characterization of a putative K-Cl cotransporter in rat brain. A neuronal-specific isoform. *J. Biol. Chem.* **271**, 16245–16252.

Petersen, O. H. (2005). Ca^{2+} signalling and Ca^{2+}-activated ion channels in exocrine acinar cells. *Cell Calcium* **38**, 171–200.

Petrukhin, K., Koisti, M. J., Bakall, B., Li, W., Xie, G., Marknell, T., Sandgren, O., Forsman, K., Holmgren, G., Andreasson, S., Vujic, M., Bergen, A. A., et al. (1998). Identification of the gene responsible for best macular dystrophy. *Nat. Genet.* **19**, 241–247.

Picollo, A. and Pusch, M. (2005). Chloride/proton antiporter activity of mammalian CLC proteins ClC-4 and ClC-5. *Nature* **436**, 420–423.

Piechotta, K., Lu, J., and Delpire, E. (2002). Cation chloride cotransporters interact with the stress-related kinases Ste20-related proline-alanine-rich kinase (SPAK) and oxidative stress response 1 (OSR1). *J. Biol. Chem.* **277**, 50812–50819.

Ponting, C. P. (1997). CBS domains in ClC chloride channels implicated in myotonia and nephrolithiasis (kidney stones). *J. Mol. Med.* **75**, 160–163.

Praetorius, H. A. and Spring, K. R. (2005). A physiological view of the primary cilium. *Annu. Rev. Physiol.* **67**, 515–529.

Pusch, M. (2004). Ca^{2+}-activated chloride channels go molecular. *J. Gen. Physiol.* **123**, 323–325.

Qu, Z., Wei, R. W., and Hartzell, H. C. (2003). Characterization of Ca^{2+}-activated Cl^- currents in mouse kidney inner medullary collecting duct cells. *Am. J. Physiol. Renal Physiol.* **285**, F326–F335.

Qu, Z., Fischmeister, R., and Hartzell, C. (2004). Mouse bestrophin-2 is a bona fide Cl^- channel: Identification of a residue important in anion binding and conduction. *J. Gen. Physiol.* **123**, 327–340.

Ravine, D., Walker, R. G., Gibson, R. N., Forrest, S. M., Richards, R. I., Friend, K., Sheffield, L. J., Kincaid-Smith, P., and Danks, D. M. (1992). Phenotype and genotype heterogeneity in autosomal dominant polycystic kidney disease. *Lancet* **340**, 1330–1333.

Raychowdhury, M. K., McLaughlin, M., Ramos, A. J., Montalbetti, N., Bouley, R., Ausiello, D. A., and Cantiello, H. F. (2005). Characterization of single channel currents from primary cilia of renal epithelial cells. *J. Biol. Chem.* **280**, 34718–34722.

Reddy, M. M. and Quinton, P. M. (2006). ENaC activity requires CFTR channel function independently of phosphorylation in sweat duct. *J. Membr. Biol.* **207**, 23–33.

Reichert, M., Muller, T., and Hunziker, W. (2000). The PDZ domains of zonula occludens-1 induce an epithelial to mesenchymal transition of Madin-Darby canine kidney I cells. Evidence for a role of beta-catenin/Tcf/Lef signaling. *J. Biol. Chem.* **275**, 9492–9500.

Reilly, R. F. and Ellison, D. H. (2000). Mammalian distal tubule: Physiology, pathophysiology, and molecular anatomy. *Physiol. Rev.* **80**, 277–313.

Reuss, L. (1983). Basolateral KCl co-transport in a NaCl-absorbing epithelium. *Nature* **305**, 723–726.

Riordan, J. R., Rommens, J. M., Kerem, B., Alon, N., Rozmahel, R., Grzelczak, Z., Zielenski, J., Lok, S., Plavsic, N., Chou, J. L., Drumm, M. L., Iannuzzi, M. C., et al. (1989). Identification of the cystic fibrosis gene: Cloning and characterization of complementary DNA. *Science* **245**, 1066–1073.

Romio, L., Musante, L., Cinti, R., Seri, M., Moran, O., Zegarra-Moran, O., and Galietta, L. J. (1999). Characterization of a murine gene homologous to the bovine CaCC chloride channel. *Gene* **228**, 181–188.

Rosenthal, R., Bakall, B., Kinnick, T., Peachey, N., Wimmers, S., Wadelius, C., Marmorstein, A., and Strauss, O. (2006). Expression of bestrophin-1, the product of the VMD2 gene, modulates voltage-dependent Ca^{2+} channels in retinal pigment epithelial cells. *FASEB J.* **20**, 178–180.

Royaux, I. E., Wall, S. M., Karniski, L. P., Everett, L. A., Suzuki, K., Knepper, M. A., and Green, E. D. (2001). Pendrin, encoded by the Pendred syndrome gene, resides in the apical region of renal intercalated cells and mediates bicarbonate secretion. *Proc. Natl. Acad. Sci. USA* **98**, 4221–4226.

Scheel, O., Zdebik, A. A., Lourdel, S., and Jentsch, T. J. (2005). Voltage-dependent electrogenic chloride/proton exchange by endosomal CLC proteins. *Nature* **436**, 424–427.

Schwiebert, E. M., Wallace, D. P., Braunstein, G. M., King, S. R., Peti-Peterdi, J., Hanaoka, K., Guggino, W. B., Guay-Woodford, L. M., Bell, P. D., Sullivan, L. P., Grantham, J. J., and Taylor, A. L. (2002). Autocrine extracellular purinergic signaling in epithelial cells derived from polycystic kidneys. *Am. J. Physiol. Renal Physiol.* **282**, F763–F775.

Scott, J. W., Hawley, S. A., Green, K. A., Anis, M., Stewart, G., Scullion, G. A., Norman, D. G., and Hardie, D. G. (2004). CBS domains form energy-sensing modules whose binding of adenosine ligands is disrupted by disease mutations. *J. Clin. Invest.* **113**, 274–284.

Sheppard, D. N. and Welsh, M. J. (1999). Structure and function of the CFTR chloride channel. *Physiol. Rev.* **79**, S23–S45.

Shindo, M., Simmons, N. L., and Gray, M. A. (1996). Characterisation of whole cell chloride conductances in a mouse inner medullary collecting duct cell line, IMCD-3. *J. Membr. Biol.* **149**, 21–31.

Simon, D. B. and Lifton, R. P. (1998). Mutations in Na(K)Cl transporters in Gitelman's and Bartter's syndromes. *Curr. Opin. Cell Biol.* **10**, 450–454.

Simon, D. B., Lu, Y., Choate, K. A., Velazquez, H., Al-Sabban, E., Praga, M., Casari, G., Bettinelli, A., Colussi, G., Rodriguez-Soriano, J., McCredie, D., Milford, D., et al. (1999). Paracellin-1, a renal tight junction protein required for paracellular Mg2+ resorption. *Science* **285**, 103–106.

Simpson, J. E., Gawenis, L. R., Walker, N. M., Boyle, K. T., and Clarke, L. L. (2005). Chloride conductance of CFTR facilitates basal Cl^-/HCO_3^- exchange in the villous epithelium of intact murine duodenum. *Am. J. Physiol. Gastrointest. Liver Physiol.* **288**, G1241–G1251.

Simske, J. S., Kaech, S. M., Harp, S. A., and Kim, S. K. (1996). LET-23 receptor localization by the cell junction protein LIN-7 during *C. elegans* vulval induction. *Cell* **85**, 195–204.

Sintchak, M. D., Fleming, M. A., Futer, O., Raybuck, S. A., Chambers, S. P., Caron, P. R., Murcko, M. A., and Wilson, K. P. (1996). Structure and mechanism of inosine monophosphate dehydrogenase in complex with the immunosuppressant mycophenolic acid. *Cell* **85**, 921–930.

Spiegel, S., Phillipper, M., Rossmann, H., Riederer, B., Gregor, M., and Seidler, U. (2003). Independence of apical Cl^-/HCO_3^- exchange and anion conductance in duodenal HCO_3^- secretion. *Am. J. Physiol. Gastrointest Liver Physiol.* **285**, G887–G897.

Steward, M. C., Ishiguro, H., and Case, R. M. (2005). Mechanisms of bicarbonate secretion in the pancreatic duct. *Annu. Rev. Physiol.* **67**, 377–409.

Stewart, G. S., Glanville, M., Aziz, O., Simmons, N. L., and Gray, M. A. (2001). Regulation of an outwardly rectifying chloride conductance in renal epithelial cells by external and internal calcium. *J. Membr. Biol.* **180**, 49–64.

Stohr, H., Marquardt, A., Nanda, I., Schmid, M., and Weber, B. H. (2002). Three novel human VMD2-like genes are members of the evolutionary highly conserved RFP-TM family. *Eur. J. Hum. Genet.* **10**, 281–284.

Stohr, H., Milenkowic, V., and Weber, B. H. (2005). VMD2 and its role in Best's disease and other retinopathies. *Ophthalmologe* **102**, 116–121.

Stokes, J. B. (1984). Sodium chloride absorption by the urinary bladder of the winter flounder. A thiazide-sensitive, electrically neutral transport system. *J. Clin. Invest.* **74**, 7–16.

Stutts, M. J., Canessa, C. M., Olsen, J. C., Hamrick, M., Cohn, J. A., Rossier, B. C., and Boucher, R. C. (1995). CFTR as a cAMP-dependent regulator of sodium channels. *Science* **269**, 847–850.

Sun, F., Hug, M. J., Lewarchik, C. M., Yun, C. H., Bradbury, N. A., and Frizzell, R. A. (2000). E3KARP mediates the association of ezrin and protein kinase A with the cystic fibrosis transmembrane conductance regulator in airway cells. *J. Biol. Chem.* **275**, 29539–29546.

Sun, H., Tsunenari, T., Yau, K. W., and Nathans, J. (2002). The vitelliform macular dystrophy protein defines a new family of chloride channels. *Proc. Natl. Acad. Sci. USA* **99**, 4008–4013.

Suzuki, M. (2006). The Drosophila tweety family: Molecular candidates for large-conductance Ca^{2+}-activated Cl^- channels. *Exp. Physiol.* **91**, 141–147.

Suzuki, M. and Mizuno, A. (2004). A novel human Cl- channel family related to Drosophila flightless locus. *J. Biol. Chem.* **279**, 22461–22468.

Tarran, R., Loewen, M. E., Paradiso, A. M., Olsen, J. C., Gray, M. A., Argent, B. E., Boucher, R. C., and Gabriel, S. E. (2002). Regulation of murine airway surface liquid volume by CFTR and Ca^{2+}-activated Cl^- conductances. *J. Gen. Physiol.* **120**, 407–418.

Tarran, R., Button, B., and Boucher, R. C. (2006). Regulation of normal and cystic fibrosis airway surface liquid volume by phasic shear stress. *Annu. Rev. Physiol.* **68**, 543–561.

Thiemann, A., Grunder, S., Pusch, M., and Jentsch, T. J. (1992). A chloride channel widely expressed in epithelial and non-epithelial cells. *Nature* **356**, 57–60.

Toda, M., Tulic, M. K., Levitt, R. C., and Hamid, Q. (2002). A calcium-activated chloride channel (HCLCA1) is strongly related to IL-9 expression and mucus production in bronchial epithelium of patients with asthma. *J. Allergy Clin. Immunol.* **109**, 246–250.

Tsunenari, T., Sun, H., Williams, J., Cahill, H., Smallwood, P., Yau, K. W., and Nathans, J. (2003). Structure-function analysis of the bestrophin family of anion channels. *J. Biol. Chem.* **278**, 41114–41125.

Tuo, B., Riederer, B., Wang, Z., Colledge, W. H., Soleimani, M., and Seidler, U. (2006). Involvement of the anion exchanger SLC26A6 in prostaglandin E2- but not forskolin-stimulated duodenal HCO_3^- secretion. *Gastroenterology* **130**, 349–358.

Van Itallie, C. M., Fanning, A. S., and Anderson, J. M. (2003). Reversal of charge selectivity in cation or anion-selective epithelial lines by expression of different claudins. *Am. J. Physiol. Renal Physiol.* **285**, F1078–F1084.

Veel, T., Buanes, T., Engeland, E., and Raeder, M. G. (1990). Colchicine inhibits the effects of secretin on pancreatic duct cell tubulovesicles and HCO_3^- secretion in the pig. *Acta Physiol. Scand.* **138**, 487–495.

Veizis, I. E. and Cotton, C. U. (2005). Abnormal EGF-dependent regulation of sodium absorption in ARPKD collecting duct cells. *Am. J. Physiol. Renal Physiol.* **288**, F474–F482.

Velazquez, H., Good, D. W., and Wright, F. S. (1984). Mutual dependence of sodium and chloride absorption by renal distal tubule. *Am. J. Physiol.* **247**, F904–F911.

Wang, S., Raab, R. W., Schatz, P. J., Guggino, W. B., and Li, M. (1998). Peptide binding consensus of the NHE-RF-PDZ1 domain matches the C-terminal sequence of cystic fibrosis transmembrane conductance regulator (CFTR). *FEBS Lett.* **427**, 103–108.

Wang, S., Yue, H., Derin, R. B., Guggino, W. B., and Li, M. (2000). Accessory protein facilitated CFTR-CFTR interaction, a molecular mechanism to potentiate the chloride channel activity. *Cell* **103**, 169–179.

Wang, S., Luo, Y., Wilson, P. D., Witman, G. B., and Zhou, J. (2004). The autosomal recessive polycystic kidney disease protein is localized to primary cilia, with concentration in the basal body area. *J. Am. Soc. Nephrol.* **15**, 592–602.

Wang, Z., Wang, T., Petrovic, S., Tuo, B., Riederer, B., Barone, S., Lorenz, J., Seidler, U., Aronson, P., and Soleimani, M. (2005). Renal and intestinal transport defects in Slc26a6-null mice. *Am. J. Physiol. Cell Physiol.* **288**, C957–C965.

Ward, C. J., Hogan, M. C., Rossetti, S., Walker, D., Sneddon, T., Wang, X., Kubly, V., Cunningham, J. M., Bacallao, R., Ishibashi, M., Milliner, D. S., Torres, V. E., *et al.* (2002). The gene mutated in autosomal recessive polycystic kidney disease encodes a large, receptor-like protein. *Nat. Genet.* **30**, 259–269.

Weinman, E. J., Steplock, D., Wang, Y., and Shenolikar, S. (1995). Characterization of a protein cofactor that mediates protein kinase A regulation of the renal brush border membrane Na^+-H^+ exchanger. *J. Clin. Invest.* **95**, 2143–2149.

Weinreich, F. and Jentsch, T. J. (2001). Pores formed by single subunits in mixed dimers of different CLC chloride channels. *J. Biol. Chem.* **276**, 2347–2353.

Wheat, V. J., Shumaker, H., Burnham, C., Shull, G. E., Yankaskas, J. R., and Soleimani, M. (2000). CFTR induces the expression of DRA along with Cl^-/HCO_3^- exchange activity in tracheal epithelial cells. *Am. J. Physiol. Cell Physiol.* **279**, C62–G71.

Wildman, S. S., Hooper, K. M., Turner, C. M., Sham, J. S., Lakatta, E. G., King, B. F., Unwin, R. J., and Sutters, M. (2003). The isolated polycystin-1 cytoplasmic COOH terminus prolongs ATP-stimulated Cl^- conductance through increased Ca^{2+} entry. *Am. J. Physiol. Renal Physiol.* **285**, F1168–F1178.

Willott, E., Balda, M. S., Fanning, A. S., Jameson, B., Van Itallie, C., and Anderson, J. M. (1993). The tight junction protein ZO-1 is homologous to the Drosophila discs-large tumor suppressor protein of septate junctions. *Proc. Natl. Acad. Sci. USA* **90**, 7834–7838.

Wilson, P. D. (1999). Cystic fibrosis transmembrane conductance regulator in the kidney: Clues to its role? *Exp. Nephrol.* **7**, 284–289.

Wilson, P. D. (2004). Polycystic kidney disease. *N. Engl. J. Med.* **350**, 151–164.

Wilson, P. D., Hovater, J. S., Casey, C. C., Fortenberry, J. A., and Schwiebert, E. M. (1999). ATP release mechanisms in primary cultures of epithelia derived from the cysts of polycystic kidneys. *J. Am. Soc. Nephrol.* **10**, 218–229.

Woods, D. F. and Bryant, P. J. (1991). The discs-large tumor suppressor gene of Drosophila encodes a guanylate kinase homolog localized at septate junctions. *Cell* **66**, 451–464.

Woods, D. F., Hough, C., Peel, D., Callaini, G., and Bryant, P. J. (1996). Dlg protein is required for junction structure, cell polarity, and proliferation control in Drosophila epithelia. *J. Cell Biol.* **134**, 1469–1482.

Xie, Q., Welch, R., Mercado, A., Romero, M. F., and Mount, D. B. (2002). Molecular characterization of the murine Slc26a6 anion exchanger: Functional comparison with Slc26a1. *Am. J. Physiol. Renal Physiol.* **283**, F826–F838.

Xu, J., Henriksnas, J., Barone, S., Witte, D., Shull, G. E., Forte, J. G., Holm, L., and Soleimani, M. (2005). SLC26A9 is expressed in gastric surface epithelial cells, mediates Cl^-/$HCO3^-$ exchange and is inhibited by NH4+. *Am. J. Physiol. Cell Physiol.* **289**, C493–C505.

Xu, J. C., Lytle, C., Zhu, T. T., Payne, J. A., Benz, E., Jr., and Forbush, B., III (1994). Molecular cloning and functional expression of the bumetanide-sensitive Na-K-Cl cotransporter. *Proc. Natl. Acad. Sci. USA* **91**, 2201–2205.

Yang, C. L., Angell, J., Mitchell, R., and Ellison, D. H. (2003). WNK kinases regulate thiazide-sensitive Na-Cl cotransport. *J. Clin. Invest.* **111**, 1039–1045.

Yap, A. S., Stevenson, B. R., Abel, K. C., Cragoe, E. J., Jr., and Manley, S. W. (1995). Microtubule integrity is necessary for the epithelial barrier function of cultured thyroid cell monolayers. *Exp. Cell Res.* **218**, 540–550.

Yun, C. H., Oh, S., Zizak, M., Steplock, D., Tsao, S., Tse, C. M., Weinman, E. J., and Donowitz, M. (1997). cAMP-mediated inhibition of the epithelial brush border Na^+/H^+ exchanger, NHE3, requires an associated regulatory protein. *Proc. Natl. Acad. Sci. USA* **94,** 3010–3015.

Yun, C. H., Lamprecht, G., Forster, D. V., and Sidor, A. (1998). NHE3 kinase A regulatory protein E3KARP binds the epithelial brush border Na^+/H^+ exchanger NHE3 and the cytoskeletal protein ezrin. *J. Biol. Chem.* **273,** 25856–25863.

Zhou, Y., Dong, Q., Louahed, J., Dragwa, C., Savio, D., Huang, M., Weiss, C., Tomer, Y., McLane, M. P., Nicolaides, N. C., and Levitt, R. C. (2001). Characterization of a calcium-activated chloride channel as a shared target of Th2 cytokine pathways and its potential involvement in asthma. *Am. J. Respir. Cell Mol. Biol.* **25,** 486–491.

Index

A

ABC proteins dimer NBDs, 156–157
4-Acetamido-4′-isothiocyanostilbene-2,2′-disulphonic acid (SITS), 185
Activated protein kinase (APKA), 117
Adenosine 50-triphosphate (ATP)-activated Cl^- channel, 5
ADPKD. See Autosomal dominant polycystic kidney disease
AE1 Cl^-/HCO_3^- exchangers, 281–284, 301, 304–306, 308–310
    cloning, 282
    expression, 282
    pathophysiology, 309–310
    physiological role, 283–284
    regulation, 284
    role in erythrocyte, 283
    topology of human AE1, 302
    transport function, 283
AE2 Cl^-/HCO_3^- exchangers, 281–282, 284–288, 301, 305, 308–311
    expression, 284–286
    pathophysiology, 309–311
    physiological role, 286–287
    regulation, 287–288
    transport function, 286
AE3 Cl^-/HCO_3^- exchangers, 281–282, 288–289, 301, 308–309, 311
    cloning, 288
    expression, 288
    pathophysiology, 309, 311
    physiological role, 288
    regulation, 289
    transport function, 288
Allopregnanolone for modulation, of GABAergic synapses, 231
Allosteric blockers, inhibitors of CFTR Cl^- channel, 133
AMP-activated kinase (AMPK), 151
Anion exchangers in epithelia, 338
Anion flow, CFTR Cl^- channel pore, 111–113
Annexins, 186
9-Anthracene-carboxylic acid (9-AC), 62, 84–85, 89, 92, 101, 185
    binding site on ClC-0 and ClC-1, 96–98
ATPase, CFTR as, 152–153
ATP-binding cassette (ABC) transporters, 110–115, 117, 124–125, 133–134, 137, 145–147, 149, 151, 341
    crystal structure, 153–156
    NBDs, 156–157
Autosomal dominant polycystic kidney disease (ADPKD), 120–123, 136, 353–356
Autosomal recessive polycystic kidney disease (ARPKD), 353–356

B

Bacterial ClC transporter. See ClC-ec1
Barbiturates for modulation, of GABAergic synapses, 231
Bartter syndrome, ClC-Kb proteins in, 46
Barttin
    ClC-K channels and, 26–27, 46
    renal functions, 26–27
Benzodiazepines modulation, of GABAergic synapses, 229–231
7,8-Benzoflavones, 126
Bestrophins, 182, 188–192. See also Calcium-dependent chloride channels (CaCCs)
    activation kinetics, 188–189
    anion permeation, 189–190
    epithelia, 344–345
    functional properties, 188–191
    ion selectivity, 189–190
    pharmacology, 190–191
    regulation, 191
    volume-sensitive Cl^- channels, 206
Brain tissue, ClC-2 proteins in, 23–24

C

Ca^{2+}-activated Cl^- channel, in epithelia, 345
Ca and $I_{Cl,swell}$, 205
Ca^{2+}-dependent Cl^+ channels. See also Calcium-dependent chloride channels (CaCCs)
    functional properties, 181–192
Caenorhabditis elegans, ClC proteins in, 44–45
Calcium-dependent chloride channels (CaCCs), 342–345, 355
    activation kinetics, 182–184, 188–189
    bestrophin family, 188–192
    blockers, 185

Calcium-dependent chloride channels
  (CaCCs) (*continued*)
    epithelia, 342–345, 355
    expression in cell types, 182
    functional properties, 182–191
    ion selectivity, 184–185, 189
    kinetics, 182–184
    molecular identity, 186–188, 192
    permeation, 184–185, 189–190
    pharmacology, 185, 190–192
    regulation, 185–187, 191
    single channel conductance, 184
Cancer, and cation-chloride cotransporter, 265–266
Carbonic anhydrase (CA) enzyme, 304, 309
Cation-chloride cotransporter, 2–3
    cancer and, 265–266
    CCC8 and CCC9, 250
    hypotension and, 266–267
    in inner ear, 256–259
    in kidney, 259–263
    molecular characterization, 242–243
    Na^+–coupled cation-chloride cotransporters, 242–247
    Na^+–independent cation-chloride cotransporters, 247–250
    in nervous system, 250–256
    physiology, 241–267
    in RBCs, 262, 264–265
Cation-chloride cotransporters (CCC), 336
CCC8, and cation-chloride cotransporter, 243, 250
CCC9, and cation-chloride cotransporter, 243, 250
Cellular physiology, CFTR Cl⁻ channel, 116–117
Cell volume homeostasis, 200–203
    RVD and RVI changes and regulation, 201–203
CF lung disease, 120
CF mutants function rescue, CFTR potentiators, 128–129
CFTR. *See also* CFTR Cl⁻ channel
    as ATPase, 152–153
    ATP role in gating, 160–169
    ClC-2 proteins and, 21–22
    Cl⁻/HCO₃⁻ exchangers activity, 292–293, 296, 305–307
    dephosphorylation regulation, 148–152
    epithelia, 341–342, 350–355
    gating, 157–172
    identification and characteristics, 110
    methods to study gating, 157–160
    MSDs, 146, 171
    NBDs of ABC proteins, 146–147, 156–157
    overview, 145–148
    permeation, 148
    phosphorylation regulation, 148–152
    R domain, 146
    structural biology of ABC transporters, 153–157
    topology, 146
    unresolved issues on gating mechanism, 169–172
CFTR Cl⁻ channel, 84. *See also* CFTR
    ADPKD disease, 120–123, 136
    anion flow through pore, 111–113
    cellular physiology, 116–117
    cystic fibrosis, 120–121
    epithelial physiology, 117–120
    F508del-CFTR rescue, 121–122, 126–128
    gating regulation, 113–114, 145–172
    hepatobiliary system, 118
    inhibitors, 130–136
    intestine, 118
    intracellular ATP, 111, 113–114
    molecular physiology, 110–116
    MSDs, 111–113, 115
    NBDs, 111, 113–115, 124–126, 132, 137
    pancreas, 118
    pathophysiology, 120–122
    PDZ domain, 116–117
    phosphorylation-dependent regulation, 111, 114–116
    physiology, 110–120
    pore properties, 112–113
    potentiators, 123–130
    R domain, 114–116
    reproductive tissues, 118
    respiratory airways, 119–120
    sweat gland, 118–119
    trans-epithelial ion transport, 116–117
CFTR potentiators, 123–130
    CF mutants function rescue, 128–129
    chemical structures, 127
    2′-Deoxy-ATP, 125
    drug therapy for CF mutation specific, 129–130
    four-point pharmacophore model, 126
    genistein, 124
    HTS identification, 125–128
    pharmacological chaperones rescue, 128
    phloxine B, 124–125
    pyrophosphate (PPi), 125
    therapeutic potential, 128–130
Chloride transport across biological membranes, 1
Chloride-transporting proteins
    gene families of, 1–2
    in mammalian organisms, 1–6

Index

Chloride transporting proteins in epithelia.
    See Epithelia
CLCAs, in epithelia, 343–344
ClC channels, 63, 65, 69–78
    blocker and gating modifier difference, 95
    Cl^- gating, 70–77
    clofibric acid derivatives muscle block
        mechanism, 91–95
    C-terminal domain role, 77–78
    fast gate, 74–77
    gating, 69–77
    H^+ gating, 70–77
    molecular identity of fast gate, 74–77
    muscle block mechanism, 91–95
    other blockers and, 101–102
    slow gate, 75–77
    structure activity study, 90
    transporters, 65–66, 78
    voltage, 70–77
ClC chloride channels and transporters, in epithelia,
        339–341
ClC-ec1. *See also* CLC proteins
    binding, 63–64
    channel gating, 69–77
    channels, 63, 65
    Cl^- gating, 70–77
    fast gate, 74–77
    foot in door mechanism, 74
    H^+ gating, 70–77
    ion selectivity, 63–64
    molecular identity fast gate, 74–77
    multi-ion pore, 64–65
    permeation, 63–64
    slow gate, 75–77
    speculative modeling, 68–70
    structure, 60–62
    transporters, 63–68
    voltage gating, 70–77
Cl^- channels
    blockers, 85
    roles in transepithelial transport, 14, 46
    transporters family, 10–12, 207–209
    unknown molecular identity, 4–5
ClC-K1. *See* ClC-Ka proteins
ClC-K2. *See* ClC-Kb proteins
ClC-Ka channels, 91–92
    expressed in kidney and inner ear, 24–27
    pharmacology, 98–101, 103
    physiological function, 24–27
ClC-Ka proteins, 10–11, 339–340
    pharmacology, 98–101, 103
    physiological function, 24–27

ClC-Kb channels
    expressed in kidney and inner ear, 24–27
    pharmacology, 98–101, 103
ClC-Kb proteins, 10–11, 339, 352
    in Bartter syndrome, 46
    pharmacology, 98–101, 103
    physiological function, 24–27
ClC-K channels, 87, 91–92, 103
    barttin and, 25–26, 46
    binding sites, 98–99
    expression, 25
    functions in inner ear, 27
    pharmacology, 98–101, 103
    renal functions, 26–27
ClC-7 KO Mice, 41
ClC-2 KO mouse, 20–21
ClC-3 KO mouse models, 28–29, 208
ClC-5 KO mouse models, 33, 34–37
ClC proteins, 2, 4–5. *See also* ClC-ec1
    binding, 63–64
    in *Caenorhabditis elegans*, 44–45
    channel gating, 69–78
    channels, 63, 65, 69–78
    Cl^- channels and transporters, 10–12, 207–209
    ClC-Ka proteins, 10–11, 24–27, 98–101, 103,
        339–340
    ClC-Kb proteins, 10–11, 24–27, 98–101, 103,
        339, 352
    ClC-0 proteins, 15–17, 96–98, 339–340
    ClC-1 proteins, 10–11, 17–19, 88–91,
        96–98, 339
    ClC-2 proteins, 10–11, 19–24, 102–103, 207,
        339–340
    ClC-3 proteins, 10–11, 27–31, 102–103,
        207–209, 340, 350
    ClC-4 proteins, 10–12, 32–40, 207, 340
    ClC-5 proteins, 10–11, 40–41, 102–103, 207,
        340, 352
    ClC-6 proteins, 10–11, 340
    ClC-7 proteins, 10–11, 41–43, 102–103, 340
    C-terminal domain role, 77–78
    endosomal-lysosomal pathway, 16
    on endosomes and synaptic vesicles, 27–31
    functions, 12–15, 46
    intracellular protein, 27–31
    ion selectivity, 63–64
    in model organism, 43–46
    mutations impact, 10–12
    permeation, 63–64
    physiological functions of, 15–43
    in plant, 45–46
    role in acidification, 15

ClC proteins (*continued*)
   structural conservation, 62–63
   structure, 10–12
   subfamilies, 10–11
   transporters, 63–70
   ubiquitously expressed channel, 17–24
   vesicular protein, 31–32, 40–41
   volume-sensitive Cl^- channels, 207–209
   in yeast, 43–44
ClC-0 proteins, 17, 62–64, 66, 72–78, 92–95, 100–102, 339–340
   9-AC and CPA binding sites on, 96–98
   gating mechanism, 96–98
   physiological functions, 17
   torpedo Cl^- channel, 15–16
ClC-1 proteins, 10–11, 62, 64, 69, 74, 76, 85, 87–88, 92–93, 339
   9-AC and CPA binding sites on, 96–98
   block by CPP, 93–95
   clofibric acid derivatives muscle block mechanism, 91–95
   gating mechanism, 96–98
   modulation by ATP, 19
   muscle membrane potential stabilization, 13
   myotonia, 17–19
   physiological functions, 17–19
   structure-activity study, 90
   targeting molecules heterologous expression, 88–91
ClC-2 proteins, 10–11, 62, 74, 76, 92, 96, 207, 339–340
   in brain tissue, 23–24
   candidate for $I_{Cl,swell}$, 22–23
   CFTR and, 21–22
   KO mouse model, 20–21
   physiological functions, 19–24
   potential blockers, 102–103
   transepithelial Cl^- transport, 21
ClC-3 proteins, 10–11, 207–209, 340, 350
   KO mouse models, 28–29, 208
   physiological functions, 27–31
   potential blockers, 102–103
   splice variants, 31
   subcellular localization, 28
   vesicular acidification, 29–30
ClC-4 proteins, 10–11, 66–67, 207, 340
   endosomal acidification, 31–32
   physiological functions, 31–32
ClC-5 proteins, 10–11, 66–67, 207, 340, 352
   cell-autonomous deficiency effect, 34–36
   Dent's disease and, 32–40, 46
   expression and subcellular localization, 33
   mutations, 39–40
   physiological functions, 32–40
   potential blockers, 102–103
   sorting and regulation, 40
ClC-6 proteins, 10–11, 340
   physiological functions, 40–41
ClC-7 proteins, 10–11, 340
   lysosomal storage, 42–43
   neurodegeneration, 42
   osteopetrosis, 41–43
   Ostm1 expression, 46
   physiological functions, 41–43
   potential blockers, 102–103
ClC proteins family, volume-sensitive Cl^- channels, 207–209
ClC transporters, 63–70
   channels and, 65–66, 78
   functions of, 66–68, 78
   mechanism, 68–70
   selectivity, 63–64
Cl^-/HCO_3^- exchange protein. *See* Cl^-/HCO_3^- exchangers
Cl^-/HCO_3^- exchangers, 2–3
   AE1, 281–284, 301, 304–306, 308–310
   AE2, 281–282, 284–288, 301, 305, 308–311
   AE3, 281–282, 288–289, 301, 308–309, 311
   CFTR and, 292–293, 296, 305–307
   disputed exchangers, 300
   HCO_3^- transport metabolon, 303–305
   human diseases associated with, 309
   inhibitors, 307–309
   kinases, 306
   NDCBE1, 281–282, 289–290, 309
   pathophysiology, 309–312
   PDZ domain-mediated interactions, 297, 306–307
   properties, 281
   regulation and inhibition, 303–309
   role in intercalated cells of kidney, 285
   SLC4, 282, 300–302, 308, 338
   SLC26, 282, 300, 303, 306–307, 346
   SLC26A3, 281–282, 290–293, 300, 305, 307–309, 311, 346
   SLC26A4, 281–282, 293–294, 300, 307, 309, 311–312, 346
   SLC26A6, 281–282, 294–297, 305–307, 309, 312, 346, 353
   SLC26A7, 281–282, 297–299, 308–309, 346
   SLC26A9, 282, 299–300, 308–309
   STAS domain, 303, 305–307
   structure, 301–303
Clofibric acid derivatives, modulators of muscle Cl^- conductance gCl, 86
congenital chloride diarrhea (CLD), 290, 311

Index

CPA, 62, 95, 102
  binding site at ClC-0 and ClC-1, 96–98
3′,5′-Cyclic adenosinemonophosphate (cAMP), 109–110, 159, 293, 305, 307, 354
Cyclic adenosine monophosphate (cAMP)-regulated Cl⁻ channel, 5, 109–110, 159, 293, 306, 307, 354
Cystathione β-synthetase (CBS) domains, 77–78, 341
Cystic fibrosis (CF) disease, 186, 352–353
  CFTR Cl⁻ channel, 120–121
Cystic fibrosis transmembrane, 3
Cystic fibrosis transmembrane conductance regulator. See CFTR

D

Dent's disease
  ClC-5 proteins and, 32–40, 46
  hypercalciuria in, 36–39
  hyperphosphaturia in, 36–39
  impaired endocytosis, 36–39
  mechanism, 34–36
  symptoms, 39
2′-Deoxy-ATP, CFTR potentiators, 125
4,4′-Diisothiocyanato-2,2′-stilbenedisulfonic acid (DIDS), 85, 100, 191, 308, 338, 343–344
  channel blocker, 102, 185
Diphenylamine-2-carboxylate (DPC), 85, 89
Diphenylamine-2-carboxylic acid (DPC), 185

E

ENaC channels, 119, 122, 342, 356
Endocytosis, Dent's disease and, 36–39
Endosomes and synaptic vesicles, ClC proteins located on, 27–31
Epithelia. See also Epithelial chloride transports
  anion exchangers, 338
  bestrophins and, 344–345
  CaCCs and, 342–345
  CFTR and, 341–342
  chloride transporting proteins, 336–347
  CLCAs, 343–344
  ClC chloride channels and transporters, 339–341
  exit mechanisms, 338–345
  KCC, 338–339
  large conductance Ca²⁺-activated Cl⁻ channel, 345
  NCC, 337–338
  NKCC, 337
  other pathways, 345–347
  uptake mechanisms, 336–338
  vectorial chloride transport, 329–356
Epithelial cells
  morphological specializations, 330–333
  structure and organization, 330–334
Epithelial chloride transports, 351–356.
  See also Epithelia
  cystic fibrosis disease, 352–353
  diseases involving, 351–356
  polycystic kidney disease, 353–356
  vectorial chloride transport, 333–334
Epithelial Na channels (ENaC).
  See ENaC channels
Epithelial physiology, CFTR Cl⁻ channel role in, 117–120

F

F508del-CFTR rescue, CFTR Cl⁻ channel, 121–122, 126–128
Flufenamic acid (FFA), 100–101
Fluoxetine, 185

G

GABA. See also GABAergic synapses
  membrane potential, 216–220
  metabolism, 224–225
  postsynaptic responses, 216–221
  receptor, 1–3, 15
  uptake, 225–227
  vesicular transporters, 222–224
GABAergic synapses
  allopregnanolone modulation, 231
  barbiturates modulation, 231
  benzodiazepines modulation, 229–231
  elements of, 227–228
  endogenous substances and drugs modulation, 228–232
  GABA metabolism, 224–225
  GABA-uptake, 225–227
  inhibitory synaptic functions, 222
  membrane potential, 216–220
  molecular composition and functions, 221–228
  neurosteroids modulation, 231
  phasic *versus* tonic inhibition, 227–228
  postsynaptic Cl⁻ equilibrium potentials, 219
  postsynaptic responses, 216–221
  signaling specificity, 232–233

GABAergic synapses (*continued*)
    synaptic inhibition, 227–228
    tetrahydrodeoxicorticosterone modulation, 231
    transmission, 215–233
    transmitter transport, 226
    vesicular GABA transporters, 222–223
    Zn^{2+} modulation, 231–232
GABA-transaminase (GABA-T) enzyme, 224
Gating of CFTR
    ATP binding in, 163–166, 169
    ATP hydrolysis in, 160–163, 167–168
    electrophysiological recording systems, 158–159
    expression systems, 157–158
    hydrolyzable nucleotide analogues, 160
    methods used to study, 157–160
    mutagenesis of NBDs, 160
    nonhydrolyzable nucleotide analogues, 160
    working model, 166–169
Gating regulation, CFTR Cl^- channel, 113–114, 145–172
Gef1. *See Saccharomyces cerevisiae* gene
Genistein, CFTR potentiators, 124
GH, modulators of muscle Cl^- conductance gCl, 86
Ghrelin, modulators of muscle Cl^- conductance gCl, 86–87
Glutamate decarboxylase (GAD) enzyme, 223–224
Glycine receptors, 1–3, 84
Glycinergic signaling specificity, 232–233
Glycinergic synapse, 215

H

H^+–activated Cl^- channel, 5
HCO_3- transporters, 280
Hepatobiliary system, CFTR Cl^- channel role in, 118
Human diseases, associated with Cl^-/HCO_3- exchangers, 309
Human Genome Organization (HUGO), 280
Hypercalciuria, in Dent's disease, 36–39
Hyperphosphaturia, in Dent's disease, 36–39
Hypotension, and cation-chloride cotransporter, 266–267

I

$I_{Cl,swell}$
    Ca and, 205
    ClC-2 proteins candidate for, 22–23
    taurine efflux and, 205
    volume-sensitive Cl^- channels, 203–206
    VSOR and, 203–206
IGF-1, modulators of muscle Cl^- conductance gCl, 86–87
Impaired endocytosis, in Dent's disease, 36–39
Inhibitors of CFTR Cl^- channel, 130–136
    allosteric blockers, 133
    chemical structure, 135
    HTS identification, 134–135
    mechanism of action, 132
    open-channel blockers, 131–132
    peptide toxin, 133–134
    therapeutic potential, 136
Inner ear, cation-chloride cotransporter in, 256–259
Intestine, CFTR Cl^- channel role in, 118
Intracellular ATP, CFTR Cl^- channel, 111, 113–114
Ionotropic GABA receptors (iGABAR), 216–217, 220, 222

J

Junctional proteins, 334–336
    zona adherens, 335–336
    zona occludens, 334–335

K

KCC, in epithelia, 338–339
KCC1 cotransporters, 243, 247–248, 255–256, 261, 264–265, 267, 338–339
KCC2 cotransporters, 243, 247–249, 252–255, 267, 339
KCC3 cotransporters, 243, 247–249, 255–256, 258–261, 264–265, 267, 339
KCC4 cotransporters, 243, 247–250, 256, 258–261, 264, 267, 339
Kidney
    cation-chloride cotransporter in, 259–263
    Cl^-/HCO_3- exchangers role in intercalated cells, 285
Kinases, 306

L

Lysosomal storage, ClC-7 proteins and, 42–43

Index

M

Mammalian organisms, chloride-transporting proteins in, 1–6
Mefloquine, 185
Membrane-spanning domains (MSDs), 110–113, 115, 146, 171
  CFTR Cl^- channel, 111–113, 115
Modulators of muscle Cl^- conductance gCl, 86–87
  clofibric acid derivatives, 86
  GH, 86
  ghrelin, 86–87
  IGF-1, 86–87
  niflumic acid, 86, 88–88, 91
  phorbol esters, 86
  statins, 86
  taurine, 86
Molecular physiology, CFTR Cl^- channel, 110–116
Myotonia, ClC-1 proteins and, 17–19

N

Na^+–coupled cation-chloride cotransporters, 243–247
  bumetanidesensitive Na^+-K^+–2Cl^- cotransporters NKCC1 (BSC2), 243
  K^+–Cl^- cotransporters, 244
  Na^+-K^+–2Cl^- cotransporters NKCC1, 243
  NCC cotransporters, 242–244, 246–247, 259, 262, 266–267
  NKCC1 cotransporters, 242–246, 250–252, 254–256, 258, 262, 265–267
  NKCC2 cotransporters, 242–244, 246, 250, 259, 261–262, 266–267
  thiazide-sensitive Na^+–Cl^--cotransporter NCC (NCCT/TSC), 243
Na^+–dependent Cl^-/HCO_3– exchanger (NDCBE1), 280
Na^+/H^+ exchanger regulatory factor isoform-1 (NHERF1), 116–117
Na^+-independent cation-chloride cotransporters, 247–250
  KCC1 cotransporters, 243, 247–248, 255–256, 261, 264–265, 267, 338–339
  KCC2 cotransporters, 243, 247–249, 252–255, 267, 339
  KCC3 cotransporters, 243, 247–249, 255–256, 258–261, 264–265, 267, 339
  KCC4 cotransporters, 243, 247–250, 256, 258–261, 264, 267, 339

Na^+–independent Cl^-/base exchangers, 280
NCC cotransporters, 242–244, 246–247, 259, 262, 266–267, 335–338, 352
  epithelia, 337–338
NDAE, 289
NDAE1, 289–290, 308
NDCBE1 Cl^-/HCO_3– exchangers, 281–282, 289–290, 309
  cloning, 289
  expression, 290
  human disease associated with, 309
  physiological role, 290
  transport function, 290
Nephrocalcinosis, 36
Nephrolithiasis, 36
Nervous system, cation-chloride cotransporter in, 250–256
Neurodegeneration
  ClC-3 proteins and, 46
  ClC-7 proteins and, 42
Neurosteroids for modulation, of GABAergic synapses, 231
Niflumic acid (NFA), 100–101
  modulators of muscle Cl^- conductance gCl, 86, 88–89, 91
5-Nitro-2-(3-phenylpropylamino)-benzoic acid (NPPB), 185
NKCC1 cotransporters, 242–246, 250–252, 254–256, 258, 262, 265–267, 336–338
NKCC2 cotransporters, 242–244, 246, 250, 259, 261–262, 266–267, 336–338, 351
NKCCs, in epithelia, 335–337, 339
Nonsteroidal anti-inflammatory drugs (NSAIDs), 88
Nucleotide-binding domains (NBDs), 110–111, 113–115, 124–126, 132, 137, 146–147, 156–157, 160
  CFTR Cl^- channel, 111, 113–115, 124–126, 132, 137

O

Open-channel blockers, inhibitors of CFTR Cl^- channel, 131–132
Osteopetrosis
  ClC-7 proteins and, 41–43
  mechanism, 42
Outwardly rectifying chloride channel (ORCC), 342

P

Pancreas
  CFTR Cl⁻ channel role in, 118
  HCO_3^- secretion in, 292
Parathyroid hormone (PTH), 37–38
Pathophysiology
  CFTR Cl⁻ channel, 120–122
  Cl^-/HCO_3^- exchangers, 309–312
2-p-Chlorophenoxy propionic acid (CPP), 85, 87–93, 98–101
PDZ domain-mediated interactions
  Cl^-/HCO_3^--exchangers, 297, 305–307, 341
PDZ proteins, 347–351
  cell-cell contact localization, 348
  cluster channels and transporters, 350–351
  diversity, 348–350
  membrane proteins stabilized by, 351
  molecular LEGO, 348–350
  nomenclature, 347–348
  transducisomes organization, 348–351
Peptide toxin, inhibitors of CFTR Cl⁻ channel, 133–134
Pharmacological chaperones rescue, CFTR potentiators, 128
Pharmacology
  CLC chloride channels and transporters, 83–103
  macroscopic skeletal muscle Cl⁻ conductance gCl, 84–88
  protein, 83–84
Phloxine B, CFTR potentiators, 124–125
Phorbol esters, modulators of muscle Cl⁻ conductance gCl, 86
Phosphorylation-dependent regulation, CFTR Cl⁻ channel, 111, 114–116
Physiological functions
  ClC-Ka channels, 24–27
  ClC-Kb proteins, 24–27
  ClC-0 proteins, 17
  ClC-1 proteins, 17–19
  ClC-2 proteins, 19–24
  ClC-3 proteins, 27–31
  ClC-4 proteins, 31–32
  ClC-5 proteins, 32–40
  ClC-6 proteins, 40–41
  ClC-7 proteins, 41–43
PIGG-pen, 67
Plant, ClC proteins in, 45–46
Plasma membrane Cl^-/HCO_3^- exchange protein. *See* Cl^-/HCO_3^- exchangers
Polycystins, 123

Postsynaptic glycine receptor, 15
Potassium-coupled chloride transporters (KCCs), 336
Protein kinase C (PKC), 87, 91, 115, 148, 150, 306, 341
Protein kinase (PKA), 115–117, 121, 124, 147–151, 159, 163, 306, 341
Pyrophosphate (PPi), CFTR potentiators, 125

R

RBCs, cation-chloride cotransporter in, 262, 264–265
Recovery volume decrease (RVD), 201–203
Recovery volume increase (RVI), 201–203
Regulatory (R) domain, 110, 115–116
  CFTR Cl⁻ channel, 114–116
Renal outer medullary potassium channel (ROMK2), 342
Reproductive tissues, CFTR Cl⁻ channel role in, 118
Respiratory airways, CFTR Cl⁻ channel role in, 119–120
Rickets, 36

S

Saccharomyces cerevisiae gene, 43
ScClC. *See Saccharomyces cerevisiae* gene
Skeletal muscle Cl⁻ conductance gCl
  chemical structure, 85
  ClC channels muscle block mechanism, 91–95
  ClC-1 targeting molecules expression, 88–91
  modulators, 86
  pharmacology, 84–88
SLC4, 2
SLC4A1. *See* AE1 Cl^-/HCO_3^- exchangers
SLC4A2. *See* AE2 Cl^-/HCO_3^- exchangers
SLC4A3. *See* AE3 Cl^-/HCO_3^- exchangers
SLC4A4. *See* NDCBE1 Cl^-/HCO_3^- exchangers
SLC4A11, 300
SLC4A9 (AE4), 300
Slc4a9 (AE4), 300
Slc26a6 (CFEX), 294–296, 312
SLC26A3 Cl^-/HCO_3^- exchangers, 281–282, 290–293, 300, 305, 307–309, 311, 346
  cloning, 290–291
  expression, 291
  pathophysiology, 309, 311
  physiological role, 291–292
  regulation, 292–293
  transport function, 291

SLC26A4 Cl⁻/HCO₃⁻ exchangers, 281–282, 293–294, 300, 307, 309, 311–312, 346
  cloning, 293
  expression, 293–294
  pathophysiology, 309, 311–312
  physiological role, 294
  regulation, 294
  transport function, 294
SLC26A6 Cl⁻/HCO₃⁻ exchangers, 281–282, 294–297, 305–307, 309, 312, 346, 353
  cloning, 294–295
  expression, 295
  pathophysiology, 309, 312
  physiological role, 295–296
  regulation, 296–297
  transport function, 295
SLC26A7 Cl⁻/HCO₃⁻ exchangers, 281–282, 297–299, 308–309, 346
  cloning, 297
  expression, 297–298
  human disease associated with, 309
  physiological role, 298
  regulation, 298–299
  transport function, 298
SLC26A8 Cl⁻/HCO₃⁻ exchangers, 300
SLC26A9 Cl⁻/HCO₃⁻ exchangers, 281–282, 299–300, 308–309
  cloning, 299
  expression, 299
  human disease associated with, 309
  physiological role, 299–300
  regulation, 300
  transport function, 299
SLC26 Cl⁻/HCO₃⁻ exchangers, 2
Sodium-coupled transporters, 336
Split-ubiquitin-system, 209
Statins, modulators of muscle Cl⁻ conductance gCl, 86
Stomach, hydrochloric acid secretion in, 287
Sulfate transporters, 280
Sweat gland, CFTR Cl⁻ channel role in, 118–119
Synaptic vesicles, ClC proteins located on, 27–31

T

Tamoxifen, 185
Taurine
  $I_{Cl,swell}$ and, 205
  modulators of muscle Cl⁻ conductance gCl, 86

Tetrahydrodeoxicorticosterone for modulation, of GABAergic synapses, 231
Therapeutic potential, CFTR potentiators, 128–130
Torpedo Cl⁻ channel, and ClC-0 proteins, 15–16
Transepithelial ion transport, CFTR Cl⁻ channel, 116–117
Transepithelial transport, roles of Cl⁻ channels in, 14

V

Vectorial chloride transport
  absorption, 333–334
  epithelia, 329–356
  junctional proteins, 334–336
  secretion, 333
γ-Vinyl-GABA, 225
Volume-regulated anion channel (VRAC), 5, 203
Volume-sensitive Cl⁻ channels
  bestrophins, 206
  candidates for VSOC, 206–209
  ClC proteins family, 207–209
  $I_{Cl,swell}$, 203–206
  taurine efflux, 205
  VSOR, 203–206
Volume-sensitive organic osmolyte/anion channel (VSOAC), 203
Volume-sensitive outwardly rectifying Cl⁻ channel (VSOR), 203–206
  $I_{Cl,swell}$, 203–206
  iconoclastic idea, 210
  molecular identity, 209–210
  stretch-activated Cl⁻ channel, 205–206

Y

Yeast in ClC proteins, 43–44

Z

Zona adherens junctional proteins, 335–336
Zona occludens junctional proteins, 334–335